3D AutoCAD 2010

One Step at a Time

Timothy Sean Sykes

Forager Publications
Spring, Texas
ForagerPub.com

ISBN 0-9794155-9-4; 978-0-9794155-9-3 (ebook)
ISBN 0-9794155-7-8; 978-0-9794155-7-9 (print)

Forager Publications
2043 Cherry Laurel
Spring, TX 77386

www.foragerpub.com

Author's Notes

Where to Find the Required Files (and Review Questions) for Using This Text:

All of the files required to complete the lessons in this text can be downloaded free of charge from this site:

http://foragerpub.com/AcadFiles/2010/2010.htm

Pick on the **Lesson Files** link for the 3D text, and save the zip file to your computer. (Windows XP and Vista will open a zipped file. If you need a zip utility, however, I suggest the free WinZip download at: http://www.winzip.com/. The evaluation version will do to get your files. You'll have to follow WinZip procedures to unzip the files.)

Once you've downloaded the zip file, follow these instructions to unzip it with Windows XP/Vista.

1. Double-click on the file. A window will display with the Step folder inside.
2. Right click on the Step folder and select **Copy** from the menu that appears.
3. Using Windows Explorer, navigate to the C-drive.
4. Select **Paste** from the Edit pull down menu.

To get the review questions and additional exercises, go to the website indicated at the end of each lesson. There you can select your lesson's link, and a PDF file will open in your Internet browser. These copyrighted files are fully printable, or you can save them to your disk if you wish. The last page of each file contains the answers to the questions in that file.

Where to Find the Software

Autodesk provides tools at their website to purchase online or to help you locate a reseller near you. Go to the Autodesk website (Autodesk.com) and select **Purchase**, then select the option that suits you.

You can also download a free, thirty day trial version of the latest software at the Autodesk website. Again, go to the Autodesk website, select **Purchase – Shop online – Product trial**. The trial version works great for short term training!

Windows Vista Users

FONTS: The 2010 edition of *3D AutoCAD: One Step at a Time* falls during a transition period for the Windows operating system – many users have upgraded to the new Vista system while many others prefer to remain with the stable XP system. AutoCAD should work with either, but the Vista system offered a new font (Calibri) which offers a much finer printed appearance. Unfortunately, it isn't readily available for XP users.

For those who use Vista and who want to take advantage of the Calibri font, use the files referenced in this text but which end with a "- v". For those who prefer the older Times New Roman font and those with XP systems, use the referenced file.

(I created most of the imagery in this text using the Calibri font.)

Contacting the Author/Publisher:

Although we tried awfully hard to avoid errors, typos, and the occasional boo boos, I admit to complete fallibility. Should you find it necessary to let me know of my blunders, or to ask just about anything about the text – or even to make suggestions about how to better the next edition, please feel free to contact me using any of the methods listed here:

http://foragerpub.com/contact/contact.htm.

I can't promise a fast response (although I'm usually pretty good about responding), but I can promise to read everything that comes my way.

An afterthought for that address: you need to leave me an email address if you want me to respond. I promise that I won't sell, give away, or in any other way distribute your address to anyone else.

Frequently Asked Questions (and responses to less-frequently-asked but more annoying questions):

I went to the web site but got a Page cannot be found *error. What gives?*

This is the most common complaint I get. The fix is simple: make sure you capitalize the "Files" in the address. The web is case-sensitive.

How do I get eBook versions of your texts? What about your other books (I hear you write other things beside AutoCAD textbooks)?

The best place to get my books at a discount is at the Forager Publications site:

http://www.foragerpub.com

But these folks also offer healthy discounts (just search for my name):

Powells.com

Amazon.com

Other sites may also offer discounts. As I learn of them I post them at the end of the specific book's page at the Forager Publications site.

I went to the web site but got a Page cannot be found *error. What gives?*

This is the most common complaint I get. The fix is simple: make sure you capitalize the "Files" in the address. The web is case-sensitive.

Why don't my graphics work? (I can't see my grips when I'm using certain visual styles.)

Turn your hardware accelerator off – enter ***3dconfig*** at the command line, pick the **Manual Tune** button, and remove the check next to **Enable hardware acceleration** in the **Hardware settings** frame.

Hardware acceleration makes a really good graphic, but it can interfere with the function of some basic stuff like viewing grips.

Contents

Lesson

Following this lesson, you will:

- ✓ *Know how to maneuver in three-dimensional space*
  - *Understand the* ***VPoint*** *command*
  - *Understand the Right-Hand Rule*
  - *Know how to use the* ***Plan*** *command*
  - *Be familiar with the View Cube*
- ✓ *Know how to draw simple three-dimensional objects*
  - *Know how to use* ***Elevation*** *and* ***Thickness*** *in your drawings*
- ✓ *Know some of the basic tricks used to view a three-dimensional drawing*
- ✓ *Know how to use Visual Styles*

"Z" Basics

Thanks, Mr. Woopie! You're the greatest.

Tennessee Tuxedo

Does anyone not remember Mr. Woopie's coveted 3D BB? How many times did he pull that tiny block from his closet, stretch it into a full-size blackboard, and help Tennessee Tuxedo devise yet another wonderful scheme?

Who would have thought – way back in those simple cartoon days – that one day you'd be learning to use your very own 3D BB?!

You're about to go where no board draftsman has gone before. You are about to enter Z-Space. (Hear that cool science fiction music playing in the background? Keep your eyes open for Dr. Who.) Z-Space is that area defined by the Z-axis. Remember the X- and Y-axes? They travel left to right and bottom to top on a sheet of paper (or drafting board). The Z-axis rises and falls into the space above and below the paper. It takes all three axes to create a three-dimensional object. (Later in this text, you'll even see where AutoCAD has taken its first tentative steps into that fourth dimension of which Einstein spoke ... that's right, AutoCAD has finally given you the rudiments of animation! We'll have to leave those other dimensions Stephen Hawking explains so well (at least to other PHDs) for a later time. *But for now, three dimensions are quite enough to keep me confused.)*

1.1 The 3D Look – AutoCAD's 3D Modeling Workspace

You'll notice when opening AutoCAD 2010 for the first time that you're facing a whole new world (see the following insert). For experienced AutoCAD users (one of which you must be or you wouldn't be reading a 3D text!), this can be unnerving.

> Your screen doesn't look any different? That's easily fixed. Select **3D Modeling** on the status bar's **Workspace** toggle.
>
> Still no different? Hmmm ... try this:
>
> 1. Start a new drawing with the ***QNew*** command.
> 2. In the Select Template dialog box, select the *acad3d.dwt* template.
> 3. Pick the **Open** button.
>
> Now you're looking at AutoCAD's 3D world!
>
> We'll work in the **3D Modeling** workspace with the *Acad3d* template for this text.

We'll spend the next few minutes getting familiar with some of the changes.

First, you notice the three-dimensional layout of the work area (Figure 1.001, p.3 – using *acad3d.dwt*).

- You have a fancier grid to use now – but you can still toggle it on or off with the F7 key or ^G. In Z-space, you may find it more helpful to leave it on.
- Your crosshairs now include an up/down line to guide you. Use this as you will the 3D UCS Icon to help keep yourself oriented within the drawing. Remember: Red = East/West, Green = North/South, and Blue = Up/Down. You'll also benefit from the Right-Hand Rule (more on that in a moment).
- You'll find the Full Navigation Wheel we discussed in our basic text; but now we'll be able to use it to its full potential. (If you don't see it, enter the ***NavsWheel*** command or use the ribbon toggle: View – Navigate – **Steering Wheels.**) We'll discuss the wheel in some detail in Lesson 2, p.48.
- Finally, you'll find the View Cube. (If you don't see it, enter the ***Cube*** command or use the ribbon's toggle: View – Views – **ViewCube.**) This phenomenal "gizmo" spans the Autodesk stable of 3D applications (you'll find it in most current releases of Autodesk software – Inventor,

Architectural Desktop, etc.) so get comfortable with it now! We'll spend some time with this tool in a moment (and throughout this text).

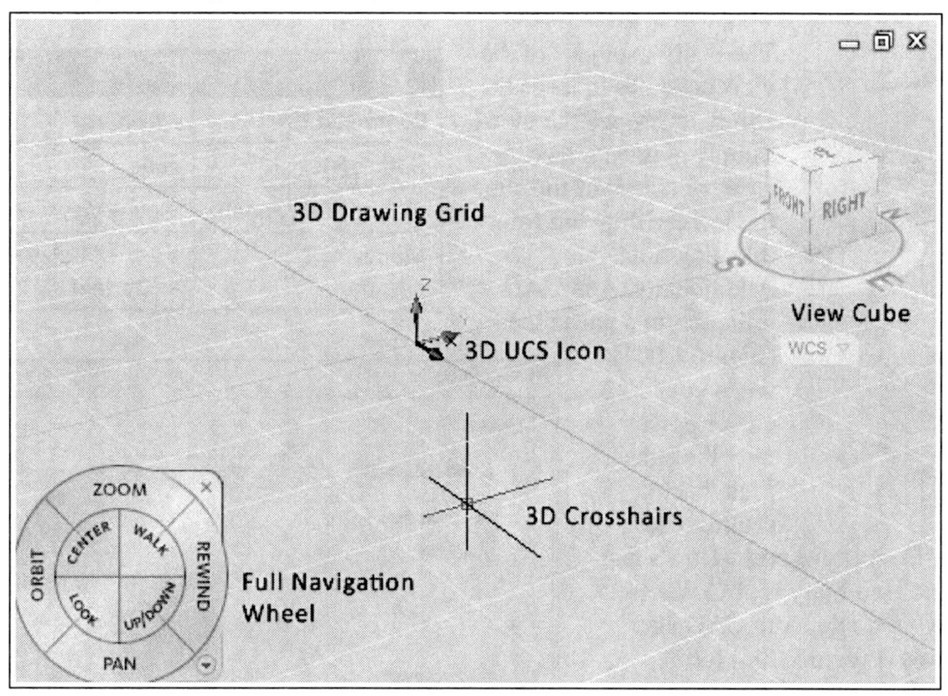

Figure 1.001

1.1.1 The View Cube

Before entering Z-Space (the third dimension), you must learn how to keep your bearings. That is, you must learn to tell up from down, top from bottom, and left from right, regardless of your orientation within the drawing (feel like an astronaut?). It's not as easy as it sounds. Place your hand over Figure 1.003 and look carefully at Figure 1.002. Are you looking at the top or bottom of the object? Now look at Figure 1.003. How'd you do?

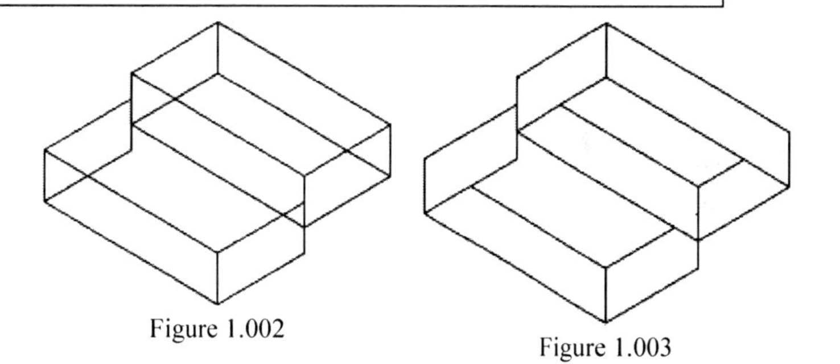

Figure 1.002

Figure 1.003

I've removed the hidden lines in Figure 1.003 (more on how I did that in Section 1.4, p.19). However, it isn't always practical to do that, so AutoCAD has provided a couple methods to determine your orientation at any time and at any place in the drawing. Let's look at the View Cube (Figure 1.001).

> In earlier editions of this text, we discussed the UCS Icon here. The View Cube assumes many of the responsibilities of this old staple of the AutoCAD world. Still, some discussion of the icon is in order. Please refer to the supplement: http://www.uneedcad.com/Files/UCS_Icon.pdf, for this information.

This extraordinary innovation deserves a new designation – we'll have to call it a *#1 Cool Tool* because it provides so much in such a simple package. It provides both a directional indicator and a view adjustor!

As a directional indicator, the cube shows the N–S–E–W directions easily using the two dimensional compass at the base of the cube. Up – or Z-space – rises from the compass while down – or negative

Z-Space – drops below it. You may find the compass's associated terms easier to follow: F–B–L–R (Front – Back – Left – Right). Up and down appear as *Top* and *Bottom* on the cube. The cube, then, should help you avoid getting lost in a 3D world!

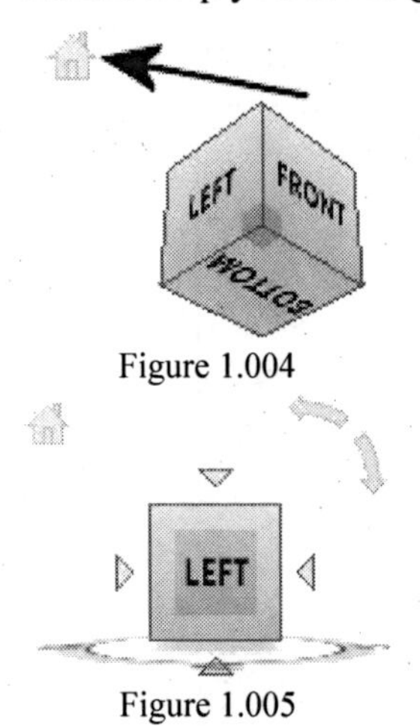

Figure 1.004

Figure 1.005

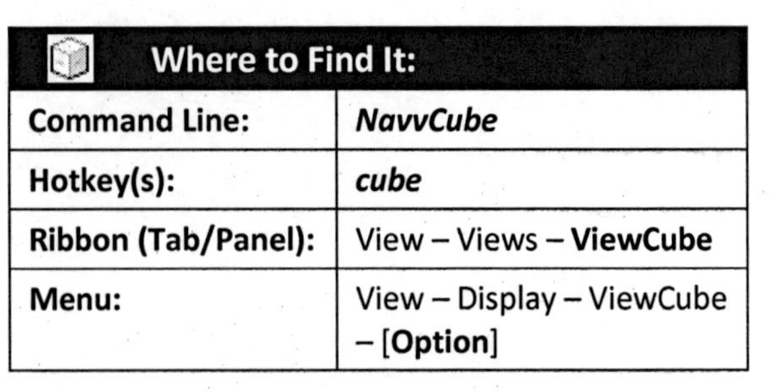

Where to Find It:	
Command Line:	***NavvCube***
Hotkey(s):	***cube***
Ribbon (Tab/Panel):	View – Views – **ViewCube**
Menu:	View – Display – ViewCube – [**Option**]

The really cool part of the view cube lies in its easy ability to *change* the view! Simply pick on a face, edge, or corner of the cube to view the drawing from that direction! Additionally, AutoCAD will present a house icon (Figure 1.004) when your cursor passes over the cube. Pick this to return to the home view from anywhere! (How's that for your first Hail Mary tool?!) When you're viewing a face, AutoCAD also presents tools (Figure 1.005) for rotating the face (curved arrows in the upper right corner) or for viewing another face (the four arrows).

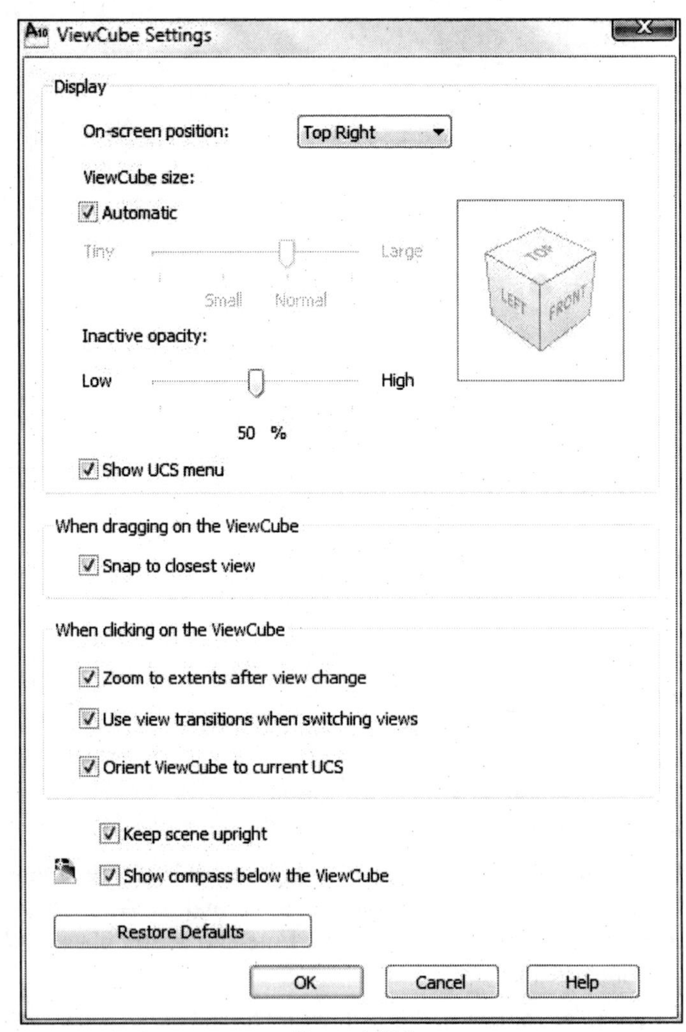

Figure 1.006

Turn the cube on or off with the ***NavvCube*** command:

> **Command: *NavvCube***
> **Enter an option [ON/OFF/Settings] <ON>:**

The **Settings** option presents the ViewCube Settings dialog box (Figure 1.006). Here you'll find some options that'll greatly enhance your use of the cube (or irritate you unmercifully)!

- The **Display** frame offers options to control:
  - the **On–screen position** of the cube. Choose a location that won't get in your way, because you'll want the cube displayed fairly consistently when you're in a three dimensional drawing.
  - the **size** of the cube. You can let AutoCAD size it **Automatic**ally, but it really doesn't need to be any larger than your ability to read the text on it. My aging eyes see the **Normal** setting just fine; you may want it a bit smaller.
  - the **Inactive opacity** – or just how nearly invisible you want it to be when you're not using it. Experienced users might make it nearly completely invisible; don't worry, it'll show quite clearly when you pass your cursor over it.

- The next frame – **When dragging on the ViewCube** – controls whether AutoCAD will smooth orbit or **snap to the closest view** when you drag the cube. Personally, I prefer the smooth orbit (remove the check next to **Snap to closest view**). I can always pick on a face, line, or corner if that's what I want.
- **When clicking on the ViewCube** offers some mind-saving options.
  - Personally, I remove the check next to **Zoom to extents after view change** (and I bless the makers and shakers at Autodesk for giving us this option!). Nothing irritates more that having your entire view change just because you need to move around a bit!
  - **Use view transitions when switching views** controls how smooth the transitions appear when orbiting the view.
  - **Orient ViewCube to current UCS** is another blessing. It controls whether the cube orients according to the WCS or the current UCS. We'll discuss the UCS in Lesson 2, p.34. But for now remember – the home icon on the view cube presents the basic three dimensional view whenever you get lost.
- **Keep the scene upright** controls whether or not you can turn the scene upside down. Using this will also help you avoid becoming lost unless you actually want to work on the bottom of an object.
- **Show compass below the ViewCube** controls whether or not the compass will appear.
- **Restore Defaults**, does just that. Use this when you get confused.

We have one last thing to examine on the View Cube – the ever-important cursor menu (Figure 1.007).

- Use the **Home** option as you do the **Home** icon.
- Use the three options in the next frame to control exactly what you see:
  - A **Parallel** view works something like an isometric. You see three sides – each with dimensional accuracy.
  - A **Perspective** view also works like an isometric, but the sides fade into a "horizon" point much like a "real world" view.
  - **Perspective with Ortho Faces** changes the view to a perspective unless the current view is a face view.
- **Set Current View as Home** does just that. After setting this, the **Home** option will return the drawing to the current view.
- **ViewCube Settings** calls the dialog box in Figure 1.006, p.4.

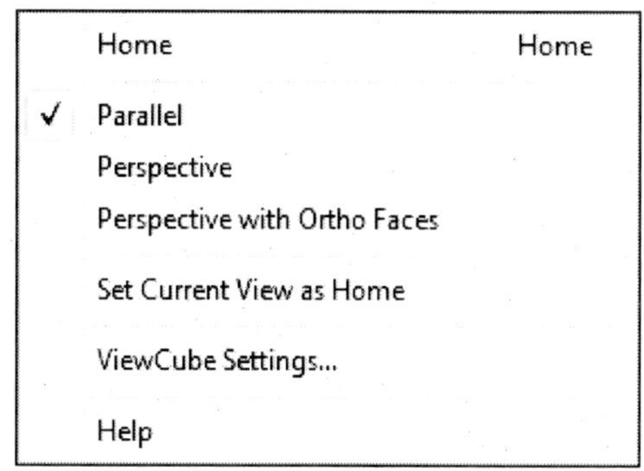

Figure 1.007

Let's try an exercise.

Do This: 1.1.1A	Manipulating the Drawing with ViewCube

I. Open the *ucs practice* file in the C:\Steps3D\Lesson01 folder. The drawing looks like the figure at right. (Don't let the funny angle of the crosshairs or UCS icon bother you – you're viewing the model from an angle in Z-Space. More on this setup in Section 1.2, p.7.)

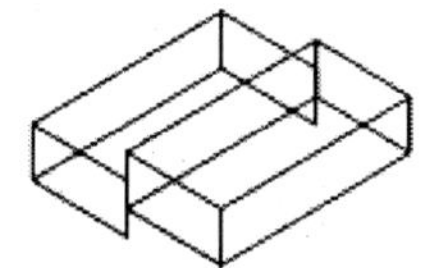

II. Pan to center the object on the screen.

III. Follow these steps.

1.1.1A: MANIPULATING THE DRAWING WITH VIEWCUBE

1. Enter the ***UCSIcon*** command.

   Command: *ucsicon*

2. Use the **ORigin** option ORigin.

   Enter an option [ON/OFF/All/Noorigin/ORigin/Properties] <ON>: *or*

Notice that the icon moves to the **0,0,0** coordinate. We'll leave it there for this exercise so you can better follow what happens.

3. Use the ***View*** command to change the view to **negative z space** (already created for you – look under **Model Views**).

   Command: *v*

The view changes to that shown here. Notice the UCS icon and the cube.

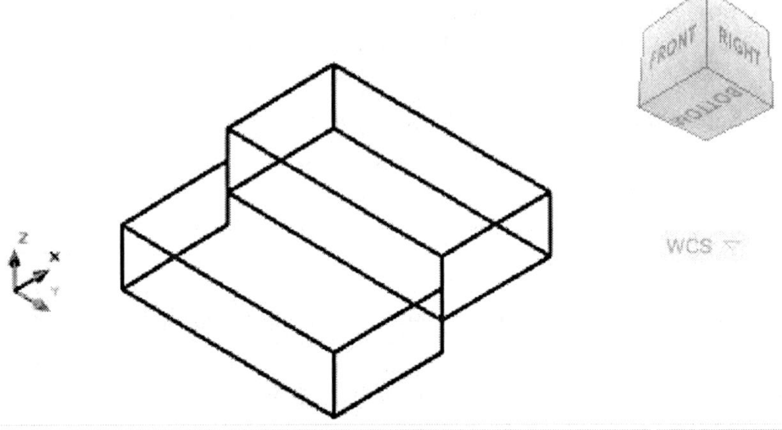

4. Pick on the top corner of the cube (FRONT – RIGHT – TOP).

Notice the change. (It's that simple to change views!)

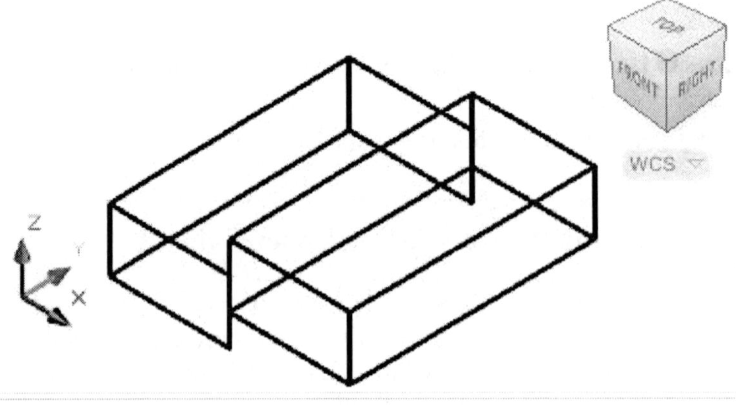

5. Zoom in on part of the model.

6. Now pick the RIGHT – BACK – BOTTOM corner of the cube.

Notice that, not only does AutoCAD adjust the position of the view, it also performs a zoom extents. Let's fix that.

1.1.1A: MANIPULATING THE DRAWING WITH VIEWCUBE

7. Pick the **ViewCube Settings...** option [ViewCube Settings...] from the cube's cursor menu. AutoCAD presents the ViewCube Settings dialog box (Figure 1.006, p.4).

8. Remove the check next to [Zoom to extents after view change].

9. **OK** [OK] the change to close the dialog box.

10. Repeat Steps 5 and 6.
Notice the difference?

These simple tricks to manipulate your drawing view will become second nature with experience.

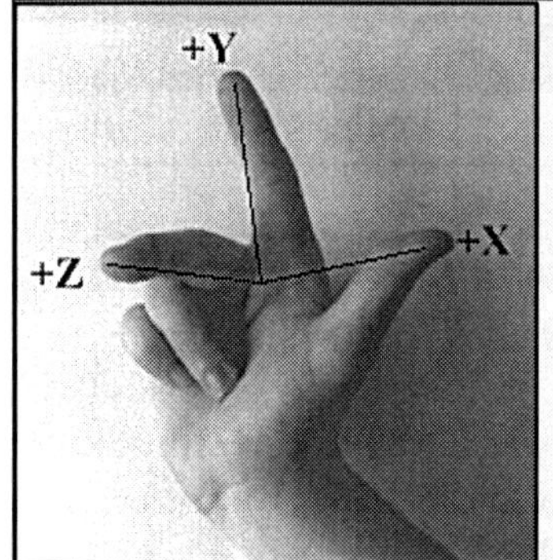

The Right-Hand Rule provides another means of navigating Z-space. It works very much like the UCS icon but is a bit more "handy." It works like this (refer to the figure at left):

1. Make a fist with your right hand.
2. Extend the index finger upward (no no – the *index* finger!).
3. Extend the thumb at a right angle to the index finger.
4. Extend the middle finger at a right angle to the index finger (pointing outward).

How's that for feeling really awkward?

Each finger serves a purpose; each indicates the positive direction of one of the XYZ axes as indicated in the picture. See how it correlates with the UCS icon? Using the right hand in this fashion, you'll always be able to determine the third axis if you know the other two. Simply orient your right hand with the appropriate fingers pointing along the known axes and the location of the unknown axis becomes clear!

1.1.2 3D Ribbon Panels

You may have notice when you opened the **3D Modeling** workspace back on page 2, that AutoCAD's ribbon added several new panels and tabs – all related to three dimensional work at one level or another. In this text, we'll examine these new panels and their specific tools as well as the three dimensional options of some of the more familiar tools.

We introduced the ribbon in Lesson 1 of *AutoCAD 2010: One Step at a Time*. If you haven't been using it before now, you might want to review that section.

I'd recommend taking advantage of the ribbons ability to hide. Pick the ribbon button to the right of the tabs until you reach **Minimize to Panel Titles** (you should see only the panel titles and the tabs). This will free up a tremendous amount of drawing area. Each panel will display when you move your cursor over the title.

1.2 Other Ways to Maneuver through Z-Space

Okay. You know how to determine your orientation in Z-Space and how to reorient yourself using the View Cube, but there are a couple other, more precise methods of which you should be aware. These include the ***VPoint*** and ***Camera*** commands.

There are three approaches to using the ***VPoint*** command: coordinate input, compass, and dialog box. Let's look at each.

1.2.1 Using Coordinates to Assign a Viewpoint

It might help your understanding if I begin by telling you that the model will not actually move or rotate. Using the ***VPoint*** command, the model holds still while you change position.

Let's use an airplane as an example. The airplane is a three-dimensional object (it has length, width, and height or thickness). To draw the top of the airplane, we'll climb into a helicopter and hover above it for a better view. To draw the front, we must fly to the front of the airplane. Side and bottom views will also require us to fly to a better vantage point.

AutoCAD's helicopter is the ***VPoint*** command (perhaps better explained as *Vantage* Point).

The ***VPoint*** command prompt looks like this:

Where to Find It:	
Command Line:	*VPoint*
Hotkey(s):	*–vp*

Command: *vpoint* (or *–vp*)

Current view direction: VIEWDIR=1.0000,-1.0000,-1.0000

Specify a view point or [Rotate] <display compass and tripod>: *[Enter a coordinate.]*

- The default response to this prompt requires you to enter the coordinates from which you wish to view the model. Does that sound simple? Only if you know the coordinates, you say? Well, it's actually easier than that! AutoCAD won't read the XYZ coordinates entered as actual coordinates but rather as a ratio. That is (keeping it simple), a coordinate of 1,-1,1 tells AutoCAD that you wish to stand – in relation to the model – a step to the right (+1 on the X-axis), back a step (-1 on the Y-axis), and up a step (+1 on the Z-axis).

  > The default viewpoint for two-dimensional drawings is **0,0,1**. This means that you view the drawing from above (+1 on the Z-axis). Draftsmen generally refer to this as the *Plan* view.

  Remembering the simple ratio approach, what do you think a coordinate of 4,-2,1 will mean?[*]

  > When the *absolute* value of all three axis entries is the same (as in 1,1,1 or 1,-1,1), you have an isometric view of the drawing. (*Iso* means one – you're using the same absolute *value* on each axis).
  >
  > When only two of the absolute values are the same (as in 2,2,1 or –2,2,1), you have a *dimetric* view. (*Di* means two.)
  >
  > What do you suppose you have when all three of the absolute values are different? Of course, it's a *trimetric* view.

  Why the ratio approach? When you change your vantage or viewpoint (your *VPoint*), AutoCAD will automatically zoom to the drawing's extents (regardless of the cube's settings)!

- The **Rotate** option of the ***VPoint*** command allows you to specify your vantage point by the angle at which you wish to see the model. The prompts look like this:

  Command: *vpoint*

  Current view direction: VIEWDIR=0.0000,-1.0000,0.0000

  Specify a view point or [Rotate] <display compass and tripod>: *r [Tell AutoCAD you wish to use the Rotate option.]*

  Enter angle in XY plane from X axis <270>: *[Indicate the two-dimensional angle (the angle along the ground) at which you wish to see the model.]*

  Enter angle from XY plane <90>: *[Indicate the three-dimensional angle (the upward angle – like the angle of the sun in the sky) at which you wish to see the model.]*

This will become clearer in our next exercise.

---

[*] It means: I want to stand twice as far to the right as I'm standing back, and twice as far back as I'm standing above the model.

You already know that a plan view is achieved when the VPoint coordinates are 0,0,1. These coordinates allow you to view the model from a step up (+Z). What do you suppose happens when you set the coordinates to 1,0,0 or 0,-1,0?

These coordinates provide elevations of the model – 1,0,0 provides a right elevation (a step to the right or +X) and 0,-1,0 provides a front elevation (a step back or –Y). Which coordinates would provide a back or left elevation?**

We'll look at the compass in Section 1.2.2, but first let's try the coordinate approach to the ***VPoint*** command.

You can access several preset viewpoints using buttons on the View toolbar or the **Current View** control box on the **View** ribbon panel (**Home** tab). Alternately, you can select the view from the View pull-down menu. Follow this path:

View – 3D Views – [selection]

Do This: 1.2.1A	The Coordinate Approach to Setting the Viewpoint

I. Open the *VPoint practice* file in the C:\Steps3D\Lesson01 folder. The drawing looks like the figure at right. It's currently in the plan view.

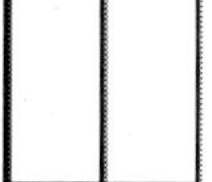

II. Follow these steps.

1.2.1A: THE COORDINATE APPROACH

1. Enter the ***VPoint*** command.

Command: *-vp*

2. AutoCAD tells you the current coordinate setting and prompts you to either **Specify a view point** or to **Rotate**. Let's use the first option. Tell AutoCAD to move to the right, back, and up as indicated (the SE Isometric View).

Current view direction: VIEWDIR=0.0000,0.0000,1.0000

Specify a view point or [Rotate] <display compass and tripod>: *1,-1,1*

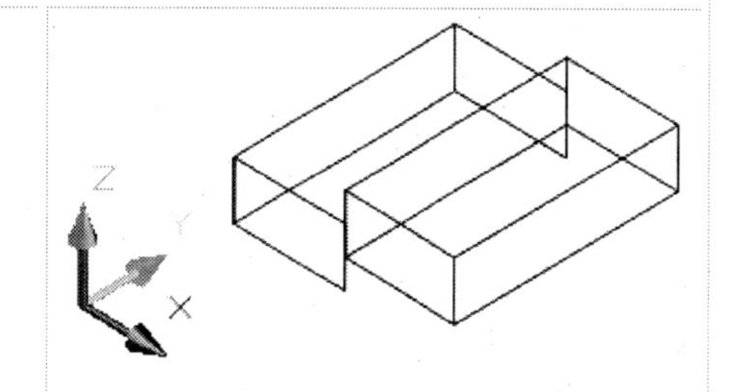

Your drawing looks like the figure at right. You can achieve the same results as Steps 1 and 2 using the option (in the View menu, toolbar, or ribbon panel).

3. Let's try the **Rotate** option [Rotate]. Repeat the ***VPoint*** command and select it.

Command: *[ENTER]*

Current view direction: VIEWDIR=1.0000,-.0000,1.0000

Specify a view point or [Rotate] <display compass and tripod>: *r*

4. AutoCAD asks for a two-dimensional angle. We'll stand in the -X,-Y quadrant of the XY plane. Enter the angle indicated.

Enter angle in XY plane from X axis <315>: *225*

** Back elevation coordinates = 0,1,0; left elevation coordinates = -1,0,0

1.2.1A: The Coordinate Approach	
5. Now AutoCAD asks for a three-dimensional angle (the angle up or down in Z-space). We'll stand above the model. Enter the angle indicated. **Enter angle from XY plane <35>:** ***45*** Your drawing now looks like the figure at right.	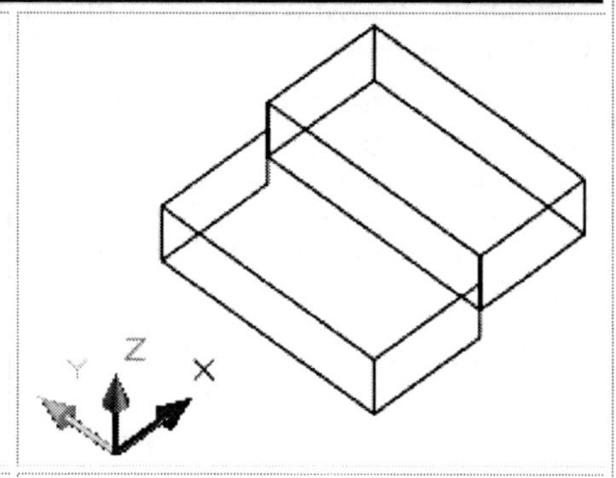
6. Repeat the ***VPoint*** command. This time, enter coordinates for a front elevation as indicated. **Command:** ***[ENTER]*** **Current view direction: VIEWDIR=-0.8660, -0.866,1.2247** **Specify a view point or [Rotate] <display compass and tripod>:** ***0,-1,0*** Your drawing looks like the figure at right. Notice the UCS icon. (*Caution:* Do *not* use the **Front View** button in the **View** panel's control box as it will also change the UCS.)	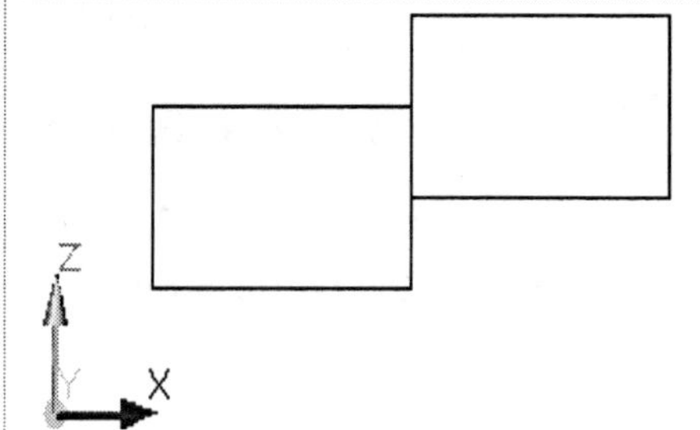

7. Experiment with different coordinate and angular entries. Use different combinations of positive and negative numbers, and watch the UCS icon and ViewCube to keep track of your orientation.

8. Save the drawing but don't exit.

Command: ***qsave***

> You may have noticed the **<display compass and tripod>** option of the ***VPoint*** command. This leads to an older compass method that is similar to but not as user–friendly as ViewCube. You can find details of how to use this older compass method here: http://www.UNeedCAD.com/Files/Compass.pdf.

1.2.2 Setting Viewpoints Using a Dialog Box

Where to Find It:	
Command Line:	***DDVPoint***
Hotkey(s):	***vp***
Menu:	View – 3D Views – **Viewpoint Presets**

Setting viewpoints using a dialog box is very similar to setting viewpoints using the **Rotate** option of the ***VPoint*** command.

Access the Viewpoint Presets dialog box (Figure 1.008) with the ***DDVPoint*** command.

The first things you'll notice on the dialog box are the two large graphics in the center. The first (on the left) looks something like a compass, and the second looks like half a compass. Use the compass

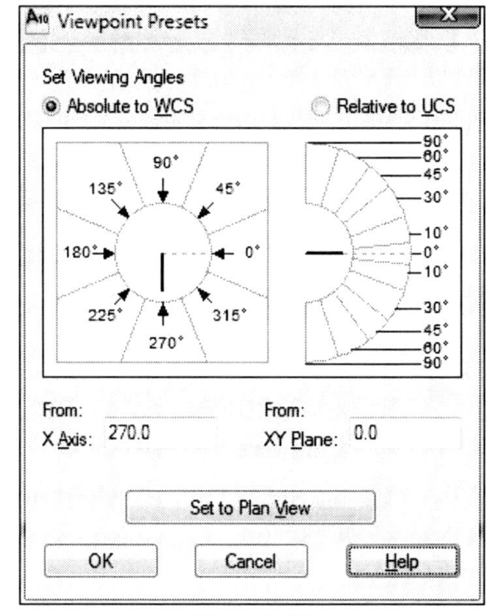

Figure 1.008

on the left to set the two-dimensional angle (*in* the XY-plane); use the compass on the right to set the three-dimensional angle (up or down *from* the XY plane).

You can set the angles in two ways – by keyboard entry in the text boxes below the compasses or by mouse pick on the compasses themselves.

At the top of the dialog box, AutoCAD asks you to identify the angle in **Absolute to WCS** or **Relative to UCS** terms. We'll learn more about the WCS (World Coordinate System) and UCS (User Coordinate System) in Lesson 2, p.34. For now, leave the viewing angles set to **Absolute to WCS**.

The long button across the bottom of the dialog box is our next "HAIL MARY" button for this text. If you read the basic text, you know to use a HAIL MARY button in an emergency. AutoCAD designed this button to return the drawing to the plan view from anywhere. Set the **Absolute to WCS** option at the top of the dialog box and then pick the **Set to Plan View** button to return to a "normal" two-dimensional view of your model. This is quite handy when you become lost in Z-space.

> The keyboard equivalent of the **Set to Plan View** "Hail Mary" button is the ***Plan*** command. It looks like this:
>
> **Command: *plan***
>
> **Enter an option [Current ucs/Ucs/World] <Current>:**
>
> Again, we'll look at the WCS and UCS in Lesson 2, p.34. Enter ***w*** for the **World** option and hit ENTER to do what the **Set to Plan View** button does – return to a normal two-dimensional view of your model.

Let's try the dialog box.

Do This: 1.2.2A	**Using the Viewpoint Presets Dialog Box**

I. Be sure you're still in the *VPoint practice* file in the C:\Steps3D\Lesson01 folder. If not, please open it now.

II. Follow these steps.

1.2.2A: Using the Viewpoint Presets Dialog Box

1. Enter the ***DDVPoint*** command.

Command: *vp*

AutoCAD presents the Viewpoint Presets dialog box (Figure 1.008).

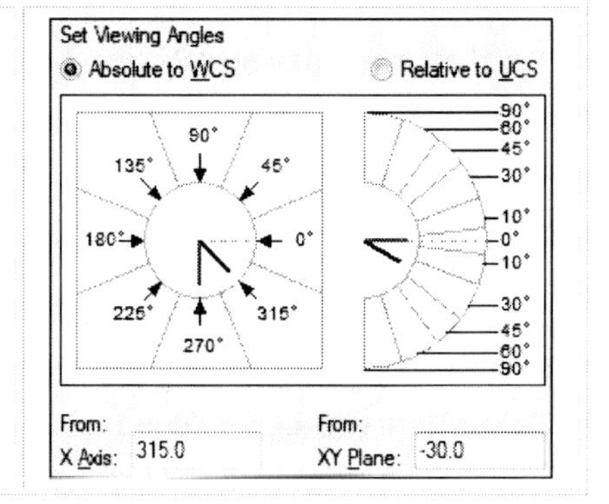

2. Set the viewpoint to **315°** on the left compass. Set the viewpoint to **-30°** on the right compass. (You can do this by typing the numbers into their respective text boxes or by picking on the numbers on the compasses themselves.)

Pick the **OK** button OK to complete the command. Your drawing looks like the following figure. Notice the UCS icon.

1.2.2A: Using the Viewpoint Presets Dialog Box

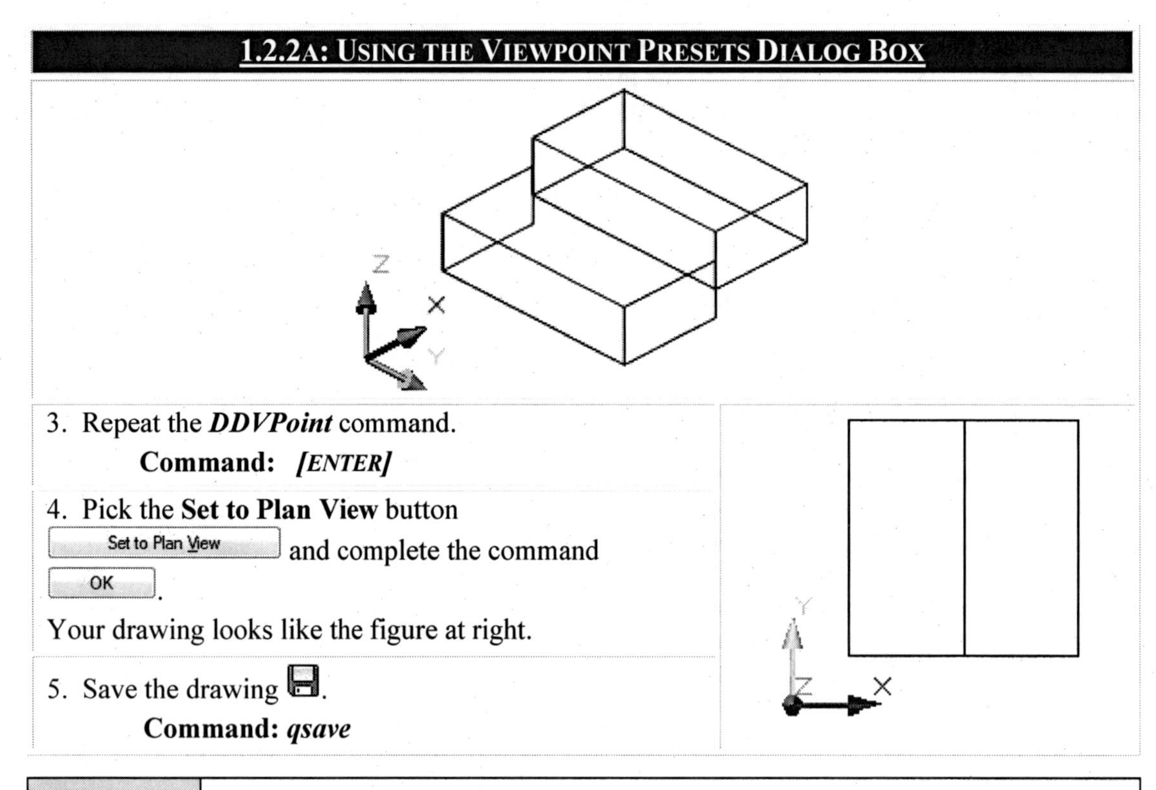

3. Repeat the ***DDVPoint*** command.

   Command: ***[ENTER]***

4. Pick the **Set to Plan View** button [Set to Plan View] and complete the command [OK].

   Your drawing looks like the figure at right.

5. Save the drawing.

   Command: ***qsave***

1.2.3 The Camera Approach

At first glance, the camera approach seems like one of the best approaches to adjusting how you view objects in a three-dimensional drawing. It works by locating the camera (your vantage point) and then identifying the position of its target. AutoCAD will automatically place a camera icon at your vantage point. You can set up more than one camera at a time. Name it, and AutoCAD automatically creates a corresponding named view. AutoCAD uses a fairly straightforward command sequence for its camera:

Where to Find It:	
Command Line:	***Camera***
Hotkey(s):	***cam***
Ribbon (Tab/Panel):	Home – View (subpanel) – [**Camera** & **Target** input boxes]
Menu:	View – **Create Camera**
Toolbar:	View – **Create Camera**

Command: ***camera***

Current camera settings: Height=0.0000 Lens Length=50.0000 mm

Specify camera location: ***[Locate the camera – your vantage point.]***

Specify target location: ***[Locate the camera's target – what you want to see.]***

Enter an option [?/Name/LOcation/Height/Target/LEns/Clipping/View/eXit]<eXit>:

Let's look at the options.

- The question mark (**?**) tells AutoCAD to display a list of cameras currently defined in your drawing.
- The **LOcation** option gives you the opportunity to set the location by coordinate rather than just picking a point on the screen.
- **Height** let's you manually set the height of the camera.
- **Target** lets you select the target.

- The **LEns** option allows you to define the lens of your camera (35mm, 120mm, etc.) Not being a photographer, I tend to let the default ride on this one.
- Clipping lets you define a clipping plane for your view. Using this, you can remove background (or even foreground) clutter. Don't you wish you could do that with a real camera?!
- **View** conveniently allows you to set the current view to the camera's settings.

Let me show you some of these cool options in action.

Do This: 1.2.3A	**Working with the Camera**

I. Open the *CameraPositionExercise* drawing file in the C:\Steps3D\Lesson01 folder. The drawing looks like the figure at right.

II. Follow these steps.

1.2.3A: WORKING WITH THE CAMERA

1. Enter the ***Camera*** command.

   Command: ***cam***

2. Select the center of the red sphere as your vantage point. (Use the center OSNAP.)

   Current camera settings: Height=0.0000 Lens Length=50.0000 mm
   Specify camera location: _cen of

3. Select the insertion point of the larger sphere block as your target.

   Specify target location: _ins of

4. Tell AutoCAD to adjust to the camera view [View].

   Enter an option [?/Name/LOcation/Height/ Target/LEns/Clipping/View/ eXit]<eXit>: ***V***
   Switch to camera view? [Yes/No] <No>: ***Y***

   AutoCAD adjusts the view.

5. From the control box on the **View** ribbon panel, select the side view. Notice the difference? AutoCAD has even adjusted the visual style to accommodate the predefined view. (More on visual styles in Section 1.4, p.19.)

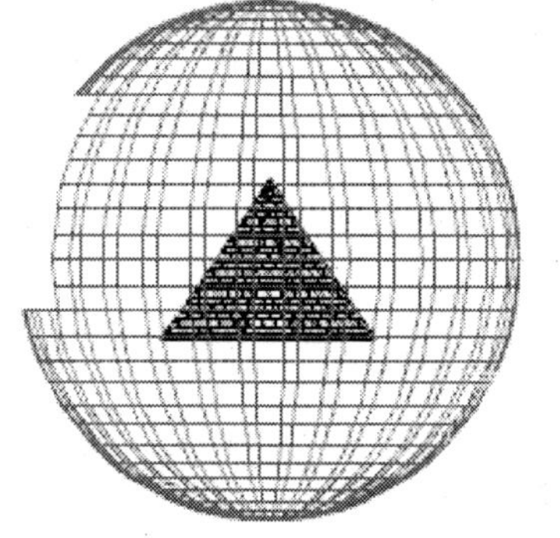

6. Take a few minutes to experiment with the different options of the ***Camera*** command.

7. Exit the drawing without saving.

   Command: ***quit***

You'll create the nifty pyramid-in-a-sphere in later lessons.

We've spent a great deal of time in this lesson on a single concept – viewpoints. I can't overemphasize the importance of being comfortable with viewpoints.

Consider the astronaut floating in space. A basic understanding of geography will tell him where he is in terms of continent, nation, or even city (XY-space). But if he doesn't understand his altitude (his Z-space position), he can't know if he's coming or going. He might wind up on the wrong planet altogether!

If necessary, repeat this lesson to this point to get comfortable with viewpoints and the VPoint command (the astronaut's navigator). Experiment with viewpoints in different viewports (each viewport can have its own viewpoint). Then proceed to the next section where we'll take our first tender steps in creating three-dimensional objects.

> Okay, I didn't cover another set of navigational tools – the Navigation Wheels. Don't fret – this just gives you something really cool to look forward to. (We'll look at these in Lesson 2, p.48.)

1.3 Drawing with the Z-Axis

I wish I could begin this section by saying that three-dimensional drafting is no different from two-dimensional drafting. However, that simply isn't the case. (I could lie if you'd prefer.) You've already seen that you'll have to master more navigation tools. Additionally, you must consider a couple things that have no meaning in a two-dimensional drawing – the Z-coordinate (or *elevation*) and the *thickness* of the object you're drawing.

Let's take a look at these.

1.3.1 Three-Dimensional Coordinate Entry

We must begin our study of three-dimensional drafting as we began our basic text – with a look at the Cartesian Coordinate System. This time, we need to understand how it works with the Z-axis. Consider the following table.

SYSTEM	ABSOLUTE	RELATIVE	POLAR/SPHERICAL/CYLINDRICAL
2D Entry	X,Y	@X,Y	@Dist<Angle
3D Entry	X,Y,Z	@X,Y,Z	@truedist<Xyangle<Zangle or @dist<2Dangle,Z-dist

The Absolute and Relative systems are easy enough to understand. Simply add the location on the Z-axis to the X and Y locations to have an XYZ-coordinate. But Polar Coordinate entry might need some explaining. (Refer to Figures 1.009a and 1.009b.)

Figure 1.009a: Spherical Coordinate Entry

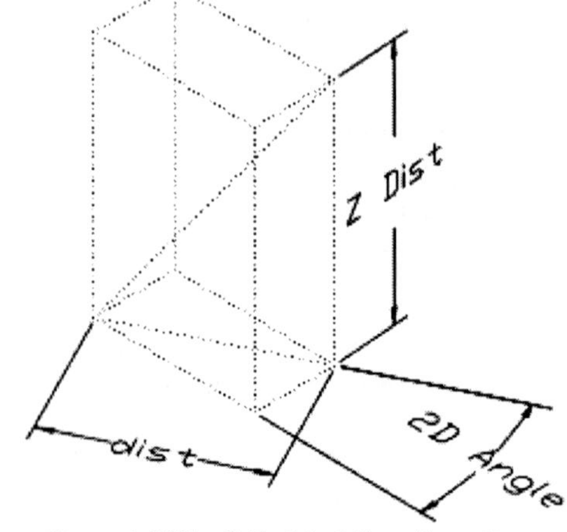

Figure 1.009b: Cylindrical Coordinate Entry

The first formula for 3D entry of polar coordinates looks like this:

@truedist<XYangle<Zangle

(Read this line as: *at a true distance of___ ... at an XY angle of___ ... and a Z angle of___*.)

We call this type of coordinate entry *Spherical* (Figure 1.009a). Begin Spherical coordinate entry with the true three-dimensional length of the line (or other object). Follow with the same two-dimensional (XY) angle you've always used for Polar Coordinates. Follow that with the three-dimensional angle (above or below the XY-plane).

The other formula for 3D entry of polar coordinates looks like this:

@dist<XYangle,Z-dist

(Read this line as: *at a distance of___ ... and an XY/2D angle of___ ... at a Z distance of___*.)

We call this type of coordinate entry *Cylindrical* (Figure 1.009b). Like the spherical coordinate entry method, begin cylindrical coordinate entry with the distance of the line *but this time, use only the XY distance*. Follow with the two-dimensional angle, but conclude with the distance *along the Z-axis* (perpendicular to the XY-plane).

We'll use these methods to draw the stick figure of a house in our next exercise.

Do This: 1.3.1A	Three-Dimensional Coordinate Entry

I. Open the *1-3D-1* file in the C:\Steps3D\Lesson01 folder. This drawing has been set up with a viewpoint of 1,-2,1, and layers have been created for you.

II. Be sure that layer **obj1** is current. (Toggle dynamic input on or off as needed.)

III. Follow these steps.

1.3.1A: THREE-DIMENSIONAL COORDINATE ENTRY

1. We'll begin using absolute coordinates. Draw a line as indicated. Don't exit the command.

   Command: *l*
   Specify first point: *1,1*
   Specify next point or [Undo]: *4,1*

Notice that a Z-coordinate entry isn't required if the value is **0**.

2. Now continue the line straight up into Z-Space.

   Specify next point or [Undo]: *4,1,2*

3. Continue the line to complete the first side of the house.

   Specify next point or [Close/ Undo]: *1,1,2*
   Specify next point or [Close/ Undo]: *c*

4. We'll use relative coordinates as indicated to draw the other side.

   Command: *[ENTER]*
   Specify first point: *1,3*
   Specify next point or [Undo]: *@3,0*
   Specify next point or [Undo]: *@0,0,2*
   Specify next point or [Close/ Undo]: *@-3,0,0*
   Specify next point or [Close/ Undo]: *c*

5. Using the Endpoint OSNAP, draw lines to connect the two walls as shown.

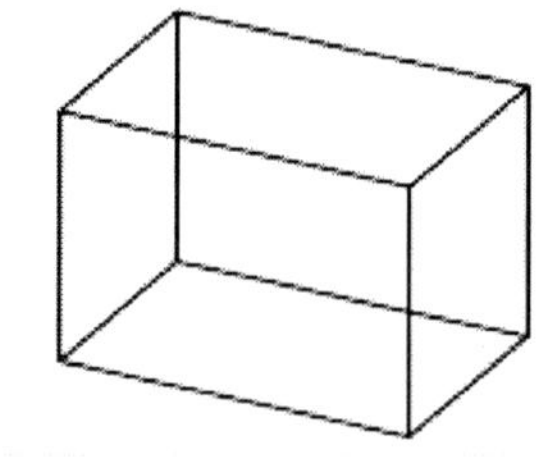

6. Remember to save occasionally. Save the drawing as *My Stick House* in the C:\Steps3D\Lesson01 folder.

   Command: *saveas*

1.3.1A: Three-Dimensional Coordinate Entry

7. Now use spherical coordinates as indicated to locate the peak of the roof. Begin at the midpoint of the upper line marking the left side of the house.

    Command: ***l***

    Specify first point: ***mid***

8. We'll want a **1.5"** line at **0°** on the XY-plane and **60°** upward.

    Specify next point or [Undo]: ***@1.5<0<60***

    Specify next point or [Undo]: ***[ENTER]***

9. Repeat Steps 7 and 8 to locate the roof peak on the opposite wall.

    Command: ***[ENTER]***

    Specify first point: ***mid***

    Specify next point or [Undo]: ***@1.5<180<60***

    Specify next point or [Undo]: ***[ENTER]***

    Your drawing looks like the figure at right.

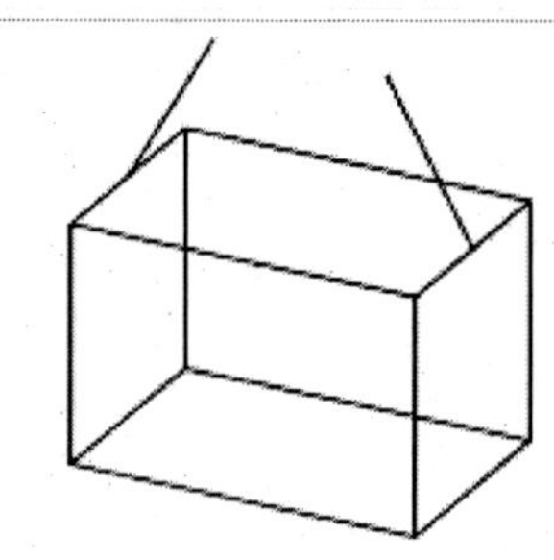

10. Using the Endpoint OSNAP, draw the roof as shown here. Erase the roof peak locators.

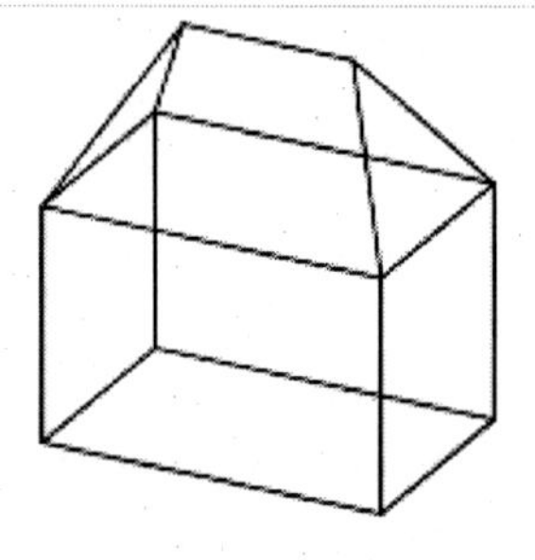

11. Save the drawing but don't exit.

    Command: ***qsave***

12. Take a few moments and examine the model using different viewpoints. Return to a location of 1,-2,1 when you've finished.

    Command: ***vp***

Congratulations! You've created your first three-dimensional drawing.

You may have noticed some three-dimensional idiosyncrasies. If not, let me list a few.

- Ortho works on all planes (X, Y, and Z).
- Object and Polar Tracking work on all planes (X, Y, and Z).
- OSNAPs work on most objects regardless of their XYZ-coordinates.
- It's impossible to locate a point in a three-dimensional drawing by arbitrarily picking a point on the screen. Remember that your screen is two-dimensional. An arbitrary point lacks a definition on one of the X-, Y-, or Z-axes, so there's no guarantee where it may actually be. *Always identify a three-dimensional point using a coordinate entry method or an OSNAP!* (Now you see why we put so much emphasis on coordinates and OSNAPs in the basic book!)
- Use three-dimensional coordinates just as you did two-dimensional coordinates when modifying the drawing (copying, moving, etc).
- Only very rarely will you want to create a three-dimensional object using a simple ***Line*** command. (After all, how often does your boss request stick figure drawings?)

Let's look next at adding a new dimension to our objects. We'll call our new dimension *thickness* and do marvelous things with it.

1.3.2	Using the Thickness and Elevation System Variables

Thickness is that property of an object that takes it from a stick figure to a true three-dimensional object. Until now, all the objects you've drawn have been stick figures. That is, they've all existed on a single, two-dimensional plane. Even the lines we used to create our stick house in the last exercise existed in two-dimensional planes (although they crossed three-dimensional space).

> Point filters are also very useful in Z-Space. Refer to Lesson 3, p.67, in the basic *One Step at a Time* text for details on the use of point filters.

Clear as milk? Let me help.

Where to Find It:	
Command Line:	***Thickness***
Hotkey(s):	***th***
Menu:	Format – **Thickness**

Consider each line created in our stick house. (Consider each line individually – not as it relates to other lines or coordinates.) Describe each in terms of length, width, and height. In each case, you can describe the line using two of the three terms mentioned. In no case can you use all three terms. Consider the first line you drew. It has a length of 3 units and a width (or lineweight) of 0.01" (AutoCAD's default). But it has no height – no *thickness*.

Drawing with thickness is as easy as setting the system variable, like this:

Command: ***thickness***

Enter new value for THICKNESS <0.0000>: ***.25***

All objects drawn while the **Thickness** system variable is set to .25 will have a thickness of .25. This brings up another very important point: *Always remember to reset the **Thickness** system variable to **0** when you've finished drawing an object.*

Let's redraw our walls using **Thickness**.

Do This: 1.3.2A	Drawing with Thickness

I. Be sure you're still in the *My Stick House* file in the C:\Steps3D\Lesson01 folder. If not, please open it now.

II. Follow these steps.

1.3.2A: DRAWING WITH THICKNESS

1. Set the **Thickness** system variable to 2.

   Command: ***th***

   Enter new value for THICKNESS <0.0000>: ***2***

2. Draw the walls as indicated.

   Command: ***l***

   Specify first point: ***6,1***

   Specify next point or [Undo]: ***@3<0***

   Specify next point or [Undo]: ***@2<90***

   Specify next point or [Close/Undo]: ***@3<180***

   Specify next point or [Close/Undo]: ***c***

   The drawing looks like the figure at right.

3. Reset the **Thickness** to ***0***.

   Command: ***th***

Notice that we used simple two-dimensional coordinate entry. The **Thickness** system variable took care of the Z requirements.

Another very useful tool to employ in three-dimensional drafting is the **Elevation** system variable. When we drew our original house, we had to enter a location on the Z-axis as part of the coordinate whenever the value of Z wasn't zero (whenever it was above or below zero-Z). The **Elevation** system variable is designed to minimize the need for entering that third number.

The command looks like this:

Command: ***elevation***

Enter new value for ELEVATION <0.0000>: ***[Enter the desired elevation.]***

Once set, AutoCAD draws all objects at the identified elevation.

> Use the ***Elev*** command to set both the **Elevation** and **Thickness** system variables at once. The command looks like this:
>
> **Command:** ***elev***
>
> **Specify new default elevation <0.0000>:** ***[Enter the elevation.]***
>
> **Specify new default thickness <0.0000>:** ***[Enter the thickness.]***

Let's try using elevation.

Do This: 1.3.2B	**Drawing with Thickness and Elevation**

I. Be sure you're still in the *My Stick House* file in the C:\Steps3D\Lesson01 folder. If not, please open it now.

II. Follow these steps.

1.3.2B: Drawing with Thickness and Elevation

1. Set the **Elevation** to *2* and the **Thickness** to *1.299*.

    Command: ***elev***

    Specify new default elevation <0.0000>: ***2***

    Specify new default thickness <0.0000>: ***1.299***

2. Draw a line as indicated.

    Command: ***l***

    Specify first point: ***6.75,2***

    Specify next point or [Undo]: ***8.25,2***

    Specify next point or [Undo]: ***[ENTER]***

    Notice that, although no Z-axis location is given, AutoCAD assumes an elevation of 2.

3. Reset the **Thickness** to *0*.

    Command: ***th***

    Enter new value for THICKNESS <1.2990>: ***0***

4. Draw the roof as shown.

    Command: ***l***

5. And now let's use the Properties palette to modify the roof. Select the roof locator line and then enter the ***Properties*** command.

    Command: ***props***

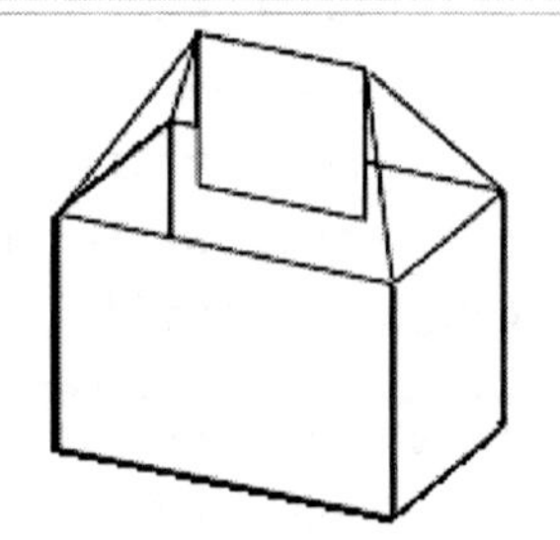

1.3.2B: Drawing with Thickness and Elevation

6. In the **Geometry** section of the Properties palette, pick on **Start Z**. Notice that a **Select objects** button appears to the right Start Z 2.0000.

7. Pick on the **Select objects** button.
AutoCAD highlights that location and allows you to redefine it by picking another point on the screen. Using the Endpoint OSNAP, pick the upper point of that same line (the western point of the roof peak).

8. Repeat Steps 6 and 7 to relocate the **End Z** point. Your drawing looks like the figure at right.

9. Now change the thickness of the line to ***0*** Thickness 0.0000. The **Thickness** property is located in the **General** section of the Properties palette.

10. Clear the grips.
Your drawing now looks like the figure at right.

11. Save the drawing.
Command: ***qsave***

> Of course, it would've been a lot easier to simply erase the roof peak line and redraw it without thickness. But I wanted to give you some experience working with the Properties palette and 3-dimensional objects.

Do the two houses look similar? They should. But don't let that fool you. There are subtle (and remarkable) differences. We'll look at those in a few moments. First, let me list some things to remember about drawing with thickness and elevation (don't you just love my lists?).

- Always remember to reset **Thickness** and **Elevation** to zero when you've finished drawing an object. Otherwise, you may have to redo the next object when it's drawn with incorrect properties.
- Use the Properties palette to change the thickness and elevation of drawn objects.
- You can enter positive or negative values for **Elevation** and **Thickness**.
- Closed objects behave differently from opened objects when drawn with thickness. (You can test this on the different objects in the *Closed Objects* drawing in the C:\Steps3D\Lesson01 folder.)

1.4 Three-Dimensional Viewing Made Easy – Visual Styles

A visual style controls how AutoCAD produces the display you see. You define the visual styles you'll use, and you can even define different styles for different viewports.

> In this section, we'll look at tools designed to help us to see our drawings more clearly. Earlier commands – ***Hide*** and ***Shademode*** –are being phased out in favor of the newer Visual Styles. I'll post the older information on ***Hide*** and ***Shademode*** at http://www.uneedcad.com/Files/Hide&Shademode.pdf for those who wish to see how they worked.

Let's take a look at what's available and how you'll use it. We'll begin with the ribbon's **Visual Styles** panel (Figure 1.010) on the **Render** tab.

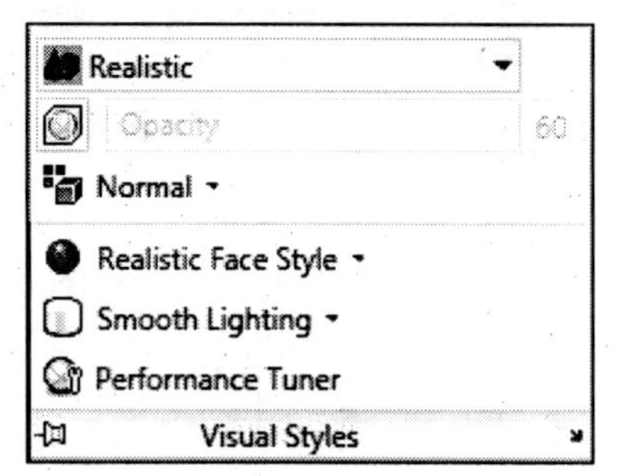

Figure 1.010

- The **Visual Styles** control box (which reads **Realistic** in our figure) offers selections of predefined visual style configurations. These include: **2D Wireframe**, **3D Hidden**, **3D Wireframe**, **Conceptual**, and **Realistic**. You'll find examples of each following. (Note: You can also find the **Visual Styles** control box on the **Home** tab's **View** panel.)
  (Don't let it bother you that **Realistic** doesn't look realistic – I'll show you how to fix that shortly.)

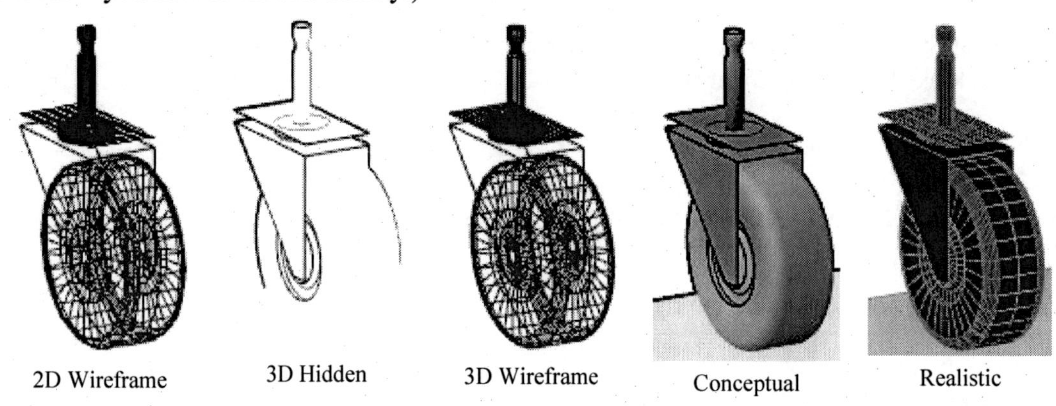

- On the **Visual Styles** panel, you'll also find several buttons (some with flyouts attached). These include:

Button	Description
Visual Styles	This button calls the Visual Styles Manager – a tool palette used to create or modify visual styles. We'll see more on this on p.23.
Realistic No Face Warm-Cool	These subpanel buttons toggle the **VSFaceStyle** system variable to control how your object's faces will appear as follows: • **No Face Style** – (setting of **0**) turns faces off and leaves you with a wireframe. (You can't have both **VSFaceStyle** and **VSEdges** off at once.) • **Realistic Face Style** – (**1**) when you turn isolines off (p.21), this presents a fairly realistic view of the objects in your drawing. • **Warm-Cool Face Style** – (**2**) uses warm/cool tones to soften the contrast.
Facet Smooth Smoothest	These subpanel buttons toggle the **VSLightingQuality** system variable to control how AutoCAD estimates the value of the colors on your model's faces and surfaces. • **Facet** (setting of **0**) computes a single color for each face. • **Smooth** (**1**) computes the face's color as a gradient between opposing vertices. • **Smoothest** (**2**) works with the Per Pixel Lighting found in the Manual Performance Tuner dialog box. When PPL is on, **Smoothest** computes colors per pixel; when off, PPL uses the **Smooth** setting.
Normal Monochrome	These buttons toggle the **VSFaceColorMode** system variable to control how AutoCAD calculates the color of your model's faces. • **Normal** (**0**) doesn't apply a modifier to the face color. • **Monochrome** (**1**) displays faces in the color specified in the **VSMonoColor** system

Button	Description
Tint Desaturated	variable. (AutoCAD uses white as the default color.) • **Tint (2)** shades the face's colors using the color specified in the **VSMonoColor** system variable. • **Desaturated (3)** reduces the saturation of the face's color by 30% to soften its appearance.
X–Ray	This button toggles the **VSFaceOpacity** system variable between 60 and -60. (The range is actually between 100 and -100.) Positive numbers indicate that objects on the screen will be opaque. Use the slider or input box control the degree of transparency.
Performance Tuner (Found in the subpanel)	Calls the Adaptive Degradation and Performance Tuning dialog box. In this dialog box, you can adjust how the AutoCAD display degrades when you reach the limit of your system's resources. My best advice here is that, if your system consistently runs short of resources – spend the money to upgrade it. AutoCAD won't get any less demanding of your system in the future!

When considering the visual quality of your display, you should also consider some tools found on the **Edge Effects** panel (Figure 1.011) also on the ribbon's **Render** tab. This panel deals primarily with the edges found on your 3D model, and most of the tools come with slider bars or control boxes to make adjustment easier.

Take a look.

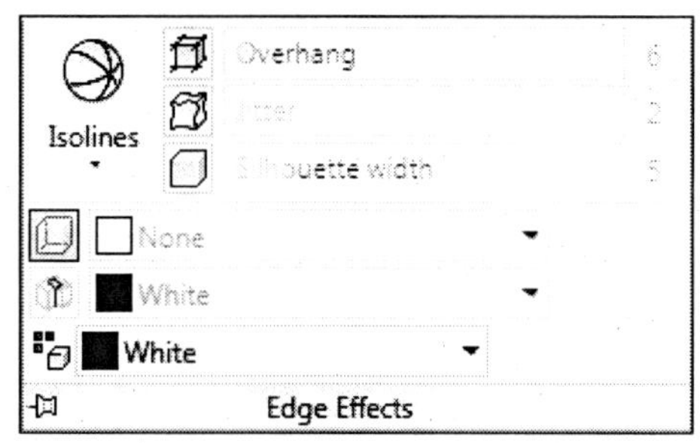

Figure 1.011

Button	Description
Isolines No Edges Faceted Edges	These three buttons toggle the **VSEdges** system variable to control the lines on your objects. • **No Edges** – (setting of **0**) turns off display of lines in objects. This creates the most realistic view of an object. • **Isolines** – (**1**) displays isolines. We'll spend some time with isolines in Lesson 8, p.177. • **Faceted Edges** – (**2**) displays faceted (sharper) edges in your objects.
Edge Overhang	Makes the drawing look semi-hand-drawn. Toggles the **VSEdgeOverhang** system variable between -6 and 6 pixels to control how much of an overhang you get. (You can manually go as high as 100 – but don't.) You can also use the slider bar to adjust the setting.
Edge Jitter	Also makes the drawing look semi-hand-drawn. Toggles the **VSEdgeJitter** system variable between -2 and 2 pixels. Settings include: 1 - Low jitter, 2 - Medium jitter and 3 - High jitter. Any number proceeded by a negative sign turns jitter off. You can also use the slider bar to adjust the setting.
Silhouette Edges	Toggles the **VSSilhEdges** system variable between **0** (off) and **1** (on) to determine whether or not silhouette edges will appear. The slider bar controls the width of the edges.

Button	Description
Obscured Edges (subpanel)	Works when you're using faceted edges to allow you to see hidden edges. Toggles the **VSObscuredEdges** system variable between **0** (off) and **1** (on). Use the control box next to the button to control the color of the obscured faceted edges.
Intersection Edges (subpanel)	Toggles the **VSIntersectionEdges** system variable on (**1**) or off (**2**) to control whether or no you will see intersecting edges. Use the control box next to the button to control the color of the faceted intersection edges.
Edge Color (subpanel)	Allows you to set the color of selected edges using a **Color** control box.

We should also consider shadows. Shadows will become considerably more important when we discuss rendering in our final lesson, but it doesn't hurt to know something about them when setting up your visual styles. You'll find them on the **Lights** panel (**Render** tab).

Button	Description
No Shadows Ground Full	These buttons work with the **VSShadows** system variable to control shadows. From the top, this flyout includes: • **No Shadows** – (setting of **0**) turns shadows off to avoid the distraction and improve system performance. • **Ground Shadows** – (**1**) Helpful for a more realistic look. Works with default lights or user-defined lights. • **Full Shadows** – (**2**) These are shadows cast by one object onto another. Full shadows only work with user-defined lights.

Once you've set things up the way you like them, you can save your style with the ***VSSave*** command. AutoCAD will prompt you for a name like this:

> **Command: *vssave***
>
> **Save current visual style as or [?]: *[Give your new style a name.]***

AutoCAD will save the style and place a visual representation in the **Visual Styles** control box (on the Visual Styles panel) for you to select it whenever you wish.

Where to Find It:	
Command Line:	***VisualStyles***
Hotkey(s):	***vsm***
Ribbon (Tab/Panel):	Render – Visual Styles – **Visual Styles Manager**
Menu:	Tools – Palettes – **Visual Styles**

I know, you've sure looked at a lot of buttons, but we have one more thing to consider before we try an exercise. Consider the Visual Styles Manager (Figure 1.012, p.23). This appears when you enter the ***VisualStyles*** command.

- In the **Available Visual Styles in Drawing** frame at the top, you'll find a panel of visual representations of those styles already defined. When you create your own styles, they, too, will appear here.
- Below the **Available Visual Styles in Drawing** panel, you'll see four buttons:
  - Use the **Create Visual Style** button to create your own style from the current setup.
  - Use the **Apply Selected Visual Style to Current Viewport** button to do just that. (Isn't that simple?)

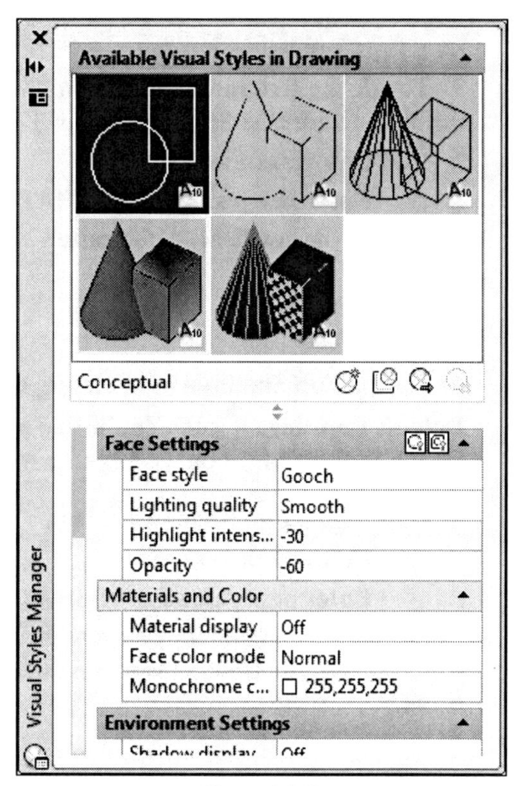

Figure 1.012

- **Export the Selected Visual Style to the Tool Palette** will place a shortcut pick for the selected style on the Visual Styles tool palette.
- Of course, **Delete the Selected Visual Style** removes the selected style. You can't remove the five AutoCAD-provided styles.

- You'll find a few expandable frames below the button. These may include one or some of the following depending upon which style you've selected in the panel above.
  - **Face Settings** – includes options to help you control how faces appear in your drawing.
  - **Materials and Color** – includes options to help you control how AutoCAD displays materials and color.
  - **Environment Settings** – options for background and shadow display.
  - **Edge Settings** – options to control how edges are displayed.
  - Several 2D frames available only when you've selected the **2D Wireframe** setup in the upper panel. Since this is a 3D text, we won't concern ourselves with these.

We've gone too long without an exercise! Let's take a look at how some of this stuff works.

> You can also find the default Visual Styles options on the View pull down menu. Follow this path:
> *View – Visual Styles – [option]*

Do This: 1.4A	Working with Visual Styles

I. Open the *caster-01* file in the C:\Steps3D\Lesson01 folder. It looks like the figure at right.

II. Follow these steps.

1.4A: WORKING WITH VISUAL STYLES

1. Let's start with the easy stuff. Pick each of the options on the **Visual Styles** control box for comparison. The results appear on p.20.

2. Toggle the **X-Ray mode** (**VSFaceOpacity** system variable) to ***60***.

Command: ***vsfaceopacity***

Enter new value for VSFACEOPACITY <-60>: ***60***

Your drawing looks like the figure at right. (Toggle back to ***-60***.)

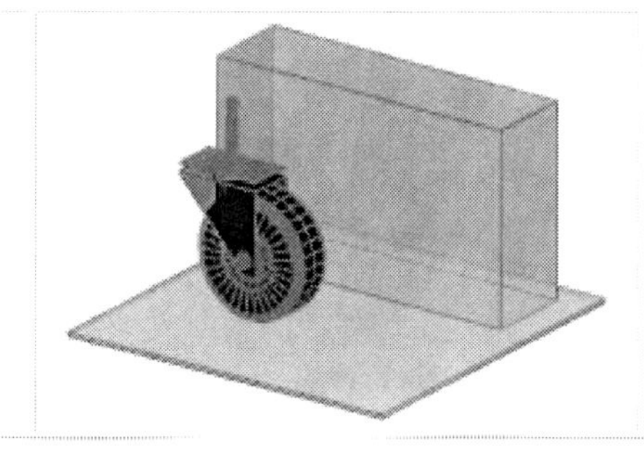

3. Toggle the **Ground Shadows** on . (You'll find the button under the **Shadows** flyout.)

Command: ***vsshadows***

Enter new value for VSSHADOWS <0>: ***1***

Notice the shadow beneath the caster.

4. For the **Full Shadows** to work properly, you must first toggle the **Sun** on and the **Default Lighting** off. You'll find a toggle for the **Sun** on the **Sun & Location** panel and for the **Default Light** on the **Lights** subpanel (**Render** tab).

5. Now toggle **Full Shadows** on.

Command: ***vsshadows***

Enter new value for VSSHADOWS <1>: ***2***

Notice the shadows now. You're viewing shadows from a user-defined spotlight. (More on those in Lesson 12, p.294.)

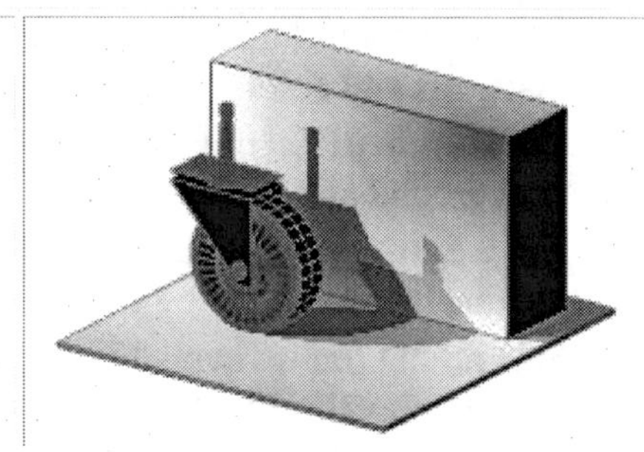

6. Freeze all layers except **Obj5**. This just leaves the box.

7. Experiment with the three edge toggles – **Edge Overhang** , **Edge Jitter** , **Silhouette Edges** . They'll produce the results shown in the following figures.

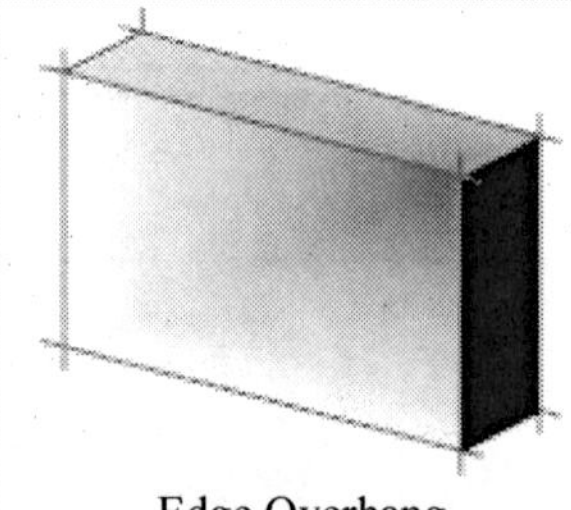

Edge Overhang

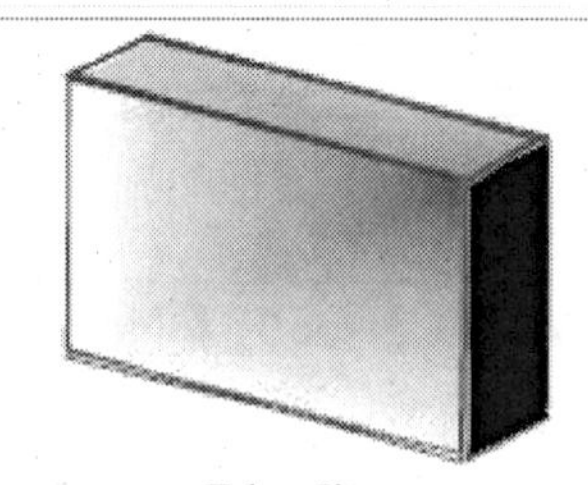

Edge Jitter

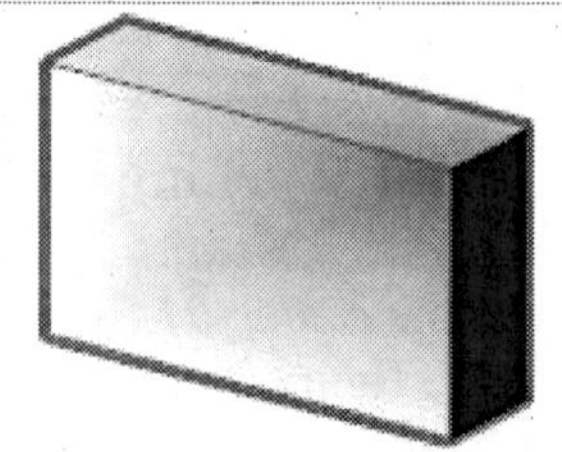

Silhouette Edges

8. Thaw all the layers.

9. Toggle the edge type buttons – **No Edges** , **Isolines** , and **Facet Edges** . The results appear in the following figures.

No Edges

Isolines

Facet Edges

1.4A: Working with Visual Styles

10. Let's set up a Visual Style of our own and save it for future use.

- Use a **Smooth**, **Realistic** face style.
- Turn the **Sun** mode off and **Default Lighting** mode on.
- Use no edges and **Ground Shadows**.

11. Now let's save our Visual Style for later recall. Call the **Visual Style Manager**.

12. In the Visual Style Manger, pick the **Create New Visual Style** button. AutoCAD asks you for a name for your new style. Call it *My First Style* as shown.
Notice that AutoCAD now displays your style in the Visual Style panel (at the top of the manager).

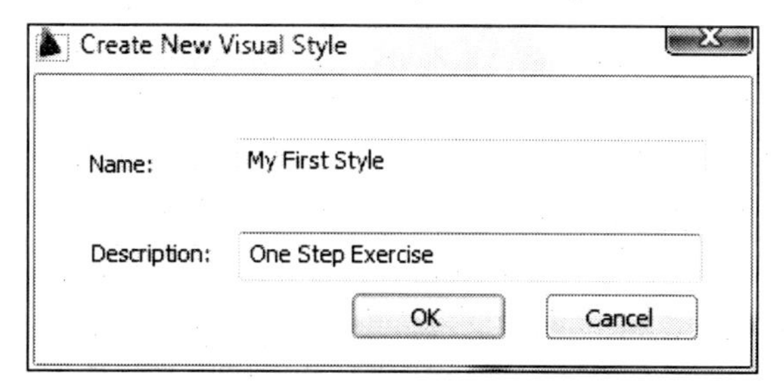

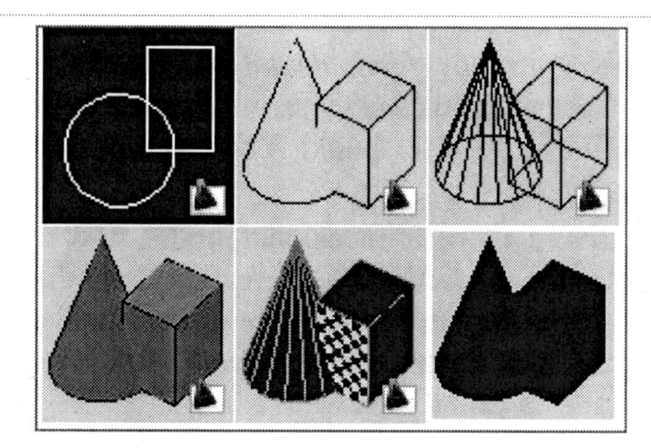

13. Exit the drawing.

1.5 Extra Steps

- Open some of the three-dimensional sample files that ship with AutoCAD (look in the \Sample subfolder of the AutoCAD folder; alternately, you can use the files in the Steps3D\Lesson12 folder). Practice your viewpoint manipulation options and visual styles.
- Print/Plot a drawing or two using Paper Space and Visual Styles.

1.6 What Have We Learned?

Items covered in this lesson include:

- *The UCS icon and the Right-Hand Rule*
- *The Camera*
- *Adjusting your view of the model*
  - *Isometric view*
  - *Dimetric view*
  - *Trimetric view*
  - *Plan view*
  - *Coordinate approach*
  - *Compass approaches*
  - *Dialog box approach*
  - *View Cube approach*
- *Coordinates in a three-dimensional drawing*

- *Cartesian Coordinate entry*
- *Spherical and Cylindrical Coordinate entry*
- *Visual Styles*
- *Commands*
  - ***Ucsicon***
  - ***VPoint***
  - ***VPoint***
  - ***Thickness***
  - ***Elevation***
  - ***Elev***
  - ***Vssave***
  - ***VSShadows***
  - ***VSFaceOpacity***
  - ***VSIntersectionEdges***
  - ***VSObscuredEdges***
  - ***VSSilhEdges***
  - ***VSEdgeJitter***
  - ***VSEdgeOverhang***
  - ***VSEdges***
  - ***VSFaceStyle***
  - ***VSLightningQuality***
  - ***VSMoveColor***
  - ***VSFaceColor***
  - ***NavvCube***
  - ***Camera***

What a list! Stop and catch your breath!

We covered a tremendous amount of new material in this lesson. But it's all fundamental to three-dimensional drafting. After some practice, you'll find this material as easy as two-dimensional work. (Well, almost.)

I must caution you, however, about just how *fundamental* this material is. You must achieve at least a small degree of expertise with these methods and those in Lesson 2 to be able to function effectively in Z-space. But that's the benefit of a good textbook! You can repeat the lesson(s) until you're comfortable.

We've seen how to maneuver around a three-dimensional model, how to improve our view for better navigation (and aesthetics), and some fundamentals of creating three-dimensional objects. But there's much more to consider! How do you draw a three-dimensional object at an angle – such as creating a solid roof instead of the stick figure outline on your stick house? Is there an easier way to rotate your viewpoint in Z-space for a better view?

We'll continue our study of Z-Basics in Lesson 2, p.33, where we'll answer some of these questions (and present some others). But first, we should practice what we've learned so far.

1.7 Exercises

1. through 8. Create the "w" drawings in Appendix B (Refer to the "su" drawings for a clearer image.) Follow these guidelines.
    1.1. Don't try to create the dimensions yet.
    1.2. Start each drawing from scratch and use the default settings.
    1.3. Adjust the viewpoint as needed to help your drawing.
    1.4. Save each drawing as *My [title]* (as in *MyB-1w*) in the C:\Steps3D\Lesson01 folder.

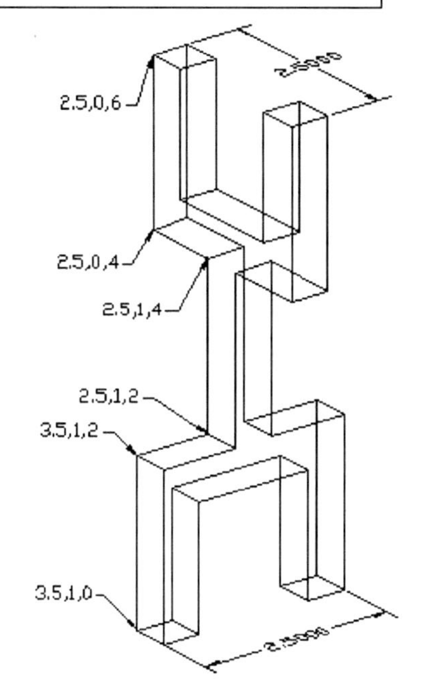

Twisted "Y"

9. Create the Twisted "Y" drawing according to the following parameters:
    9.1. Start the drawing from scratch and use the default settings.
    9.2. Set the viewpoint to 8.5,11,5.75.
    9.3. Don't attempt to draw the dimensions yet.
    9.4. Use a thickness setting of 0.
    9.5. The depth of the piece is ½".
    9.6. Save the drawing as *My Twisted Y* in the C:\Steps3D\Lesson01 folder.
10. Create the 3-Dimensional Block drawing (p.29), according to the following parameters:
    10.1. Start the drawing from scratch and use the default settings.
    10.2. Don't attempt to draw the dimensions yet.
    10.3. Text is in Paper Space and is 3/16" and 1/8".
    10.4. Title block text is ¼", 3/16", and 1/8".
    10.5. The viewpoint for each viewport is indicated.
    10.6. Use lines with thickness.
    10.7. Use the 3D Wireframe visual style for all views except the diametric; use the 3D Hidden visual style to achieve the diametric view.
    10.8. Save the drawing as *My Block* in the C:\Steps3D\Lesson01 folder.
11. Create the Grill drawing (p.29) according to the following parameters.
    11.1. Start the drawing from scratch and use the default settings.
    11.2. Use polylines with 1/16" width for everything except the handle.
    11.3. Use a thickness of 1/16" for all objects.
    11.4. I used a scale of 1:4 in the plan and elevation views but no scale in the isometric.
    11.5. Don't attempt to draw the dimensions yet.
    11.6. Use the 3D Wireframe visual style for all views.
    11.7. Text is in Paper Space and is 3/16" and 1/8".
    11.8. Title block text is ¼", 3/16", and 1/8".
    11.9. Save the drawing as *MyGrill* in the C:\Steps3D\Lesson01 folder.
12. Create the Bookend drawing (p.30) according to the following parameters:
    12.1. Start the drawing from scratch and use the default settings.
    12.2. I used a scale of 1:2 for all views.
    12.3. Don't attempt to draw the dimensions yet.

12.4. Text is in Paper Space and is 3/16” and 1/8”

12.5. Title block text is ¼”, 3/16”, and 1/8”.

12.6. You’ll find it easier to fillet the corners rather than drawing arcs.

12.7. Use the 3D Wireframe visual style for all views.

12.8. Save the drawing as *MyBookEnd* in the C:\Steps3D\Lesson01 folder.

13. Create the Magazine Rack drawing (p.30) according to the following parameters:

13.1. Start the drawing from scratch and use the default settings.

13.2. Use 0.6mm lineweights for all lines.

13.3. Don’t attempt to draw the dimensions yet.

13.4. Use the 3D Wireframe visual style for all views.

13.5. Text is in Paper Space and is 3/16” and 1/8”

13.6. Title block text is ¼”, 3/16”, and 1/8”.

13.7. I used ½” diameter circles with ½” thickness for the legs.

13.8. Save the drawing as *MyMagRack* in the C:\Steps3D\Lesson01 folder.

14. Create the Anchor Stop drawing (p.31) according to the following parameters:

14.1. Start the drawing from scratch and use the default settings.

14.2. Don’t attempt to draw the dimensions yet.

14.3. Text is in Paper Space and is 3/16” and 1/8”

14.4. Title block text is ¼”, 3/16”, and 1/8”.

14.5. The viewpoint for each viewport is indicated. Use the 3D Wireframe visual style for all views.

14.6. Use lines without thickness except when drawing the slot.

14.7. Save the drawing as *MyAnchorStop* in the C:\Steps3D\Lesson01 folder.

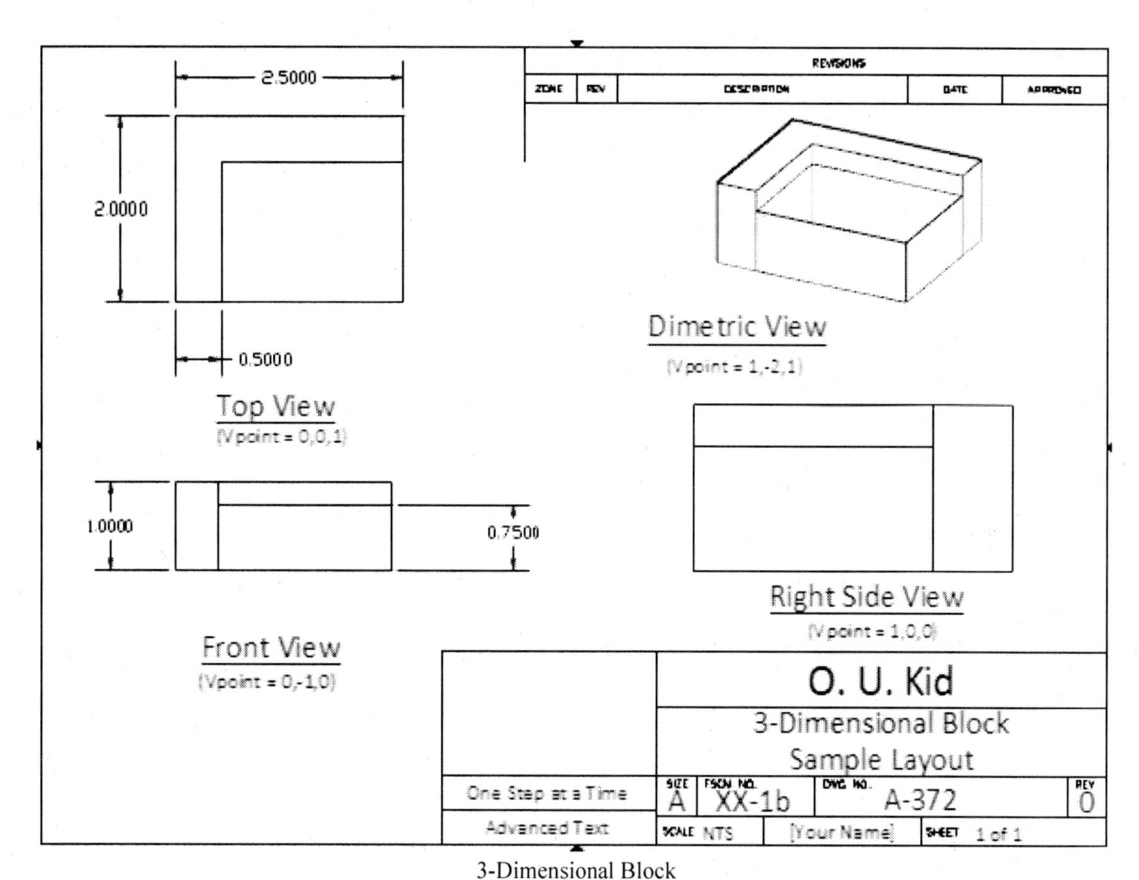

REVISIONS
ZONE
REV
DESCRIPTION
DATE
APPROVED
2.5000
2.0000
0.5000
Top View
(Vpoint = 0,0,1)
Dimetric View
(Vpoint = 1,-2,1)
1.0000
0.7500
Front View
(Vpoint = 0,-1,0)
Right Side View
(Vpoint = 1,0,0)
O. U. Kid
3-Dimensional Block
Sample Layout
One Step at a Time
Advanced Text
SIZE A
FSCM NO. XX-1b
DWG NO. A-372
REV 0
SCALE NTS
[Your Name]
SHEET 1 of 1

3-Dimensional Block

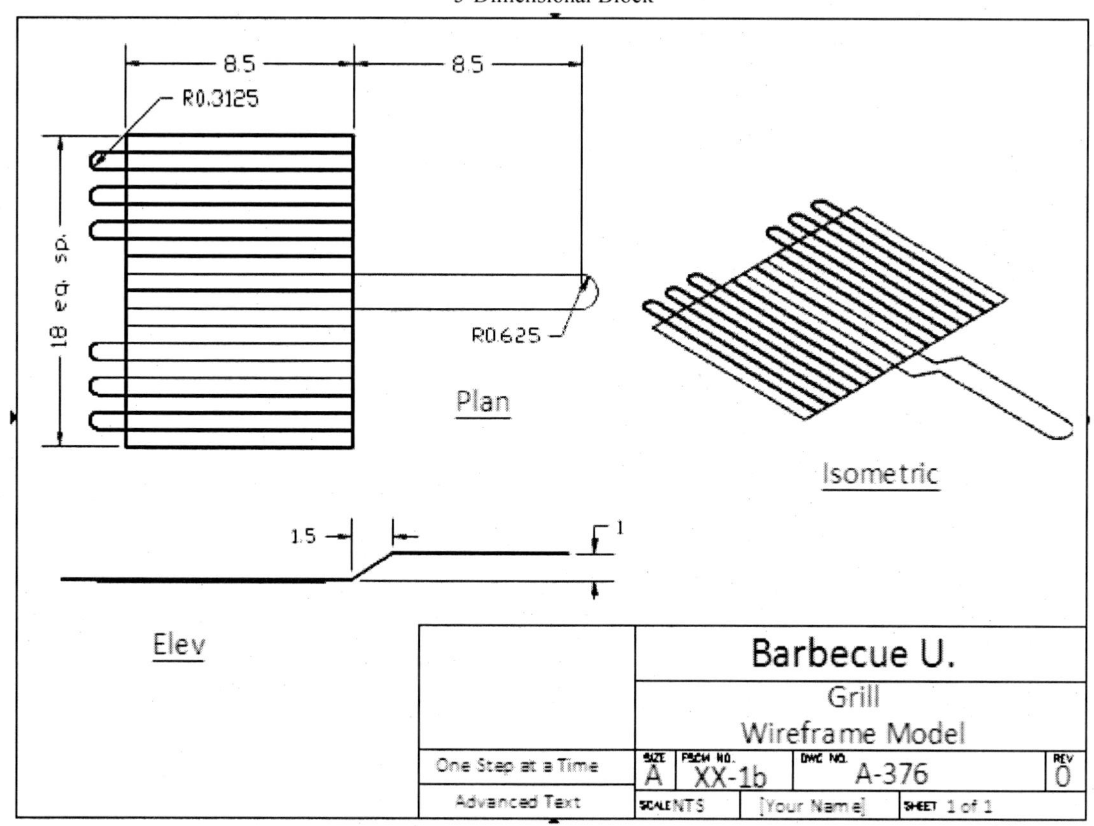

8.5
8.5
R0.3125
18 eq. sp.
R0.625
Plan
Isometric
1.5
1
Elev
Barbecue U.
Grill
Wireframe Model
One Step at a Time
Advanced Text
SIZE A
FSCM NO. XX-1b
DWG NO. A-376
REV 0
SCALE NTS
[Your Name]
SHEET 1 of 1

Grill

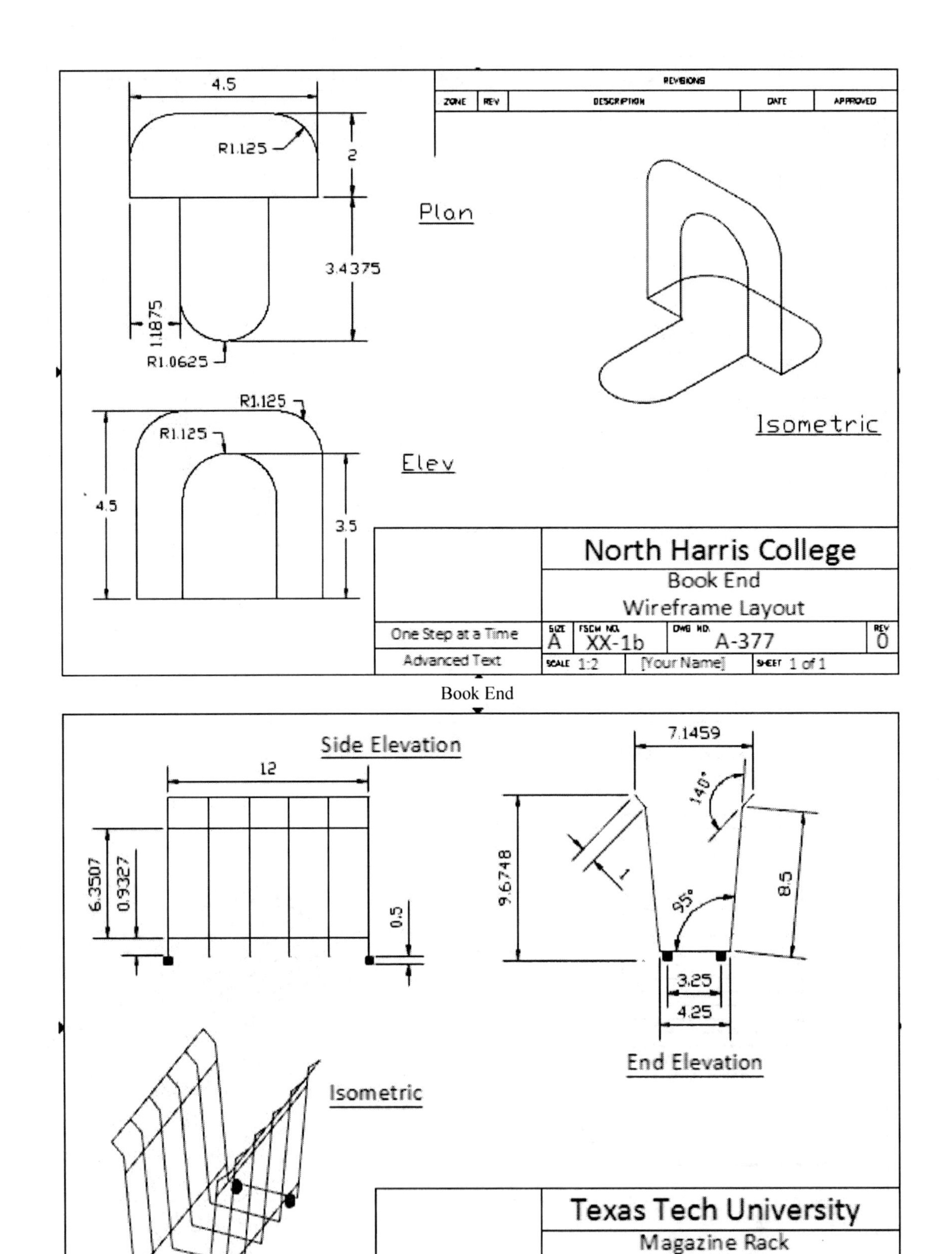

REVISIONS
ZONE
REV
DESCRIPTION
DATE
APPROVED
4.5
R1.125
2
Plan
3.4375
1.1875
R1.0625
R1.125
R1.125
Elev
4.5
3.5
Isometric
North Harris College
Book End
Wireframe Layout
One Step at a Time
Advanced Text
SIZE A
FSCM NO. XX-1b
DWG NO. A-377
REV 0
SCALE 1:2
[Your Name]
SHEET 1 of 1

Book End

Side Elevation
12
6.3507
0.9327
0.5
7.1459
140°
9.6748
1
95°
8.5
3.25
4.25
End Elevation
Isometric
Texas Tech University
Magazine Rack
Wireframe Layout
One Step at a Time
Advanced Text
SIZE A
FSCM NO. XX-1b
DWG NO. A-378
REV 0
SCALE NTS
[Your Name]
SHEET 1 of 1

Magazine Rack

15. Here's a challenge. Create the Chess Queen drawing (p.32) according to the following parameters:
    15.1. Start the drawing using *template #1* found in the C:\Steps3D\Lesson01 folder.
    15.2. Don't attempt to draw the dimensions yet.
    15.3. Text is in Paper Space and is 3/16" and 1/8".
    15.4. Title block text is ¼", 3/16", and 1/8".
    15.5. The crown is made up of solids drawn with thickness.
    15.6. The round pieces are donuts drawn with thickness.
    15.7. Use the 3D Wireframe visual style for all views except the diametric; use the Realistic visual style for the diametric.
    15.8. Save the drawing as *MyQueen* in the C:\Steps3D\Lesson01 folder.

16. Here's another challenge. Create the Scientific Ring Stand drawing (p.32) according to the following parameters:
    16.1. Start the drawing using *template #1* found in the C:\Steps3D\Lesson01 folder.
    16.2. Don't attempt to draw the dimensions yet.
    16.3. Text is in Paper Space and is 3/16" and 1/8"
    16.4. Title block text is ¼", 3/16", and 1/8".
    16.5. Polylines have a width of 1/16".
    16.6. Anchors are donuts with an outer diameter of 9/16" and an inner diameter of 3/16".
    16.7. The nut is a six-sided polygon inscribed in a radius of ¼"; its width is 1/16".
    16.8. Use the 3D Wireframe visual style for all views except the diametric; use the Realistic visual style for the diametric.
    16.9. Save the drawing as *MyRingStand* in the C:\Steps3D\Lesson01 folder.

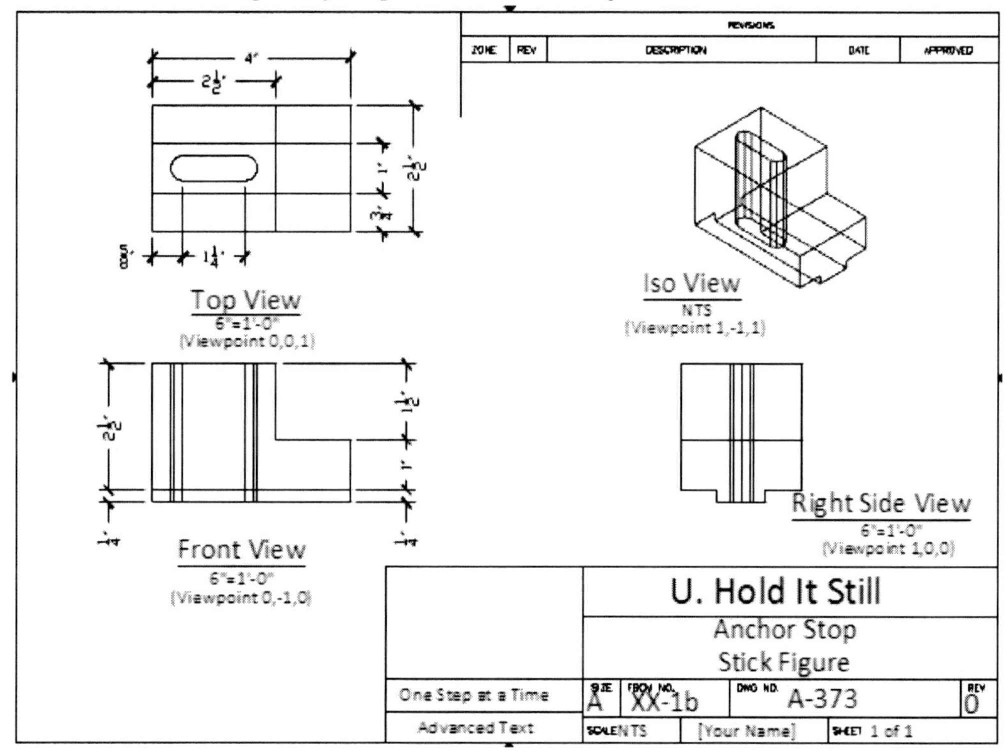

Anchor Stop

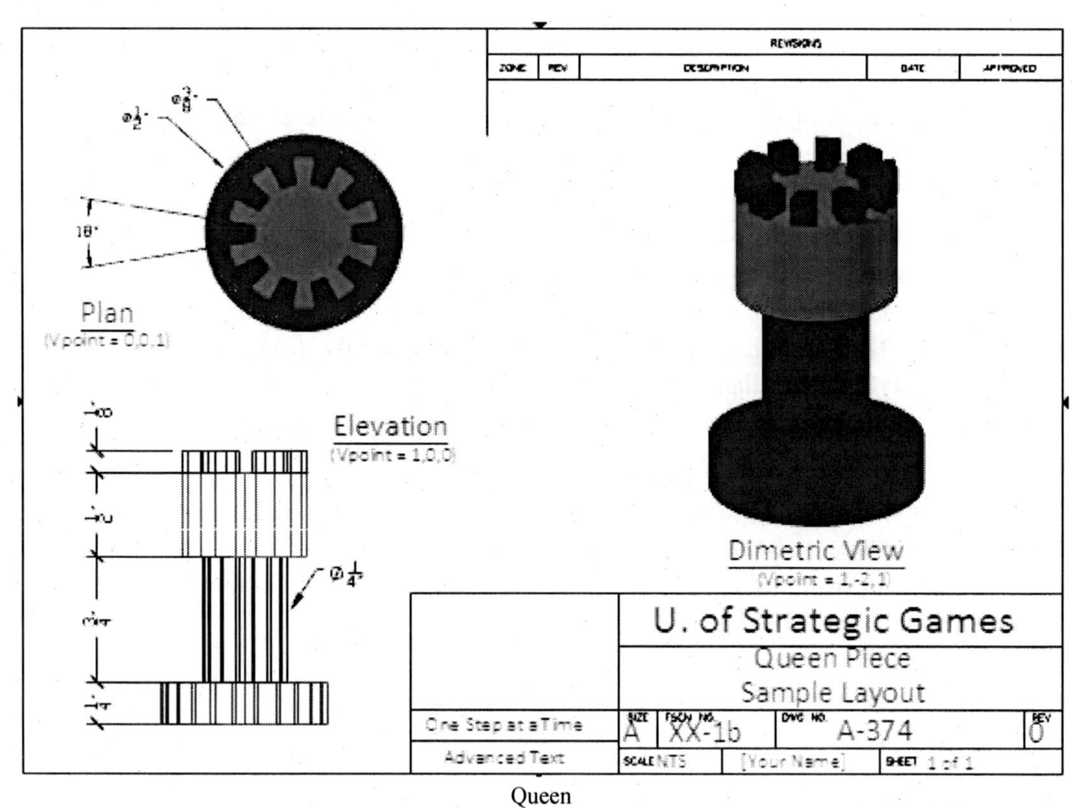

Queen

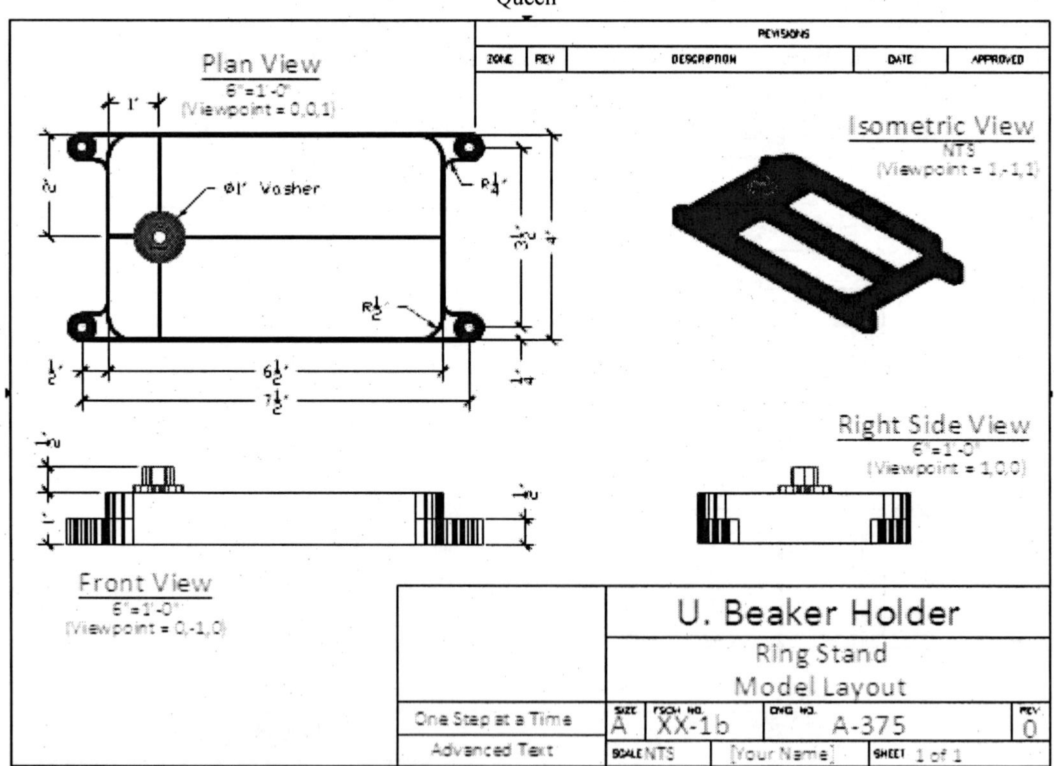

Ring Stand

1.8	**For Web-Based Review Questions, visit: http://foragerpub.com/AcadFiles/2010/2010.htm**

Lesson

Following this lesson, you will:

✓ *Understand the differences between the UCS and the WCS*

- *Know how to use working planes and the dynamic UCS*
- *Know how to use the UCS Dialog Box (Manager)*
- *Know how to dimension a three-dimensional drawing*

✓ *Be familiar with some other viewing techniques*

- *Be able to use the Navigation Wheels*
- *Be able to use the* ***3DDistance*** *and* ***3DSwivel*** *commands*

✓ *Be able to use the* ***3DClip*** *command*

More of Z Basics

When we first drew our three-dimensional stick figure house, it had no walls and only sticks for a roof. Later we saw that, by using the Thickness and Elevation system variables, we could make our walls solid. But what about the roof? There must be an AutoCAD tool for making it solid as well.

We also discussed the need for point entry precision in Z-space. What about text and dimensions? Must we use coordinates to place them? (Have you tried to dimension a three-dimensional drawing yet?) And what if you want to place text along a slope (like the roof)?

Believe it or not, all of these questions have the same answer! The answer lies in a tool called the User Coordinate System – the UCS.

In Lesson 1, you learned how to create simple three-dimensional objects and how to view those objects from different angles. In this lesson, we'll discuss the tools with which you'll work on the different faces of your three-dimensional objects. Then we'll look at some more viewing tools.

Let's start with the UCS.

2.1 WCS vs. UCS

Although you may not be aware of it, you're already familiar with the UCS. You've been using it since you began your study of AutoCAD. However, it's always been aligned with the World Coordinate System – the WCS – so you never noticed it. So what's the difference? You need to understand a little about how AutoCAD works to really understand the UCS.

To avoid any chance of damage to the WCS, AutoCAD placed it out of your reach.

In the basic text, you learned that all objects in an AutoCAD drawing are defined by information stored in that drawing's database. When you regenerate a drawing, AutoCAD reads this information and restores the drawing accordingly. When you create an attributed block, the attribute information is also stored in the database.

Part of the information stored in a drawing's database is the location and orientation of each object. For consistency (and to avoid a programming nightmare), AutoCAD developed a coordinate system that remains the same throughout the life of the drawing. Thus, the definition of point 0,0,0 will remain the same, and the point will always be in the same place. The X, Y, and Z directions will never change.

This coordinate system is the WCS.

AutoCAD uses the WCS point 0,0,0 much as a mariner uses the North Star. Unchanging in the night sky, Polaris shows the sailor the way home. Unchanging in the WCS, point 0,0,0 shows AutoCAD how to orient any object in the drawing.

The User Coordinate System – or UCS – is what the CAD operator uses to determine up from down and left from right for the immediate task at hand. Until now, the UCS has always been aligned with the WCS – our mariner's ship has always been pointed toward the North Star. Until now, the front of our ship has always been north; the masts have always pointed upward. With a compass, our sailor could find anything on the ship. Until now, we had only two dimensions with which to create a drawing – our sailor had only a single deck on which to work. Until now, the CAD operator had no need to know about the UCS or WCS. Until now…

Sigh. It used to be so easy …

Now our ship has changed directions and acquired new decks. Our sailor with the compass is completely lost. To make it easier for him to understand where things are on the ship, he must change his reference point from the North Star to something on the ship itself. This way, he can always find his way regardless of the direction in which the ship is sailing.

He'll use the mainsail as his reference point (0,0,0). His compass directions will now reference points on the ship – the bow becomes north, the stern becomes south, starboard and port become east and west. The mainsail will always point upward from the main deck (regardless of how violently the sea rocks the ship). So now, using the North Star, our mariner's ship will never be lost, and using his new Mariners Coordinate System, he'll never be lost on the ship.

CAD operators must adjust for Z-space as our mariner adjusted for a change in the ship's direction. Using the 0,0,0 coordinate of the WCS as our North Star, we can always find our way home. And using the User Coordinate System as our sailor uses his Mariner's Coordinate System, we can work on any surface – or working plane – of a three-dimensional model (as our mariner could work on any deck of his ship).

So how do you use the UCS? Where do you begin?

> Mercifully, AutoCAD's UCS is dynamic – that is, it will change automatically when you pass over a face of a 3-dimensional solid object. Unfortunately, not all objects are solid, so we'll spend some time learning to manually change the UCS.

Where to Find It:	
Command Line:	***UCS***
Ribbon:	View – Coordinates – **UCS**
Menu:	Tools – New UCS – [**option**]
Toolbar:	UCS – **UCS**

Let's begin by accessing the ***UCS*** command. What do you suppose the command would be?

To keep it simple, AutoCAD calls the command ***UCS***! The command line approach looks like this:

> **Command: *UCS***
> **Current ucs name: *WORLD***
> **Specify origin of UCS or [Face/NAmed/OBject/Previous/View/World/X/Y/Z/ZAxis] <World>:** ***[Enter an option or specify (locate) the origin.]***

If you **Specify origin**, AutoCAD prompts

> **Specify point on X-axis or <Accept>:** ***[Tell AutoCAD through which point the X-axis will go.]***
> **Specify point on the XY plane or <Accept>:** ***[Tell AutoCAD in which direction you want the y-axis to go.]***

With this procedure, AutoCAD will determine the direction for Z from the locations of the X & Y axes.

Let's look at the other options. We'll begin with the options listed at the command prompt and their corresponding buttons on the ribbon's **UCS** panel (**View** tab), then look at some which don't show up with the other prompts.

- The **Face** option allows you to define the new UCS by selecting a face on an existing three-dimensional solid object.
- The **NAmed** option allows you to **Save** a new UCS (define a new 0,0,0 and new orientations for the X-, Y-, and Z-axes), **Restore** a named UCS, or **Delete** a named UCS. You can even have AutoCAD list the named UCSs with the question mark (**?**).

  It prompts

  > **Enter an option [Restore/Save/Delete/?]:**
- The **OBject** option allows you to define the new UCS by selecting an existing three-dimensional object. (Remember, a line without thickness is two-dimensional regardless of its orientation.) AutoCAD aligns the new UCS with that object.
- **Previous** restores the previous UCS.
- **View** sets the UCS flat against the screen (the X-axis parallel to the bottom of the screen and the Y axis parallel to the left side of the screen). The 0,0,0 coordinate remains where it is currently located.

- The **World** option is the most important (so important, in fact, it deserves "Hail Mary" status). It tells AutoCAD to restore the UCS to match the WCS's orientation. In other words, when you're lost, use the World option to reorient yourself.
- The **X**, **Y**, **Z** options allow you to rotate the UCS around the selected axis. They prompt

  Specify rotation angle about X *[or Y or Z]* axis <90>:
- Using the **Z Axis Vector** option, you'll define the new UCS by identifying a location for 0,0,0, and then a point on the Z-axis.

Some options appear on the ribbon panel and toolbars that don't appear on the command prompt, but they're well worth examination.

- The **3point** option is the same as the default **Specify origin** option.
- The **Origin** button provides the opportunity to change the UCS origin without changing the XYZ planes. If you accept the defaults at the **Specify origin** option of the command line, you'll accomplish the same thing.

Some other options include:

- **Move** (entering ***M*** at the UCS prompt) allows you to move the UCS without changing its orientation. It prompts

  Specify new origin point or [Zdepth]<0,0,0>:

  Respond by picking a new origin point or by selecting the **Zdepth** option. The **Zdepth** option allows you to move the origin along the Z-axis.
- The **orthoGraphic** option (entering ***G*** at the UCS prompt) offers a quick way to change the UCS to one of the standard orthographic projections without moving 0,0,0 from the WCS location. It prompts

  Enter an option [Top/Bottom/Front/BAck/Left/Right]<Top>:

Most of this will become clearer with practice. Let's try a get-acquainted exercise.

Do This: 2.1A	Manipulating the UCS

I. Open the *ucs practice2* file in the C:\Steps3D\Lesson02 folder. The drawing looks like the figure at right.

II. Follow these steps.

2.1A: MANIPULATING THE UCS
1. Our first step is one of the most important steps to remember when manipulating the UCS. Set the UCS icon to **ORigin**. This way, you'll always know where 0,0,0 is. **Command: *ucsicon*** **Enter an option [ON/OFF/All/Noorigin/ORigin Properties] <ON>: *or***
2. Now we'll tell AutoCAD to use the UCS (our mariner must navigate within the ship). Enter the ***UCS*** command. **Command: *ucs***
3. Our first UCS will simply relocate 0,0,0. We'll use the unlisted **Move** option. **Current ucs name: *WORLD*** **Specify origin of UCS or [Face/NAmed/OBject/Previous/View/World/X/Y/Z/ZAxis] <World>: *m***

2.1A: MANIPULATING THE UCS

4. Specify the FRONT-BOTTOM-LEFT corner of the object …

Specify new origin point or [Zdepth] <0,0,0>:

Notice that the UCS icon moves. It's locating 0,0,0 as we told it to do in Step 1.

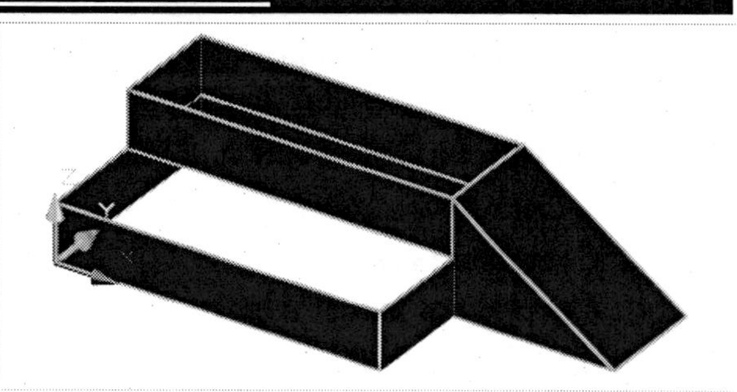

5. Now we'll save this UCS for later retrieval. Repeat the ***UCS*** command.

Command: ***[ENTER]***

6. AutoCAD responds by asking what you would like to do. Select the **NAmed** option NAmed.

Current ucs name: *WORLD*

Specify origin of UCS or [Face/NAmed/OBject/Previous/View/World/X/Y/Z/ ZAxis] <World>: ***na***

7. Tell AutoCAD you wish to save Save the UCS …

Enter an option [Restore/Save/Delete/?]: ***s***

8. … and call it something appropriate.

Enter name to save current UCS or [?]: ***lower left base***

9. Let's create another UCS. Repeat the command. (Alternately, the **3 Point UCS** button will work.)

Command: ***ucs***

10. Select the points indicated.

Current ucs name: lower left base

Specify origin of UCS or [Face/NAmed/OBject/Previous/View/ World/X/ Y/Z/ZAxis] <World>:

Specify point on X-axis or <Accept>:

Specify point on the XY plane or <Accept>:

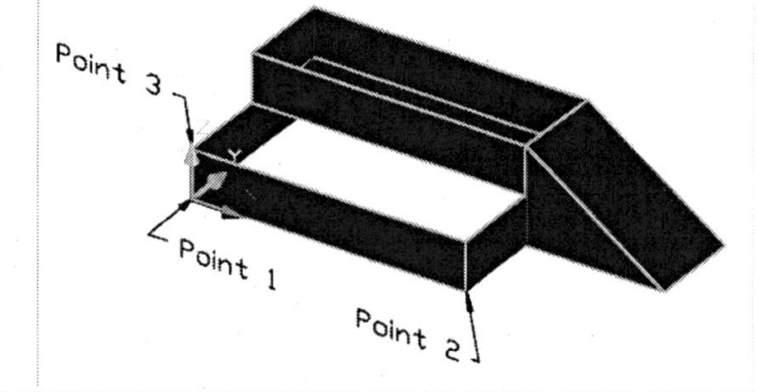

11. The UCS icon now looks like the figure at right.

Before continuing, stop and ask yourself where the X-, Y-, and Z-axes are located. Use the UCS icon, the ViewCube, and the Right-Hand Rule to help answer that question.*

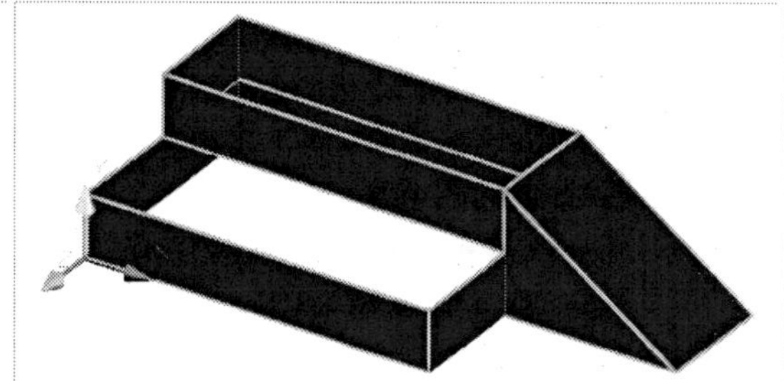

12. Repeat Steps 5 through 8 to save this UCS as **MyFront**.

* The UCS icon indicates the X-, and Y-, and Z-axes.

13. Now let's look at some other ways to set the UCS. You can select the option of the ***UCS*** command for the next several steps, or you can simply pick the button indicated.

Let's begin with the **Object** option. Enter the sequence shown or pick the **Object UCS** button on the **Coordinates** panel.

Command: *ucs*

Current ucs name: MyFront

Specify origin of UCS or [Face/NAmed/OBject/Previous/View/World/X/Y/Z/ ZAxis] <World>: *ob*

14. Select the front of the object closest to the lower-left corner of the screen.

Select object to align UCS:

The UCS icon now looks like the figure at right. The 0,0,0 coordinate of the UCS matches the line's end point closest to where you picked. The Y- and Z-axes also line up according to the line's definition.

15. Restore the **Previous** UCS.

Command: *ucs*

Current ucs name: *NO NAME*

Specify origin of UCS or [Face/NAmed/OBject/Previous/View/World/X/Y/Z/ ZAxis] <World>: *p*

16. Let's look at the dynamic UCS. (The wedge piece is a three-dimensional solid – an item we'll discuss in Lesson 8, p.182. It was necessary to include it here for demonstration purposes.)

Enter the ***Line*** command and move your cursor over the angled face of the solid to see how the dynamic UCS works. (Be sure the DUCS toggle on the status bar is lit.) Notice that the crosshairs change to indicate the dynamic UCS.

17. Begin the line around the center of the angled face. Notice that the UCS icon moves and changes to reflect the UCS of the face.

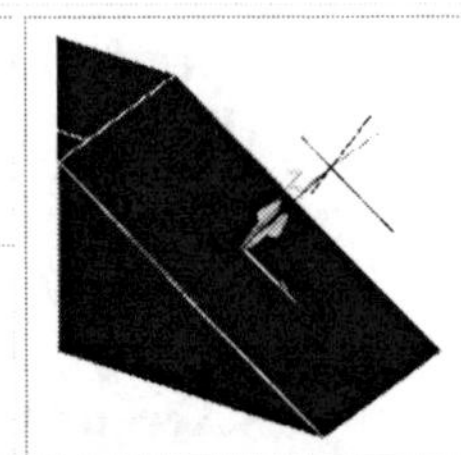

18. Cancel the command Esc.

19. Now set the UCS according to the current view. Enter the sequence shown or pick the **View UCS** button. Notice the new orientation of the UCS icon. Notice also that, although it changes orientation, it doesn't change position.

Command: *ucs*

Current ucs name: MyFront

Specify origin of UCS or [Face/NAmed/OBject/Previous/View/World/X/Y/Z/ ZAxis] <World>: *v*

20. Let's experiment with the X/Y/Z options. Watch the UCS icon as we rotate the UCS.

Select the **X** button.

2.1A: MANIPULATING THE UCS

21. Accept the **90°** default and watch the UCS icon.

    Specify rotation angle about X axis <90>: ***[ENTER]***

It can be difficult to follow axial rotations, but that's where the ViewCube comes in handy.

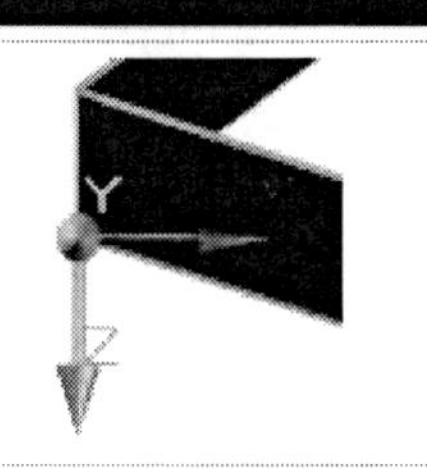

22. Repeat Steps 20 and 21 for the **Y** and **Z** options.

23. Using any of the tools just discussed, create the new UCS setups identified in the following figures. Save the setups as indicated.

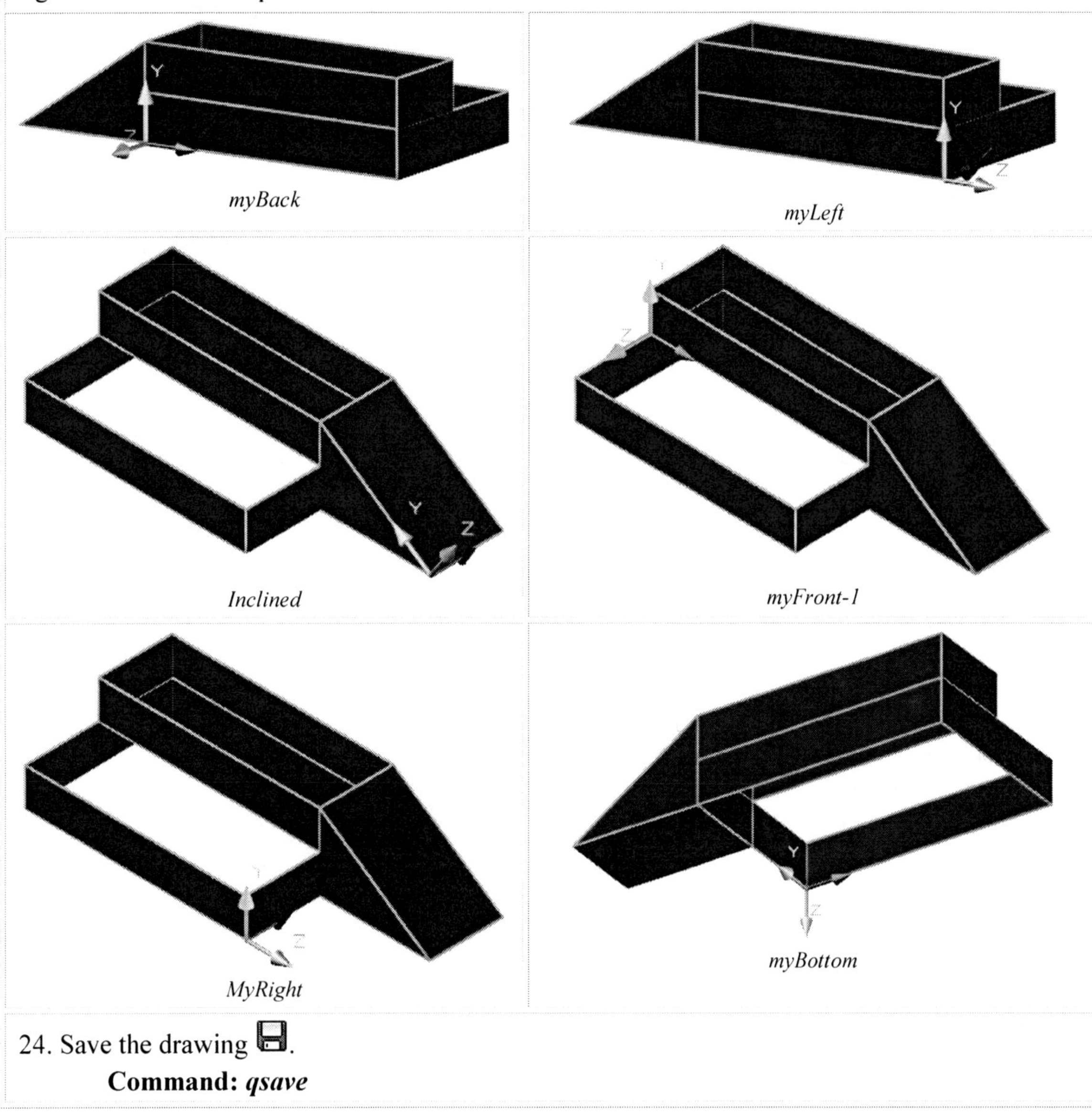

myBack

myLeft

Inclined

myFront-1

MyRight

myBottom

24. Save the drawing.

    Command: ***qsave***

You've created several UCS setups using a variety of methods. It's okay to be a bit confused at this point. Let's pause for a moment, catch our breath, and review what we've learned. Here are some important things to remember.

- Despite their similarities, viewpoints, viewports, and the UCS are three different things. Remember:

- Viewpoints are *points* (where you stand) *from which you view* the model.
- Viewports are like *port*holes in a ship *through which you view* the model.
- The UCS orients you *on* the model itself.

COOL STUFF

- Points used to define viewpoints will always reference the WCS (*not* the UCS). This will make it easier for you to get your bearings.

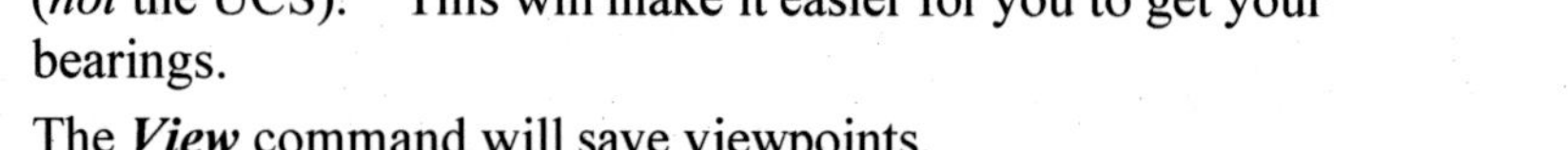

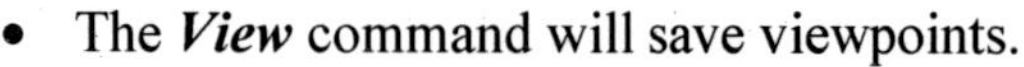

- The ***View*** command will save viewpoints.
- Each viewport can have a unique viewpoint and/or UCS assigned to it.
- Each – viewpoint, viewport, and UCS – works independently of the other two, but all should be considered as a team to assist you when you work on a three-dimensional model.

Let's take a moment to look at a tool that might make UCS management easier. Then we'll use the working planes to create some lines, text, and dimensions on our stick house.

2.2 The UCS Dialog Box (Manager)

Over the course of creating a drawing – particularly a larger drawing – you may find it necessary to create several UCSs. You've already seen how to save these setups for later retrieval, but where do you keep the list of names you've assigned the UCSs?

Where to Find It:	
Command Line:	***UCSMan***
Hotkey(s):	***uc***
Ribbon (Tab/Panel):	View – Coordinates – **Named**
Menu:	Tools – **Named UCS**
Toolbar:	UCS II – **Named UCS**

AutoCAD makes it simple with the UCS Manager (Figure 2.001). This provides a dialog box approach to keeping track of the various setups as well as some additional tools. Access the UCS Manager with the ***UCSMan*** command.

Let's take a look.

- The first tab presents a list of **Named UCSs** as well as a **World** option. To set a UCS current, either double click on its name or select the name and pick the **Set Current** button. Then pick the **OK** button to complete the procedure. It's that simple – no need to remember or to store a list of names! (But it's always a good idea when assigning the names to make them self-explanatory.)

  The **Details** button presents a dialog box (Figure 2.002, p.41) that indicates the origin location, as well as the orientation of the X-, Y-, and Z-axes for the currently selected UCS. You view the data in relation to the WCS or any other existing UCS by selecting the coordinate system from the **Relative to** control box.

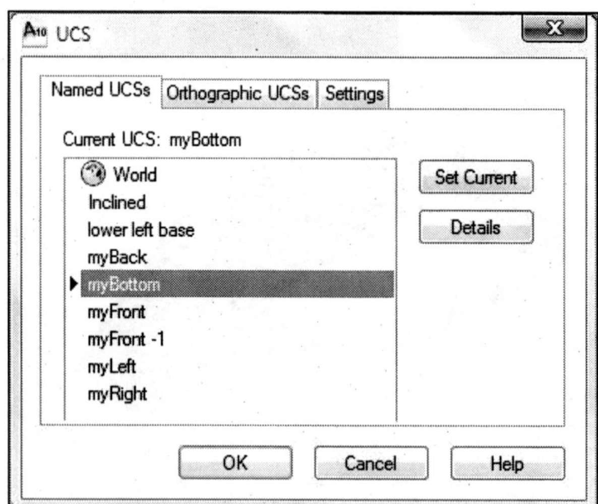

Figure 2.001

- The middle tab (Figure 2.003) – **Orthographic UCSs** – lists several standard UCSs (the same ones available using the **orthoGraphic** option of the ***UCS*** command). Use the same procedure you used on the **Named UCSs** to set one current.

  Notice the **Relative to** control box on this tab. You can set one of the orthographic UCSs relative to the WCS or relative to one of the user-defined UCSs. I suggest leaving this control set relative to the WCS at least for now. Any other setting might make it difficult to orient yourself.
- The **UCS icon settings** frame of the **Settings** tab (Figure 2.004) is a dialog box interface for the ***UCSIcon*** command.

  Use the **UCS settings** frame to **Save UCS with viewport** (the default). This means that each viewport can have its own UCS setting. Clear this check and each viewport will reflect the UCS settings of the current viewport.

  A check next to **Update view to Plan when UCS is changed** will regenerate the viewport in a plan view of the current UCS whenever the UCS settings are changed. I suggest leaving this box clear since you're more likely to want to keep the view even if you change the UCS.

We'll use the UCS dialog box in our next exercise to help us see how using different UCS settings can benefit us.

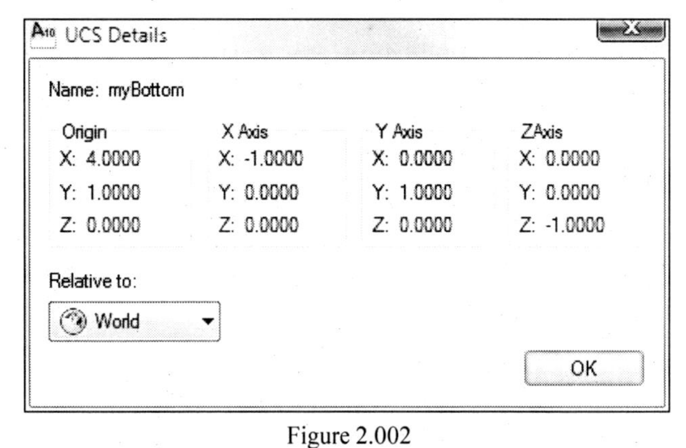

Figure 2.002

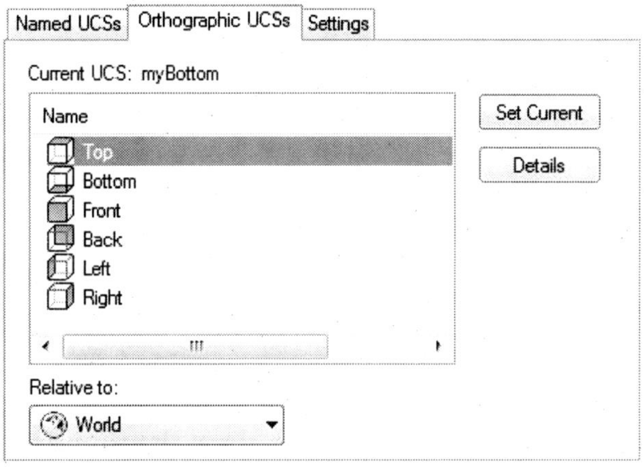

Named UCSs | Orthographic UCSs | Settings

Current UCS: myBottom

Name

Top

Bottom

Front

Back

Left

Right

Set Current

Details

Relative to:

World

Figure 2.003

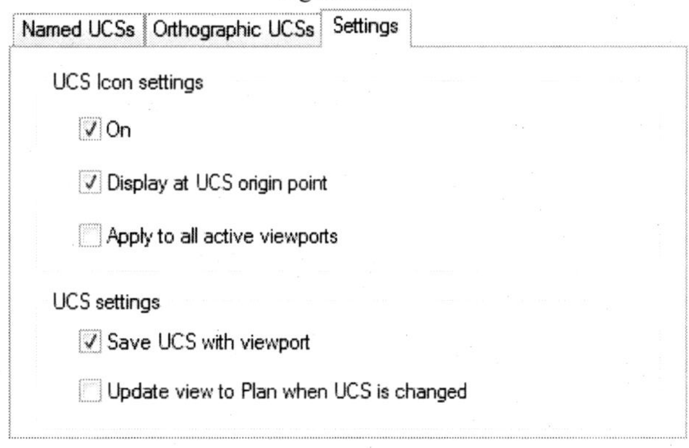

Figure 2.004

2.3 Using Working Planes

We've spent many pages learning to set up User Coordinate Systems. But as yet, we haven't seen how to use the UCS once it's set up. We haven't seen the answers to the questions that began our lesson – How do we make a solid roof? How do we place text and dimensions in a three-dimensional drawing?

Let's do an exercise to put UCSs to practical use.

Do This: 2.3A	Manipulating the UCS (Cont.)

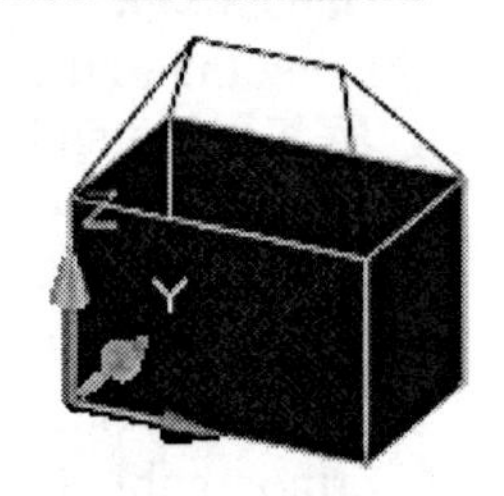

I. Open the *Stick House 2* file in the C:\Steps3D\Lesson02 folder. The drawing looks like the figure at right.

This drawing has already been set up for you with several UCSs. The house is a stick figure (wireframe) structure with thickness assigned to the lower lines. We'll create a solid roof, add a few dimensions, and then set up some Paper Space viewports.

II. Be sure **Obj2** is the current layer and set the **Annotation Scale** to **1:8**.

III. Follow these steps.

2.3A: Manipulating the UCS (Cont.)

1. Try drawing a solid using the points indicated.

Command: *so*

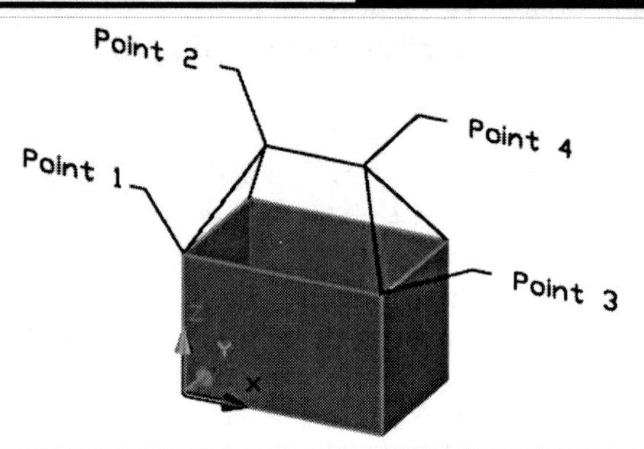

2. Notice (right) that the solid was drawn two-dimensionally in the current UCS. To use a solid to form the roof, the UCS must be aligned to the side of the roof you wish to draw.

Erase the solid.

Command: *e*

3. Call the UCS Manager.

Command: *ucsman*

4. Double click on the **south roof** UCS to make it current.

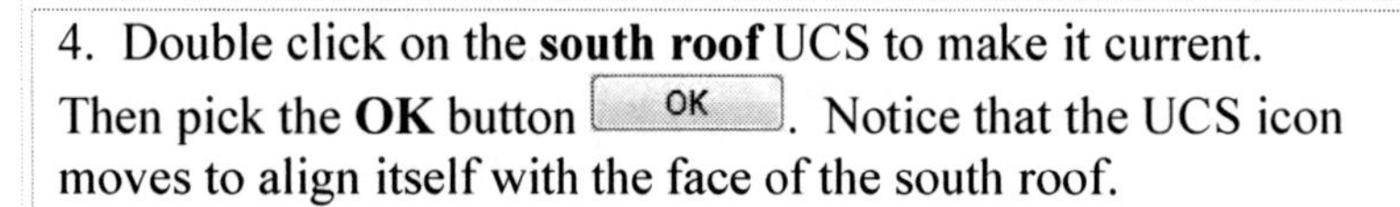

Then pick the **OK** button. Notice that the UCS icon moves to align itself with the face of the south roof.

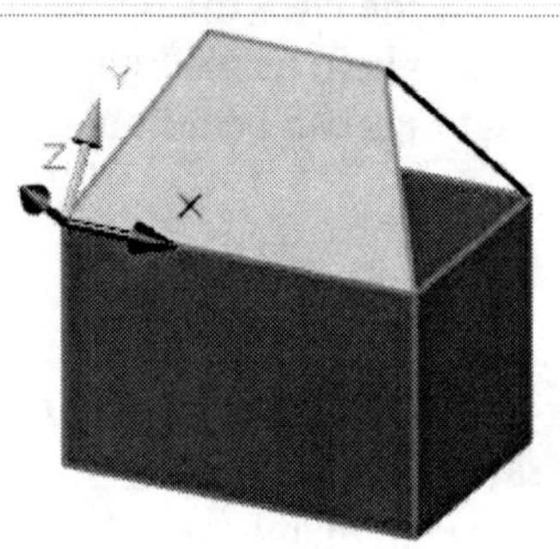

5. Now that the UCS has been properly set, repeat Step 1.

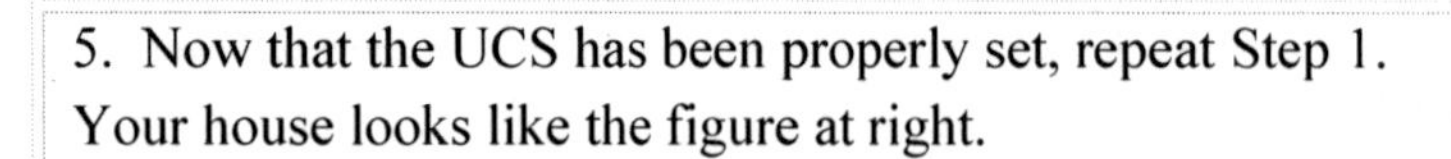

Your house looks like the figure at right.

2.3A: Manipulating the UCS (Cont.)

6. Using the techniques seen in this exercise, complete the roof using solids (adjust the viewpoint as necessary, then return to the current setting of 1,-2,1).
Your drawing looks like the figure at right.

7. Set the current UCS to **south west floor**.

8. Now we'll add some dimensions in Model Space. Be sure the **Stick House** dimstyle and the **dims** layer are current.

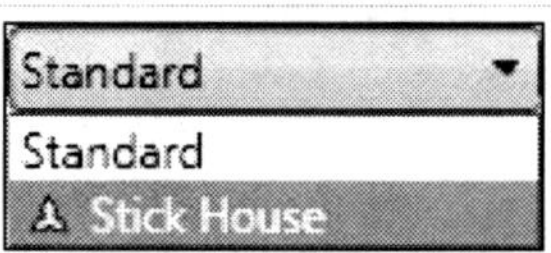

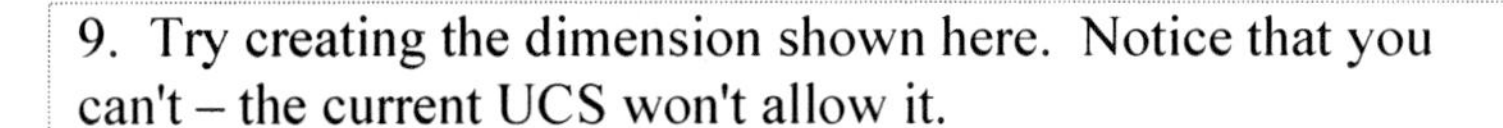

9. Try creating the dimension shown here. Notice that you can't – the current UCS won't allow it.

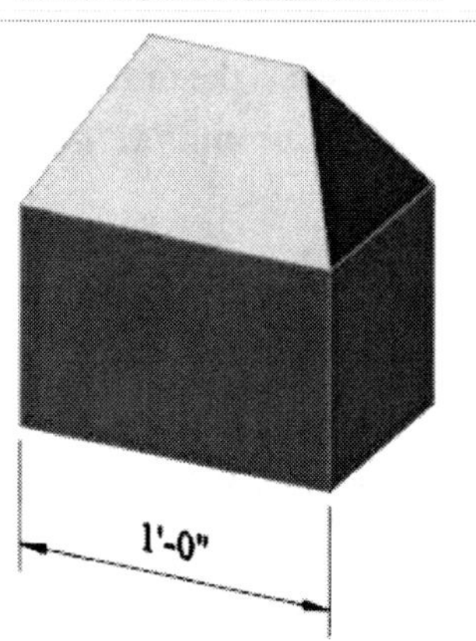

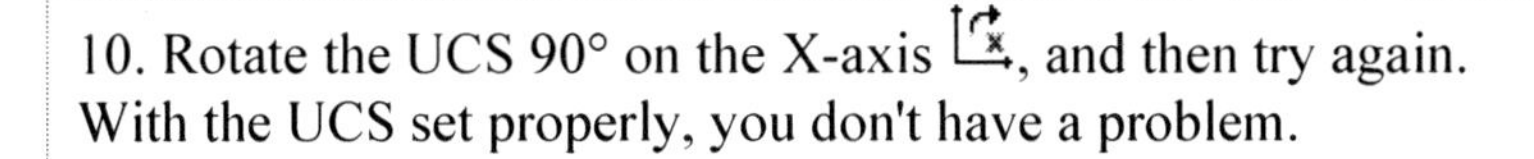

10. Rotate the UCS 90° on the X-axis, and then try again. With the UCS set properly, you don't have a problem.

11. Adjust the UCS position and orientation as necessary to create the remaining dimensions shown here.
So you see that you can dimension only in the current UCS.

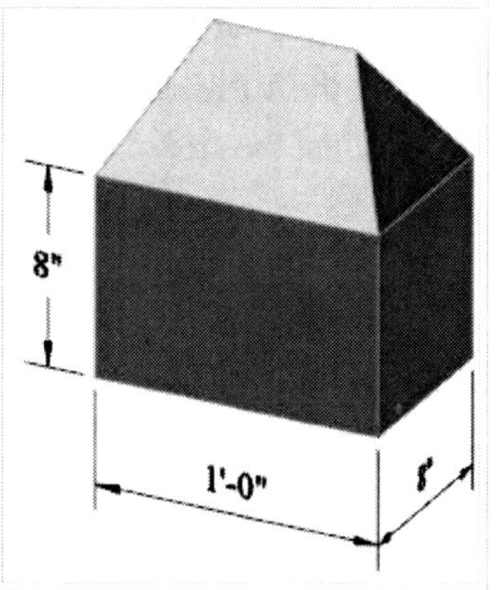

12. Let's get fancy – we'll add some shingles to our roof. Restore the **south roof** UCS, freeze the **dims** layer, and set the **obj1** layer current.

13. In the Hatch dialog box, set the options that appear in the following figure.

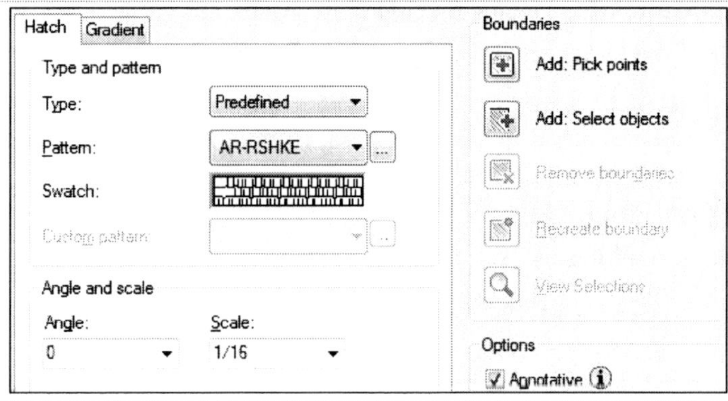

2.3A: Manipulating the UCS (Cont.)

14. Use the **Add: Pick points** button to put the hatching on the southern roof as indicated.

15. Repeat this procedure to add shingles to the entire roof.

Pretty cool, huh?

16. Save the drawing.

Command: ***qsave***

17. Now we'll set up some viewports on our layout to reflect some different UCSs. Activate the **Layout1** tab. (We've already started the layout.)

18. Activate the upper-left viewport and do the following:

- Create a WCS plan view (set the viewpoint to 0,0,1).
- Set the **VP** and **Annotation Scales** for this viewport to 1:8.

19. Activate the lower-left viewport and do the following:

- Create a WCS front view (set the viewpoint to 0,-1,0).
- Set the **VP** and **Annotation Scales** for this viewport to 1:8.
- Adjust the size of the viewport and the position of the house as necessary to see the entire house.

20. Activate the lower-right viewport and do the following:

- Create a WCS right side view (set the viewpoint to 1,0,0).
- Set the **VP** and **Annotation Scales** for this viewport to 1:8.
- Adjust the size of the viewport and the position of the house as necessary to see the entire house.

21. Activate the upper-right viewport and do the following:

- Set the **VP** and **Annotation Scales** for this viewport to 1:8.
- Adjust the size of the viewport and the position of the house as necessary to see the entire house.

22. Using the ***MVSetup*** command to align the plan, front, and side views. Adjust the position of the viewports as necessary for aesthetics.

Command: ***mvsetup***

Your drawing looks something like the following figure.

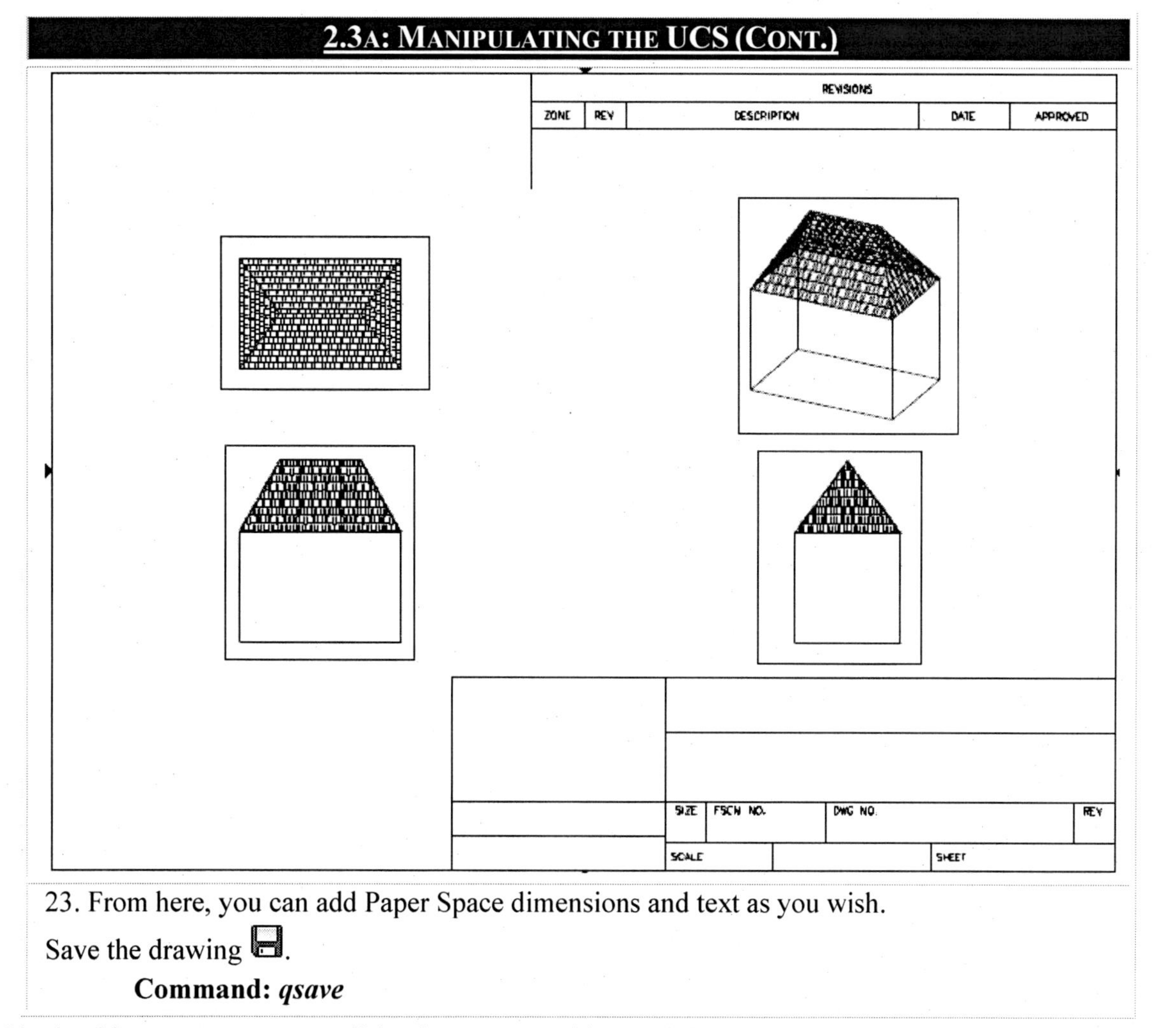

23. From here, you can add Paper Space dimensions and text as you wish.

Save the drawing .

Command: ***qsave***

Oh, the things you can accomplish when you stand in the right place! Remember, you can work on your model through each of the layout viewports – with the UCS designated for that viewport!

2.4 Other Viewing Tools

In previous editions of this text, we discussed *Advanced Viewing Techniques* in this section. These included the ***3DFOrbit***, ***3DCOrbit***, and ***3DOrbit*** commands. The addition of the new ViewCube has made these tools redundant so we've removed their discussion here. You can, however, review the material at: http://www.UNeedCAD.com/Files/OtherViewingTools.pdf. (I recommend viewing this supplement regardless of which release you'll be using.)

We'll begin this section with a look at some older tools (and some with which you're already familiar – sort of) that can help you better see your model. Then we'll take another look at the navigation wheels we discussed in our basic text.

2.4.1 Tools for Better Viewing

You can find most of the tools we'll discuss in this section on a cursor menu (Figure 2.005, p.46) which appears when any of the tools has been engaged. Otherwise, refer to the WTFI table for the best ways to begin the command.

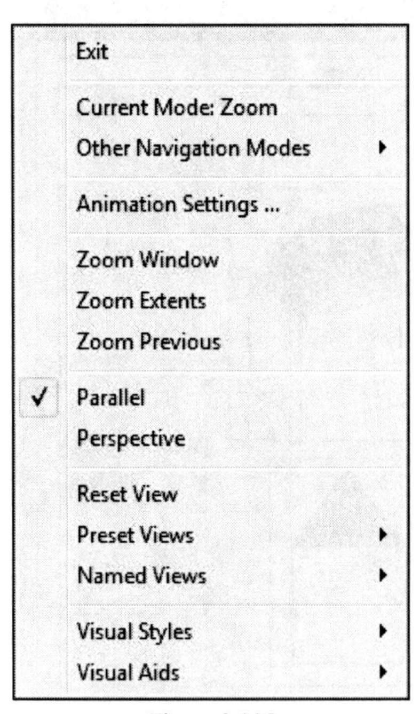

Figure 2.005

- ***3DZoom*** works very much like a realtime zoom. In fact, when you're in a *parallel* view, it *is* a realtime zoom. When in a *perspective* view, however, ***3DZoom*** actually works as though you're working with a zoom lens on your camera. (The position of the camera doesn't change, but the zoom factor does.) Parallel and perspective views change how you see your drawing (as we discussed in Section 1.1.1, p.3,). You can toggle back and forth using the cursor menu (Figure 2.005).

Where to Find It:	
Command Line:	***3DZoom***
Toolbar:	3D Navigation – **3DZoom**

- As with the ***3DZoom*** command, ***3DPan*** provides panning capabilities while you're in a perspective view.

Where to Find It:	
Command Line:	***3DPan***
Toolbar:	3D Navigation – **3DPan**

- ***3DDistance*** appears to do the same thing that the ***3DZoom*** command does. The difference is that 3***DDistance*** actually changes the distance between you and the model. No distortion results from the ***3DDistance*** command as it may from the ***3DZoom*** command. The ***3DDistance*** command even has its own cursor .

Where to Find It:	
Command Line:	***3DDistance***
Menu:	View – Camera – **Adjust Distance**
Cursor Menu:	Other Navigation Modes – **Adjust Distance**
Toolbar:	3D Navigation – (Swivel flyout) **Adjust Distance**

- The ***3DSwivel*** command adjusts the view as though a camera, although stationary, is revolving on its tripod. This tool comes in handy when viewing an architectural, structural, or piping drawing from inside the model. It also has its own cursor .

Where to Find It:	
Command Line:	***3DSwivel***
Menu:	View – Camera – **Swivel**
Cursor Menu:	Other Navigation Modes – **Swivel**
Toolbar:	3D Navigation – **Swivel**

- **Walk** calls the ***3DWalk*** command and the Position Locator tool palette. **Fly** calls the ***3DFly*** command. It also calls the Position Locator tool palette. These tools work together and we'll look at them together in Lesson 13, p.313.

The cursor menu we saw in Figure 2.005 contains some other tools – some you're already familiar with, but some others might interest you.

- **Animation Settings** calls the Animation Settings dialog box. We'll see the options in that one in Lesson 13, p.322, (as part of the Motion Path Animation dialog box).
- **Reset View** is your "Hail Mary" while you're using these viewing tools. Use it to set the view back to what it was when you began.

- **Preset Views** presents a flyout list of the basic preset views in AutoCAD (Figure 2.006). You'll find these on the **Views** panel of the ribbon's **View** and **Home** tabs as well.
- **Named Views** does the same thing with any user-named views in the drawing.
- **Visual Styles** presents a flyout list of the visual styles in the drawing. You'll find these in the **Visual Styles** panel (**Render** tab) or the **View** panel (**Home** tab) as well.

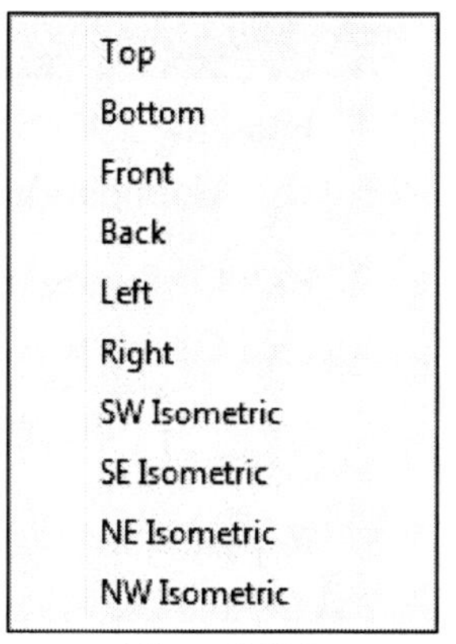

Figure 2.006

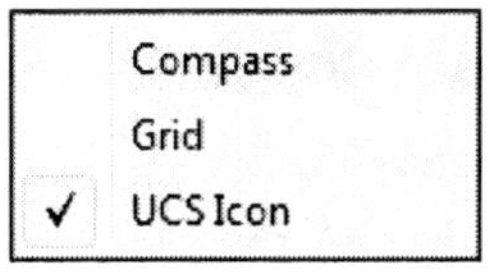

Figure 2.007

- Finally, **Visual Aids** calls the menu in Figure 2.007. I can't recommend using the **Compass** called by this option. It tends to clutter the screen. The **Grid** option simply toggles the grid on. This can be useful to help orient you just as the **UCS Icon** can.

2.4.2 Clipping 3D Objects for Better Viewing

The ***3DClip*** command presents the Adjust Clipping Planes dialog box (Figure 2.008) with the model shown at 90° to the current 3D Orbital display. The lines through the center of the window are the clipping planes. Pick and drag to adjust their locations.

Figure 2.008

Control what you see with the five buttons along the top of the dialog box. These are (from the left):

- **Adjust Front Clipping** allows you to move the front clipping plane up or down.
- **Adjust Back Clipping** allows you to move the back clipping plane up or down.
- **Create Slice** allows you to move both clipping planes together.
- **Pan** calls the ***Pan*** command within the Adjust Clipping Plane window.
- **Zoom** calls the ***Zoom*** command within the Adjust Clipping Plane window.
- **Front Clipping On/Off** toggles front clipping on or off. When **On**, AutoCAD won't display anything in front of (below) the front clipping plane.
- **Back Clipping On/Off** toggles back clipping on or off. When **On**, AutoCAD won't display anything behind (above) the clipping plane.
- Toggle both clippings off to view the entire model.

Let's look at the orbiter's clipping planes.

Do This: 2.4.2A	Using Clipping Planes

I. Reopen the *UCS Practice 2*[or *2a*] file in the C:\Steps3D\Lesson02 folder.

II. Follow these steps.

2.4.2A: Using Clipping Planes

1. Enter the ***3DClip*** command.

   Command: *3dclip*

2. AutoCAD presents the **Adjust Clipping Planes** window with **Adjust Front Clipping** toggled **On** (refer to Figure 2.008, p.47). Toggle the **Front Clipping** On.

3. Move the front clipping plane up and down (pick and drag). Watch the display as you move the plane.

4. Repeat Steps 2 and 3 using the **Back Clipping** plane. (Be sure to toggle **Back Clipping** On and use the **Adjust Back Clipping** toggle .)

5. Let's create a slice. Move both clipping planes fairly close to each other, then pick the **Create Slice** button .

6. Move the clipping planes up and down (pick and drag). Watch the display. You're cutting everything before and behind the clipping planes from your view.

7. Close the Adjust Clipping Planes dialog box.

8. Rotate the object as you did in the previous exercise and watch the effect clipping has on your view.

9. Repeat the ***3DClip*** command and toggle both the clipping planes off and .

2.4.3 A Cool New Tool – The Navigation Wheel

I repeat much of this section from Lesson 4 of *AutoCAD 2010: One Step at a Time* where I first introduced it. I couldn't cover some of the material in the basic text, however, as it related to a 3D world.

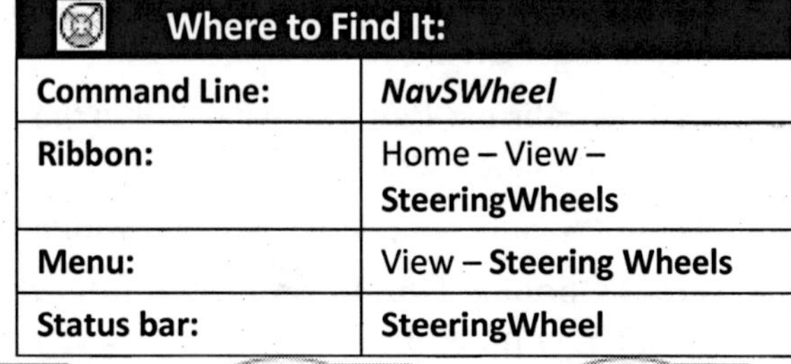

Where to Find It:	
Command Line:	***NavSWheel***
Ribbon:	Home – View – **SteeringWheels**
Menu:	View – **Steering Wheels**
Status bar:	**SteeringWheel**

It's probably not fair to refer to the Navigation Wheel as a single feature – AutoCAD actually provides three *steering* wheels – the **Full Navigation Wheel** (the default), the **View Object Wheel**, and the **Tour Building Wheel** (see the following figures).

Use one of the methods indicated in the **Where To Find It** table to display the default wheel. It may follow your cursor around the screen. You may find this a bit disconcerting at first, but AutoCAD had its reasons for this as you'll soon see.

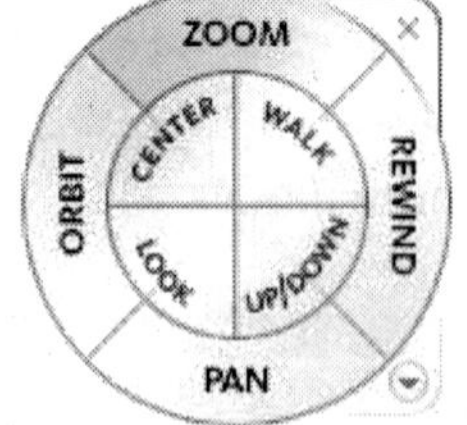

Full Navigation Wheel

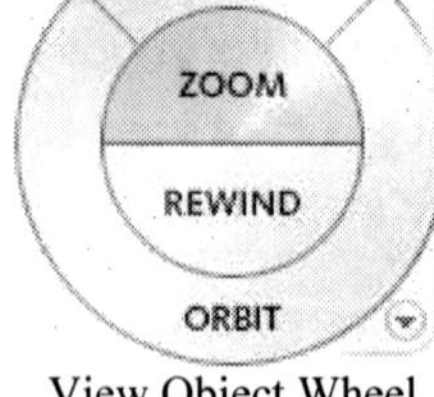

View Object Wheel

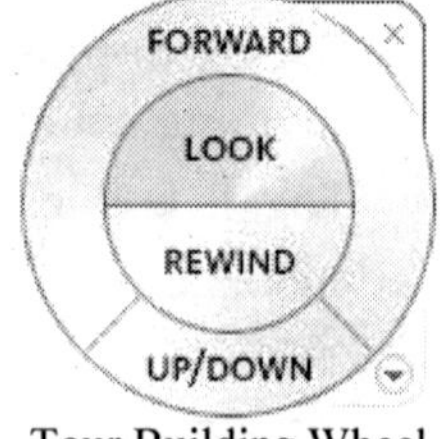

Tour Building Wheel

To change from one wheel to another, right click anywhere on the displayed wheel, and select either **Full Navigation Wheel** or **Basic Wheels** from the menu (more on this menu shortly). The latter will produce a menu giving you the option of either **View Object Wheel** or the **Tour Building Wheel**. Of course, you can also display the mini wheels as well – we'll look at those shortly.

You'll find many of the tools redundant among the various wheels – that is, they exist on more than one wheel. Let's look at the specifics.

- The **Zoom** button (Full Navigation Wheel and View Object Wheel) on the wheels manages to cover several of the command line options. Holding the button down allows you to zoom in and out about the pivot point – similar to the way you did with realtime zoom. But you can also click the button to zoom in and out (also about the pivot point) just as you do with the **Zoom In** option on the Zoom menu.
- You won't find the **Rewind** option (all wheels) anywhere else – and it's a winner! When you pick (and hold) the **Rewind** button, AutoCAD presents a "reel" of frames from previous views (Figure 2.009). You can move the selection frame (within the corner brackets) back and forth until you've found the display you want. AutoCAD will rewind dynamically – that is, it'll rewind the actual display as you move the selection frame. (Very cool!)

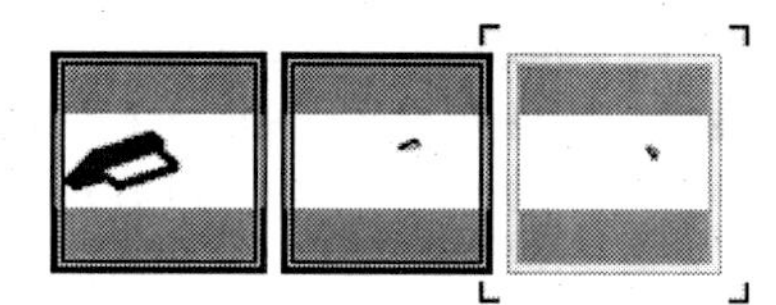

Figure 2.009

- **Pan** (Full Navigation Wheel) works just like realtime pan.
- **Orbit** (Full Navigation Wheel and View Object Wheel) allows you to orbit around the model just as you did when you held down the mouse button while dragging over the ViewCube in Lesson 1, p.3.
- **Center** (Full Navigation Wheel and View Object Wheel) allows you to center a drawing point on your screen, but more importantly, it's also what you use to locate the pivot point about which you'll orbit. Hold down the **Center** button of the wheel (left mouse button) and move your cursor to an object you wish to center or use as a pivot point. AutoCAD lets you know when it finds something with a green *pivot point* (right). When the location satisfies you, release the button and AutoCAD will center that point.

PIVOT

- When you press and hold **Walk** (Full Navigation Wheel) AutoCAD will display a directional icon (right) and you'll move in that direction (the actual view moves in the opposite direction). Hold down the SHIFT key to move Up/Down while in the Walk procedure.

Directional Icon

- **Up/Down** (Full Navigation Wheel and Tour Building Wheel) displays a slider (right) which you can use to move your display up or down.

TOP

BOTTOM

Up/Down Slider

- **Look** (Full Navigation Wheel and Tour Building Wheel) works as though you're standing in one place and turning your head. This can be a useful tool when performing a "walk–thru" of a building or plant area. Look also uses its own icon (right).

Look Icon

- **Forward** (Tour Building Wheel only) uses another slider that allows you a bit more control over moving forward or backward in your display.

SURFACE

START

Forward Slider

> Don't forget that ***Plan*** command! When all else fails, it's a lifesaver!
>
> **Command: *plan***
>
> **Enter an option [Current UCS/UCS/World] <Current>: *w***

You'll find the last two items on the wheel on the corners that extend to the right. The upper corner contains the usual "X" button that closes a window (or in this case, a wheel). The lower right corner contains an arrow that opens a menu (Figure 2.010).

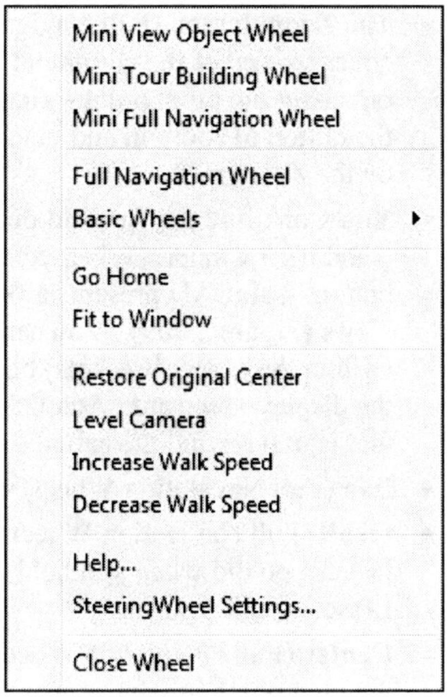

Mini View Object Wheel
Mini Tour Building Wheel
Mini Full Navigation Wheel
Full Navigation Wheel
Basic Wheels ▸
Go Home
Fit to Window
Restore Original Center
Level Camera
Increase Walk Speed
Decrease Walk Speed
Help...
SteeringWheel Settings...
Close Wheel

Figure 2.010

We've already discussed the **Full Navigation Wheel** and the **Basic Wheels** (View Object Wheel and Tour Building Wheel), but what about the rest of the options?

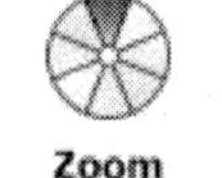

Full Navigation Mini Wheel

- The **Mini ... Wheel** options call miniaturized versions of the steering wheels. The primary benefit of the mini–wheels lies in the way they function without being nearly as obtrusive in your display. But they're also a bit more intuitive; that is, they don't display the names of buttons until you move your cursor in the button's direction (left).
- **Go Home** restores the same home view you'll find when you pick the **Home** icon on the ViewCube.
- **Fit to Window** adjusts the view to see all the objects in the model.
- **Restore Original Center** removes any work you may have done moving the pivot point and returns it to the model's extents.
- **Level Camera** rotates the view so that the view is relative to the XY plane of the current UCS.
- **Increase** or **Decrease Walk Speed** allows you some control over how fast you move when you use the **Walk** button on the wheel.

The remaining options need no explanation. You should, however, become familiar with the SteeringWheels Settings dialog box (Figure 2.011, p.51) accessed from this menu. It offers option frames with tools for controlling the **Size** and **Opacity** of both the **Big Wheels** and the **Mini Wheels**.

You can also adjust the size and opacity of the wheels with these commands:

- **NavSWheelSizeBig** – size of the full size wheels (0 to 2; default is 1)
- **NavSWheelOpacityBig** – opacity of the full size wheels (25 – 90; default is 50)
- **NavSWheelSizeMini** – size of the mini wheel
- **NavSWheelOpacityMini** – opacity of the mini wheels

- The **Display** frame allows some control over showing tool messages and/or tooltips, but it also contains the control for **pin**[ning] the **wheel at startup**. With this box checked, AutoCAD will anchor the wheel in the lower left corner of the screen. Unfortunately, it follows the cursor once you use it.
- The only **Zoom tool** available is an option to **Enable single click incremental zoom**. With this enabled, AutoCAD will zoom into the display 25% when you click once on the **Zoom** button of the Full Navigation Wheel.
- You'll find three frameless check boxes below the Zoom tool.
  - When using the **Look** tool, by default, when you move the cursor up, the view moves downward. **Invert vertical axis for Look tool** switches this so that moving the cursor up causes the view to move upward. This doesn't affect the left/right movement of the cursor/view, however, and can cause some confusion.
  - **Maintain Up direction for Orbit tool** controls whether or not you can turn the model upside down when orbiting.

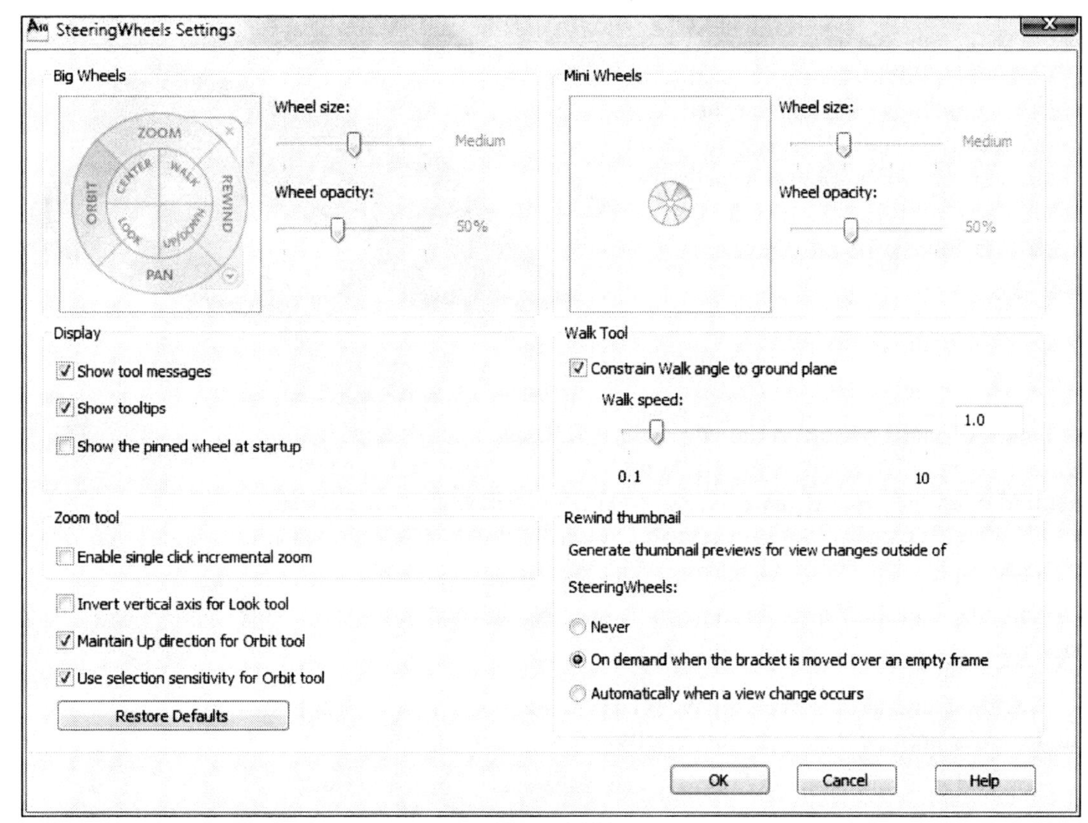

Figure 2.011

  - **Use selection sensitivity for Orbit tool** controls whether or not AutoCAD will use the objects selected *before* you open the wheel as the pivot point of an orbit.
- The **Walk Tool** frame includes a couple tools.
  - A check next to **Constrain Walk angle to ground plane** makes sure you actually walk through your model (as opposed to flying).
  - Use the **Walk speed** slider and input box to control how fast you move when using the **Walk** tool.
- Use the options in the **Rewind thumbnail** frame to control whether or not AutoCAD will display an image in the frame during a rewind. Naturally, displaying images takes greater system resources, so consider the power of your system before changing this setting.

Okay, time for an exercise.

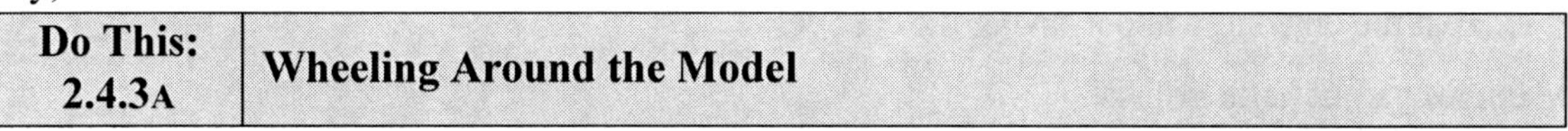

Do This: 2.4.3A	Wheeling Around the Model

I. If the *UCS Practice 2*[or *2a*] file isn't already open, please open it now. It's in the C:\Steps3D\Lesson02 folder.

II. Zoom extents.

III. Follow these steps.

2.4.3A: WHEELING AROUND THE MODEL
1. Open the default wheel.   **Command: *navswheel***   The wheel appears next to your cursor.

2.4.3A: Wheeling Around the Model

2. Pick and hold the **Center** button while you move the wheel over the model area indicated. Release the button when the pivot point appears as shown here.

AutoCAD centers the area in your display.

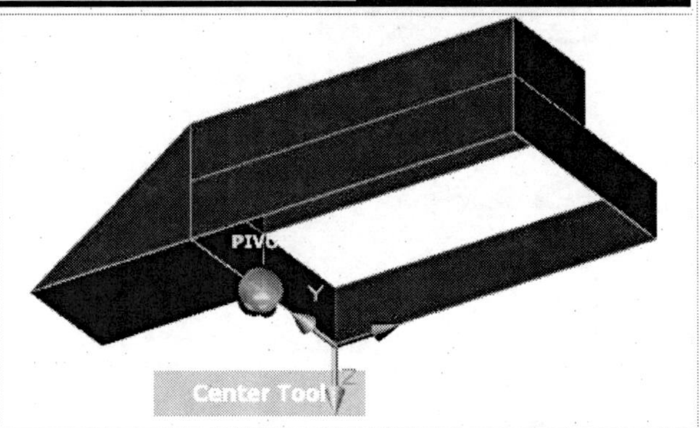

3. Pick the **Zoom** button on the wheel several times. Notice that, with each pick, AutoCAD zooms in. (If AutoCAD doesn't zoom in, use the menu to open the SteeringWheels Settings dialog box and place a check next to **Enable single click incremental zoom.**) Notice also that the center of the zoom is where your wheel was when you began the zoom.

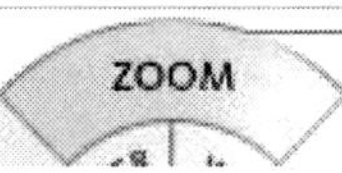

4. Pick and drag the **Zoom** button downward. Notice that you're zooming back out.

5. Pick the **Rewind** button on the wheel a couple times. Notice that you "undo" your previous zoom incrementally.

6. Pick and hold the **Rewind** button. Notice the reel of previous zoom displays. Move the frame slowly to the left and watch the display. Stop when you like what you see.

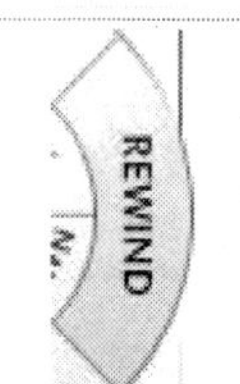

7. Right click on the wheel and pick **Mini Full Navigation Wheel** from the menu. Notice the new wheel.

8. Move your mouse to the left until **Orbit** appears on the mini wheel as indicated.

9. Orbit the model. Notice how it orbits around a pivot point.

10. Use the mini wheel to change the pivot point (**Center**) and repeat Step 9.

11. Experiment with the rest of the tools you'll find on the mini wheel.

12. Experiment with the other wheels' tools.

13. Exit the drawing without saving the changes.

Sometimes ya just hafta grin ☺!

2.5 Extra Steps

I don't always encourage students to study the "olde" ways AutoCAD did things, but in this case, I must.

Many of the 3D orbiting tools were introduced to AutoCAD within the past few years and still offer a lot to the CAD designer – even though we find better ways of doing almost everything in our current release. I encourage you to download the OtherViewingTools.pdf file from the website (http://www.UNeedCAD.com/Files/OtherViewingTools.pdf) and examine the orbiting tools. Pay particular attention to the **3DOrbit** tools; you'll be glad you did!

2.6 What Have We Learned?

Items covered in this lesson include:

- *The differences between the UCS and the WCS*
- *How to use the UCS and different working planes*
- *How to use the UCS Dialog Box (Manager)*
- *How to dimension a three-dimensional drawing*
- *How to use AutoCAD's navigation wheels*
- *Commands:*
  - ***UCS***
  - ***UCSMan***
  - ***3DPan***
  - ***3DZoom***
  - ***3DDistance***
  - ***3DSwivel***
  - ***3DClip***
  - ***Plan***

My chief Grammar and Usage Editor will say this was another full lesson!

But pat yourself on the back! Having made it to this point is no slight accomplishment. In these two lessons, you've mastered the basics for working in Z-Space. Let's take a minute and think about what you can do now that you couldn't do before beginning this text.

- You can maneuver in a drawing from front to back, side to side, and up and down (Spherical and Cylindrical Coordinate Systems, Point Filters).
- You can see your drawing from any point in the universe (***VPoint*** and ***Plan***).
- You can work on any surface as though it were lying flat on your desk (***UCS***).
- You have a host of new viewing tools to help you see your model from any angle or several angles at one time (***VPoint***, ***VPorts***, Steering Wheels).
- You can create three-dimensional stick figures (wireframes) and even draw with thickness and elevation.

You've come a long way in a short period of time, but there's still far to go. (Oh, the sights still to see ...) Most of what you learn about Z-space from here will be tools and techniques to make three-dimensional drawing easier, faster, and prettier!

As always, we should practice what we've learned before continuing our study. Do the exercises and answer the questions. Then proceed to our study of Wireframe and Surface Modeling techniques.

2.7 Exercises

1. to 8. Dimension the drawings you created in Exercises 1 through 8 of Section 1.7 (in Lesson 1). Refer to the drawings in Appendix B as a guide. If these drawings are not available, use the corresponding drawing in the C:\Steps3D\Lesson02 folder.

9. [Refer to Section 1.7 of Lesson 1 – Exercise #9, p.27.] Open the *My Twisted Y* file in the C:\Steps3D\Lesson01 folder. [If the *My Twisted Y* file isn't available, use the *Twisted Y* file found in the C:\Steps3D\Lesson02 folder.]
   - 9.1. Place the dimensions shown in the *Twisted Y* drawing.
   - 9.2. Save the drawing to the C:\Steps3D\Lesson02 folder.
10. [Refer to Section 1.7 of Lesson 1 – Exercise #10, p.27.] Open the *My Block* file in the C:\Steps3D\Lesson01 folder. [If the *My Block* file isn't available, use the *Block* file found in the C:\Steps3D\Lesson02 folder.]
    - 10.1. Place the dimensions shown in the *3-Dimensional Block* drawing.
    - 10.2. Save the drawing to the C:\Steps3D\Lesson02 folder.
11. [Refer to Section 1.7 of Lesson 1 – Exercise #11, p.27.] Open the *MyGrill* file in the C:\Steps3D\Lesson01 folder. [If the *My Grill* file isn't available, use the *Grill* file found in the C:\Steps3D\Lesson02 folder.]
    - 11.1. Place the dimensions shown in the *Grill* drawing.
    - 11.2. Save the drawing to the C:\Steps3D\Lesson02 folder.
12. [Refer to Section 1.7 of Lesson 1 – Exercise #12, p.27.] Open the *MyBookEnd* file in the C:\Steps3D\Lesson01 folder. [If the *MyBookEnd* file isn't available, use the *BookEnd* file found in the C:\Steps3D\Lesson02 folder.]
    - 12.1. Place the dimensions shown in the *Bookend* drawing.
    - 12.2. Save the drawing to the C:\Steps3D\Lesson02 folder.
13. [Refer to Section 1.7 of Lesson 1 – Exercise #13, p.28.] Open the *MyMagRack* file in the C:\Steps3D\Lesson01 folder. [If the *MyMagRack* file isn't available, use the *MagRack* file found in the C:\Steps3D\Lesson02 folder.]
    - 13.1. Place the dimensions shown in the *Magazine Rack* drawing.
    - 13.2. Save the drawing to the C:\Steps3D\Lesson02 folder.
14. [Refer to Section 1.7 of Lesson 1 – Exercise #14, p.28.] Open the *My Anchor Stop* file in the C:\Steps3D\Lesson01 folder. [If the *My Anchor Stop* file isn't available, use the *Anchor Stop* file found in the C:\Steps3D\Lesson02 folder.]
    - 14.1. Place the dimensions shown in the *Anchor Stop* drawing.
    - 14.2. Save the drawing to the C:\Steps3D\Lesson02 folder.
15. [Refer to Section 1.7 of Lesson 1 – Exercise #15, p.31.] Open the *My Queen* file in the C:\Steps3D\Lesson01 folder. [If the *My Queen* file isn't available, use the *Queen* file found in the C:\Steps3D\Lesson02 folder.]
    - 15.1. Place the dimensions shown in the *Chess Queen* drawing.
    - 15.2. Save the drawing to the C:\Steps3D\Lesson02 folder.
16. [Refer to Section 1.7 of Lesson 1 – Exercise #16, p.31.] Open the *My Ring Stand* file in the C:\Steps3D\Lesson01 folder. [If the *My Ring Stand* file isn't available, use the *Ring Stand* file found in the C:\Steps3D\Lesson02 folder.]
    - 16.1. Place the dimensions shown in the *Ring Stand* drawing.
    - 16.2. Save the drawing to the C:\Steps3D\Lesson02 folder.

17. Create the *Angle Blocks* drawing according to the following parameters:
    17.1. Start the drawing using *template #2* found in the C:\Steps3D\Lesson02 folder.
    17.2. The Paper Space text is 3/16" and 1/8".
    17.3. Title block text is ¼", 3/16", and 1/8".
    17.4. Text on the blocks uses either AutoCAD's standard text style or a type using the Times New Roman or Calibri font. Text heights are ¼".
    17.5. Save the drawing as *My Angled Blocks* in the C:\Steps3D\Lesson02 folder.

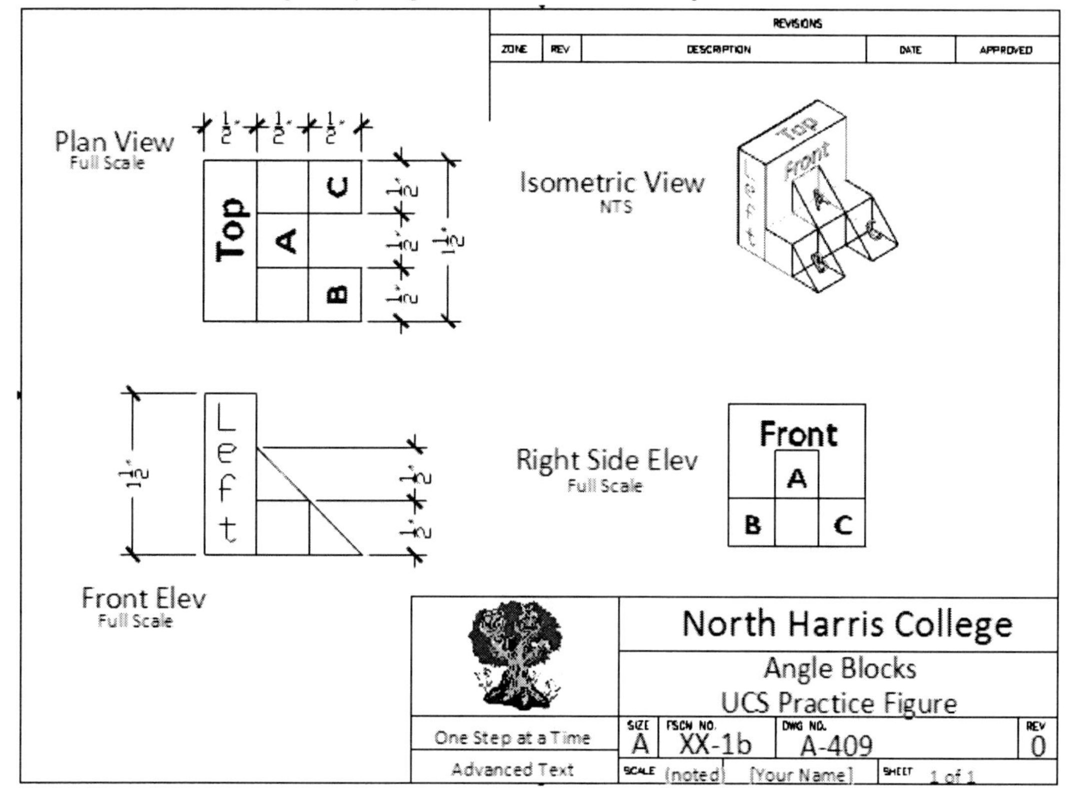

Angle Blocks

18. Create the *Corner Steps* drawing according to the following parameters:
    18.1. Start the drawing using *template #1* found in the C:\Steps3D\Lesson01 folder.
    18.2. The Paper Space text is 3/16" and 1/8".
    18.3. Title block text is ¼", 3/16", and 1/8". The font is Times New Roman or Calibri.
    18.4. Text on the blocks uses AutoCAD's standard text style. Text height is 3/16".
    18.5. Save the drawing as *My Corner Steps* in the C:\Steps3D\Lesson02 folder.

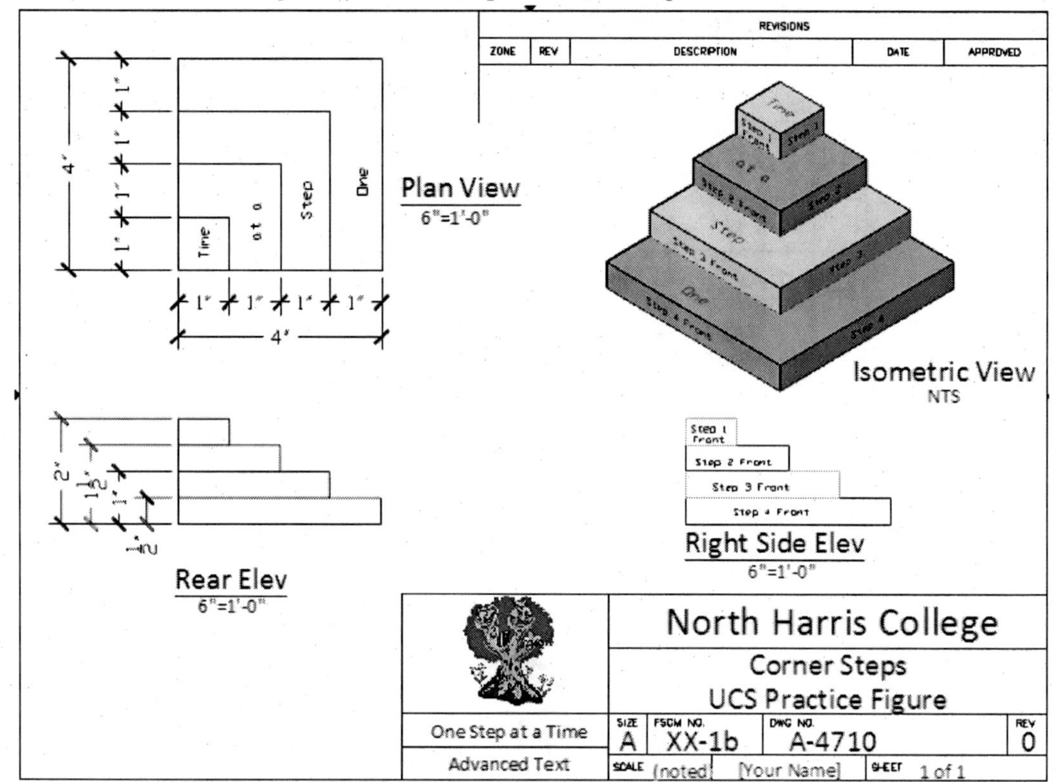

Corner Steps

19. The *table* drawing is a game of UCS manipulation and the ***Array*** and ***Mirror*** commands. Create it according to the following parameters:
    19.1. The tabletop is a 5" width x 5" long polyline drawn with a ½" thickness.
    19.2. Each of the eight legs is made up of eight lines. Each line is 5" long. The original was drawn at 30° in the XY-lane and 60° from the XY-plane. The lines were then arrayed in a 1/16" circle.
    19.3. The eight "feet" are ¼" radius circles. Again, they were arrayed (eight circles in each array).
    19.4. The center ball (at the intersection of the legs) is also an arrayed circle (eight in all) – this one with a ½" radius.
    19.5. The upper legs are mirrored from the bottom.
    19.6. Have fun!
    19.7. Save the drawing as *My Table* in the C:\Steps3D\Lesson02 folder.

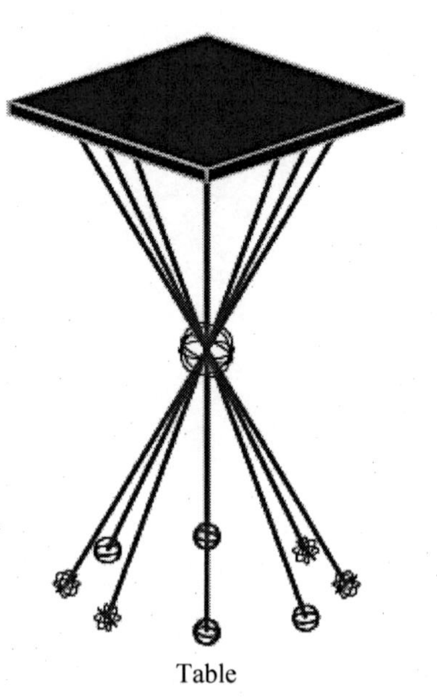

Table

20. Create the *table2* drawing according to the following parameters:
    20.1. This is a B-size (11" x 17") layout. Use the appropriate AutoCAD title block/border.
    20.2. The Paper Space text is 3/16" and 1/8".
    20.3. Title block text is ¼", 3/16", and 1/8". The font is Times New Roman or Calibri.
    20.4. The top is a single polyline drawn with ½ thickness.
    20.5. Save the drawing as *My Other Table* in the C:\Steps3D\Lesson02 folder.

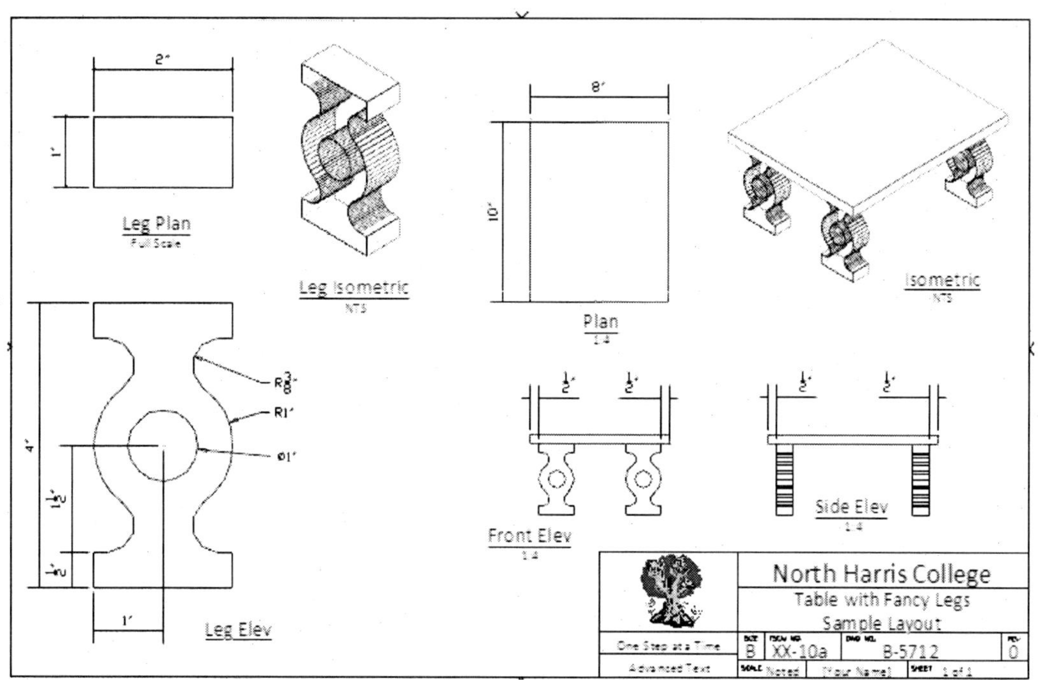

Table2

21. Create the *Corner Bracket* drawing (p.58) according to the following parameters:
    21.1. This is a B-size (11" x 17") layout. Use the appropriate AutoCAD title block/border.
    21.2. The Paper Space text is 3/16" and 1/8".
    21.3. Title block text is ¼", 3/16", and 1/8". The font is Times New Roman or Calibri.
    21.4. This is a wireframe drawing – use thickness only on the arcs/circles.
    21.5. Save the drawing as *My Corner Bracket* in the C:\Steps3D\Lesson02 folder.
22. Create the *Phone Plug* drawing (p.58) according to the following parameters:
    22.1. This is an A-size (8½" x 11") layout. Use the appropriate AutoCAD title block/border.
    22.2. The Paper Space text is 3/16" and 1/8".
    22.3. Title block text is ¼", 3/16", and 1/8". The font is Times New Roman or Calibri.
    22.4. This is a wireframe drawing – don't use thickness.
    22.5. You'll need to use the ***3DClip*** procedures to clean up the views.
    22.6. Save the drawing as *My Phone Plug* in the C:\Steps3D\Lesson02 folder.

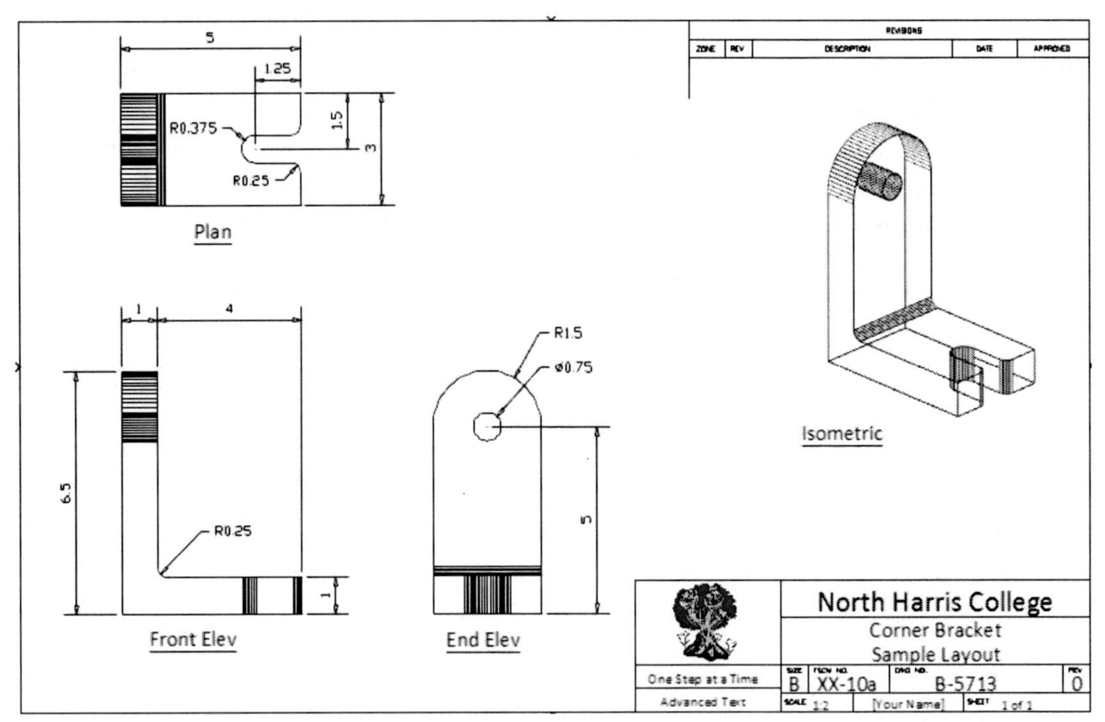

Corner Bracket

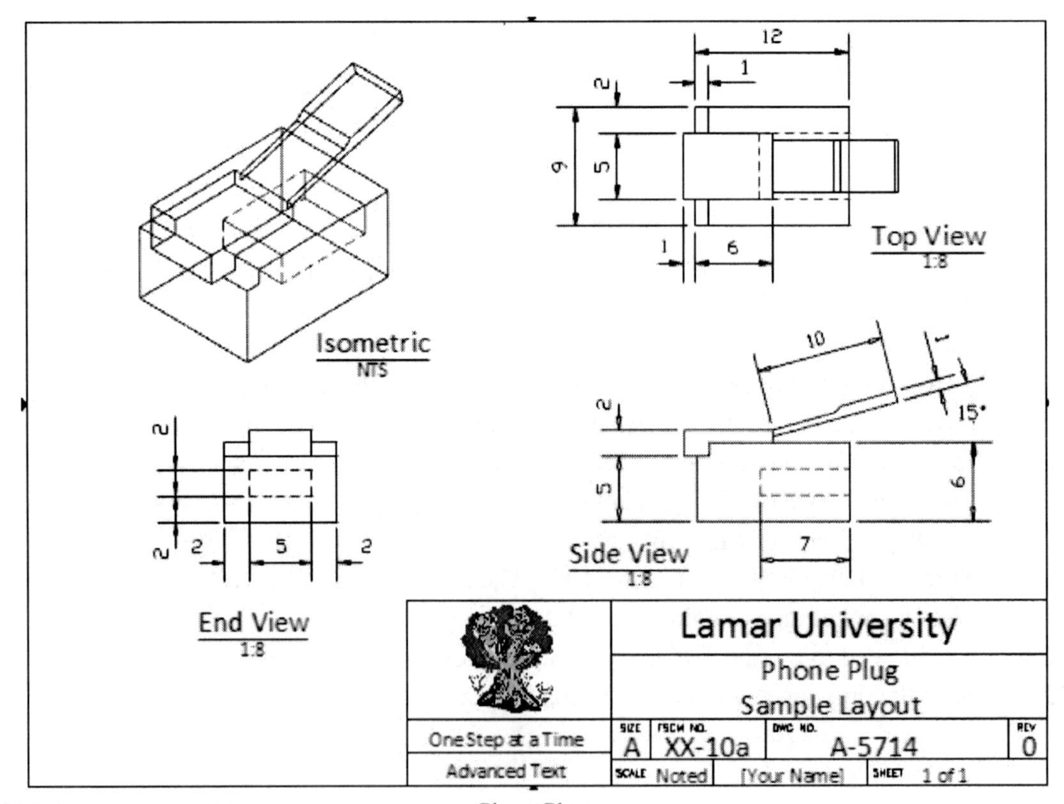

Phone Plug

23. Create the *Corner Bracket #2* drawing according to the following parameters:
    23.1. This is a B-size (11" x 17") layout. Use the appropriate AutoCAD title block/border.
    23.2. The Paper Space text is 3/16" and 1/8".
    23.3. Title block text is ¼", 3/16", and 1/8". The font is Times New Roman or Calibri.
    23.4. This is a wireframe drawing – use thickness only on the arcs/circles.
    23.5. You'll need to use the ***3DClip*** procedures to clean up the views.
    23.6. Save the drawing as *My Other Corner Bracket* in the C:\Steps3D\Lesson02 folder.

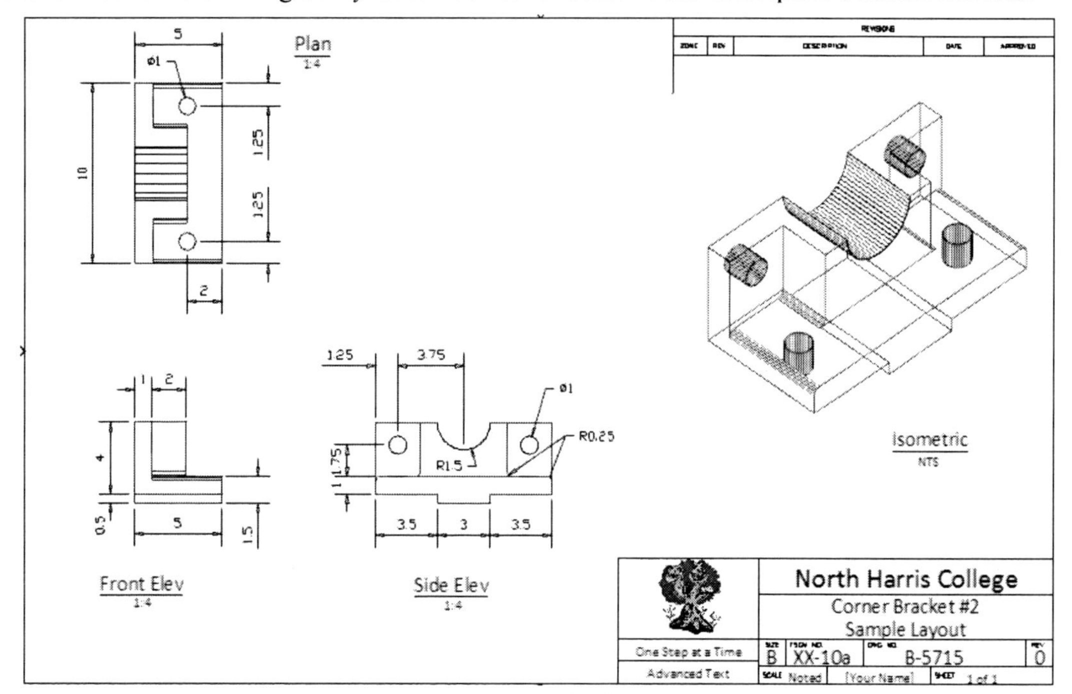

Corner Bracket #2

2.8 For Web-Based Review Questions, visit: http://foragerpub.com/AcadFiles/2010/2010.htm

Lesson 3

Following this lesson, you will:

✓ *Know the differences between a polyline and a three-dimensional polyline*

✓ *Know how to project a curved surface in three dimensions*

✓ *Know how to create a three-dimensional face (**3DFace**)*

- *Know how to control the visibility of edges*

✓ *Know the differences between solids and regions*

- *Know how to create a region with the **Region** command*
- *Know how to create a region with the **Boundary** command*

✓ *Know how to use the **Subtract** command to remove one region from another*

Wireframes and Surface Modeling

Most textbooks separate Wireframe Modeling and Surface Modeling into two distinct chapters. But the inevitable result is confusion. The two are so closely related that distinguishing between them often causes more bewilderment than just teaching them as they are – two sides of the same coin.

Let me make the distinction as simple as possible.

A wireframe model (what we've called a stick figure up until now) is a skeleton drawing. It has all the necessary parts – but no flesh. Drawing a wireframe model is relatively fast (compared to a surface model), but it provides little more than an outline of the model. We've seen that visual styles have no effect on wireframe models. When used, wireframes generally lead the three-dimensional design process (in the layout stage).

Fleshing out the wireframe – turning it into a surface model – comes when the layout is accepted, and you want to turn the skeleton into a production or display drawing. A surface model essentially stretches some skin over the skeleton making it possible to see surfaces (hence, the name).

In this lesson, we'll learn how to create more complex wireframe models. Then we'll look at some ways to create surfaces.

A quick note about surface modeling before we begin:

I should point out the differences between surface modeling and the newer "mesh" modeling. Both provide essentially "paper" thin surfaces over a model, and you can create primitives (boxes, cones, etc) with each, but meshes offer infinitely better results once you know how to use them. We'll discuss surface modeling here and mesh modeling primitives in Lesson 4. Then, we'll concentrate on mesh modeling as most animation packages (Lightwave, 3DS Max, Maya, Poser, and others) rely heavily on meshes.

3.1 *3DPoly* vs. *PLine*

Now that you're working with Z-coordinates, you may have noticed a certain limitation in polylines – polylines are two-dimensional creatures. True, you can give a polyline thickness and elevation, but you can't draw a polyline using different points on the Z-axis. That is, when prompted to **Specify next point**, your selection will use the same point on the Z-axis as the first point you identified regardless of any three-dimensional coordinate you give it!

But that doesn't mean that you have to sacrifice the benefits of a polyline … well, not entirely anyway. AutoCAD provides a three-dimensional version of the polyline called the *3DPoly*. When you need a multi-segmented polyline drawn in Z-space, simply use the ***3DPoly*** command instead of the ***PLine*** command.

Where to Find It:	
Command Line:	***3DPoly***
Hotkey(s):	***3p***
Ribbon (Tab/Panel):	Home – Draw – **3D Polyline**
Menu:	Draw – **3D Polyline**

There are, however, some restrictions to the 3DPoly. Chief among these is that a 3D polyline can't contain width. AutoCAD hasn't added this useful property yet. Additionally, you can't draw a 3D polyline using arcs or linetypes other than continuous.

Bear in mind that, while use of the polyline is restricted in Z-space, use of the spline isn't. You can use the ***Spline*** command when you want to draw curved lines in three dimensions. However, most surfaces created for a surface model will have flat edges and won't be able to lie flat against a spline.

The benefits of the 3DPoly include the ability to draw it in three dimensions, to edit it with the ***PEdit*** command, and to spline it.

We'll use the ***3DPoly*** command in our first exercise.

3.2 Drawing in Three Directions at Once – Point Projection

You may think that wireframe modeling is a fairly easy thing to do. After all, a wireframe model is just stick figures, right?

Of course, you're absolutely right. Stick figures are quite simple to draw – as long as the model you want to draw uses nice straight sticks. But consider the curved panel roof in Figure 3.001. Using the UCS procedures you learned in Lesson 2, you can easily draw the arcs and rooflines. But how would you draw the joint between the roofs? (Uh, oh! Here we go with the hard questions again.)

To draw in three directions at once as this joint requires – (front to back, side to side, and up and down – means that you must identify a series of points where the two roofs intersect. To do this, you must project points – that is, you must identify points by intersection of lines in Z-space.

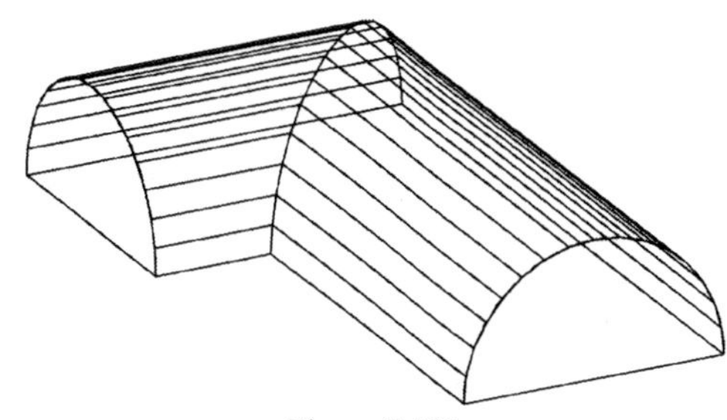

Figure 3.001

This isn't as difficult as it sounds. In fact, it's a lot like duck hunting (or skeet shooting for those with weaker stomachs). To hit a moving target, you must lead it a bit so that your bullet and the bird (or clay pigeon) arrive at the same place at the same time. What you do when you project points is simply identify where the bird and the bullet will meet. You do this by projecting one line along the bird's flight path and another along the barrel of your rifle. The intersection of the two lines is your actual target – or in drafting terms, your projection point.

Let's see how this works. We'll use our projection technique to identify the intersecting arc of the two roofs and then draw the arc using the 3D polyline.

> Remember that all three-dimensional drafting requires precise point identification using one of the methods we've already discussed. You can't pick an "about here" point and get away with it as you might have in two-dimensional drafting.

Do This: 3.2A	Projections and 3D Polylines

I. Open the *Cabin* file in the C:\Steps3D\Lesson03 folder. The drawing looks like the figure at right.

II. Follow these steps.

3.2A: PROJECTIONS AND 3D POLYLINES

1. Use the ***Divide*** command to divide each of the arcs into 16 segments. (If the divisions aren't clearly marked, set the **PDMode** system variable to **3** and regenerate the drawing.)

    Command: ***div***

    Select object to divide:

    Enter the number of segments or [Block]: ***16***

Notice that the nodes appear in the UCS that was current when the arcs were drawn.

2. Set the **roof** layer current.

3.2A: Projections and 3D Polylines

3. Using OSNAPs, draw lines between the corresponding nodes and endpoints as shown. (The lines represent the duck's flight path and the barrel of our rifle.)

(Hint: You may find it easier to draw one line in each direction and then copy it to each of the nodes/endpoints.)

Command: *l*

4. Freeze the **Marker2** layer.

5. Carefully trim the extra portions of the lines. Start with the nearest intersection – lowest north-south line with lowest east-west line – and work back. (Hint: You may fine it easier to do this from the plan view. But watch the first and last endpoints.)

Command: *tr*

Return to this view (VPoint 2,-1,1) when you've finished. Your drawing looks like this.

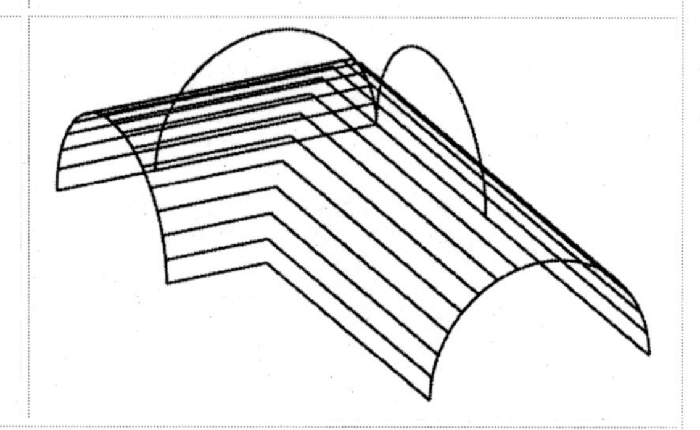

6. Erase the arcs in the back.

Command: *e*

7. Draw a 3D polyline connecting the intersections of the extension lines as shown. Use OSNAPs! (It might be easier to do this in plan view.)

Command: *3dpoly*
Specify start point of polyline:
Specify endpoint of line or [Undo]:
Specify endpoint of line or [Close/Undo]:

8. Save the drawing but don't exit.

Command: *qsave*

Look closely at the 3D polyline you created. Notice that it's a series of straight lines, not curved like the roof. We might have used a spline instead of the 3D polyline and achieved a nice soft curve, but the tool we'll use to "stretch the skin around our skeleton" doesn't allow for curves. So we're better off using the straighter 3D polyline (as we'll soon see).

3.3 Adding Surfaces – Regions, Solids, and 3D Faces

AutoCAD provides three methods for creating surface models – the ***Region***, ***Solid***, and ***3DFace*** commands. The three are so closely related that it's often difficult to tell the difference:

- You'll draw each as a two-dimensional object.
- Each becomes a 3D solid when extruded (more on the ***Extrude*** command in Lesson 8, p.177).
- Each creates an opaque (or solid) surface.

But despite their similarities, each has its place.

- The ***Region*** command converts a closed object (polygon, circle, two-dimensional spline, etc.) into a surface. A region can't have thickness.
- The ***Solid*** command fills an area only in the current UCS, but it can have thickness.
- The ***3DFace*** command draws a true three-dimensional surface (in all of the X-, Y-, and Z-planes). A 3D face can't have thickness.

When would you use one instead of the other two? Let's take a look at each and see.

3.3.1 Creating 3D Faces

Of the three surfacing methods, the ***3DFace*** command is the most versatile. However, it's also more difficult to use when cutouts are involved. The ***3DFace*** command makes no allowance for removal of part of the surface (as the ***Trim*** command allows you to remove part of a line or circle). You can, however, draw a 3D face without concern for the current UCS (provided coordinate entry is precise).

Where to Find It:	
Command Line:	***3DFace***
Hotkey(s):	***3f***
Menu:	Draw – Modeling – Meshes – **3D Face**

The command sequence looks like this:

Command: *3DFace*

Specify first point or [Invisible]: *[Select the first corner point.]*

Specify second point or [Invisible]: *[Select the second corner point.]*

Specify third point or [Invisible] <exit>: *[Select the third corner point.]*

Specify fourth point or [Invisible] <create three-sided face>: *[Select the fourth point or hit ENTER to create a 3D face from the three points already selected.]*

Specify third point or [Invisible] <exit>: *[You can continue selecting points or hit ENTER to exit the command.]*

The only option available – **Invisible** – isn't one you really want to use. When creating 3D faces, it'll occasionally be necessary to hide one of the edges (or make it invisible). (This will become apparent in the next exercise.) The command line procedure for doing this involves typing an ***I*** before the first point selection that defines the edge to be hidden. The edge drawn between the two points that follow the ***I*** will be invisible. This becomes a real chore when two or more edges must be hidden. We'll look at an easier approach to hiding the edges of 3D faces following the next exercise.

First, let's use the ***3DFace*** command to place a surface – with a window in it – on one of the walls of our cabin.

Do This: 3.3.1A	Surfaces with 3D Faces

I. Be sure you're still in the *Cabin* file in the C:\Steps3D\Lesson03 folder. If not, please open it now.

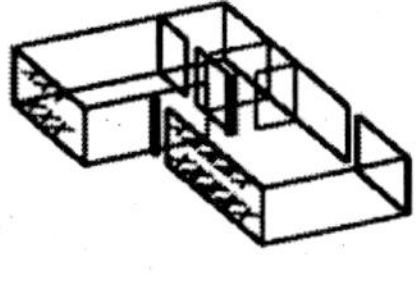

II. Set UCS = WCS.

III. Set the **3D Wireframe** visual style for this exercise.

IV. Thaw the **WALLS** and **MARKER** layers; set **WALLS** current and freeze the **roof** layer. The drawing looks like the figure at right.

V. Follow these steps.

3.3.1A: Surfaces with 3D Faces

1. Zoom in around the protruding part of the cabin – the wall with six nodes (refer to the Step 3 figure).

    Command: *z*

2. The front wall consists of two lines drawn with thickness. We can't put a window in these objects, so we'll remove the thickness and replace the lines with 3D faces.

    Begin by changing the thickness of both lines (inner and outer walls) to **0**. (Use the Properties palette.)

    Command: ***props***

3. Begin the ***3DFace*** command. Pick Point 1 as your first point.

    Command: ***3f***

    Specify first point or [Invisible]: ***[Select Point 1.]***

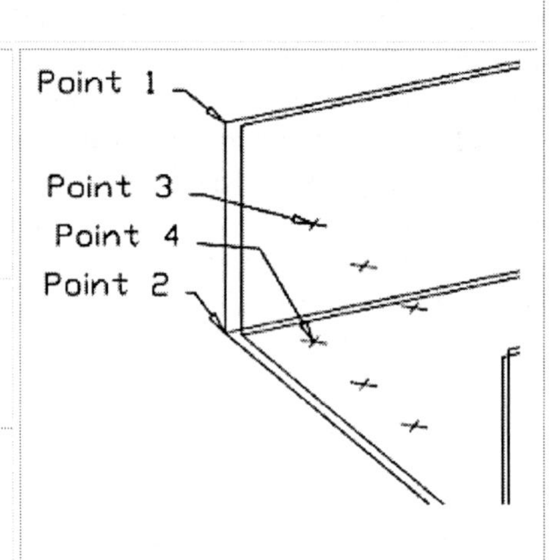

4. Pick Point 2 as the second point.

    Specify second point or [Invisible]: ***[Select Point 2.]***

5. We'll use point filters to locate the third point. Tell AutoCAD to use the XY values of Point 3 …

    Specify third point or [Invisible] <exit>: ***.xy***

6. … and the Z value of Point 2.

    of (need Z): ***[Select Point 2.]***

7. Again, we'll use point filters to locate the fourth point. Tell AutoCAD to use the X and Y values of Point 4 …

    Specify fourth point or [Invisible] <create three-sided face>: ***.xy***

8. … and the Z value of Point 1.

    of (need Z): ***[Select Point 1.]***

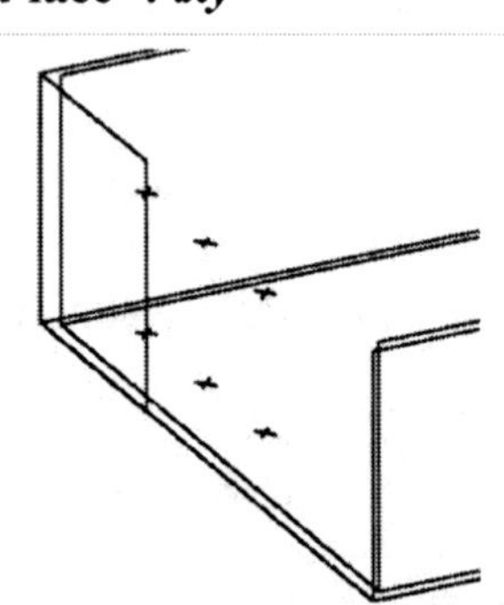

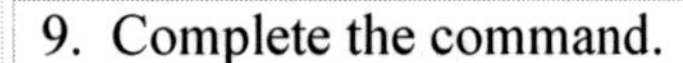

9. Complete the command.

    Specify third point or [Invisible] <exit>: ***[ENTER]***

    Your drawing looks like the figure at right.

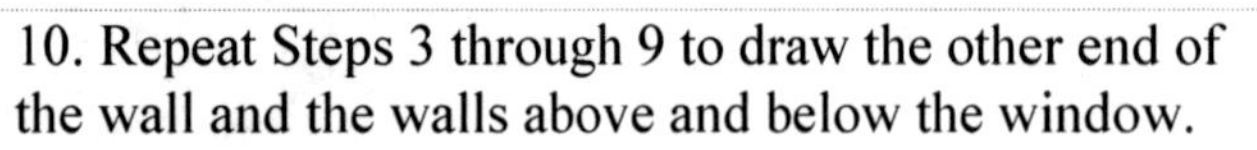

10. Repeat Steps 3 through 9 to draw the other end of the wall and the walls above and below the window.

    Command: ***3f***

    Set the **3D Hidden** visual style current. Your drawing will look like the figure at right.

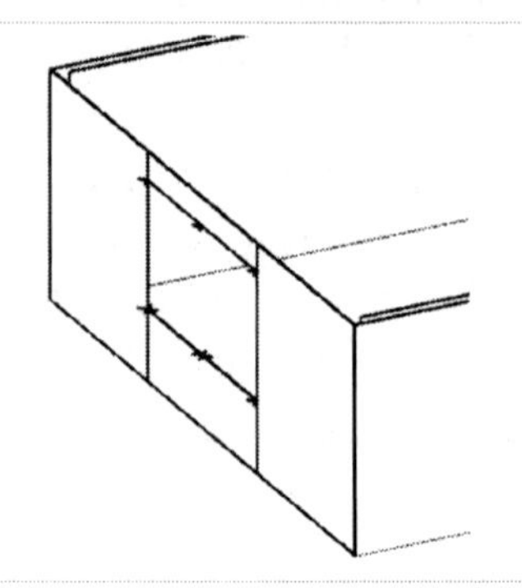

11. Save the drawing but don't exit.

    Command: ***qsave***

You should've noticed three things about the last exercise.

- Although quite useful, the skeleton (wireframe model) isn't required when drawing a surface model.
- The ***3DFace*** command has left lines (edges) above and below the window that don't normally appear on a model (or a wall).
- The current UCS didn't affect placement of the 3D faces (although it may affect the points you use to create the faces).

Let's take a look at those edges.

3.3.2 Invisible Edges of 3D Faces

Although you can make 3D face edges invisible as you draw them, it's a tedious, time-consuming, and error-prone task. You'll find it much easier to draw the 3D face and then use the Properties palette (or the more complicated, less reliable ***Edge*** command) to hide the edges that you don't want to see.

To display all of the 3D face edges in a drawing, whether visible or not, set the **SPLFrame** system variable to **1**.

Let's make the necessary edges invisible on our new surfaces.

Do This: 3.3.2A — Invisible Edges

I. Be sure you're still in the *Cabin* file in the C:\Steps3D\Lesson03 folder. If not, please open it now.

II. Freeze the **Marker** layer.

III. Follow these steps.

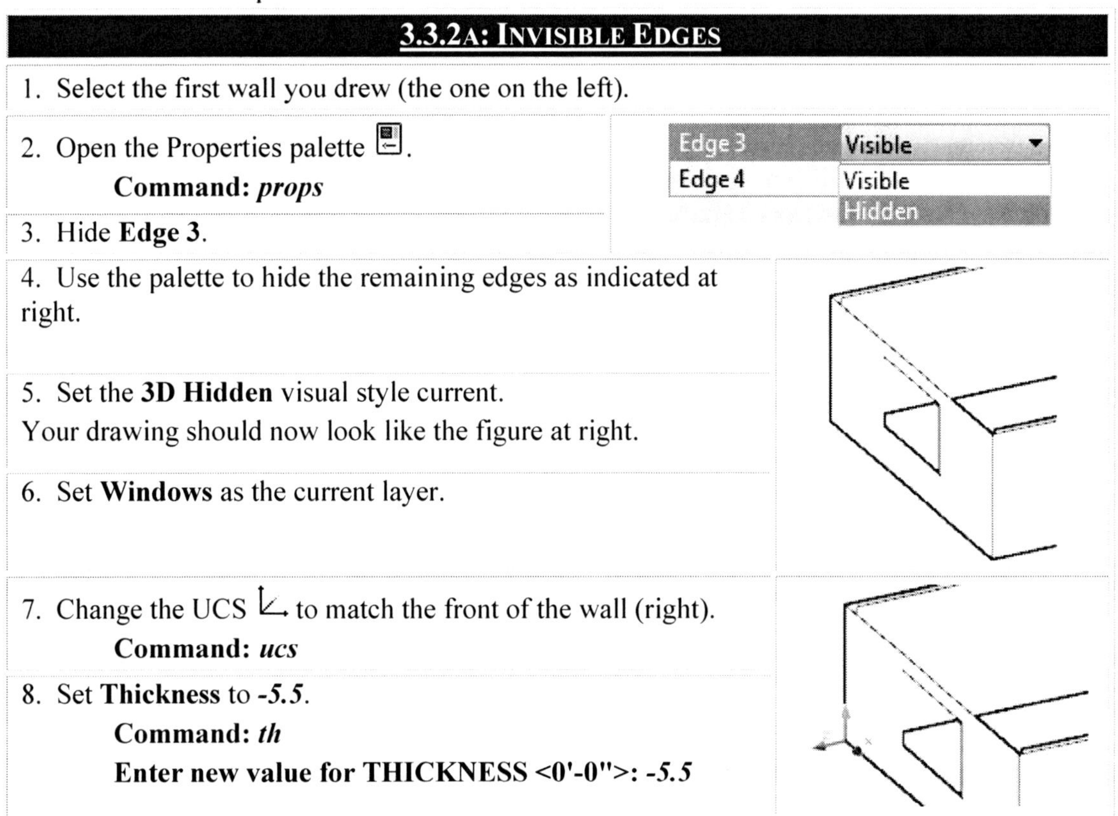

3.3.2A: INVISIBLE EDGES

1. Select the first wall you drew (the one on the left).
2. Open the Properties palette.
   Command: *props*
3. Hide **Edge 3**.
4. Use the palette to hide the remaining edges as indicated at right.
5. Set the **3D Hidden** visual style current.
   Your drawing should now look like the figure at right.
6. Set **Windows** as the current layer.
7. Change the UCS to match the front of the wall (right).
   Command: *ucs*
8. Set **Thickness** to ***-5.5***.
   Command: *th*
   Enter new value for THICKNESS <0'-0">: *-5.5*

3.3.2A: INVISIBLE EDGES

9. Set the **SPLFrame** system variable to 1 so you can see the outline of the window. (You may need to regenerate the drawing.)

Command: *splframe*

Enter new value for SPLFRAME <0>: *1*

10. Draw the window frame (trace the opening.)

Command: *l*

11. Return the **SPLFrame** system variable to zero and regenerate the drawing.

The front window looks like the figure at right.

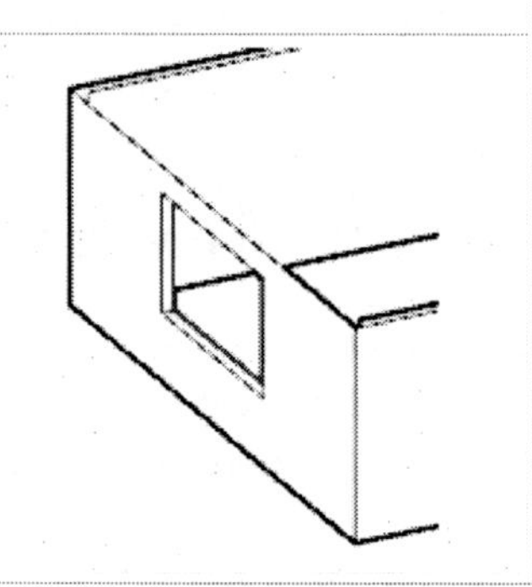

12. Save the drawing but don't exit.

Command: *qsave*

Of course, you've only drawn the outer surface of the wall. If you'd like, you can repeat the exercise to draw the inner surface as well. Then we can look at another method of Surface Modeling – Regions.

3.3.3 Solids and Regions

Some confusion inevitably arises between the terms *solid* and *3D solid*. Let me clarify the distinction.

A *solid* is a two-dimensional, filled polygon. Use solids when you need to highlight a particular object, building, or area on a drawing. Create a solid using the ***Solid*** command.

AutoCAD has no ***3DSolid*** command. The term 3D solid refers to a family of objects (including spheres, cones, boxes, and more). Although similar objects can be created using surface modeling techniques, those techniques won't create *solid* objects. We'll discuss techniques to create 3D solid objects in Lessons 8 and 9.

There are two real differences between solids and regions. The first is that a solid is a filled two-dimensional polygon (although it can have thickness) while a region is an actual surface. The second is that you can create a solid from scratch while creating a region requires an existing object.

A general rule of thumb to follow when considering solids and regions is this: use a region for a surface and a solid to fill in a two-dimensional object.

Since we looked at solids in the basic text and will spend several lessons on solid modeling, we'll concentrate here on regions.

What is a region? Without getting too technical, a region is a two-dimensional surface. It looks very much like a 3D face, but you don't have to hide the edges. You'd use a region anywhere an arc, circle, or hole is required. (Imagine trying to show a round hole with the ***3DFace*** command. Remember that you're restricted to straight edges!)

Let's compare 3D faces with regions.

3D FACE	REGION
Can be drawn from scratch in three dimensions.	Selected objects must be coplanar (share the same UCS). The ***Boundary*** command will only create regions in the current UCS and from existing geometry.
Can't use to show curves or arcs.	Can use to show curves and arcs.
May need to make some edges invisible.	No need to worry about edges.

3D FACE	**REGION**
Can be extruded into a 3D solid.	Can be extruded into a 3D solid.
Create using the ***3DFace*** command.	Create using either the ***Region*** command or the ***Boundary*** command.

As you can see, there are two ways to create a region. Let's consider both.

- The ***Region*** command converts an existing closed object into a region. The objects you can convert include: closed lines, polylines, arcs, circles, splines, and ellipses. Once converted, the original object(s) exists as a region – it's no longer a line, polyline, etc. AutoCAD will create a region from a selected object regardless of the object's relation to the current UCS.

Where to Find It:	
Command Line:	***Region***
Hotkey(s):	***reg***
Ribbon (Tab/Panel):	Home – Draw (subpanel) – **Region**
Menu:	Draw – **Region**
Toolbar:	Draw – **Region**
Tool Palette:	Draw – **Region**

  The ***Region*** command looks like this:

  Command: ***region***

  Select objects: ***[Select the closed object you want to convert.]***

  Select objects: ***[Hit ENTER to complete the selection.]***

  1 Region created. ***[AutoCAD tells you how many regions it has created.]***

- The ***Boundary*** command uses boundaries to create a region. (We discussed boundaries as part of hatching in the basic text.) No objects are lost or converted with this approach and AutoCAD uses a dialog box to assist you. For the ***Boundary*** command to work properly, the objects forming the boundary must be on the zero coordinate of the Z-axis (or the ground – so-to-speak) in the current UCS.

We'll use both of these methods to add some more windows in our cabin.

Do This: 3.3.3A	Creating Regions

I. Be sure you're still in the *Cabin* file in the C:\Steps3D\Lesson03 folder. If not, please open it now.

II. Set the viewpoint to 1,-1,1 (the SE Isometric view).

III. Follow these steps.

3.3.3A: CREATING REGIONS

1. Zoom in around the other front wall as shown.

   Command: *z*

2. Change the **Thickness** to zero for the lines forming the inner and outer walls (on both sides of the door opening).

   Command: ***props***

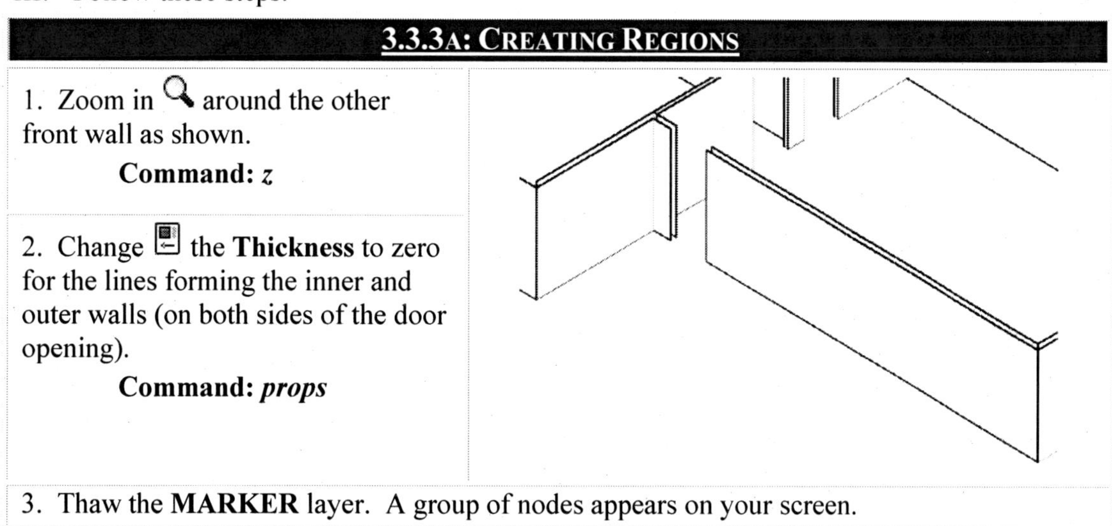

3. Thaw the **MARKER** layer. A group of nodes appears on your screen.

3.3.3A: Creating Regions

4. Use the nodes as a guide to draw four closed-polyline windows as shown. (The thickness should already be set to –5.5 and the UCS should already be the SE Isometric. If not, please make these corrections before doing this step.)

   Command: ***pl***

5. Reset **Thickness** to zero.

   Command: ***th***

   Enter new value for THICKNESS <-5.5000>: ***0***

6. Set the **Walls** layer current.

7. Use a polyline to draw the outline of the outer walls as shown (the height of the door opening is 6'-8"). Be sure to close the polyline.

   Command: ***pl***

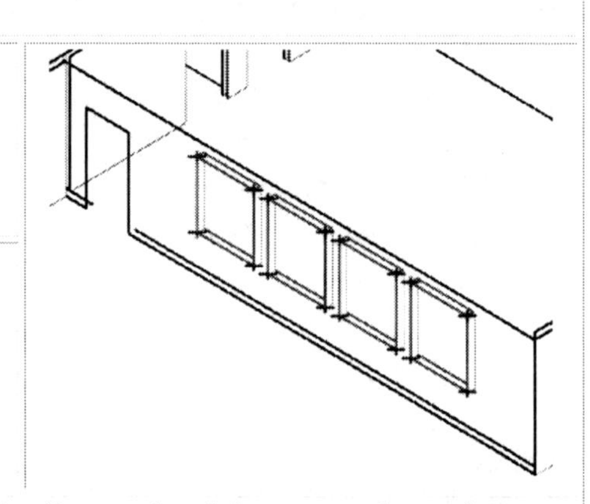

8. Now we'll create our first region. Enter the ***Region*** command. (If you haven't entered the command previously, it may take a moment to load.)

   Command: ***reg***

9. Select the polyline that defines the wall. AutoCAD tells you that it has created a region.

   Select objects:

   Select objects: ***[ENTER]***

   1 loop extracted.

   1 Region created.

10. Freeze the **Marker** layer. Your drawing looks like the figure at right.

Notice that you can't see through the windows. This is because the wall is a region and there are, as yet, no openings for the windows. We'll deal with that now, first by creating window regions and then by removing the window regions from the wall region.

11. Move the current UCS to the lower left corner of the wall as shown.

   Command: ***ucs***

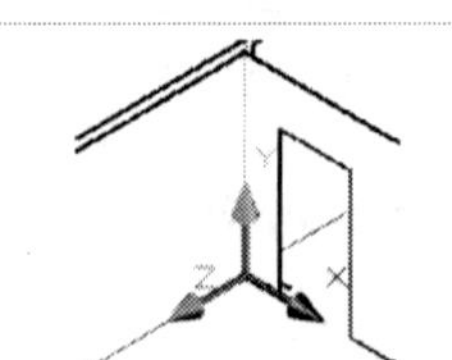

3.3.3A: Creating Regions

12. Enter the ***Boundary*** command.

Command: ***bo***

AutoCAD presents the Boundary Creation dialog box (see Step 13). You're familiar with the options from your study of hatching in the basic text.

13. Set the control box in the **Object type** frame to **Region** as indicated. Then pick the **Pick Points** button. AutoCAD returns to the graphics screen.

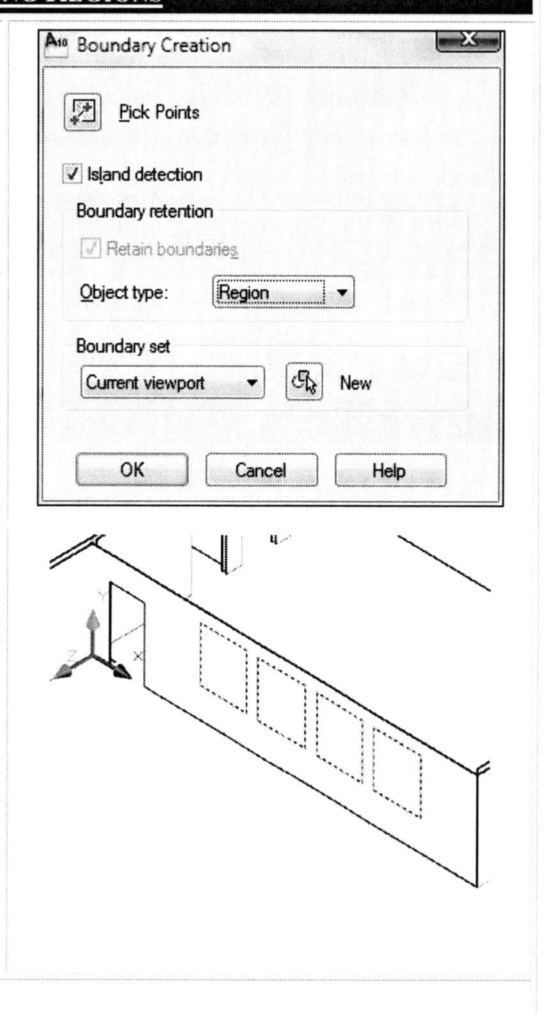

14. Select points inside each of the four windows as shown.

Pick internal point:

15. Complete the command. You can see that the new regions were created without converting the polylines. (Regions are on the **Walls** layer while the polylines are still on the **Windows** layer).

Pick internal point: ***[ENTER]***

4 loops extracted.

4 Regions created.

BOUNDARY created 4 regions

16. Save the drawing but don't exit.

Command: ***qsave***

You've used two methods to create your regions. The first – the ***Region*** command – converted the polyline outlining the wall to a region. That polyline doesn't exist anymore. The second – the ***Boundary*** command – created regions within the defined boundaries without changing the boundaries themselves (the polylines defining the windows).

But what you haven't done is to use the regions created with your window boundaries to cut holes in the wall for the windows. Right now, you simply have four window regions sitting on top of a wall region. Let's take a look at how we can cut those holes.

We'll use a tool with which we'll become considerably more familiar when we study solid modeling in Lesson 9, p.201. In fact, the tool is one of several modifying tools shared by regions and solid models. The tool we'll use here is the ***Subtract*** command. It looks like this:

Command: ***subtract***

Select solids, surfaces, and regions to subtract from..

Select objects: ***[Select the region or solid***

Where to Find It:	
Command Line:	***Subtract***
Hotkey(s):	***su***
Ribbon (Tab/Panel):	Home – Solid Editing – **Subtract**
Menu:	Modify – Solid Editing – **Subtract**
Toolbar:	Modeling – **Subtract**

from which you'll subtract – in our exercise, this would be the wall.]
Select objects: ***[Hit ENTER to complete the selection.]***
Select solids, surfaces, and regions to subtract..
Select objects: ***[Select the regions you wish to remove.]***
Select objects: ***[ENTER to complete the command.]***

Let's finish our wall.

Do This: 3.3.3B	Using Regions to Create Holes

I. Be sure you're still in the *Cabin* file in the C:\Steps3D\Lesson03 folder. If not, please open it now.

II. Follow these steps.

3.3.3B: USING REGIONS TO CREATE HOLES
1. Enter the ***Subtract*** command.   **Command:** ***su***
2. AutoCAD needs to know from which surface you'll subtract. Select the wall.   **Select solids, surfaces, and regions to subtract from ..**   **Select objects:**
3. Now AutoCAD needs to know what to subtract. Select the windows. (Be sure to select the regions and not the polylines.)   **Select solids, surfaces, and regions to subtract ..**   **Select objects:**   Your drawing looks like the following figure. (Starting to look pretty good, don't you think?)
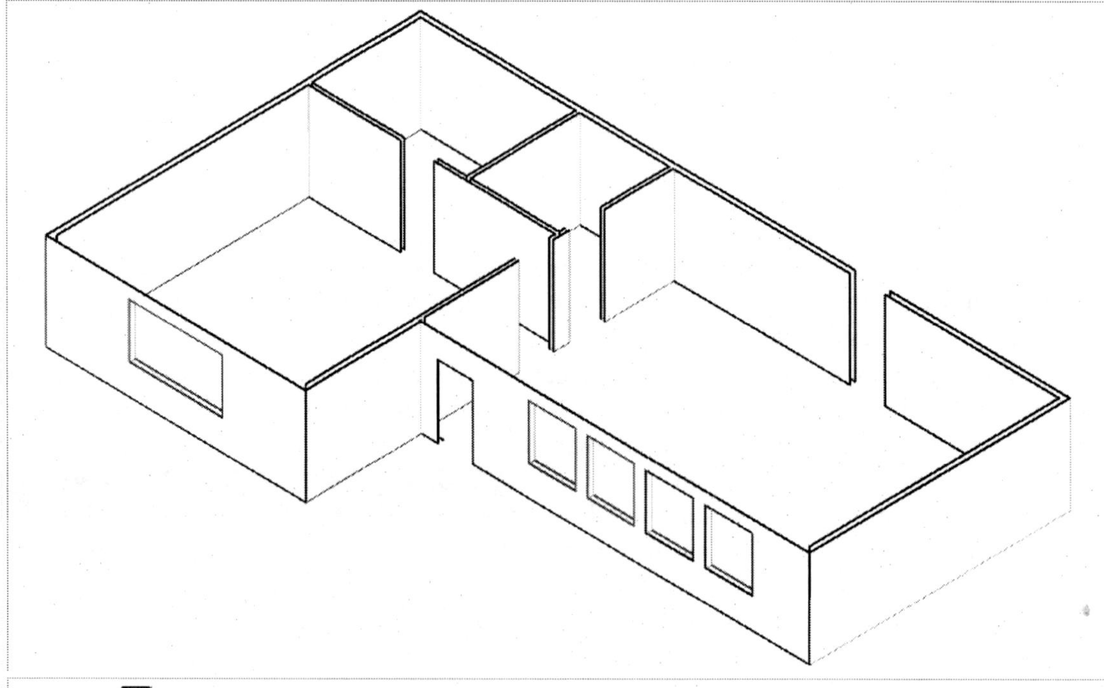
4. Save and close the drawing.   **Command:** ***qsave***

3.3.4 Which Method Should You Use?

Which method of Surface Modeling do you prefer – 3D Face or Regions? Believe it or not, you'll need both.

Consider the model in Figure 3.002. Take a moment to consider each surface. Ask yourself which method of surface modeling you would use to create it. Then (more importantly) ask yourself why you'd use that method.

Once you've examined each surface, continue to the following explanations.

Figure 3.002

- You can easily draw **Surface "A"** using lines with thickness. This brings us to the first rule of three-dimensional work: *Never draw a surface when a line will do.* Surfaces are complex objects and take up more drawing memory than simple lines.

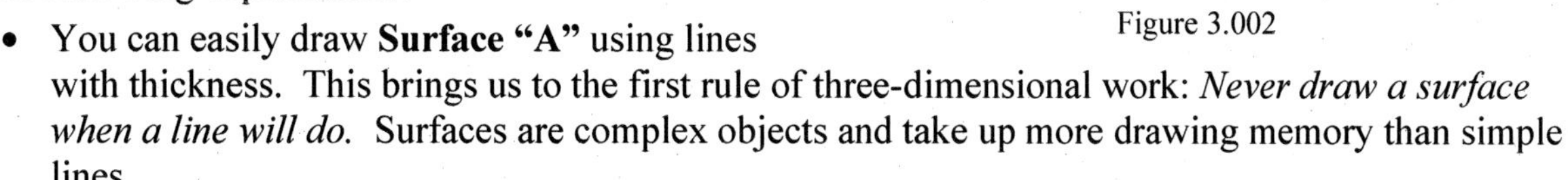

- **Surface "B"** has a round hole in it. 3D faces create flat edges so they won't work here. A region will serve best, but which method should you use? Using the **Boundary** method to create a region, you can pick a point inside the rectangular area (but outside the circle) and let AutoCAD do the rest. This is the best approach.

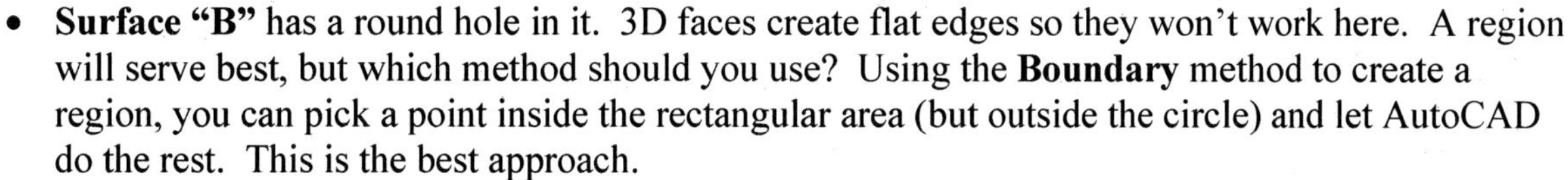

- **Surface "C"** is the inside surface of the hole. The only method we've seen to create this surface is **Thickness**. But you can't use a circle to create a hole. That will result in a closed cylinder (a drum). Use two arcs with thickness to create the hole.
- **Surface "D"** looks very much like the roof of our cabin. Indeed, it's the same type of construction. You can use a region or solid to create the surfaces over the wireframe, but that would involve setting the UCS flat against the surface to be created for each roof section. There are 16 sections, so you'd have to set the UCS and then draw the surface 16 times. Alternately, you can draw 16 3D faces.

  But look at the surface again. There's an easier way that will provide a more rounded surface. How about drawing an arc with thickness in the proper UCS? This will work on this surface, as there are no intersections with which you have to contend. (Rule #2 of three-dimensional work then might be: *Think about it twice; draw it once.*)
- **Surface "E"** has an odd shape to it. Like Surface "B", the odd shape gives away the answer. Another rule of three-dimensional surface work is: *When faced with an unusual shape or holes in a surface, use the boundary approach to create a region.*

Use Figure 3.002 as a guide in your first steps toward creating three-dimensional surface models. (We'll draw Figure 3.002, p.73, in our exercises. Then you can plot it and hang it on your monitor!) Memorize the explanation of each surface and consider each point when determining how to draw a surface on your model.

> AutoCAD includes another command – ***XEdges*** – for those who wish to create 3D wireframes by extracting the edges from existing 3D solids, meshes, regions, and surfaces.

3.4 Extra Steps

Did you use the **Subtract** button on the **Solid Editing** panel in Exercise 3.3.3B, p.72? If you did, you might have noticed that it was grouped with two other buttons – **Union** and **Intersect**. These three buttons work on regions as well as 3D solids. Can you tell from their symbols what they'll do?

We'll discuss them in more detail in Lesson 9, but that doesn't mean that you can't experiment with them now.

Open the *solids & regions* file in the C:\Steps3D\Lesson03 folder. (It looks like Figure 3.003). Experiment with each of these commands on the objects shown. When you've finished, see Figure 3.004, p.74, for the results.

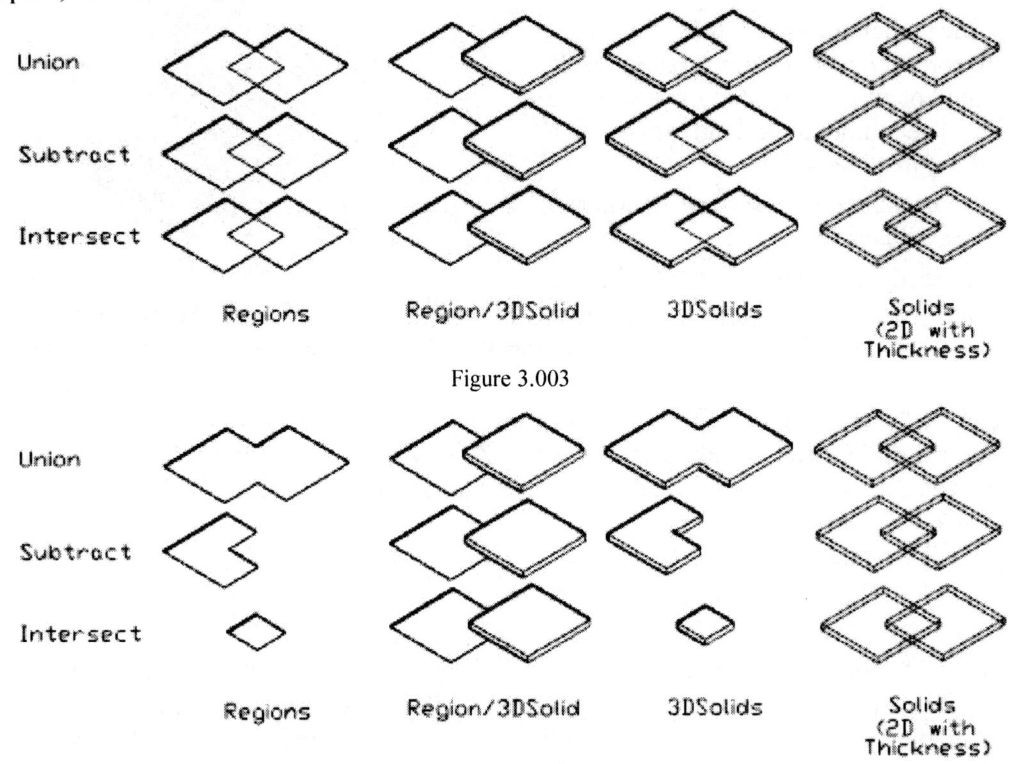

Figure 3.003

Figure 3.004

You'll notice that these commands work on regions or 3D solids. You can't, however, use them to subtract or join (and so forth) different types of objects with each other. Additionally, you'll notice that they don't work at all on simple solids. This is why you should generally use regions and 3D faces when creating surface models.

> In earlier releases, regions only shaded on top when viewed with a Realistic or Conceptual visual style. This had some unpleasant results – we had to draw bottom regions upside down to see them! The programmers have corrected this problem. (Praise the programmers!)

3.5 What Have We Learned?

Items covered in this lesson include:

- *The differences between 3D faces, solids, 3D solids, and regions*
- *The differences between polylines and 3D polylines*
- *Projecting points in three dimensions*
- *The two approaches to creating regions*
  - ***Region***
  - ***Boundary***
- *Commands:*
  - ***3DPoly***
  - ***3DFace***
  - ***Region***
  - ***Edge***
  - ***SPLFrame***
  - ***Boundary***
  - ***Subtract***
  - ***Xedges***

It's taken a few lessons to get comfortable with the basics of Z-space wireframe and surface modeling. At this point, you're either anxious to continue or feeling somewhat overwhelmed by it all (probably a little of both).

I can't overemphasize the importance of practice – if you're uncomfortable with the material thus far, go back and do it again. You shouldn't feel that you're the only person who ever found Z-space difficult to master. But that's the benefit of computer labs and a good textbook! (If you've already changed the files that came from the web, just reload them to start over!)

Are you ready for an easy lesson? Our next chapter – "Predefined Mesh Models" – will show you how to create some more complex objects easily. So do the problems that follow ... review as necessary to get comfortable ... then forward – ever forward!

3.6 Exercises

1. through 8. Add surfaces to the drawings you created in Exercises 1 through 8 of Section 1.7 (in Lesson 1, p.27). Refer to the drawings in Appendix B as a guide. If these drawing aren't available use the corresponding drawing in the C:\Steps3D\Lesson\Steps\Lesson03 folder.
9. Open the *My Twisted Y* file you created in Section 1.7 – Exercise 9, p.27, (C:\Steps3D\Lesson02 folder). (If that file isn't available, use the *Twisted Y-3* file in the C:\Steps3D\Lesson03 folder.) Convert the drawing into a surface model by placing thickness, 3D faces, or regions on the necessary surfaces. See the Twisted Y figure for the completed drawing.
10. Open the *My Block* file you created in Section 1.7 – Exercise 10, p.27, (C:\Steps3D\Lesson02 folder). (If that file isn't available, use the *Block-3* file in the C:\Steps3D\Lesson03 folder.) Convert the drawing into a surface model by placing thickness, 3D faces, or regions on the necessary surfaces. See the Block figure for the completed drawing.

Twisted Y

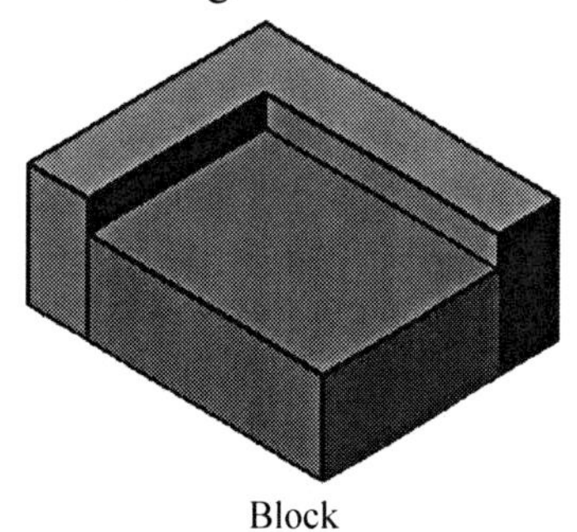

Block

11. Open the *projection* file in the C:\Steps3D\Lesson03 folder. Create the saddle tee shown here. Save the drawing as *My Saddle Tee* in the C:\Steps3D\Lesson03 folder.

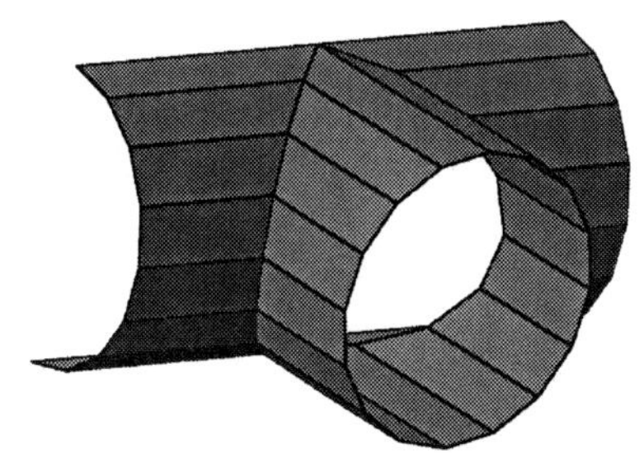

Saddle Tee

12. Open the *MyBookEnd* file you created in Section 1.7 – Exercise 12, p.27, (C:\Steps3D\Lesson02 folder). (If that file isn't available, use the *Book End-3* file in the C:\Steps3D\Lesson03 folder.) Convert the drawing into a surface model by placing thickness, 3D faces, or regions on the necessary surfaces. See the Book End figure for the completed drawing.
13. Open the *My Anchor Stop* file you created in Section 1.7 – Exercise 14, p.28, (C:\Steps3D\Lesson02 folder). (If that file isn't available, use the *Anchor Stop-3* file in the C:\Steps3D\Lesson03 folder.) Convert the drawing into a surface model by placing thickness, 3D faces, or regions on the necessary surfaces. See the Anchor Stop figure for the completed drawing.

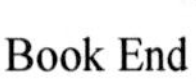

Book End

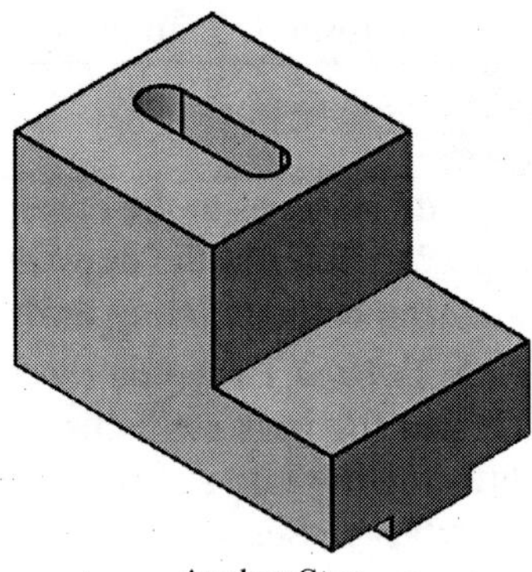

Anchor Stop

14. Open the *MyOtherTable* file you created in Section 2.7 – Exercise 20, p.57, (C:\Steps3D\Lesson02 folder). (If that file isn't available, use the *Other Table-3* file in the C:\Steps3D\Lesson03 folder.) Convert the drawing into a surface model by placing thickness, 3D faces, or regions on the necessary surfaces. See the Table figure for the completed drawing.
15. Open the *My Corner Bracket* file you created in Section 2.7 – Exercise 21, p.57, (C:\Steps3D\Lesson02 folder). (If that file isn't available, use the *corner bracket-3* file in the C:\Steps3D\Lesson03 folder.) Convert the drawing into a surface model by placing thickness, 3D faces, or regions on the necessary surfaces. See the Corner Bracket figure for the completed drawing.

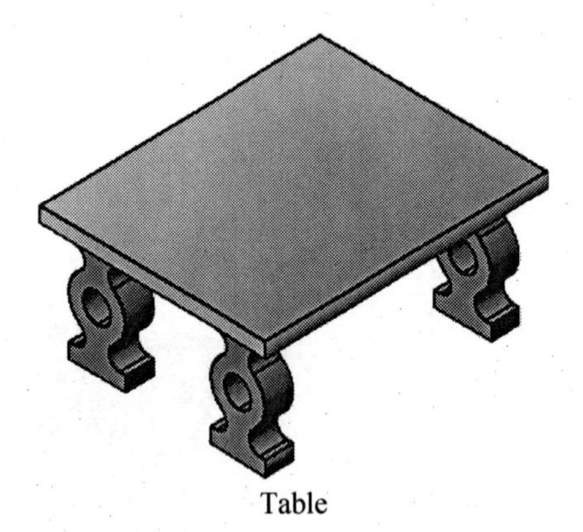

Table

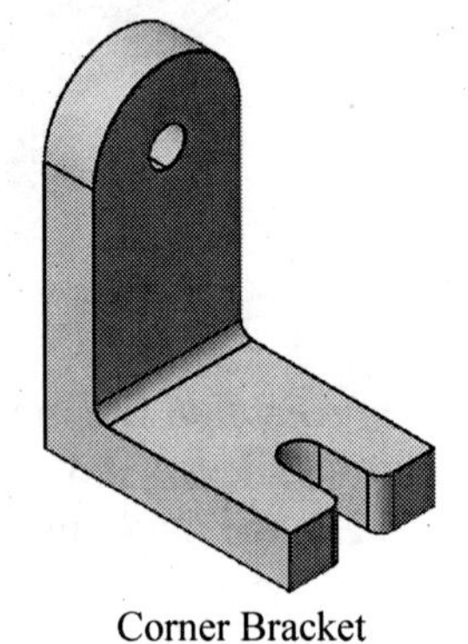

Corner Bracket

16. Open the *My Phone Plug* file you created in Section 2.7 – Exercise 22, p.57, (C:\Steps3D\Lesson02 folder). (If that file isn't available, use the *phone plug-3* file in the C:\Steps3D\Lesson03 folder.) Convert the drawing into a surface model by placing thickness, 3D faces, or regions on the necessary surfaces. See the Phone Plug figure for the completed drawing.
17. Open the *My Other Corner Bracket* file you created in Section 2.7 – Exercise 23, p.59, (C:\Steps3D\Lesson02 folder). (If that file isn't available, use the *other corner bracket-3* file in the C:\Steps3D\Lesson03 folder.) Convert the drawing into a surface model by placing thickness, 3D faces, or regions on the necessary surfaces. See the Other Corner Bracket figure for the completed drawing.

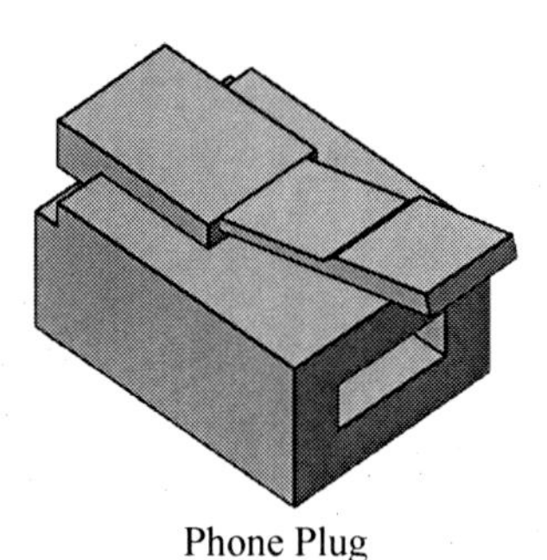

Phone Plug

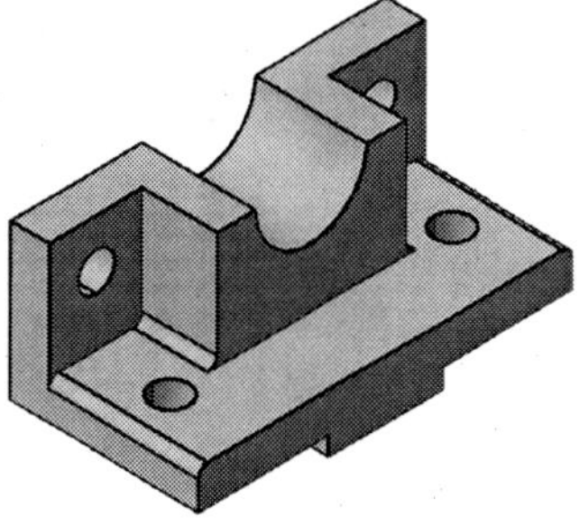

Other Corner Bracket

18. Create the cinder block shown. The following details will help:
    18.1. The overall dimensions of a cinder block are 8" x 8" x 16".
    18.2. The holes are 7" x 6½" and are evenly spaced.
    18.3. The fillets in the holes have a 1" radius.
    18.4. Color 164 looks like a cinder block.
    18.5. Remember that a region can only be seen from what was the positive Z-axis direction at the time of its creation.
    18.6. Save the drawing as *My Cinder Block* in the C:\Steps3D\Lesson03 folder.

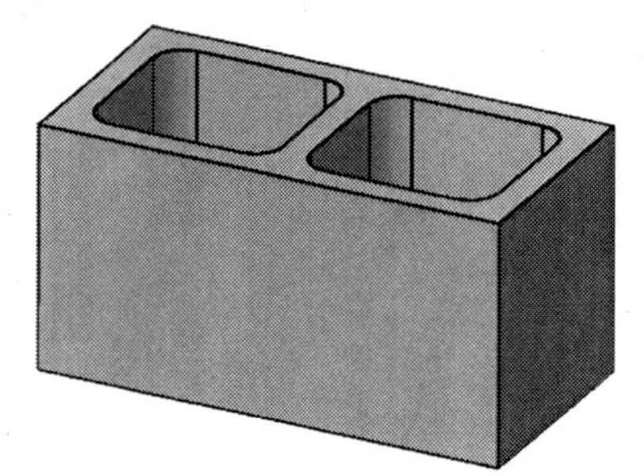

Cinder Block

19. Draw the demo model shown (refer to Figure 3.002, p.73). Use the following details as a guide.
    19.1. Label the types of surfaces as explained in Section 3.3.4.
    19.2. Save the drawing as *Surfaces Model* in the C:\Steps3D\Lesson03 folder.
    19.3. Plot the drawing to fit on about one quarter of a standard 8½" x 11" sheet of paper. Cut it out and tape it to the side of your monitor as a reference.
    19.4. You don't have to dimension the drawing.

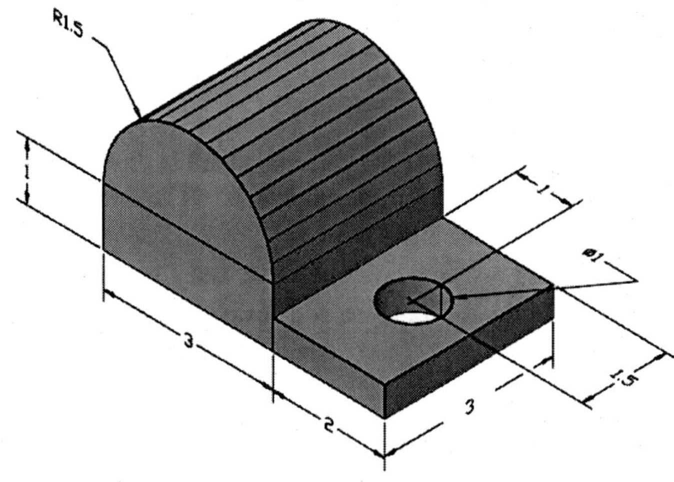

Demo Model

20. Finish the Cabin drawing we started in this lesson. The final drawing is shown in the Cabin figure. Use these details as a guide.
    20.1. All the windows are the same size.
    20.2. All the windows are 12" from the top of the walls.
    20.3. Spaced windows are 12" apart.
    20.4. Change the layer for all inner walls to a new **Inner Walls** layer. Then freeze the layer for clarity.
    20.5. Use the ***3DClip*** command for clarity on the elevations.
    20.6. The scale in each viewport is 1/16"=1'-0" (this layout is set up for an 8½" x 11" sheet).
    20.7. Use the title block of your choice, and plot the drawing.
    20.8. Save the drawing in the C:\Steps3D\Lesson03 folder.

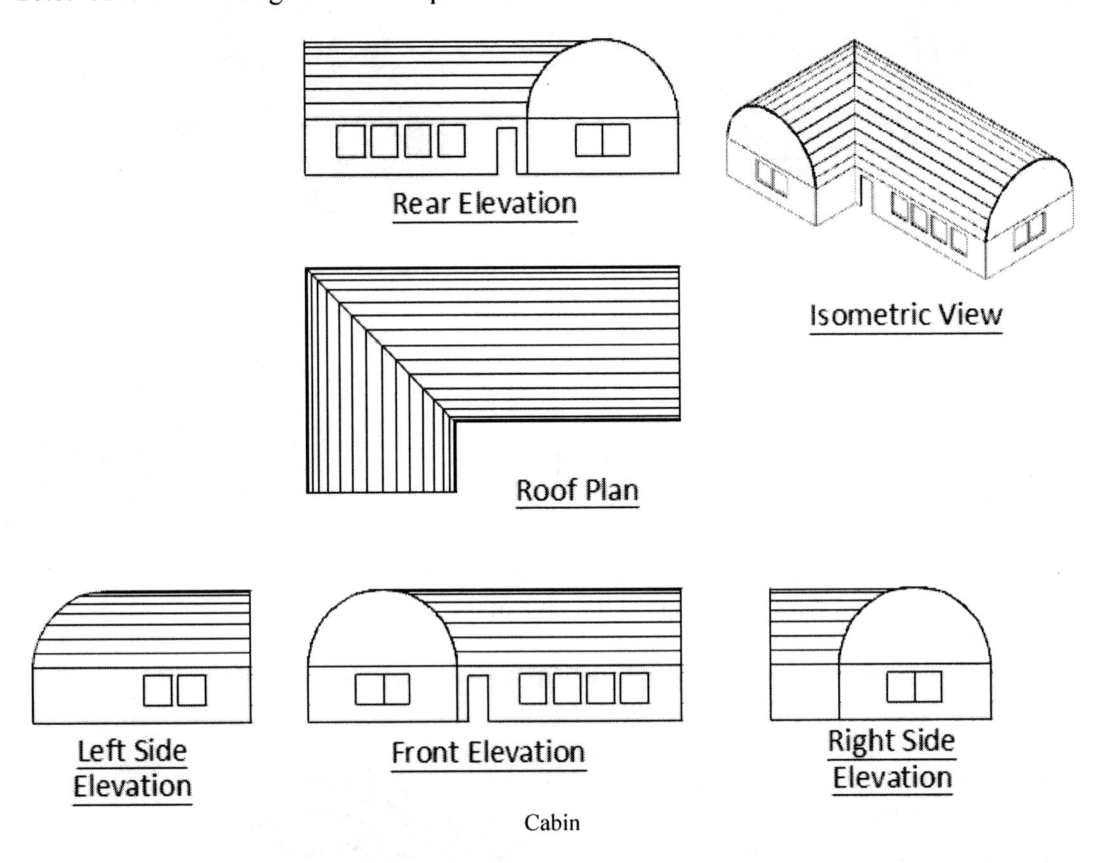

Cabin

21. Reopen the *Cabin* file you completed in Exercise 20 (above). We'll add a different roof and a chimney. Follow these guidelines. (The final drawing is shown here).
    21.1. Freeze the **Roof** layer.
    21.2. Add two new layers – **Roof2** and **Chimney**.
    21.3. On the **Roof2** layer, add hexagons at the end walls of the roofs.

21.4. Add the 3D polyline and 3D faces to complete the roof.

21.5. Place a 4' x 2' rectangle, with a thickness of 24', at coordinates 33',22',0.

21.6. Use construction lines and UCS manipulation to locate where the chimney penetrates the roof. Draw a polyline around the penetration.

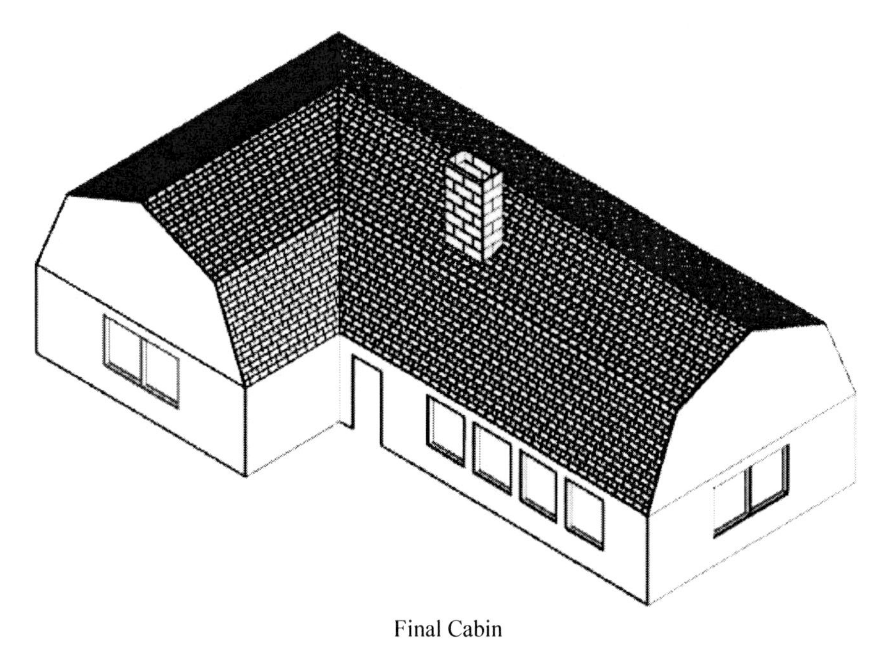

Final Cabin

21.7. Finish drawing the chimney using wireframe techniques. Erase the rectangle.

21.8. Add 3D faces to the chimney.

21.9. Hatch everything as shown.

21.10. Save and plot the drawing.

3.7 For Web-Based Review Questions, visit: http://foragerpub.com/AcadFiles/2010/2010.htm

Lesson

Following this lesson, you will:

✓ *Know how to build and use AutoCAD's predefined mesh models*

- ***Box***
- ***Wedge***
- ***Pyramid***
- ***Cone***
- ***Cylinder***
- ***Sphere***
- ***Dome***
- ***Dish***
- ***Torus***

✓ *Know how to use the **Mesh** command*

Predefined Meshes

In our next few lessons, we'll look at some tools that should greatly simplify your work with mesh models.

First, in this lesson, we'll learn to use AutoCAD's predefined mesh models. Draftsmen at all levels of development can quickly and easily learn to use these remarkable timesaving devices. In fact, I'd be surprised if you hadn't discovered them in your three-dimensional explorations already. You must become familiar with these tools if you wish to pursue a career in any type of three-dimensional work, but you'll find them irreplaceable for any future animation work (3DS Max, Pixar, Lightwave, Poser, and so forth). Unlike their earlier surface cousins, predefine mesh models form the basis of most mesh work.

> *Most of the mesh tools have been introduced in AutoCAD 2010; only a couple are hold-overs from older surface tools.*

Then in Lesson 5, p.103, you'll discover some surprisingly simple tools that you can use to create elaborate, non-uniform mesh and surface models. Then finally, after an introduction to Z-space editing in Lesson 6, p.125, we'll see what meshes can really do in Lesson 7, p.149.

So Let's begin with an overview of the predefined mesh modeling tools AutoCAD has provided.

4.1 What Are Predefined Mesh Models?

Simply put, predefined mesh models are standard geometric shapes that AutoCAD creates for you with a minimal amount of user input. The shapes include (Figure 4.001) a box, a wedge, three types of pyramid, two types of cone, a cylinder, a dome, a dish, a sphere, and a torus. User input for each commonly includes length, width, height, radius, and rotation angle definitions.

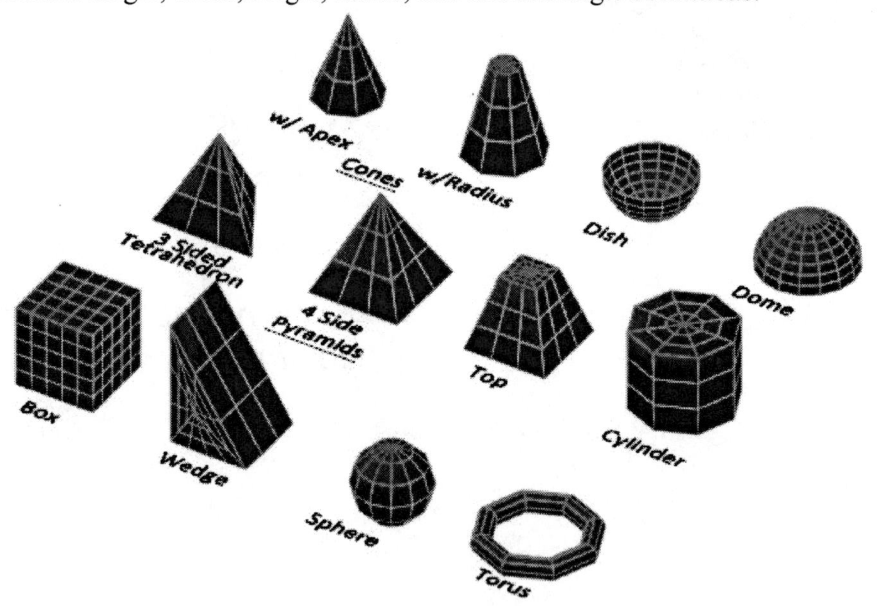

Figure 4.001

You can create each of these models with little effort – simply follow AutoCAD's prompts.

AutoCAD creates each model as a 3D mesh – a collection of 3D faces. We'll discover a host of editing tools designed specifically for meshes in Lesson 7, p.149, which we'll come to love! Additionally, you can explode the mesh (or surface) into a series of 3D faces. This enables you to remove part of the model (with the ***Erase*** command) or hide some of the edges (Properties palette) without affecting the rest of the faces.

4.2 Drawing Predefined Mesh Models

Where to Find It:	
Command Line:	***Mesh [option]***
Ribbon (Tab/Panel):	Mesh Modeling – Primitives – **[Mesh flyout]**
Menu:	Draw – Modeling – Meshes – Primitives – **[Option]**
Toolbar:	Smooth Mesh Primitives – **[option]**

AutoCAD has included a command that enables you to access predefined mesh models without difficulty. The command is, simply enough, ***mesh***. It looks like this:

Command: ***mesh***

Current smoothness level is set to : 0

Enter an option [Box/Cone/CYlinder/Pyramid/Sphere/Wedge/Torus/SEttings]<Box>:

Typically, you'll respond with the type of object you wish to draw and proceed with the appropriate prompts. But what's that "smoothness level" setting?

Well, **smoothness level** refers to roundness of the mesh objects. You can control it by selecting the **SEttings** option. AutoCAD will prompt:

Specify level of smoothness or [Tessellation] <0>:

You can enter a number from 1-5 (higher numbers mean rounder meshes), or you can select the **Tessellation** option. This opens the Mesh Primitive Options dialog box (Figure 4.002). This powerful tool begs for a few paragraphs.

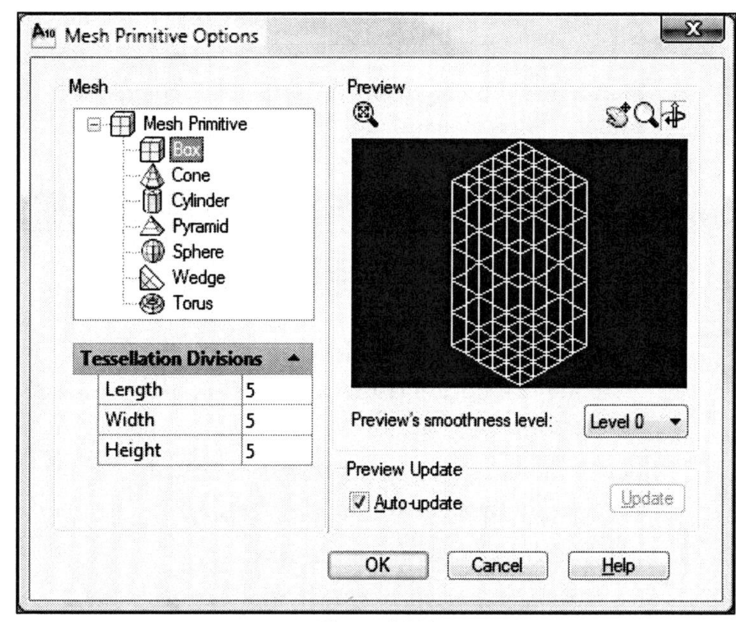

Figure 4.002

- The **Mesh** frame on the left contains a selection box where you'll find the available mesh primitives listed for your selection. Below that, you can set the number of **Tessellation Divisions** (the number of faces AutoCAD will use to create the mesh) you wish for the selected mesh.
- You'll find the **Preview** frame in this dialog box considerably more dynamic than most. Use the buttons above the preview to (from the left): **Zoom Extents**, **Pan**, **Zoom Dynamically**, and **Orbit** the display. Below the preview, you can set the default smoothness level for the selected mesh. (You can change the smoothness level on the Properties palette or using tools on the ribbon's **Mesh Modeling** tab, **Mesh** panel.)
- The **Preview Update** frame offers options to control whether or not the preview automatically or manually updates as you change the other options in the dialog box.

> You can also open the Mesh Primitive Options dialog box by entering the ***MeshPrimitiveOptions*** command on the command line or picking the **Options** button on the **Primitives** panel's title bar (ribbon's **Mesh Modeling** tab).

Let's examine the procedures for drawing each of the predefined surface models.

Note: Unless noted otherwise, we show all the exercises in this section with a **Realistic** visual style, a **Realistic** face style, and showing **Isolines**.

4.2.1	Box

Use the **Box** option of the ***Mesh*** command to draw any six-sided box whose sides, top, and bottom are parallel or perpendicular to the current UCS. The command sequence looks like this:

> **Command:** ***mesh***
>
> **Current smoothness level is set to : 0**
>
> **Enter an option [Box/Cone/CYlinder/Pyramid/Sphere/Wedge/Torus/SEttings] <Box>:** ***b***
>
> **Specify first corner or [Center]:** ***[Identify the first corner of the box.]***
>
> **Specify other corner or [Cube/Length]:** ***[Identify the opposite, same plane corner of the box or tell AutoCAD to prompt for Length (if you use the*** **Length** ***option, AutoCAD will also prompt for*** **Width*).]***
>
> **Specify height or [2Point]:** ***[Tell AutoCAD how tall to make the box; AutoCAD won't use this prompt if you opted for a Cube at the last prompt.]***

Let's try one.

Do This: 4.2.1A	Creating a Mesh Box

I. Start a new drawing using the *lesson 04 template* file located in the C:\Steps3D\Lesson04 folder.

II. Follow these steps.

4.2.1A: CREATING A MESH BOX

1. Begin the **Mesh Box** command.

   Command: ***mesh***

   Current smoothness level is set to : 0

   Enter an option [Box/Cone/CYlinder/Pyramid/Sphere/Wedge/Torus/SEttings] <Sphere>: ***b***

2. Specify the start point of the box as indicated.

   Specify first corner or [Center]: ***1,1***

3. Specify the **other corner** of the box.

   Specify other corner or [Cube/Length]: ***@4,2***

4. And give the box a **height** of ***1***.

   Specify height or [2Point] <14.3922>: ***1***

Your box looks like the figure at right.

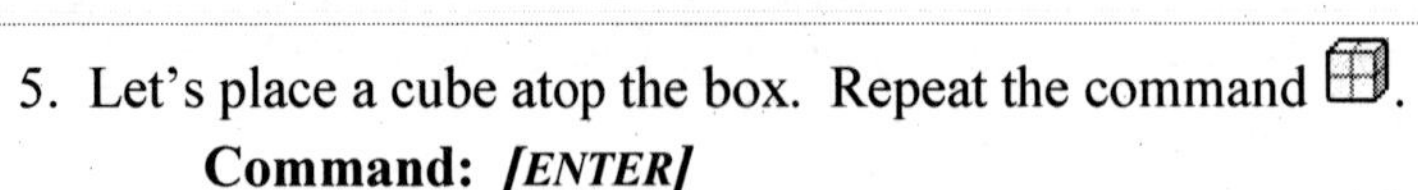

5. Let's place a cube atop the box. Repeat the command.

   Command: ***[ENTER]***

6. Place the first corner atop the first corner of the box.

   Specify first corner or [Center]: ***1,1,1***

7. Select the **Cube** option Cube…

   Specify other corner or [Cube/Length]: ***c***

4.2.1A: Creating a Mesh Box

8. (Use Ortho and orient the box toward the right while proceeding.) Give our cube a **length** of 2 …

Specify length: ***2***

Your drawing looks like the figure at right.

9. Save the drawing as *MyBoxes* in the C:\Steps3D\Lesson04 folder.

Command: ***save***

10. Perform the ***List*** command on the upper box.

Command: ***li***

Notice (following figure) that AutoCAD identifies the box as a MESH object.

```
                    MESH        Layer: "obj1"
                                Space: Model space
                     Handle = 123
             Watertight mesh = YES
       Initial mesh face count = 150
     Initial mesh vertex count = 152
            Mesh Smooth Level = 0
        Level 0 mesh face count = 150
      Level 0 mesh vertex count = 152
              Bounding box: Lower bound X = 1.0000    , Y = 1.0000    , Z = 1.0000
                            Upper bound X = 3.0000    , Y = 3.0000    , Z = 3.0000
```

11. Explode the upper box.

Command: ***x***

12. Repeat Step 10.

Command: ***li***

Notice (following figure) the difference. Now you can remove faces, hide edges, and so forth.

```
              3D FACE    Layer: "obj1"
                         Space: Model space
               Handle = 125
          first point, X=    1.0000  Y=    1.4000  Z=    3.0000
         second point, X=    1.4000  Y=    1.4000  Z=    3.0000
          third point, X=    1.4000  Y=    1.8000  Z=    3.0000
         fourth point, X=    1.0000  Y=    1.8000  Z=    3.0000
```

13. Save and close the drawing.

Command: ***qsave***

Wasn't that easier than drawing a stick figure and stretching skin around it? (I love easy!) Let's try another predefined mesh model.

4.2.2 Wedge

You won't find many differences between the ***Mesh*** command's **Wedge** and **Box** tools. In fact, except for the outcome, the two are identical! But let's draw a wedge for practice.

Do This: 4.2.2A — Creating a Mesh Wedge

I. Start a new drawing using the *lesson 04 template* file located in the C:\Steps3D\Lesson04 folder.

II. Follow these steps.

4.2.2A: Creating a Mesh Wedge

1. Enter the **Mesh Wedge** command.

   Current smoothness level is set to : 0

   Enter an option [Box/Cone/CYlinder/Pyramid/Sphere/Wedge/Torus/SEttings] <Sphere>: ***w***

2. Identify the starting point of the wedge as indicated.

   Specify first corner or [Center]: ***1,1***

3. Specify the values of the wedge as indicated.

   Specify other corner or [Cube/Length]: ***@4,2***

   Specify height or [2Point] <1.6663>: ***1***

   Your drawing looks like the figure at right.

4. Save the drawing as *MyWedge* in the C:\Steps3D\Lesson folder.

   Command: ***save***

Remember that you can change the UCS prior to creating the wedge to help control the direction of the slope. You can even use negative numbers!

Let's look at something more complex than simple boxes and wedges. Let's look at the **Pyramid** tool.

4.2.3 Pyramid

As you saw in Figure 4.001, p.82, AutoCAD provides for three types of mesh pyramid. The earlier, surface approach provided for a fourth – a *ridged* pyramid – which isn't available as a mesh. If you'd like to know how to create a ridged, surface pyramid, I've placed this section of our '09 text on the web. Go to: http:/www.uneedcad.com/Files/Surface_Pyramids.pdf.

Did you know that there's more than one kind of pyramid?

Technically speaking, a pyramid is a polyhedron – a multi-triangular structure (and you thought pyramid was hard to spell!). The common idea of a pyramid comes from the Egyptian model with a rectangular base (or four triangular sides on a rectangular base). Most of the Egyptian pyramids come to a point at the top.

But there are other pyramids. In fact, the largest pyramid in the world isn't Egyptian at all! Look for it just outside Mexico City. And it has a flat top!

Some pyramids even have triangular bases – or three triangular sides on a triangular base. These are called tetrahedrons. (Sounds like something that ate the bad guys in *Jurassic Park XXIV*.)

You probably didn't realize how complicated the world of pyramids was! But not to worry – AutoCAD provides for drawing each within one simple ***Mesh*** command option – **Pyramid**. Its command sequence looks like this:

> **Command:** ***mesh***
>
> **Current smoothness level is set to : 0**
>
> **Enter an option [Box/Cone/CYlinder/Pyramid/Sphere/Wedge/Torus/SEttings] <Wedge>:** ***p***
>
> **4 sides Circumscribed** ***[AutoCAD lets you know what type of pyramid it will build.]***
>
> **Specify center point of base or [Edge/Sides]:** ***[AutoCAD draws a pyramid base in much the same manner as it draws a polygon. By default, you'll draw from the center. Use the***

Edge ***option to draw from an edge; use the*** **Sides** ***option to change the number of sides your pyramid will have.]***

Specify base radius or [Inscribed]: ***[Tell AutoCAD around how large a circle it should place the pyramid's base. Use the Inscribed option to place the base within the circle rather than outside it.]***

Specify height or [2Point/Axis endpoint/Top radius] <1.0000>: ***[Define your pyramid. Enter a point in Z-space to identify the point at the top of the pyramid (*****Axis endpoint*****), tell AutoCAD to define the height from two selected points (*****2Point*****), or use the*** **Top radius** ***option to create a flat top within a defined circle.]***

It really isn't as complicated as it looks. Let's draw some pyramids and see.

Do This: 4.2.3A	Creating Mesh Pyramids

I. Start a new drawing using the *lesson 04 template* file located in the C:\Steps3D\Lesson04 folder.

II. Follow these steps.

4.2.3A: CREATING MESH PYRAMIDS

1. Enter the **Mesh Pyramid** command.

    Current smoothness level is set to : 0

    Enter an option [Box/Cone/CYlinder/Pyramid/Sphere/Wedge/Torus/SEttings] <Sphere>: ***p***

2. We'll start with a simple, four-sided pyramid that comes to a point at the top. Enter the base figures indicated.

    Specify center point of base or [Edge/Sides]: ***0,0***

    Specify base radius or [Inscribed]: ***2***

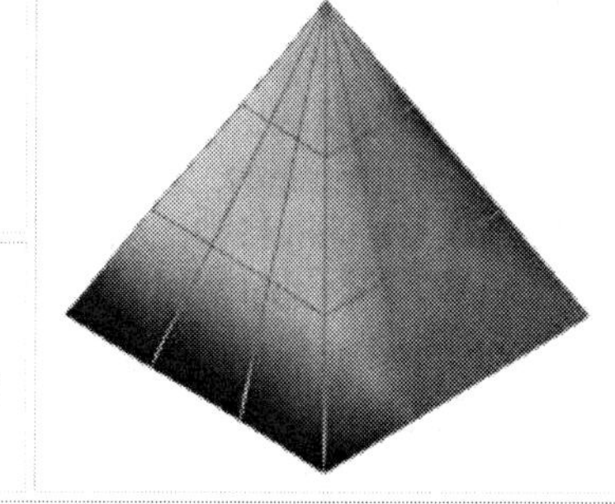

3. An apex point (sharp point at the top) is the default, so enter the height indicated.

    Specify height or [2Point/Axis endpoint/Top radius]: ***4***

    Your first pyramid looks like the figure at right.

4. Let's draw a three-sided pyramid – a tetrahedron. Repeat the command.

5. Select the **Sides** option and tell AutoCAD to create a three-sided pyramid.

    Specify center point of base or [Edge/Sides]: ***s***

    Enter number of sides <4>: ***3***

6. We'll draw this one next to the first. Enter the figures indicated for the base.

    Specify center point of base or [Edge/Sides]: ***6,0***

    Specify base radius or [Inscribed]: ***1.5 [Using ortho, move your cursor toward the back of the drawing.]***

7. And give it a height of 4".

    Specify height or [2Point/Axis endpoint/Top radius]: ***4***

    Your pyramid looks like the figure at right.

8. Now let's draw a four-sided pyramid with a flat top. (Erase the previous pyramids or freeze their layer and set a different one current.) Repeat the command.

4.2.3A: CREATING MESH PYRAMIDS

9. First, draw the base just as you did for the first pyramid.

   Specify center point of base or [Edge/Sides]: ***0,0***

   Specify base radius or [Inscribed]: ***2***

10. But instead of identifying an apex point, tell AutoCAD to draw a **Top** on the pyramid.

    Specify height or [2Point/Axis endpoint/Top radius]: ***t***

11. Give the pyramid's top a radius of .75" …

    Specify top radius <0.0000>: ***.75***

12. … and make it 4" tall.

    Specify height or [2Point/Axis endpoint]: ***4***

Your pyramid looks like the figure at right.

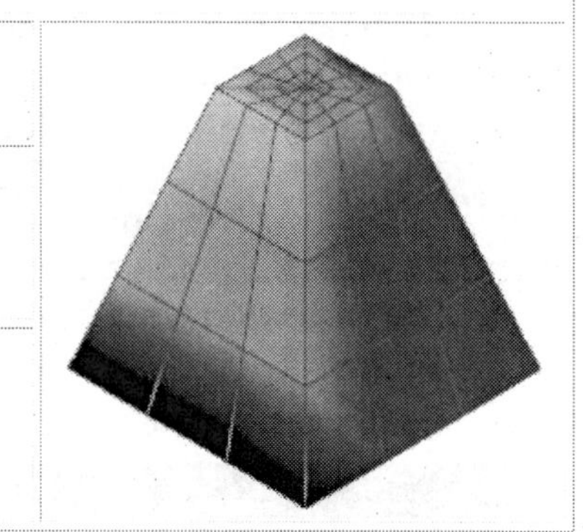

13. Save the drawing as *MyPyramids* in the C:\Steps3D\Lesson04 folder.

    Command: ***save***

Now let's look at pyramids with round bottoms – let's look at cones.

4.2.4 Cone

Like pyramids, cones are smaller on the top than on the bottom (generally speaking). The top can be pointed like the pyramid's apex point, or it can be flat. But the similarities end there.

The predefined mesh models we've examined so far have all had sides that loaned themselves easily to 3D faces. Beginning with cones, we'll look at several predefined shapes that incorporate circles or arcs in their structures. Because these structures are 3D meshes (and will convert to 3D faces when exploded), we should tell AutoCAD how many faces to use when creating the surfaces of the circles or arcs. We'll do that with the Mesh Primitive Options dialog box we saw in Figure 4.002 (p.83).

The command sequence for cones is:

> **Command:** ***mesh***
>
> **Current smoothness level is set to : 0**
>
> **Enter an option [Box/Cone/CYlinder/Pyramid/Sphere/Wedge/Torus/SEttings] <Pyramid>:** ***c***
>
> **Specify center point of base or [3P/2P/Ttr/Elliptical]:** ***[Use standard*** **circle** ***command options to locate the base of the cone.]***
>
> **Specify base radius or [Diameter] <2.5634>:** ***[Use standard*** **circle** ***command options to size the base of the cone.]***
>
> **Specify height or [2Point/Axis endpoint/Top radius] <4.2992>:** ***[This step duplicates the pyramid's approach. See p.87.]***

We'll draw a couple cones for practice.

Do This: 4.2.4A	Creating Mesh Cones

I. Start a new drawing using the *lesson 04 template* file located in the C:\Steps3D\Lesson04 folder.

II. Follow these steps.

4.2.4A: Creating Mesh Cones

1. Enter the **Mesh Cone** command.

   Command: ***mesh***

   Current smoothness level is set to : 0

   Enter an option [Box/Cone/CYlinder/Pyramid/Sphere/Wedge/Torus/SEttings] <Pyramid>: *c*

2. Identify the **center point** and the **radius** of the base as indicated.

   Specify center point of base or [3P/2P/Ttr/Elliptical]: ***0,0***

   Specify base radius or [Diameter] <2.8901>: ***2***

3. We'll draw a pointed cone first. Specify the height of the cone as indicated.

   Specify height or [2Point/Axis endpoint/Top radius] <5.5656>: ***4***

Your drawing looks like the figure at right.

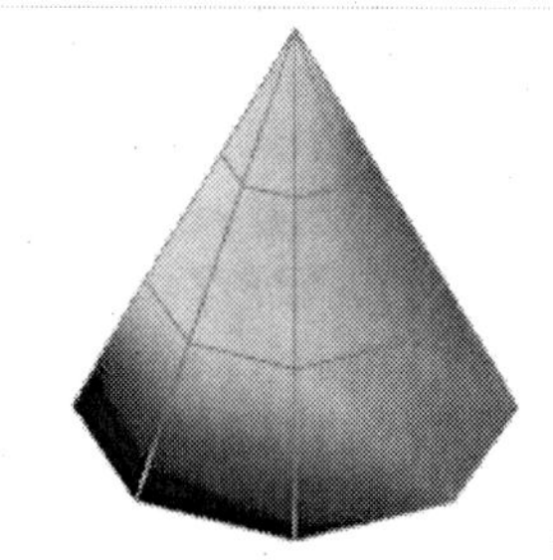

4. Now let's draw a cone with a flat top. Repeat the command.

5. Locate the base and specify the radius as you did in Step 2.

   Specify center point of base or [3P/2P/Ttr/Elliptical]: ***6,0***

   Specify base radius or [Diameter] <2.8901>: ***2***

6. But this time, tell AutoCAD you wish to use a **Top radius**.

   Specify height or [2Point/Axis endpoint/Top radius] <4.0000>: ***t***

7. Give the top a radius of .5".

   Specify top radius <0.0000>: ***.5***

8. And make the cone 4" tall.

   Specify height or [2Point/Axis endpoint] <4.0000>: ***4***

Your cone looks like the figure at right.

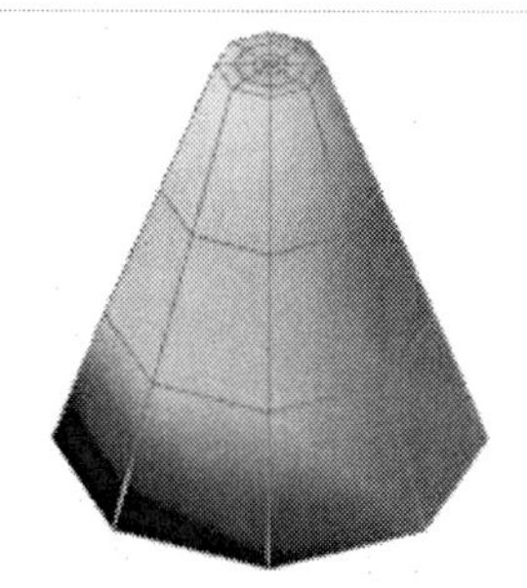

9. Have you noticed that the cone doesn't appear to be very round. Let's fix that by increasing the number of faces.

   Open the Mesh Primitive Options dialog box (Figure 4.002, p.83).

10. Select **Cone** in the **Mesh** selection box and increase the **Tessellation Divisions** as shown, and close the dialog box.

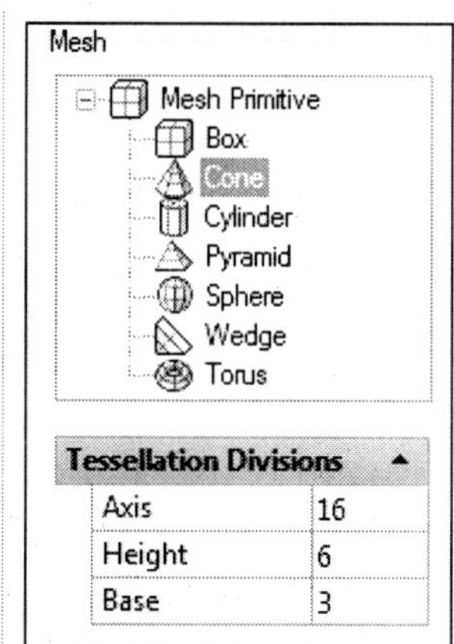

4.2.4A: Creating Mesh Cones

11. Create the cone indicated and compare it (following figure) with your previous cones.

Command: ***mesh***

Current smoothness level is set to : 0

Enter an option [Box/Cone/CYlinder/Pyramid/Sphere/Wedge/ Torus/SEttings] <Pyramid>: ***c***

Specify center point of base or [3P/2P/Ttr/Elliptical]: ***3.5,-3.5***

Specify base radius or [Diameter] <2.0000>: ***[ENTER]***

Specify height or [2Point/Axis endpoint/Top radius] <4.0000>: ***[ENTER]***

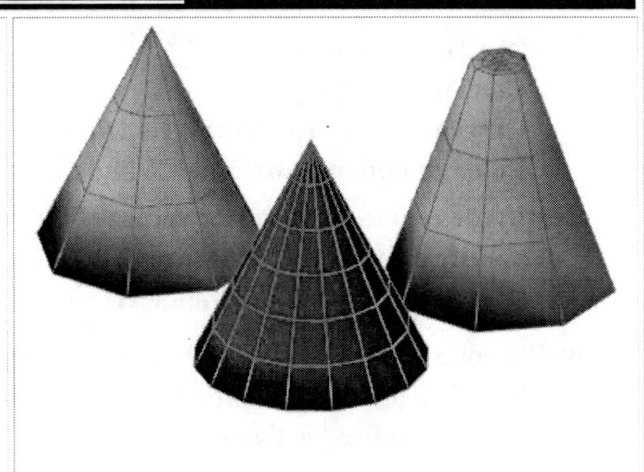

12. Still not quite round enough for you? Try this. Select the last cone you created and open the Properties palette.

13. Under the **Geometry** header, set the **Smoothness** to **Level 3** as indicated.

Smoothness	None
Vertex	None
Vertex X	Level 1
Vertex Y	Level 2
	Level 3
Vertex Z	Level 4

14. Clear the grips. Now compare the cones (following figure). Cool, huh?

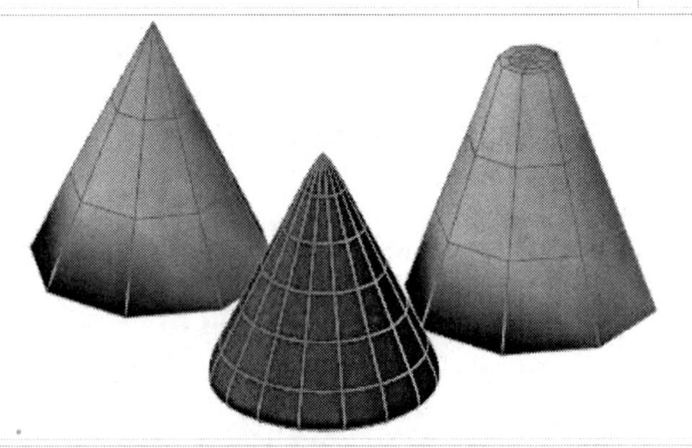

15. Save the drawing as *MyCones* in the C:\Steps3D\Lesson04 folder.

Command: ***save***

The last few steps in this exercise reinforces what you've learned about meshes consisting of flat faces. Remember, for a mesh to appear round, you must increase the number of tessellation divisions and/or the **Smoothness** factor!

4.2.5 Cylinder

You'll find shifting from cones to cylinders simple; you just drop the **Top Radius** height options! The prompts look like this:

Command: ***mesh***

Current smoothness level is set to : 0

Enter an option [Box/Cone/CYlinder/Pyramid/Sphere/Wedge/ Torus/SEttings] <Pyramid>: ***cy***

Specify center point of base or [3P/2P/Ttr/Elliptical]: *[Use standard* **circle** *command options to locate the base of the cylinder.]*

Specify base radius or [Diameter] <2.0000>: *[Use standard* **circle** *command options to size the cylinder.]*

Specify height or [2Point/Axis endpoint] <4.0000>: *[This step duplicates the pyramid's approach without the* **Top Radius** *option.* *See p.87.]*

Let's draw a cylinder.

Do This: 4.2.5A	Creating a Mesh Cylinder

I. Start a new drawing using the *lesson 04 template* file located in the C:\Steps3D\Lesson04 folder.

II. Follow these steps.

4.2.5A: CREATING A MESH CYLINDER

1. Enter the **Mesh Cylinder** command.

 Command: *mesh*

 Current smoothness level is set to : 0

 Enter an option [Box/Cone/CYlinder/Pyramid/Sphere/Wedge/Torus/SEttings] <Cone>: *cy*

2. Create the cylinder as indicated.

 Specify center point of base or [3P/2P/Ttr/Elliptical]: *0,0*

 Specify base radius or [Diameter] <0.0000>: *2*

 Specify height or [2Point/Axis endpoint] <0.0000>: *4*

 Your cylinder looks like the figure at right.

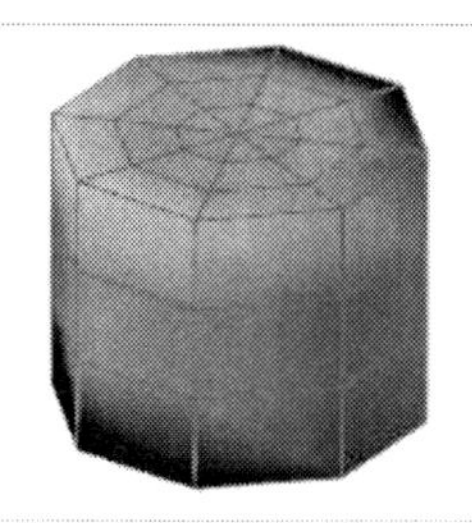

3. Save the drawing as *MyCylinder* in the C:\Steps3D\Lesson04 folder.

 Command: *save*

4. Experiment with tessellation and smoothness settings for some other cylinders.

Well, it's about time we got around to spheres.

4.2.6 Sphere

A sphere also undergoes enhanced appearance with increased tessellation and smoothness. And it's even easier than creating a cylinder.

Command: *mesh*

Current smoothness level is set to : 0

Enter an option [Box/Cone/CYlinder/Pyramid/Sphere/Wedge/Torus/SEttings] <Cylinder>: *s*

Specify center point or [3P/2P/Ttr]: *[Use standard* **circle** *command options to locate the center of the sphere.]*

Specify radius or [Diameter] <1.7127>: *[Use standard* **circle** *command options to size the sphere.]*

Let's draw a sphere.

Do This: 4.2.6A	Creating a Mesh Sphere

I. Start a new drawing using the *lesson 04 template* file located in the C:\Steps3D\Lesson04 folder.

II. Follow these steps.

4.2.6A: CREATING A MESH SPHERE

1. Enter the **Mesh Sphere** command.

   Command: *mesh*

   Current smoothness level is set to : 0

   Enter an option [Box/Cone/CYlinder/Pyramid/Sphere/Wedge/Torus/SEttings] <Cylinder>: *s*

2. Specify the **center point** of the sphere. Be sure to use a three-dimensional coordinate as indicated.

   Specify center point or [3P/2P/Ttr]: *3,3,3*

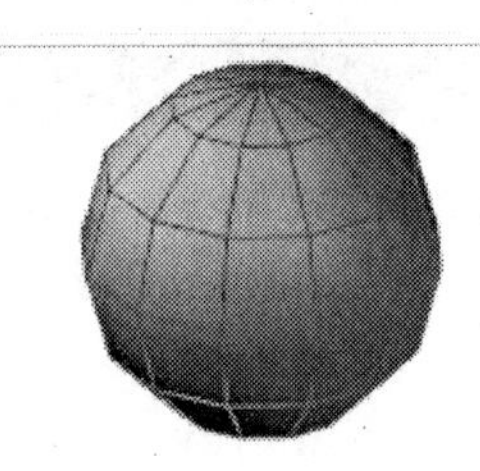

3. Identify the **radius** as indicated.

   Specify radius or [Diameter] <1.7127>: *2*

   Your drawing looks like the figure at right.

4. In the Mesh Primitive Options dialog box, set the **Tessellation Divisions** for spheres to 36 on the Axis and 18 on the **Height** as indicated.

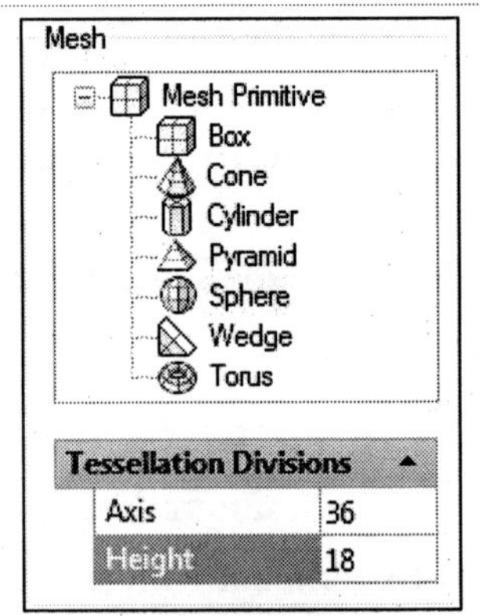

5. Erase the sphere you drew, and recreate it with the new settings (Steps 1 – 3).

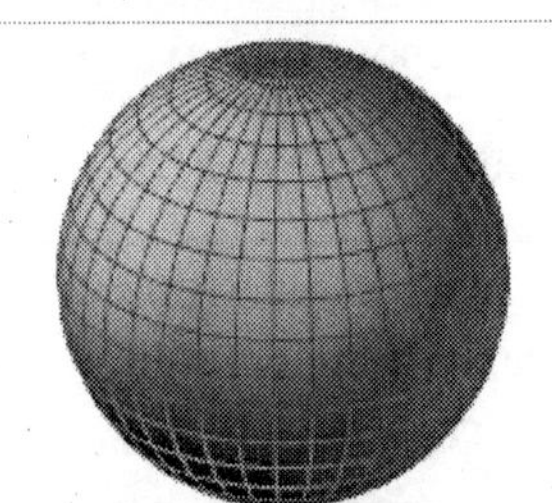

6. Change the **Smoothness** level to **4**.

   Your sphere looks like the figure at right. How does that compare with the one you had after Step 3?

7. Now for a real treat, toggle your isolines off **(Render** tab). Your drawing looks like the figure at right. Way cool!

8. Save the drawing as *MySphere* in the C:\Steps3D\Lesson04 folder.

   Command: *save*

What do you mean you wanted to use the *right half* or *left half* of the sphere?! Some people just have to be difficult!

Well, I'll show you how to do that in Section 4.3, p.94. But for now, let's finish off the mesh tools with a three-dimensional donut. Let's create a torus.

4.2.7	Torus

A torus looks like the inner tube you used at the beach or lake when you were a kid. (Okay, some of you are still young enough to enjoy that type of activity without cracking bones that saw the breakup of Pangaea!) Although it isn't difficult (it's a lot like the last several meshes you've drawn), drawing a torus requires a double effort. When drawing a torus, you must identify both the radius of the *torus* and the radius of the *tube*.

> Manipulating the various radii of a torus can lead to some startling (and nifty) results as you'll discover when working with 3D solids in Lesson 8, p.185.

The command sequence looks like this:

Command: ***mesh***

Current smoothness level is set to : 0

Enter an option [Box/Cone/CYlinder/Pyramid/Sphere/Wedge/Torus/SEttings] <Sphere>: ***t***

Specify center point or [3P/2P/Ttr]: ***[Use standard* circle *command options to locate the center of the torus.]***

Specify radius or [Diameter] <1.7127>: ***[Use standard* circle *command options to size the torus. This radius will be from the* center *of the torus to the* center *of the tube.]***

Specify tube radius or [2Point/Diameter]: ***[Use standard* circle *command options to size the torus tube.]***

Give it a try.

Do This: 4.2.7A	Creating a Mesh Torus

I. Start a new drawing using the *lesson 04 template* file located in the C:\Steps3D\Lesson04 folder.

II. Set the **Tessellation Divisions** for your torus to **Radius: 32** and **Sweep path: 16**. This way, you not only get a better torus but you can tell the difference between the radius and the sweep path.

III. Follow these steps.

4.2.7A: CREATING A MESH TORUS
1. Enter the **Mesh Torus** command. **Command:** ***mesh*** **Current smoothness level is set to : 0** **Enter an option [Box/Cone/CYlinder/Pyramid/Sphere/Wedge/Torus/SEttings] <Sphere>:** ***t***
2. Locate the torus using a three-dimensional coordinate. **Specify center point or [3P/2P/Ttr]:** ***5,5,.5***

4.2.7A: Creating a Mesh Torus	
3. Specify the radii of the torus and the tube as indicated. **Specify radius or [Diameter] <1.7127>:** *4* **Specify tube radius or [2Point/Diameter]:** *1* Your torus looks like the figure at right.	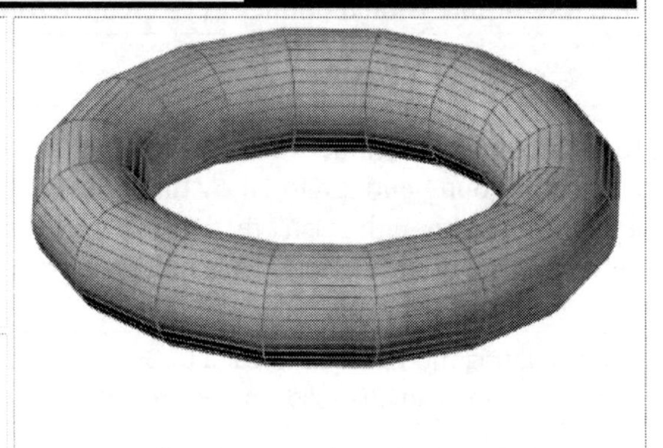
4. Save the drawing as *MyTorus* in the C:\Steps3D\Lesson04 folder. **Command:** ***save***	

That covers AutoCAD's mesh primitives. But a couple other tools merit some attention, even though they technically belong with the older, surface tools. These include a dish and a dome – halve spheres that you just can't draw any other way. Let's look at these next.

4.3 A Couple Legacy Surface Meshes: Domes and Dishes

> The older surface command – ***3D*** – may look familiar after studying meshes, but you really don't want to use it as the results can't compare with meshes. The command looks like this:
>
> **Command: *3d***
>
> **Enter an option [Box/Cone/DIsh/DOme/Mesh/Pyramid/Sphere/Torus/Wedge]:**
>
> Still, you won't find dish and dome tools anywhere else – and both actually produce meshes!

So you only need half a sphere, huh? Well, you could draw a sphere, explode it, and then erase what you don't need. But that's too much work; and let's face it, that's not what drafting is all about. Besides, once you explode a mesh, you can't use any of the mesh editing tools on it! (You'll learn about those in Lesson 7, p.149.)

A dome is the upper half of a sphere; a dish is the lower half. The command sequences for both the **Dome** and **Dish** options is:

Command: *3d*

Enter an option [Box/Cone/DIsh/DOme/Mesh/Pyramid/Sphere/Torus/Wedge]: *do* **[or** *di***]**

Specify center point of dome: ***[Specify the center point of the dome/dish.]***

Specify radius of dome or [Diameter]: ***[Specify the radius or diameter of the dome/dish.]***

Enter number of longitudinal segments for surface of dome <16>: ***[How many divisions will you want from side to side? You won't use tessellation divisions for dishes and domes; use longitudinal and latitudinal segments instead.]***

Enter number of latitudinal segments for surface of dome <8>: ***[How many divisions will you want from top to bottom?]***

Let's draw one of each.

Do This: 4.3A	Creating 3D Surface Domes and Dishes

I. Start a new drawing using the *lesson 04 template* file located in the C:\Steps3D\Lesson04 folder.

II. Be sure isolines are toggled on.

III. Follow these steps.

4.3A: CREATING 3D SURFACE DOMES AND DISHES

1. Enter the ***3D*** command and select the **Dome** option [DOme].

   Command: ***3d***

   Enter an option [Box/Cone/DIsh/DOme/Mesh/Pyramid/Sphere/Torus/Wedge]: ***do***

2. Use a three-dimensional coordinate for the **center point** as indicated.

   Specify center point of dome: ***3,3,5***

3. Specify the **radius**.

   Specify radius of dome or [Diameter]: ***2***

4. Accept the default number of **longitudinal** and **latitudinal segments**.

   Enter number of longitudinal segments for surface of dome <16>: ***[ENTER]***

   Enter number of latitudinal segments for surface of dome <8>: ***[ENTER]***

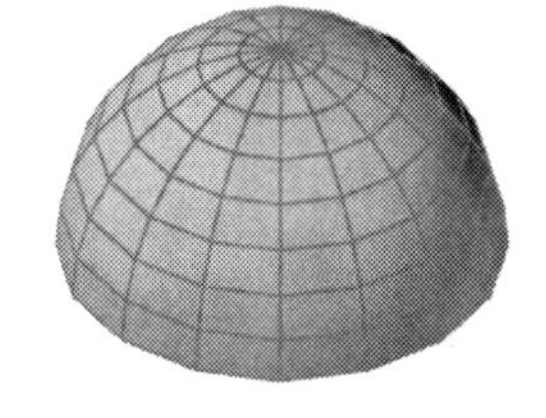

Your drawing looks like the figure at right.

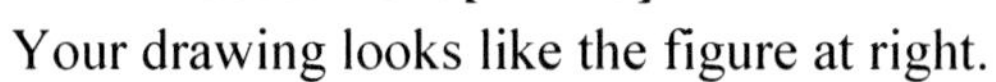

5. We'll draw the dish below the dome. Enter the ***3D*** command and select the **Dish** option [DIsh].

   Command: ***3d***

   Enter an option [Box/Cone/DIsh/DOme/Mesh/Pyramid/Sphere/Torus/Wedge]: ***di***

6. Follow the sequence indicated.

   Specify center point of dish: ***3,3,2***

   Specify radius of dish or [Diameter]: ***2***

   Enter number of longitudinal segments for surface of dish <16>: ***[ENTER]***

   Enter number of latitudinal segments for surface of dish <8>: ***[ENTER]***

Your drawing looks like the figure at right.

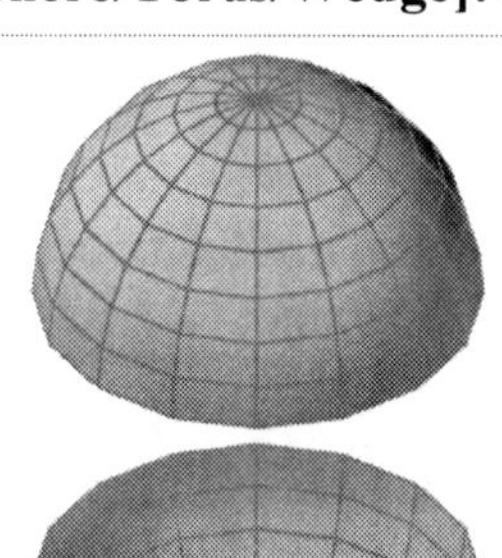

7. Set the **Smoothness** level for both to **4**. Notice the difference (right)?

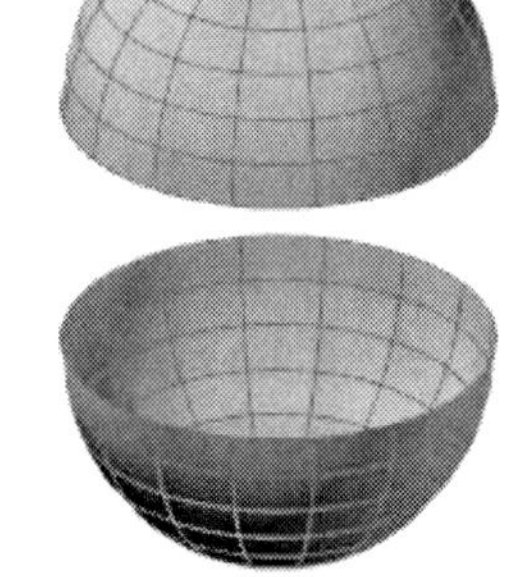

8. Save the drawing as *MyDome* in the C:\Steps3D\Lesson04 folder.

   Command: ***save***

4.4 Understanding the Limitations of Mesh Models

By now you may be thinking how wonderful predefined mesh models are. And you're right to think so. But remember that you've yet to consider their limitations.

- You can explode the models discussed in this lesson, but remember that you can't easily modify 3D faces. You can't trim or extend them, nor can you fillet, chamfer, break, lengthen, or offset them. (You can, however, stretch, mirror, array, rotate, and copy them.)
- Neither 3D faces nor meshes have wall thickness. They are, essentially, two-dimensional objects existing in three-dimensional space.
- While you can use these predefined shapes to build many things, you can't combine them as you can solids. (The ***Union***, ***Subtract***, and ***Intersect*** commands won't work on meshes.)

But take heart, AutoCAD has several other commands (that you'll see in our next lesson) to enable you to draw shapes that aren't predefined. Then in Lessons 6 and 7, we'll look at some editing tools that do work!

4.5 Extra Steps

You've probably noticed that there are similar predefined models on the ribbon's **Home** tab – **Modeling** panel. These create 3D solid objects. We've spent this lesson studying predefined *mesh* objects. Take a few minutes and compare the two. Are they the same? What are the differences? Do they make the same types of models available?

In a new drawing created with the *Lesson 04 Template*, create a box using the mesh approach and then create another box using the solid approach. Notice the similarities and differences in the command sequences. Repeat this procedure for each of the predefined models.

It's early to be studying Solid Modeling, yet the similarities between solids and meshes are too tempting to ignore. As you continue your study of Mesh Modeling, bear in mind that each procedure probably has a solid modeling equivalent. Then, when we get to Solid Modeling, you'll be a step a head of the game!

4.6 What Have We Learned?

Items covered in this lesson include:

- *Commands*
  - ***Mesh***
  - ***MeshPrimitiveOptions***
  - ***3d***
- *AutoCAD's predefined mesh models*
  - *Box*
  - *Wedge*
  - *Pyramid*
  - *Cone*
  - *Sphere*
  - *Cylinder*
  - *Dome*
  - *Dish*
  - *Torus*

This has been a fun lesson. It's nice to know that they're not all difficult!

We've seen several predefined objects designed by AutoCAD to make three-dimensional drafting move more quickly. But you shouldn't limit these objects to their obvious uses. Whenever you have a complex object to build with mesh models, think about these predefined tools as basic – and often elastic – building blocks.

Practice the exercises at the end of this lesson for experience. Remember to use the UCS and visual styles to assist you. Then try to draw different objects around your desk – how about your mouse or the keyboard?

Our next lesson will cover more complex, user-defined shapes created as meshes. So what you can't draw yet, you'll soon be able to!

4.7 Exercises

1. through 8. Recreate the "su" drawing found in Appendix B using the tools found in this lesson to help you. You'll need to explode many of the primitives to get 3D faces to manipulate, but the drawing should be faster. Use cylinders rather than arcs to line the holes.
9. Create the flying saucer drawing. The following hints will help:
   - 9.1. Use the *lesson 04 template* to begin.
   - 9.2. Change the layer colors as needed.
   - 9.3. Use these tools: cone, dish, dome, and torus.
   - 9.4. Use clipping planes to create the section.
   - 9.5. Use the direct hatch approach to hatch the section.
   - 9.6. While you're still in Model Space, use the continuous orbit tool (refer to the www.uneedcad.com/Files/OtherViewingTools.pdf supplement) to make the saucer fly back and forth across the screen. (Eat your heart out, Marvin the Martian!)
   - 9.7. Save the drawing as *MySaucer* in the C:\Steps3D\Lesson04 folder.

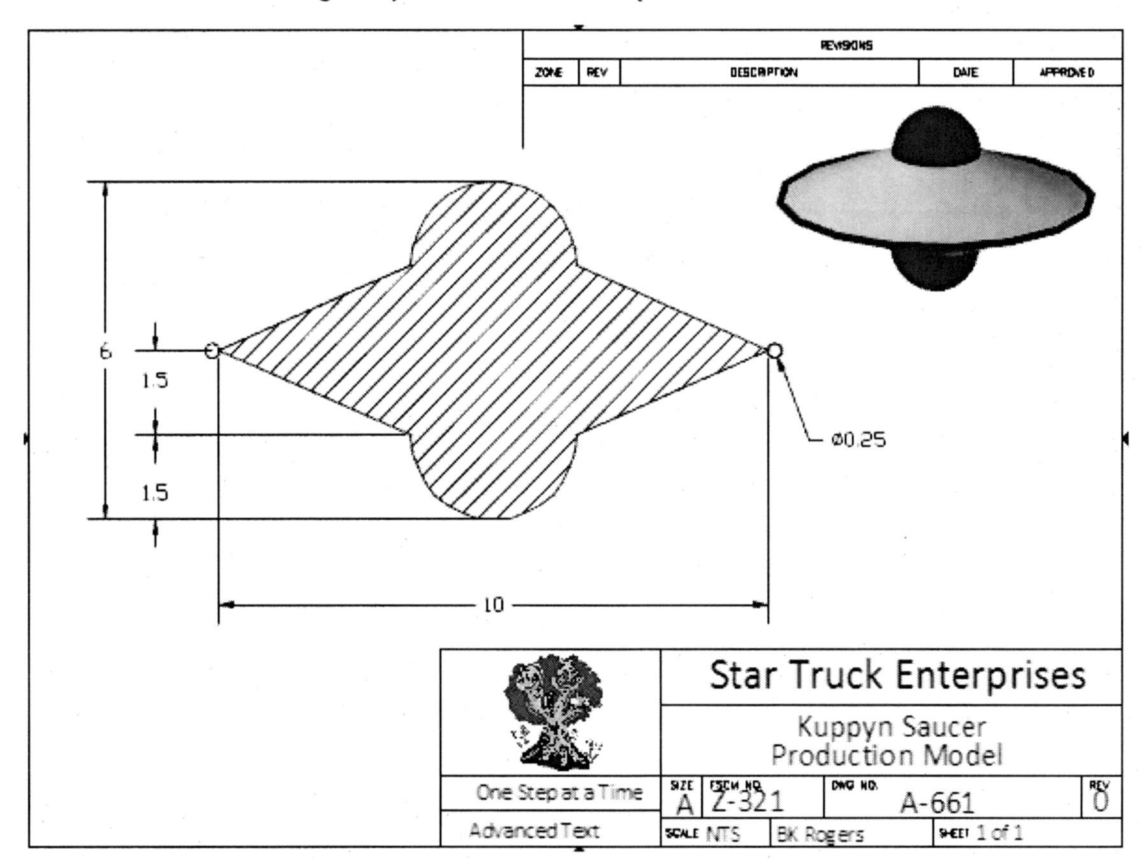

Flying Saucer

10. Create the aquarium aerator drawing. The following hints will help:
    10.1. Use the *lesson 04 template* to begin.
    10.2. Change the layer colors as needed.
    10.3. Use these tools: box, cylinder, and dish.
    10.4. Viewport scales are a uniform 1:1.
    10.5. Text uses the Times New Roman or Calibri font.
    10.6. Save the drawing as *MyAerator* in the C:\Steps3D\Lesson04 folder.

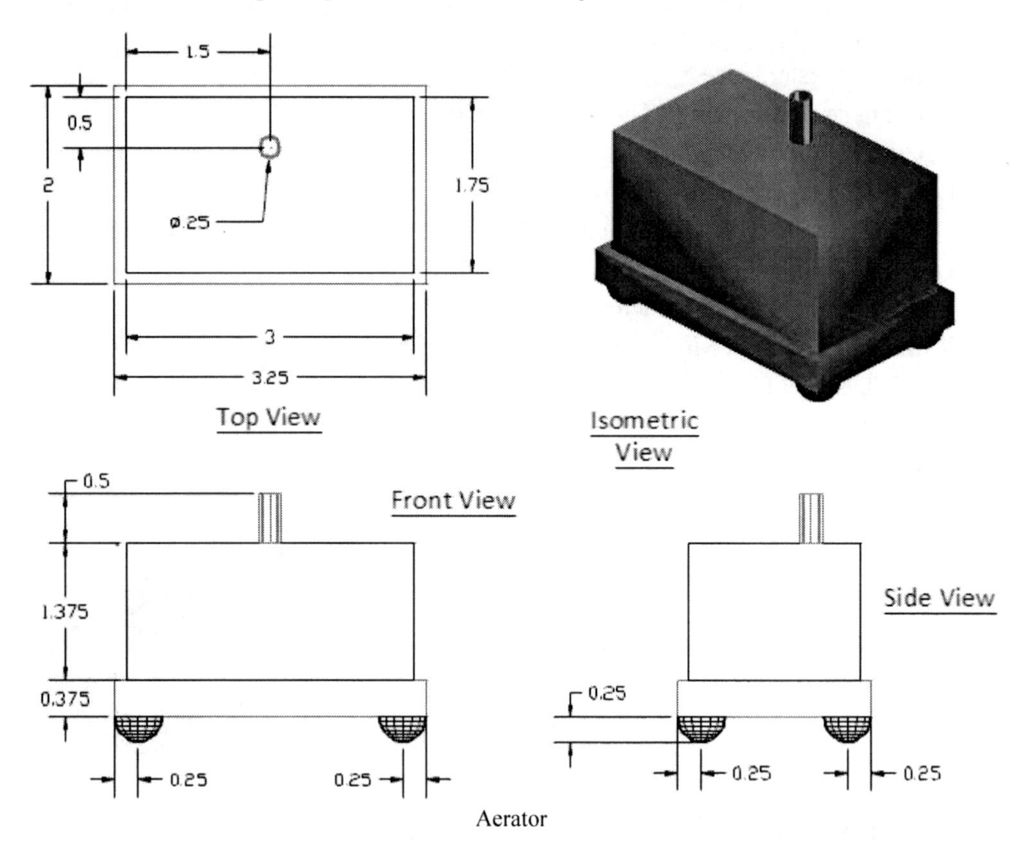

Aerator

11. Create the coffee table drawing. The following hints will help:
    11.1. Use these tools: cone, torus, and region.
    11.2. The torus is 36" diameter and the tube is 1" diameter.
    11.3. The legs are 21" long and extend 2" below the bottom shelf.
    11.4. Save the drawing as *MyCoffeeTable* in the C:\Steps3D\Lesson04 folder.

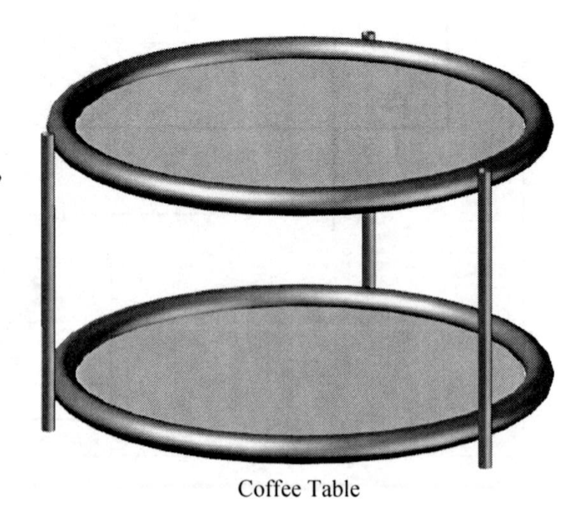
Coffee Table

12. Create the trailer light drawing shown in the following figure. The following hints will help:
    12.1. Use the *lesson 04 template* to begin.
    12.2. Change the layer colors as needed.
    12.3. Use these tools: box, cylinder, pyramid, and cone.
    12.4. Viewport scales are a uniform 1:2.
    12.5. Text uses the Times New Roman font.
    12.6. Save the drawing as *MyLight* in the C:\Steps3D\Lesson04 folder.

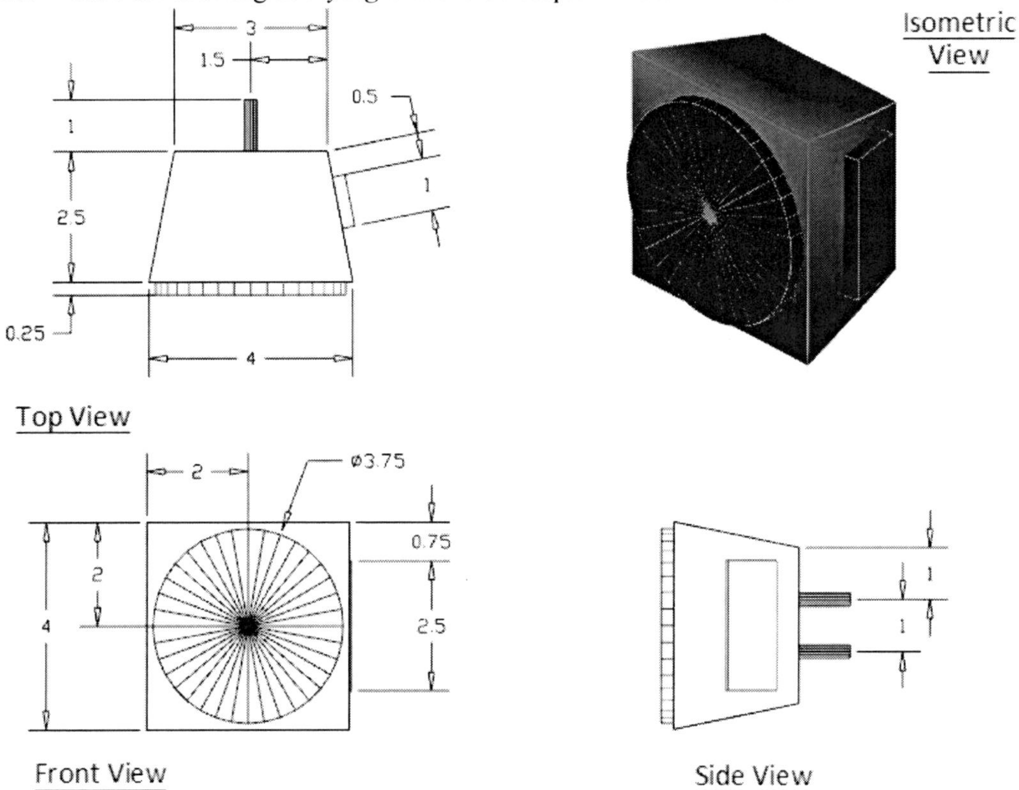

Trailer Light

13. Create the microphone drawing at right. The following hints will help:
    13.1. Use the *lesson 04 template* to begin.
    13.2. Change the layer colors as needed.
    13.3. Use these tools: box, cone, sphere, and wedge.
    13.4. The UCS is tricky on this one. Align it with the cone to draw the button. Use the UCS to help you rotate the microphone so that it sits atop the wedge.
    13.5. The wedge is ¼" wide.
    13.6. Save the drawing as *MyMicrophone* in the C:\Steps3D\Lesson04 folder.

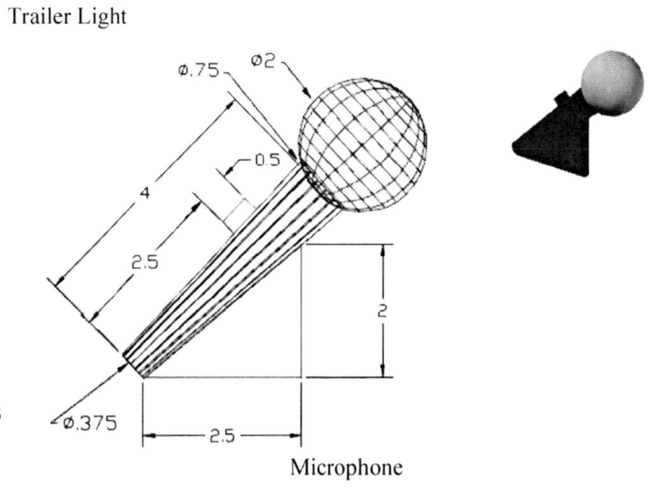

Microphone

14. Create the wagon drawing. The following hints will help:
    14.1. Use these tools: cone, torus, box, and dome.
    14.2. The wheel tori are 4" diameter with 1" diameter tubes.
    14.3. The axles are 1" diameter x 8" long cones.
    14.4. The axles are 12" apart.
    14.5. The wagon floor is 18" x 6" x ½".
    14.6. The outer boards are 2½" wide and spaced ½" apart.
    14.7. The inner boards are 1½" wide.
    14.8. Save the drawing as *MyWagon* in the C:\Steps3D\Lesson04 folder.

Wagon

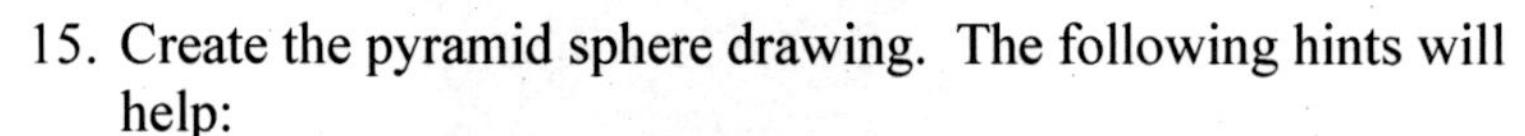

15. Create the pyramid sphere drawing. The following hints will help:
    15.1. Use these tools: sphere, torus, and pyramid.
    15.2. The sphere is 16" diameter. (You'll need to explode it and erase some of the 3D faces.)
    15.3. The pyramid is 5" squared on the base x 5" high.
    15.4. The torus is 10" diameter with a 1" diameter tube.
    15.5. The pyramid is hatched to create the brick pattern.
    15.6. Save the drawing as *MyPyrSph* in the C:\Steps3D\Lesson04 folder.

Pyramid Sphere

16. Create the temple drawing. The following hints will help:
    16.1. Use these tools: wedge, box, and pyramid.
    16.2. The sphere is the drawing we created in Exercise 15 (above). I inserted it as a block at 1/8 scale. (Be sure the UCS in both drawings is equal to the WCS. We'll discuss more on three-dimensional blocks in Lesson 10.)
    16.3. The ramps are 8" x 2" x 2".
    16.4. The center box is 4" x 4" x 2".
    16.5. The pyramid is 4" x 4" x 2" (with a 2" x 2" top).
    16.6. The torus is 10" diameter with a 1" diameter tube.
    16.7. The pyramid is hatched to create the brick pattern.
    16.8. Save the drawing as *MyTemple* in the C:\Steps3D\Lesson04 folder.

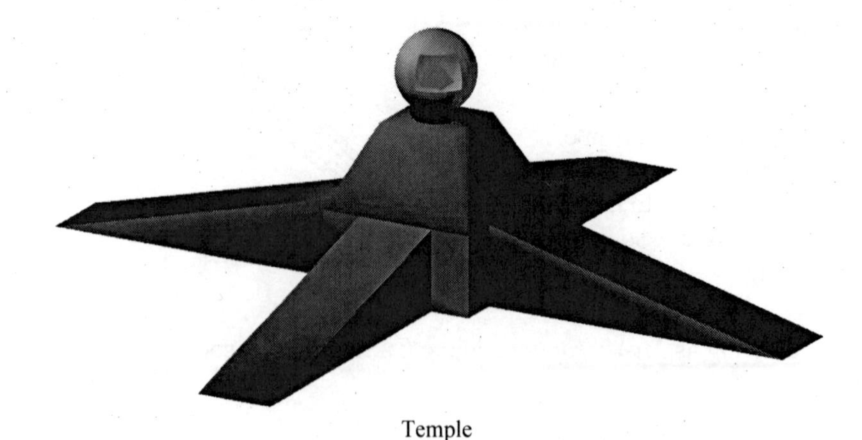

Temple

17. Create the lamp drawing. The following hints will help:
    17.1. Use these tools: cone.
    17.2. The center pole is 6' tall x 1" diameter.
    17.3. The base is 12" diameter on the bottom and 1" diameter on the top. It is 2" high.
    17.4. The top is 1" diameter on the bottom and 12" diameter on the top. It is 6" high.
    17.5. The switch is ½" diameter x ½" high. It's located halfway up the center pole.
    17.6. Save the drawing as *MyLamp* in the C:\Steps3D\Lesson04 folder.

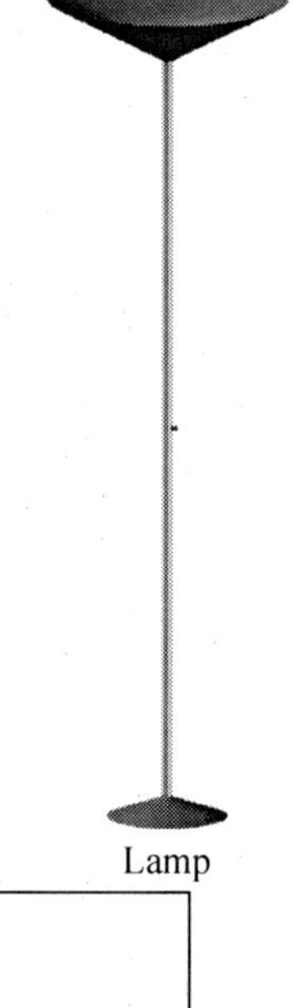
Lamp

4.8 **For Web-Based Review Questions, visit: http://foragerpub.com/AcadFiles/2010/2010.htm**

Lesson 5

Following this lesson, you will:

✓ *Know how to build more complex mesh models*

- ***Rulesurf***
- ***Revsurf***
- ***Tabsurf***
- ***Edgesurf***
- ***3DMesh***
- ***Planesurf***
- ***ConvToSurface***
- ***Loft***

✓ *Know how to control the number of tessellation divisions used to draw a mesh model with:*

- ***Surftab1***
- ***Surftab2***

Complex Mesh Models

In Lesson 4, you discovered some simple tools that you can use to create mesh models. But what if you need a model that doesn't easily translate into one of the predefined mesh models? For example, suppose you need to draw an I-Beam, a piping elbow, or an ornate lamp. Which predefined model would you use?

The answer of course, is that none of the predefined tools would help. Well then, would you have to draw a wireframe model (a stick figure) and stretch faces over it? Or would you just give up and wait for Autodesk to develop some more predefined shapes? (Sorry, I don't think any more are coming.)

Oh, take heart, AutoCAD provides other tools that'll help you deal easily with more intricate designs.

In this lesson, we'll examine tools for creating complex mesh models. Here, we'll conclude our study of mesh model creation techniques with a look at the procedures needed to put the razzle-dazzle in your three-dimensional drawing.

Let's get started.

5.1 Controlling the Number of Surfaces – Surftab1 and Surftab2

When you drew the dome, and dish in our last lesson, AutoCAD asked you for the number of segments you wanted to use in defining the object. Remember that we defined longitudinal and latitudinal segments differently so that you could see the distinction.

That approach worked well for a predefined shape. But when you draw complex shapes, AutoCAD can't know what you're doing. So, it can't ask you for the number of segments (aka. tabulations or tessellation divisions) you'll need to define the object. Still, that information will be required for the complex object to take the shape you want.

For this reason, AutoCAD established two system variables to define the number of faces it'll use to create an object. The first – **Surftab1** – defines the number of tessellation divisions AutoCAD will use to create a linear object, or the number of surfaces it will use to create the circumference (axial direction) of a round (or arced) object. The second – **Surftab2** – defines the number of tessellation divisions AutoCAD will use to create latitudinal sections (along the path of rotation) of an object.

Consider the drawing in Figure 5.001. I created this tube (using the ***Tabsurf*** command) in two steps – the top with a **Surftab1** setting of 20, the bottom with a **Surftab1** setting of 6.

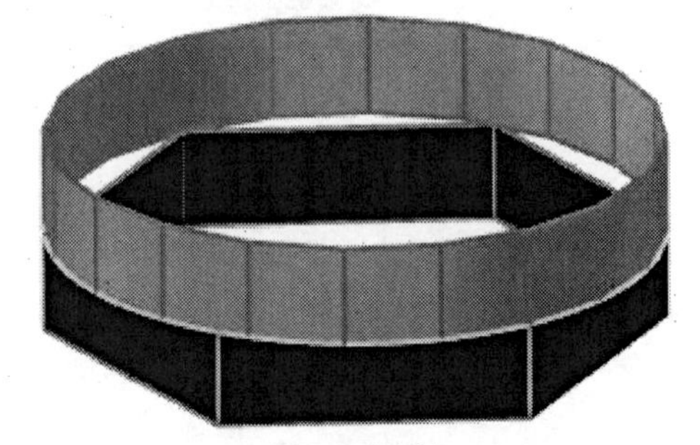

Figure 5.001

Notice the difference. I created both top and bottom using the same circle object. The only difference is the **Surftab1** setting. On the top, AutoCAD used twenty faces to define the object; on the bottom, AutoCAD used 6. Here you see another important aspect of the **Surftab** system variables – *their settings affect the shape of the object being drawn.*

> Do you remember how you would make the tube round? That's right; you'd change its **Smoothness** property to **Level 1** (or 2, 3, or 4). With the introduction of the **Smoothness** property, the importance of the **Surftab** system variables diminishes slightly. Still, you should be aware of them as the number of faces used to create a mesh will affect its ultimate size and shape.

We'll set **Surftab1** and **Surftab2** as needed throughout the exercises in this lesson.

5.2 Different Approaches for Different Goals

AutoCAD provides basic and advanced commands to handle the drawing of complex shapes. You'll probably enjoy the basic commands – they're lots of fun! The advanced commands, however, are more challenging and may take some time to master.

The basic commands all have one thing in common – they all use something two dimensional as a guide to create a three-dimensional object. But remember, a two-dimensional object can exist in Z-space. That is, an object can have any two of the three properties required for a three-dimensional object (length, width, and height). The command you use to create the three-dimensional object will provide the third property. This'll become clear once you've used the commands.

We'll discuss the basic commands in this section and save the advanced commands for the next one. Let's get started.

5.2.1 Follow the Path – The *Tabsurf* Command

Where to Find It:	
Command Line:	***Tabsurf***
Ribbon (Tab/Panel):	Mesh Modeling – Primitives – **Tabsurf**
Menu:	Draw – Modeling – Meshes – **Tabulated Mesh**

Tabsurf is an easy command. You'll need a basic shape – circles, arcs, polylines, and splines make great shapes – and something to indicate a path. AutoCAD will expand the shape into three dimensions along the path you identify.

The command sequence looks like this:

Command: ***tabsurf***

Current wire frame density: SURFTAB1=6 ***[AutoCAD reports the Surftab1 settings.]***

Select object for path curve: ***[Select the object that'll give shape to your three-dimensional object.]***

Select object for direction vector: ***[Select an object that'll tell AutoCAD the direction in which to expand the shape.]***

This will become clearer with some practice. (This is a fun project. We'll use the basic commands for drawing complex shapes to create a three-dimensional toy train. So put on your Engineer's hat and let's get started!)

Do This: 5.2.1A — Using *Tabsurf* to Create a Three-Dimensional Object

I. Open the *train* file in the C:\Steps3D\Lesson05 folder. The drawing looks like the figure at right.

II. Set **Tank** as the current layer.

III. Follow these steps.

5.2.1A: Using *Tabsurf*

1. Set the **Surftab1** system variable to **20** for a more defined shape.

   Command: ***surftab1***

   Enter new value for SURFTAB1 <6>: ***20***

2. Enter the ***Tabsurf*** command.

   Command: ***tabsurf***

5.2.1A: USING *TABSURF*

3. AutoCAD wants you to select the object for path curve – the *shape*. Select the circle indicated.

Current wire frame density: SURFTAB1=20

Select object for path curve: ***[Select the circle.]***

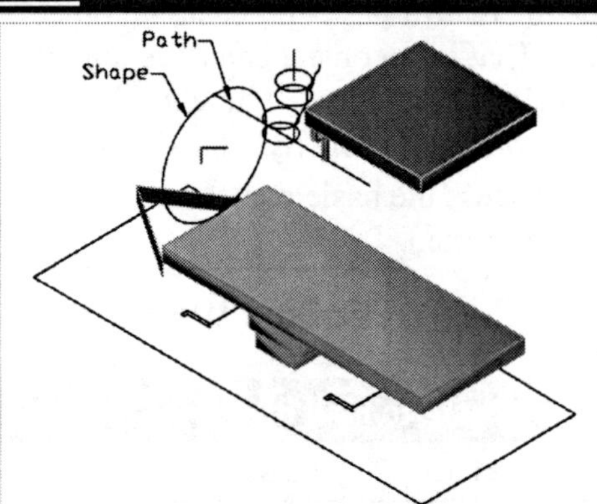

4. AutoCAD wants to know in which direction to expand the shape – the *path*. Select the line indicated. (Select next to where the arrow points in the previous figure.)

Select object for direction vector: ***[Select the line.]***

Your drawing looks like the figure at right.

[Note: First AutoCAD determined the path by the selected object. Then it determined the direction for the three dimensional object by where you select on the object for direction vector. Try repeating this step but select the other end of the line. Then undo until your drawing again looks like this.]

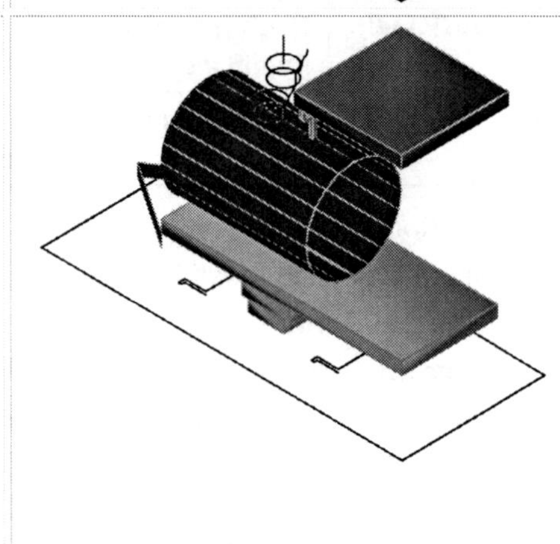

5. Zoom in around the smokestack (the stacked circles atop the tank) and set the **stack** layer current.

Command: *z*

6. Use the ***Tabsurf*** command to expand the circles along the lines. Notice that the path is independent of the UCS.

Command: ***tabsurf***

7. Set the **Smoothness** property to **Level 1** for each of the tabsurfed meshes. (Use the Properties palette.) Your drawing looks like the figure at right.

8. Save the drawing but don't exit.

Command: ***qsave***

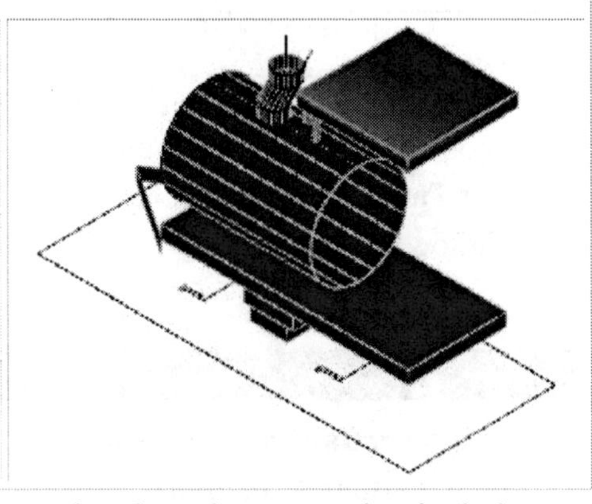

Some things you may have noticed about the ***Tabsurf*** command and need to remember include:

- The layer of the new object is based on the current layer at the time it is drawn and not on the layer of the original shape or path.
- The original shape and path remain intact after the three-dimensional object is drawn – put them on a separate layer that can be frozen later.
- The part of the path AutoCAD uses to define the new object is the endpoint – *a curved path won't produce a curved object.*
- Where you pick on the path can affect the direction of the expansion.

- The UCS doesn't affect the creation.

We used circles in our exercise, but **Tabsurf** will work on any predefined shape provided the shape is a single object. Polylines and splines are particularly useful in creating shapes for **Tabsurf** expansion.

Let's take a look at another complex surface modeling command.

5.2.2 Add a Surface between Objects – The *Rulesurf* Command

Use the ***Rulesurf*** command when you have two existing objects and want to place a surface between them. It's particularly useful when creating surfaces between uneven objects or objects of different size.

Where to Find It:	
Command Line:	***Rulesurf***
Ribbon (Tab/Panel):	Mesh Modeling – Primitives – **Rulesurf**
Menu:	Draw – Modeling – Meshes – **Ruled Mesh**

The command sequence is:

Command: *rulesurf*

Current wire frame density:

SURFTAB1=20 *[AutoCAD reminds you of the current Surftab1 setting; Surftab2 doesn't affect Rulesurf.]*

Select first defining curve: *[Select the first edge of the surface to be created.]*

Select second defining curve: *[Select the opposite edge of the surface to be created.]*

There are some similarities between ***Rulesurf*** and ***Tabsurf***, including:

- The layer of the new object is based on the current layer at the time it's drawn and not on the layer of the original shape or path.
- The original edges remain intact after the three-dimensional object is drawn.
- Where you pick on the path can affect the expansion. You should pick in the same vicinity on both edges (toward the same endpoints).
- The UCS doesn't affect the expansion.

Let's try the ***Rulesurf*** command on our train's cow catcher.

Do This: 5.2.2A	Using *Rulesurf* to create a Three-Dimensional Object

I. Be sure you're still in the *train* file in the C:\Steps3D\Lesson05 folder. If not, please open it now.

II. Restore the **–111** view.

III. Set the **tank** layer current.

IV. Follow these steps.

5.2.2A: Using Rulesurf
1. Turn the circle in the front of the tank into a region. (This'll improve viewing later.) **Command: *reg***
2. Set the **cow catcher** layer current.
3. Enter the ***Rulesurf*** command. **Command: *rulesurf***

5.2.2A: Using Rulesurf

4. AutoCAD needs to know where to draw the surface. Select the edges indicated.

If the surface appears crossed, erase it and try again. When you select the edges, pick in the same general location of each.

(If you have trouble selecting the edges, hold down the CTRL key and select until the edge is found.)

Current wire frame density: SURFTAB1=20

Select first defining curve: ***[Select the first edge.]***

Select second defining curve: ***[Select the second edge.]***

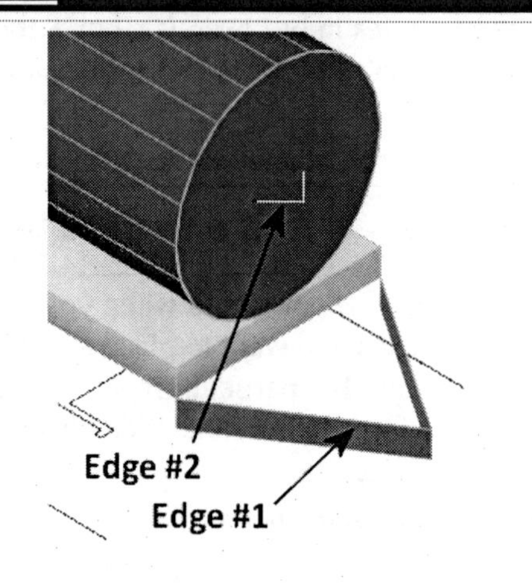

5. Repeat Steps 3 and 4 for the other side of the cow catcher.

Command: ***rulesurf***

Your drawing looks like the figure at right.

6. Save the drawing but don't exit.

Command: ***qsave***

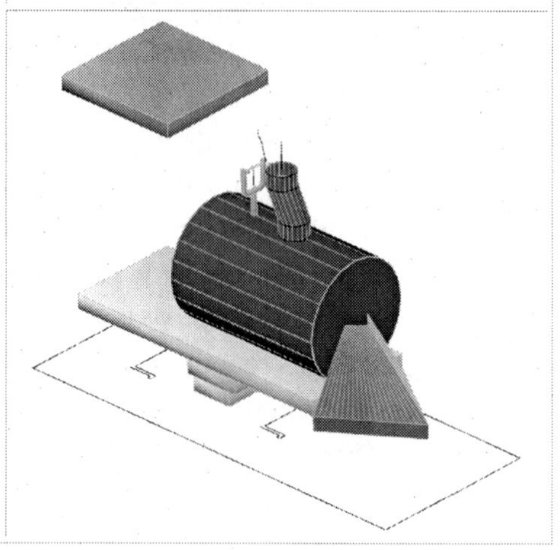

Like ***Tabsurf***, any type of object will do for a ***Rulesurf*** edge – the fancier the original object, the fancier the results!

Speaking of fancy, let's look at the ***Revsurf*** command!

5.2.3 Creating Circular Surfaces – The *Revsurf* Command

Revsurf is one of the more popular of AutoCAD's mesh modeling commands. That could be because ***Revsurf*** is so simple to use; but it's more likely because of the nifty gizmos you can draw with it!

Where to Find It:	
Command Line:	***Revsurf***
Ribbon (Tab/Panel):	Mesh Modeling – Primitives – **Revsurf**
Menu:	Draw – Modeling – Meshes – **Revolved Mesh**

Like the ***Tabsurf*** command, ***Revsurf*** begins with an object that'll define its basic shape and another object that'll define its path (or, in the case of ***Revsurf***, its axis). Devote some time and care when creating the basic shape since the final object will reflect a well-defined shape. Although other objects may occasionally be required (like a circle as a basic shape to define a piping elbow), I'd use polylines or splines almost exclusively. They tend to produce some truly professional results!

The ***Revsurf*** command sequence looks like this:

Command: ***revsurf***

Current wire frame density: SURFTAB1=20 SURFTAB2=6 ***[AutoCAD reports the current settings for both Surftab1 and Surftab2. Both will be needed for this procedure.]***

Select object to revolve: ***[Select the object that defines the basic shape of the object you wish to create.]***

Select object that defines the axis of revolution: ***[Select an object that defines the axis around which you'll revolve the shape.]***

Specify start angle <0>: ***[specify a starting angle.]***

Specify included angle (+=ccw, -=cw) <360>: ***[Tell AutoCAD if you want a fully or partially revolved shape. The default – 360°– defines a full revolution. For less than a full revolution, enter the degrees that define the arc you wish to fill.]***

Like ***Tabsurf*** and ***Rulesurf***, the UCS doesn't affect ***Revsurf***. However, you might need to adjust it when creating the basic shape you intend to revolve.

Let's take a look at the ***Revsurf*** command.

Do This: 5.2.3A	Using *Revsurf* to Create a Three-Dimensional Object

I. Be sure you're still in the *train* file in the C:\Steps3D\Lesson05 folder. If not, please open it now.

II. Restore the **bell** view and set the **Stack** layer current.

III. Follow these steps.

5.2.3A: Using *Revsurf*

1. Enter the ***Revsurf*** command.

   Command: ***revsurf***

2. AutoCAD reports the **Surftab1** and **Surftab2** settings.

   Current wire frame density: SURFTAB1=20 SURFTAB2=6

   These suit our purpose, so we'll continue. Select the spline rising from the top of the smokestack as the object to revolve.

   Select object to revolve:

3. Select the line rising from the center of the smokestack as your **axis of revolution**.

   Select object that defines the axis of revolution:

4. Accept the **start angle** and **included angle** defaults.

   Specify start angle <0>: ***[ENTER]***

   Specify included angle (+=ccw, -=cw) <360>: ***[ENTER]***

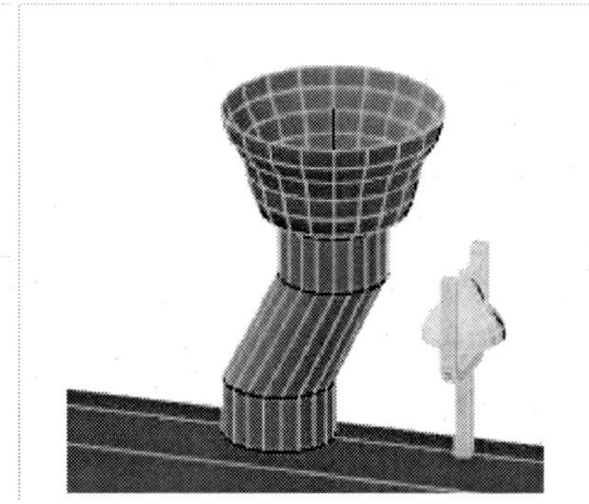

5. Repeat Steps 1 to 4 for the bell assembly (be sure to use the **Bell** layer).

   Command: ***revsurf***

   The top of your train looks like the figure at right.

6. Thaw the **MARKER** layer. Notice the nodes, but also notice the lines that appear as axes for the wheel shapes.

   Set the **wheels** layer current.

7. Repeat Steps 1 to 4 for both wheels using the lines on the **MARKER** layer as the axes. (Refreeze the **MARKER** layer when you've finished.)

   Command: ***revsurf***

5.2.3A: USING *REVSURF*	
8. Set the **Smoothness** property to **Level 1** for bell, stack, and wheels. Your drawing looks like the figure at right.	
8. Save the drawing but don't exit. **Command:** ***qsave***	

Wasn't that fun?

But remember that I'm providing the basic shapes for these exercises. It's a bit more involved (although not difficult) when you draw them yourself. But just imagine the sense of accomplishment that'll give you!

Our train is starting to take shape (as is our expertise with some cool new commands). Have you tried viewing the model with the isolines toggled off? It looks very nice (especially if you freeze layer **0**).

But we still have to draw the cab and the ground beneath the wheels. Let's not waste a moment – full *steam* ahead!

5.2.4 Using Edges to Define a Surface Plane – The *Edgesurf* Command

Edgesurf is actually one of the simplest of the complex surface commands. But I've placed it at the end of the basic commands because it makes a good transition to the more advanced commands.

Where to Find It:	
Command Line:	***Edgesurf***
Ribbon (Tab/Panel):	Mesh Modeling – Primitives – **Edgesurf**
Menu:	Draw – Modeling – Meshes – **Edge Mesh**

Edgesurf creates a surface plane. That doesn't mean that it creates a surface along the X-, Y-, or Z-planes but rather a plane like an open field. It doesn't have to be flat, although it's not the tool for very complex surfaces.

The ***Edgesurf*** command uses four edges to define a plane. The edges can be parallel to one another or skewed in any direction.

The command sequence simply asks for the four edges:

Command: ***edgesurf***

Current wire frame density: SURFTAB1=20 SURFTAB2=6 ***[AutoCAD reminds you of the Surftab1 and Surftab2 settings. Like Revsurf, both will be needed here.]***

[Use the next four options to define the edges of the surface.]

Select object 1 for surface edge:

Select object 2 for surface edge:

Select object 3 for surface edge:

Select object 4 for surface edge:

We'll use ***Edgesurf*** to create the ground beneath our train's wheels.

Do This: 5.2.4A	Using *Edgesurf* to Create a Three-Dimensional Object

I. Be sure you're still in the *train* file in the C:\Steps3D\Lesson05 folder. If not, please open it now.

II. Remain in the **bell** view but zoom out so you can see the ground.

III. Set the **ground** layer current, and freeze the **MARKER** layer.

IV. Follow these steps.

5.2.4A: USING *EDGESURF*

1. Enter the ***Edgesurf*** command.

   Command: *edgesurf*

2. AutoCAD reports the current **Surftab1** and **Surftab2** settings and then asks you to select the edges of your object. Select the four lines that form the boundary of the ground (pick one of the shorter lines first).

   Current wire frame density: SURFTAB1=20 SURFTAB2=6

   Select object 1 for surface edge:

   Select object 2 for surface edge:

   Select object 3 for surface edge:

   Select object 4 for surface edge:

   Your drawing looks like the figure at right.

3. You've seen how ***Edgesurf*** works. But let me show you what it can do with a little imagination. Thaw the **Flag** layer and set it current. Notice the outline of the flag atop the roof.

4. Now do an ***Edgesurf*** using the four sides (two lines and two splines) of the flag as your edges.

   Command: *edgesurf*

5. Give the flag a **Smoothness** property of **Level 3**. Your flag looks like the figure at right.

6. Save the drawing but don't exit.

   Command: *qsave*

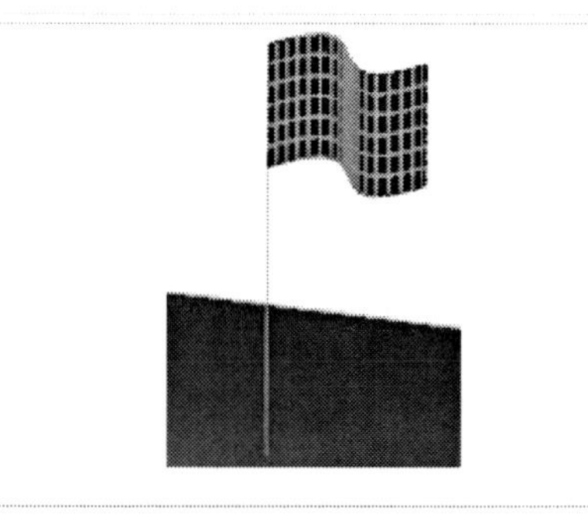

How's that for a bit of razzle-dazzle?!

This concludes the basic Surface Modeling commands. But there're a couple parts of our train we have yet to draw – the hill in front of the train, and the cab. For drawing these surfaces, let's look at some more complex tools.

5.3 More Complex Meshes & Surfaces

There are actually several advanced mesh and surface commands – ***3DMesh***, ***Planesurf***, ***Loft***, and ***3DFace***. I don't recommend ***3DMesh*** or ***3DFace*** for the faint of heart! (In fact, we won't even cover ***3DFace*** in this text!) The others, however, can be a lot of fun.

AutoCAD invented these commands for you to create objects similar to the predefined surface models so you can manipulate many surfaces as a single object (as in a single sphere rather than dozens of faces). But be forewarned: creating an object with the ***3DMesh*** command means manually specifying *every vertex on the object*! Besides that, you must identify the vertices *in a specific order*! Does that sound like a lot of work? It is! But in AutoCAD's defense, I must add that these commands are actually better suited for Lisp routines or other third-party programs. You might say that you're on the border where CAD operation ends and CAD programming begins.

> A third-party program is one that's designed to use AutoCAD as a base. In other words, AutoCAD becomes something like a CAD operating system (as Windows is your computer's operating system). The third-party program builds on AutoCAD by providing shortcuts toward a specific end. Popular third-party programs include *Mechanical Desktop, Architectural Desktop, Propipe, ProISO*, and many others.

We'll start with the ***3DMesh*** command.

5.3.1 Creating Meshes with the *3DMesh* Command

The ***3DMesh*** command is similar to the ***Edgesurf*** command. (The similarity is akin to that between a sculptor and a whittler – both will give you a carving. But what the sculptor produces will embarrass the whittler.) Both work well in creating that open field look. But where four edges define the ***Edgesurf*** object, you have no limit to the number of defined vertices with the ***3DMesh*** command. As I've mentioned, however, it requires tedious effort to identify each vertex involved.

The command sequence looks like this:

> **Command: *3dmesh***
>
> **Enter size of mesh in M direction: *[Tell AutoCAD how many lines are required to define the columns of faces you'll need.]***
>
> **Enter size of mesh in N direction: *[Tell AutoCAD how many lines are required to define the rows of faces you'll need.]***
>
> **Specify location for vertex (0, 0): *[This prompt will repeat for each vertex (intersection) of lines defining the rows and columns.]***

Notice that AutoCAD defines the rows and columns in terms of **M** and **N**. This helps avoid any confusion with the X-, Y-, and Z-axes. ***3DMesh*** works independently of the UCS.

There's an interesting point to remember when defining the size of the mesh in terms of rows and columns. Notice that AutoCAD doesn't ask for the number of rows and columns, but rather, it asks for the number of lines required to define the rows and columns. *You'll be working with the vertices that create the surfaces, not the spaces between the vertices.* The easiest way to determine the number of lines required is simply to add one to the number of rows and columns you want.

This'll become clearer in our next exercise. Let's draw a hill in front of our train (we'll call it *the little engine that could*).

Do This: 5.3.1A — Using *3DMesh* to Create a Three-Dimensional Object

I. Be sure you're still in the *train* file in the C:\Steps3D\Lesson05 folder. If not, please open it now.

II. Thaw the **MARKER** layer and set the **ground** layer current.

III. Begin in the **Bell** view, but adjust so you can see the nodes at the front of the train.

IV. Set your running OSNAPs to **Node**; clear all other settings.

V. Follow these steps.

5.3.1A: Using *3DMesh*

1. Enter the ***3DMesh*** command.

   Command: *3dmesh*

2. (Refer to the figure at right.) Tell AutoCAD you want 5 rows and 5 columns. (Remember that you need 6 lines to define 5 rows or columns.)

   Enter size of mesh in M direction: *6*

   Enter size of mesh in N direction: *6*

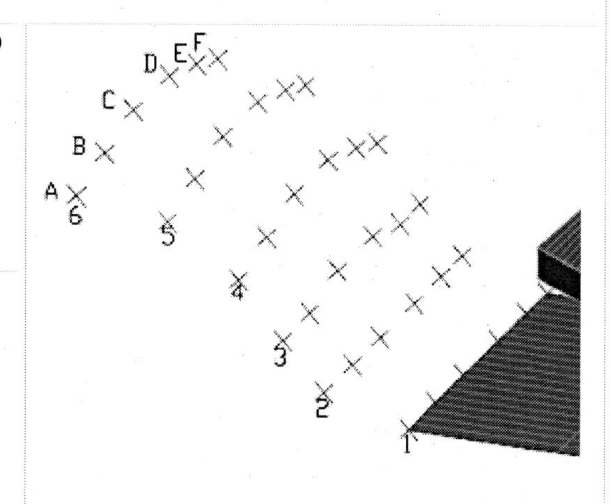

3. Now we'll identify the vertices. Select nodes A-1 through A-6 sequentially. (Hint: Use running OSNAPs.)

   Specify location for vertex (0, 0):

   Specify location for vertex (0, 1):

   Specify location for vertex (0, 2):

   Specify location for vertex (0, 3):

   Specify location for vertex (0, 4):

   Specify location for vertex (0, 5):

4. Follow Step 3 for rows B through F. (Be sure to select the nodes sequentially.)

   Specify location for vertex (1, 0):

5. Freeze the **MARKER** layer. Your drawing looks like the figure at right.

6. Save the drawing but don't exit.

   Command: *qsave*

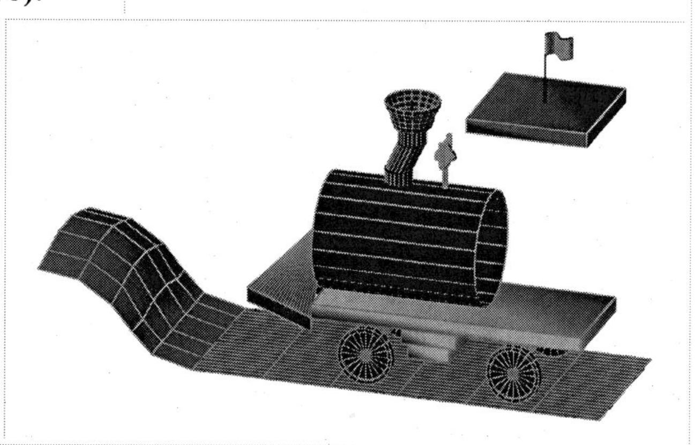

> Note that, once the basic shape of our hill has been defined by using the ***3DMesh*** command, you can add or remove faces using the ***PEdit*** command. See the supplement at http://www.uneedcad.com/Files/PEditing_a_3DMesh.pdf for more details.

Did you find it tedious having to select a node to define each of the 36 vertices? Imagine what it's like defining a large area! Bear in mind that I provided the nodes as a guide for this exercise. Normally, you'll identify the location and place the nodes as well (or enter coordinates at the **Specify location** prompt). Do you see now why I described this command as the edge between CAD operating and CAD programming? You'll often create a 3D mesh using a lisp routine or a Visual Basic program in conjunction with coordinates defined in a database or spread sheet. This is really one of the best ways to create large surfaces (like mountains and valleys).

But wait! There is an easier way!

5.3.2 Creating Meshes with the *Loft* Command

We have another command designed to make creation of the ground surface much easier than the procedure we just went through with the ***3DMesh*** command. The new command is ***Loft***, and it makes drawing uneven surfaces a snap.

The sequence looks like this:

Command: ***loft***

Select cross sections in lofting order: ***[Select the lines that will define the curve.]***

Enter an option [Guides/Path/Cross-sections only] <Cross-sections only>: ***[Hit ENTER here or select an option – if you accept the default, AutoCAD will draw the mesh and present the Loft Settings dialog box (Figure 5.002).]***

Where to Find It:	
Command Line:	***Loft***
Ribbon (Tab/Panel):	Home – Modeling – (Extrude flyout) **Loft**
Menu:	Draw – Modeling – **Loft**
Toolbar:	Modeling – **Loft**

Okay, it looks deceptively simple, but the real work – drawing the "cross-sections" or defining lines will have already been done by the time you begin the ***Loft*** command. These lines can be just about anything – lines, splines, circles, polylines, etc. Further, they can be either open or closed.

> The **DelObj** system variable controls whether or not AutoCAD will delete the original objects once it's drawn the mesh. A setting of **0** tells AutoCAD to delete the objects; **1** tells AutoCAD to leave them.

Let's consider the options.

- **Guides** are guidelines you can use (create) to further define the shape of your wireframe mesh. AutoCAD limits these to lines that intersect each of the defining shapes (cross-sections) and begin and end at the first and last defined shape.
- **Path** defines a single line (spline, etc.) to shape the mesh. While you can use several guides, you're limited to a single path.
- **Cross-sections only** displays the Loft Settings dialog box (Figure 5.002), which allows you some control of the mesh.
  - o **Ruled**, **Smooth Fit**, and **Normal to** control how sharp the mesh will appear.
    - ▪ **Ruled** forces the spaces between the defining lines to be flat and sharp.
    - ▪ **Smooth Fit** allows a smoother appearance.
    - ▪ **Normal to** controls how the mesh behaves at the intersections with the defining lines. **Normal to** options (available in the control box) include:
      - ❖ **All cross sections**
      - ❖ **Start cross sections**
      - ❖ **End cross sections**
      - ❖ **Start and end cross sections**

      Accept the default for a smooth surface.

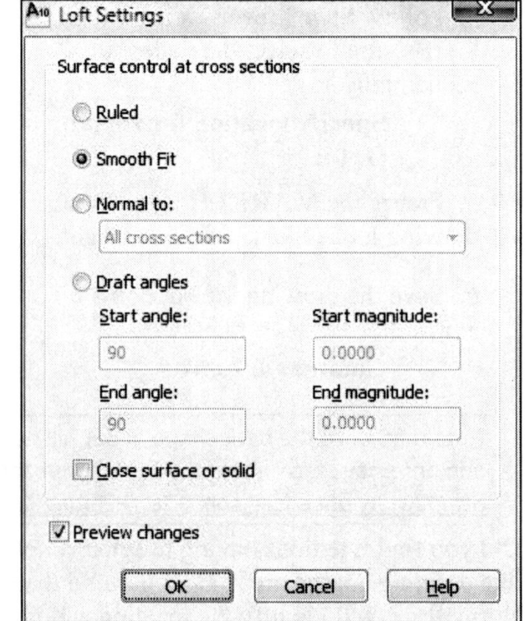

Figure 5.002

  - o **Draft angles** work something like the start and end vectors of a spline. They'll allow you to control the start and end angles of the mesh.
  - o **Close surface or solid** does just that (much like the **Close** option of the ***Line*** or ***PLine*** commands).
  - o **Preview chances**, of course, allows you to see the changes as you define them. It's usually a good idea to leave this box checked.

Let's use the ***Loft*** command to create the same surface in front of the train that we created with the ***3DMesh*** command.

Do This: 5.3.2A	Using *Loft* to Create a Three-Dimensional Mesh

I. Be sure you're still in the *train* file in the C:\Steps3D\Lesson05 folder. If not, please open it now.
II. Erase the mesh in front of the train.
III. Thaw the **Loft Lines** layer, set the bell view current and adjust the view so you can see the loft lines clearly. Freeze the **Marker** layer and set the **ground** layer current.
IV. Set the UCS to World.
V. Follow these steps.

5.3.2A: LOFTING

1. Enter the ***Loft*** command.
   Command: *loft*

2. Select the LEFT-RIGHT splines as your cross-sections. (Hit ENTER to complete the selection process.)
   Select cross sections in lofting order:
   Select cross sections in lofting order: *[ENTER]*

3. AutoCAD wants to know what to do next. Hit ENTER to accept the **Cross-sections only** option, and then pick the **OK** button [OK] on the Loft Settings dialog box (Figure 5.002, p.114) to accept the default.
   Enter an option [Guides/Path/Cross-sections only] <Cross-sections only>: *[ENTER]*
   Your mesh looks like the figure at right.

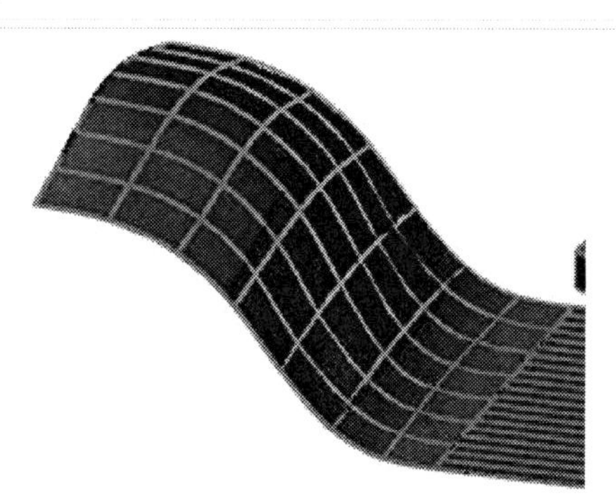

4. Undo the changes. Let's try this again using a path.
   Command: *u*

5. Repeat the ***Loft*** command.
   Command: *loft*

6. Select the first and last LEFT-RIGHT splines.
   Select cross sections in lofting order:
   Select cross sections in lofting order: *[ENTER]*

7. Tell AutoCAD you want to use a **Path** [Path] this time.
   Enter an option [Guides/Path/Cross-sections only] <Cross-sections only>: *P*

8. And select the center (FRONT-BACK) spline as your path.
   Select path curve:
   Your mesh looks like the figure at right.
   Notice how the mesh pulls away from the flat ground. You don't have quite as much control using the **Path** option.

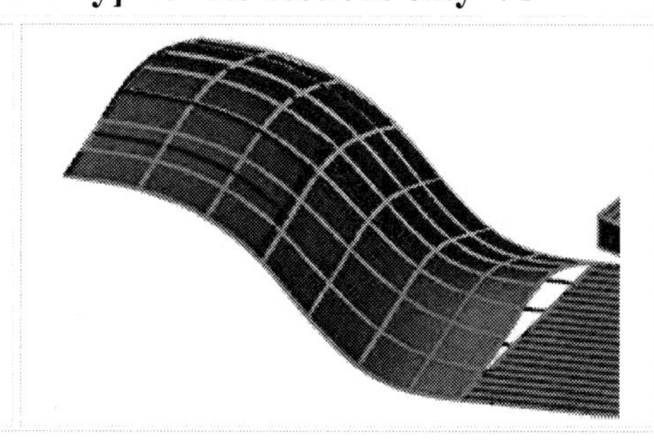

9. Undo the changes. Let's try this again using guides.
Command: *u*

10. Repeat the ***Loft*** command.
Command: *loft*

11. Select only the first and last LEFT-RIGHT splines.
Select cross sections in lofting order:
Select cross sections in lofting order: *[ENTER]*

12. Tell AutoCAD you'd like to use **Guides** [Guides] this time.
Enter an option [Guides/Path/Cross-sections only] <Cross-sections only>: *G*

13. Select the FRONT-BACK lines as your guides.
Select guide curves:
Select guide curves: *[ENTER]*
Your mesh looks like the figure at right. Is there a difference? Which was easier? Faster?

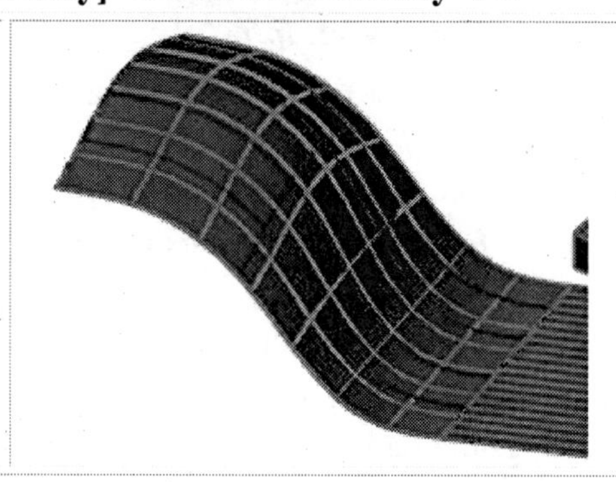

14. Save the drawing but don't exit.
Command: *qsave*

5.3.3 Creating a 3D Surface with *ConvToSurface* and *Planesurf*

In previous editions of this text, I included a section here that introduced the ***PFace*** command. This remarkable tool works great for programmers who have a great deal of coordinates to enter and the expertise to create the required programming. For everyday users, however, it's a nightmare.

We'll use ***Planesurf*** and ***ConvToSurface*** (newer tools) to accomplish the same thing quite easily. But for those of you who demand to know the hard way, I'll post the 2006 section on ***PFace*** on the web at http://www.uneedcad.com/Files/PFace.pdf. Enjoy!

ConvToSurface converts 2D solids, regions, polylines (with thickness but no width), lines, ellipses and arcs with thickness, and planar 3d faces to planar surfaces. (That's a fancy way of saying that it creates a two-dimensional surface, but most surfaces are two-dimensional.) The selected object doesn't have to be in the XY plane of the current UCS.

Where to Find It:	
Command Line:	***ConvToSurface***
Ribbon (Tab/Panel):	Home – Solid Editing (subpanel) – **Convert to Surface**
	Mesh Modeling – Convert Mesh – **Convert to Surface**
Menu:	Modify – 3D Operations – **Convert to Surface**

It's really quite simple; the sequence looks like this:
Command: *convtosurface*
Mesh conversion set to: Smooth and optimized.
Select objects: *[Select the object(s) to convert.]*

Planesurf also converts objects to planar surfaces, or it can create a planar surface from scratch (much as you would create a rectangle).

Where to Find It:	
Command Line:	***PlaneSurf***
Ribbon (Tab/Panel):	Home – Modeling – **Planar Surface**
Menu:	Draw – Modeling – **Planar Surface**
Toolbar:	Modeling – **Planar Surface**

The sequence looks like this:

> **Command:** ***planesurf***
>
> **Specify first corner or [Object] <Object>:** ***[It's just like drawing a rectangle; pick the first corner …]***
>
> **Specify other corner:** ***[…then pick the opposite corner.]***

This may be the easiest surface you'll draw!

Okay, things to remember about ***Planesurf***:

- ***Planesurf*** creates surfaces in the XY-plane of the current UCS.
- The **Object** option provides the best approach for using this command. You can convert an existing object to a planar surface in much the same way you can convert a closed object to a region. Convert either closed objects or select multiple objects that form a closed area. Possible object include: lines, circles, arcs, ellipses, elliptical arcs, polylines, planar 3D polylines, planar splines, and regions. This way, you can create a planer surface with a hole in it or with an oddball shape (by selecting a region – you'll see this in our next exercise).
- When using the **Object** option, the surface to be converted does not have to be in the XY-plane of the current UCS.

> It's important to remember that these two commands – ***Planesurf*** and ***ConvToSurface*** – create surfaces, *not* meshes. Once you've created surfaces, you can convert them to meshes using the ***MeshRefine*** command we'll discuss in Lesson 7, p.163. Converting surfaces to meshes, however, can produce some undesirable (bizarre) results.

Let's take a look.

Do This: 5.3.3A	**Using *Planesurf* and *ConvToSurface* to Create a Planar Surface**

I. Be sure you're still in the *train* file in the C:\Steps3D\Lesson05 folder. If not, please open it now.

II. Freeze the **Loft Lines** and **Marker** layers; thaw the **Cabin** layer and set it current; set the **1-11** view current and adjust the view so you can see all three walls of the cabin clearly.

III. Set the UCS to World.

IV. Follow these steps.

5.3.3A: Using Planesurf and ConvToSurface

1. Enter the ***Planesurf*** command.

   Command: ***planesurf***

2. Tell AutoCAD to use the **Object** option [• Object].

   Specify first corner or [Object] <Object>: ***o***

5.3.3A: Using PlaneSurf and ConvToSurface

3. Select two walls of the cabin.

   Select objects:

   Select objects: ***[ENTER]***

Your train looks something like the figure at right.

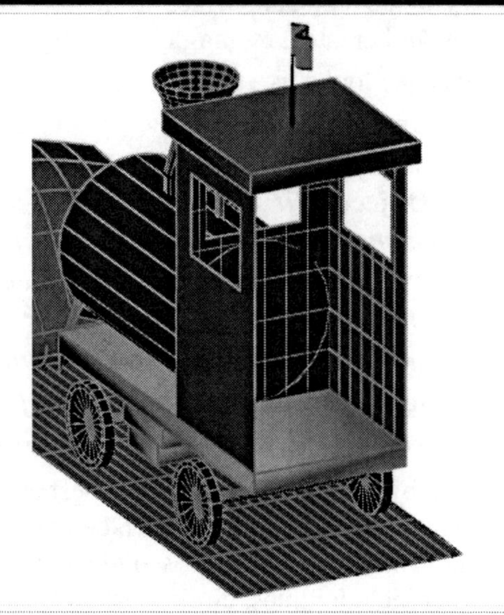

4. Enter the ***ConvToSurface*** command.

   Command: ***convtosurface***

5. Select the remaining cabin wall.

   Mesh conversion set to: Smooth and optimized.

   Select objects:

   Select objects: ***[ENTER]***

6. Restore the -1,1,1 view, and adjust the orbit to improve the view. Remove the isolines and freeze layer **0**. (You may have to thaw the **Cabin** layer again.) How do you like this look (following figure)?

7. Save the drawing and exit.

   Command: ***qsave***

(That exercise took many more very tedious pages with the ***PFace*** command! Don't you just love progress?!)

5.4 Extra Steps

Try to incorporate all you've learned thus far into the train drawing.

- Adjust the views so you can see it from all sides.
- Freeze the **ground** layer and use the Continuous Orbiter to make the train revolve about the screen. (For details on the Continuous Orbiter, see the supplement: http://www.uneedcad.com/Files/OtherViewingTools.pdf.)

- Set up the train drawing for plotting:
  - Show it in three views and an isometric.
  - Dimension it.
  - Put it on a title block with your school/business name.

5.5 What Have We Learned?

Items covered in this lesson include:

- *Controlling the number of surfaces on a surface model*
- *Basic and Advanced Surface Modeling Commands*
  - ***Surftab1***
  - ***Surftab2***
  - ***Rulesurf***
  - ***Tabsurf***
  - ***Revsurf***
  - ***Edgesurf***
  - ***3DMesh***
  - ***Loft***
  - ***Planesurf***
  - ***ConvToSurface***

Congratulations! You've finished AutoCAD's Wireframe and Mesh Modeling creation commands. You've really come a long way in five lessons!

The decisions you'll now face concern which procedure or method you'll need to create the objects that you want to create. The best help you can get for that is *practice*! So work through the problems at the end of this lesson until you're comfortable with your new abilities.

When you've finished with the exercises, go on to Lesson 6, p.125. There we'll discuss the three-dimensional aspects of several editing tools you already know … and some new ones that are specific to three-dimensional drawings. Then in Lesson7, p.149, we'll look at some really cool tools designed for mesh editing. After that, we'll start a whole new ballgame – we'll learn how to create solid models!

So do the problems, pat yourself on the back for having come this far, and then move onward … ever onward!

5.6 Exercises

1. Create the *Toothpaste* tube drawing. The following information will help.
   1.1. The main tube is 7¼" long x 2" wide on the flat end. The round end has an additional ¼" bubble.
   1.2. The round end is 1¼" diameter.
   1.3. The cap is ½" long, 5/8" diameter at the base and ½" diameter at the top.
   1.4. Use ***Edgesurf*** to create the body of the tube.
   1.5. Use ***Revsurf*** to create the cap and the bubble-end of the tube.
   1.6. Create layers as needed.
   1.7. Save the drawing as *MyToothpaste* in the C:\Steps3D\Lesson05 folder.

Toothpaste Tube

2. Create the *Window Guide* drawing. The following information will help.
   2.1. Use a title block of your choice.
   2.2. Use the Times New Roman or Calibri font – 3/16" and 1/8".
   2.3. Adjust the dimstyle as needed.
   2.4. Create layers as needed.
   2.5. The object is a tabulated mesh model created from a polyline shape.
   2.6. Use either a region or a 3D face to close the ends.
   2.7. Use the **Conceptual** visual style for the isometric figure.
   2.8. Save the drawing as *MyWinGuide* in the C:\Steps3D\Lesson05 folder.

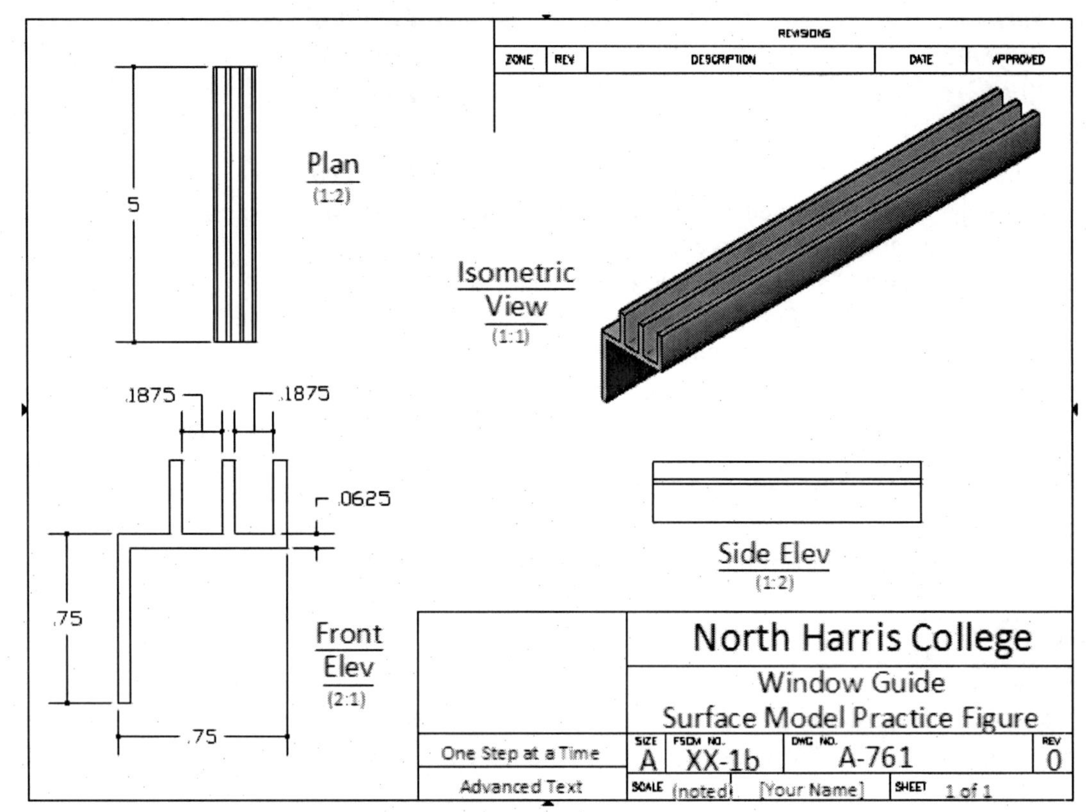

Window Guide

3. Create the *Lighter* drawing. The following information will help.
   3.1. The main bottle is 2¼" tall and is based on an ellipse that is 1" x ½".
   3.2. The top is based on half the bottle's ellipse and is ½" tall.
   3.3. The button is also based on half the bottle's ellipse, is 1/16" thick, and is at an angle of 15°.
   3.4. The striker wheel is 5/16" wide and ¼" diameter.
   3.5. Use ***Revsurf*** to create the striker wheel.
   3.6. Use ***Edgesurf*** to create the bottle and the top.
   3.7. Use regions where needed.
   3.8. Save the drawing as *MyLighter* in the C:\Steps3D\Lesson05 folder.

Lighter

4. Create the *Light Fixture* drawing. The following information will help.
    4.1. Use a title block of your choice.
    4.2. Use the Times New Roman or Calibri font – 3/16" and 1/8".
    4.3. Adjust the dimstyle as needed.
    4.4. Create layers as needed.
    4.5. I used a **Surftab1** setting of **32** and a **Surftab2** setting of **6**.
    4.6. The object is a simple revolved mesh model. I created the shape on one layer and the revolved surface model on another. I did the cross section by adjusting the viewport and freezing unnecessary layers.
    4.7. Save the drawing as *MyFixture* in the C:\Steps3D\Lesson05 folder.

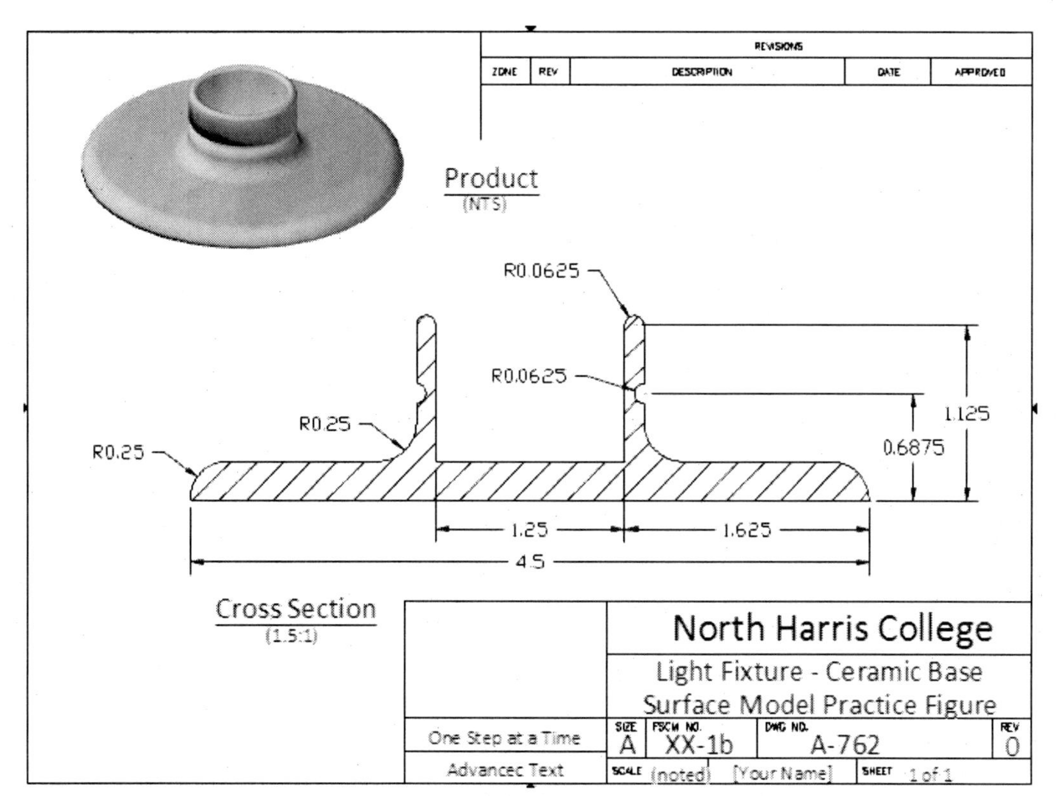

Light Fixture

5. Create the *Alan Wrench* drawing. The following information will help.
   5.1. Use a title block of your choice.
   5.2. Use the Times New Roman or Calibri font – 3/16" and 1/8".
   5.3. Adjust the dimstyle as needed.
   5.4. Create layers as needed.
   5.5. I used the default settings for both surftabs.
   5.6. This object is made up of two tabulated meshes and a revolved mesh. I used six-sided polygons as my basic shapes.
   5.7. Save the drawing as *MyWrench* in the C:\Steps3D\Lesson05 folder.

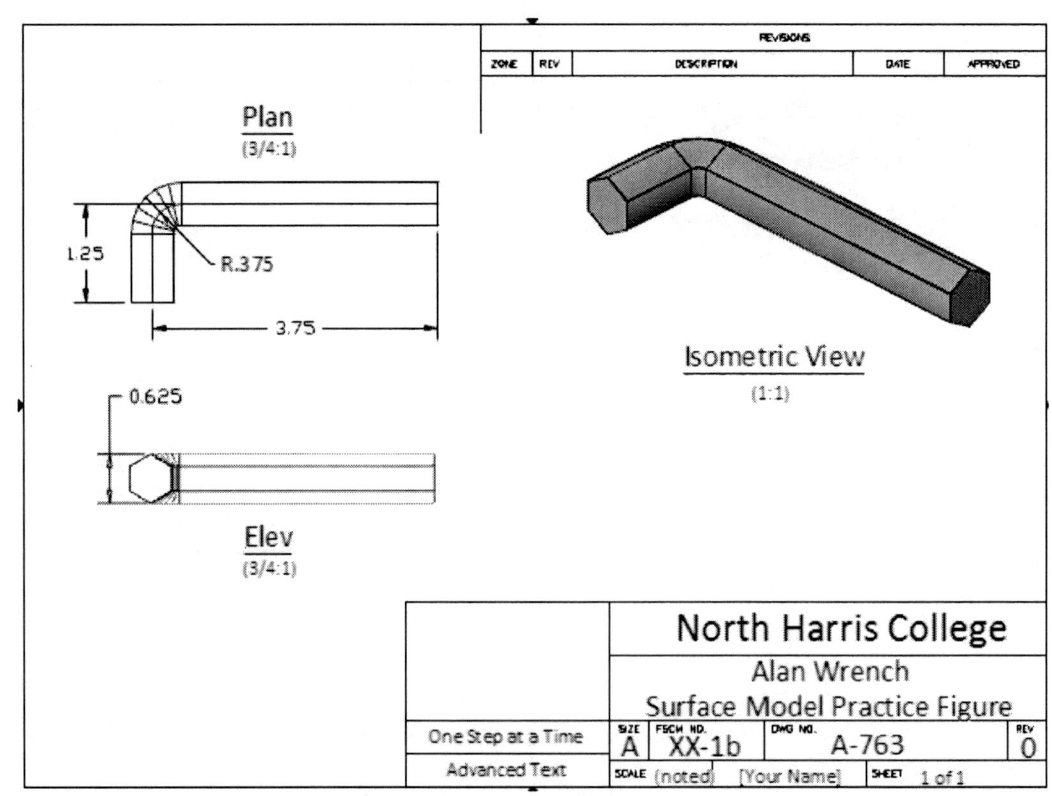

Alan Wrench

6. Create the *Caster* drawing. The following information will help.
    6.1. Use the Times New Roman font – 3/16" and 1/8".
    6.2. Adjust the dimstyle as needed.
    6.3. Create layers as needed (you may need more than you think).
    6.4. I used 16 as my setting for both surftabs.
    6.5. The spindle is a revolved mesh; the upper place is an edged mesh.
    6.6. The axle is a revolved mesh.
    6.7. The wheel is a revolved mesh with a hole in it for the axle.
    6.8. The ball bearings are 1/8" diameter mesh spheres.
    6.9. To create the sections, use the same technique you used in Exercise 2.
    6.10. Save the drawing as *MyCaster* in the C:\Steps3D\Lesson05 folder.

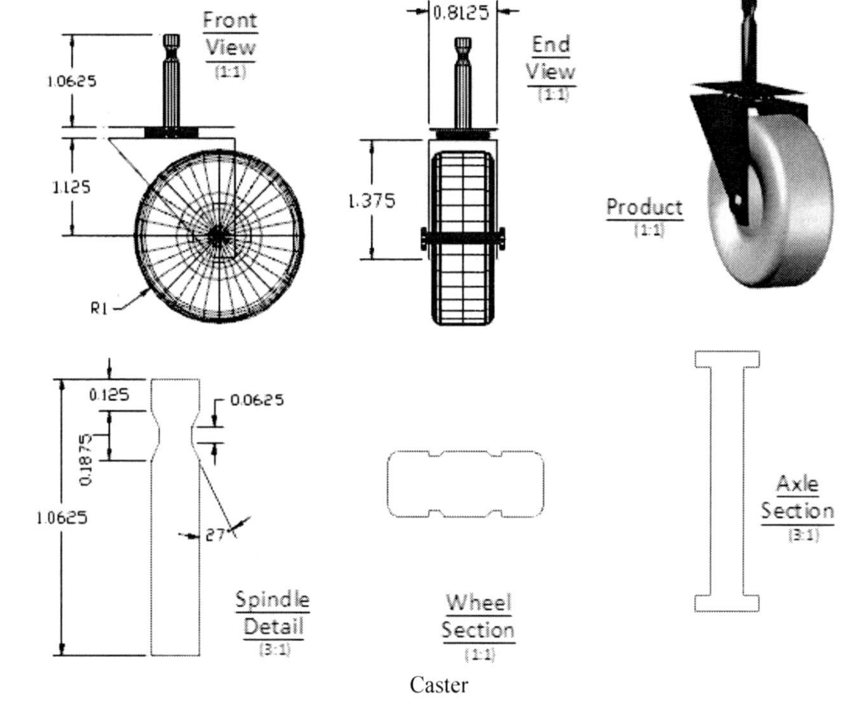

Caster

7. Create the *Remote* drawing. The following information will help.
    7.1. The main face is 5¾" long x 2" wide.
    7.2. There's an additional ½" molded arc on the sides and ends.
    7.3. The thickness of the instrument is ½"
    7.4. Round buttons are 3/8" diameter and ¼" diameter domes.
    7.5. Ellipse buttons are ½" x ¼".
    7.6. Other buttons are either pyramids or tabsurfed polylines with regions closing the tops. These are 1/8" high.
    7.7. Use the Times New Roman font at a 1/8" text height.
    7.8. Save the drawing as *MyRemote* in the C:\Steps3D\Lesson05 folder.

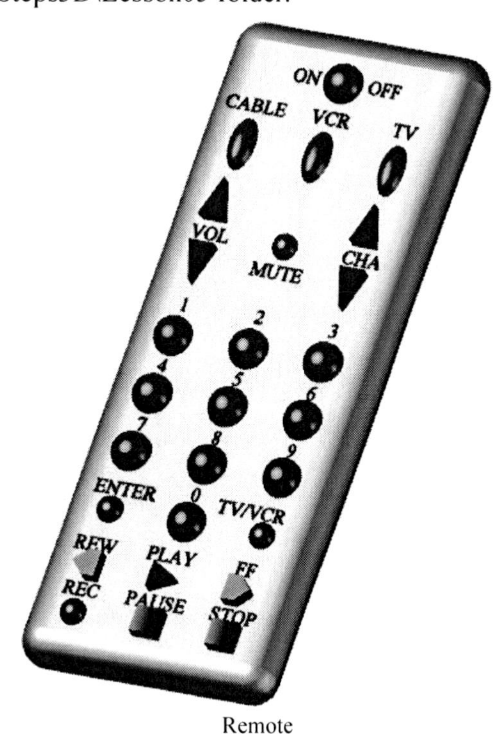

Remote

8. Create the *Racer* drawing. The ½" grid figure will help. Save the drawing as *MyRacer* in the C:\Steps3D\Lesson05 folder.

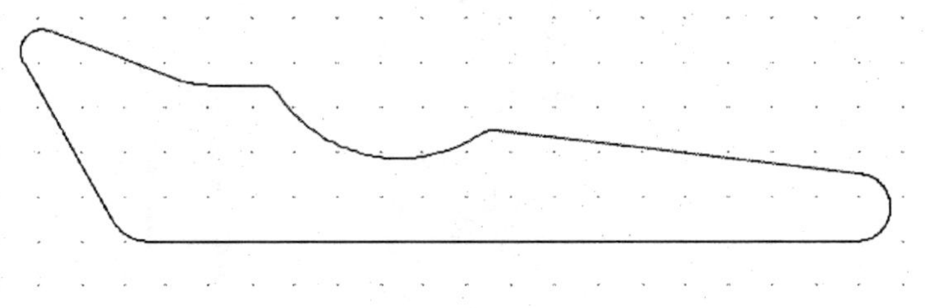

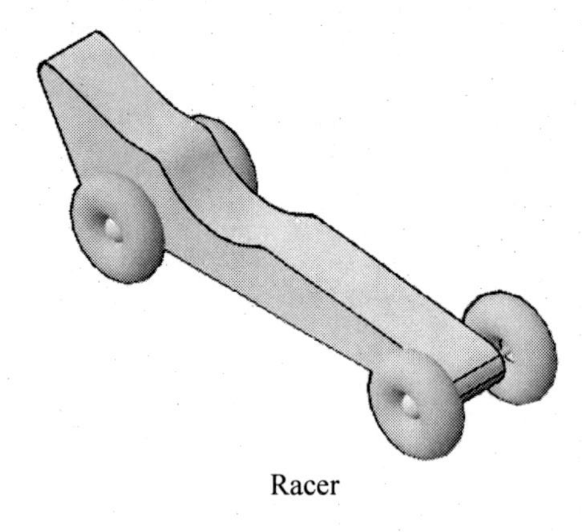

Racer

5.7	**For Web-Based Review Questions, visit: http://foragerpub.com/AcadFiles/2010/2010.htm**

Lesson 6

Z-Space Editing

Following this lesson, you will:

✓ *Know how to use AutoCAD's basic editing tools in Z-Space*

- ***Trim and Extend***
- ***3DMove***
- ***3DRotate***
- ***3DScale***
- ***Mirror3D***
- ***3DArray***
- ***3DAlign***

It's time to look at how our editing tools work both on three-dimensional objects and on two-dimensional objects drawn in Z-Space. You'll find little difference in the way some tools and procedures work, but the differences in others may unsettle you. You'll find still others to be brand new (although strangely familiar).

Regardless of their differences or newness, however, you'll find each tool we discuss in this lesson to be invaluable in your three-dimensional efforts. As they did in the two-dimensional world, editing tools will enhance your speed and drawing ability in Z-Space.

We'll divide our editing tools into those that will work on all objects in Z-space, which we'll discuss in this lesson, and mesh specific editing tools, which we'll look at in Lesson 7, p.149.

Let's get started.

6.1 Trimming and Extending in Z-Space

As we studied editing tools in our basic text, we occasionally came across a prompt that I said would appear in the 3D text. AutoCAD designed these prompts to allow you to continue using some of the more common (and useful) tools when you made the transition into Z-Space. The tools – ***Trim*** and ***Extend*** – weren't difficult to learn in the basic book. And now that you're familiar with them, covering their three-dimensional functions will be a snap!

Trim and ***Extend*** in Z-Space begin very much like ***Trim*** and ***Extend*** in two-dimensional space. In fact, the commands are the same, so all you must learn is the option required to control what gets trimmed/extended in a three-dimensional view.

Remember how the **Edgemode** system variable controlled the **Edge** option in both ***Trim*** and ***Extend*** commands? It still holds true in Z-Space, but there's an additional system variable to consider now – **Projmode**. Like **Edgemode**, **Projmode** affects both the ***Trim*** and ***Extend*** commands. But where **Edgemode** controls your ability to trim/extend to an imaginary extension of the selected cutting edge/boundary, **Projmode** controls how the ***Trim*** and ***Extend*** commands behave in three-dimensional space.

There are two ways to set the **Projmode** system variable – by selecting the **Project** option at the **[Fence/Crossing/Project/Edge/eRase/Undo]** prompt of either command, or by entering ***Projmode*** at the command prompt. The command prompt requires that you enter a number code for the option you wish to use; the **Project** option of the ***Trim/Extend*** command presents the available settings. The number codes and their corresponding settings follow.

CODE	SETTING	FUNCTION
0	**None**	This is the *True 3D* setting. It requires that both the cutting edge/boundary and the object to trim/extend be in the same plane. That is, they must actually intersect or, using the **Edgemode** system variable, intersect at an imaginary extension.
1	**UCS**	(Default Setting) This setting projects the cutting edge/boundary and object to trim/extend onto the XY plane of the current UCS and then performs the task as though the objects exist in two-dimensional space.
2	**View**	This setting causes AutoCAD to project the cutting edge/boundary and object to trim/extend onto the current view as though your monitor's screen is the XY plane. It then performs the task as though the objects exist in two-dimensional space.

Let's see how these settings work in a couple exercises.

Do This: 6.1A	Trimming in Z-Space

I. Open the *trim* file in the C:\Steps3D\Lesson06 folder. The drawing looks like the figure at right.

II. Be sure the **Edgemode** system variable is set to 1.

III. Follow these steps.

6.1A: Trimming in Z-Space

1. Let's begin by setting the **Projmode** system variable to zero (the *True 3D* setting).

    Command: *Projmode*

    Enter new value for PROJMODE <1>: *0*

2. Now begin the ***Trim*** command and select the line indicated as your cutting edge.

    Command: *tr*

    Current settings: Projection=None, Edge=Extend

    Select cutting edges ...

    Select objects or <select all>:

    Select objects: *[ENTER]*

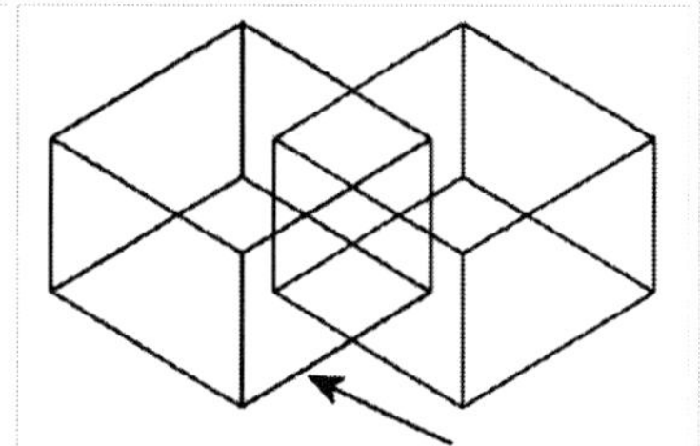

3. Now select the lines indicated to trim. Select the bottom line first.

    Select object to trim or shift-select to extend or [Fence/Crossing/Project/Edge/eRase/Undo]:

Notice that the lower line trims, but the upper line doesn't. With the **Projmode** system variable set to zero, the lines must be on the same plane (as I explained in the chart). The upper line doesn't intersect the cutting edge, so it didn't trim.

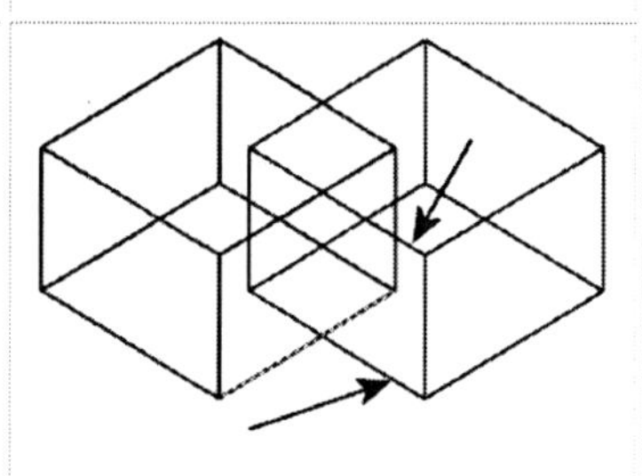

4. Without leaving the command, change **Projmode** Project to the **UCS** setting Ucs as indicated.

    Select object to trim or shift-select to extend or [Fence/Crossing/Project/Edge/eRase/Undo]: *p*

    Enter a projection option [None/Ucs/View] <None>: *u*

5. Now trim the line that wouldn't trim previously.

    Select object to trim or shift-select to extend or [Fence/Crossing/Project/Edge/eRase/Undo]:

It trims now because AutoCAD projects the cutting edge and object to trim against the XY plane of the current UCS (as if both lines were drawn in 2D space).

6. Complete the command.

    Select object to trim or shift-select to extend or [Fence/Crossing/Project/Edge/eRase/Undo]: *[ENTER]*

Your drawing looks like the figure at right.

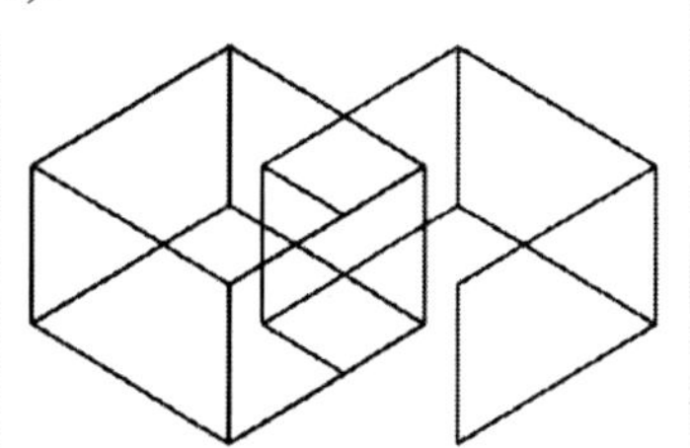

7. Repeat Step 2.

    Command: *[ENTER]*

6.1A: TRIMMING IN Z-SPACE

8. Select the line indicated to trim.

   Select object to trim or shift-select to extend or [Fence/Crossing/Project/Edge/eRase/Undo]:

Notice that the line won't trim. The **Projmode** system variable setting of **1** (**UCS**) means that the lines must intersect in a two-dimensional projection of the current UCS. The line doesn't intersect the cutting edge, so it didn't trim.

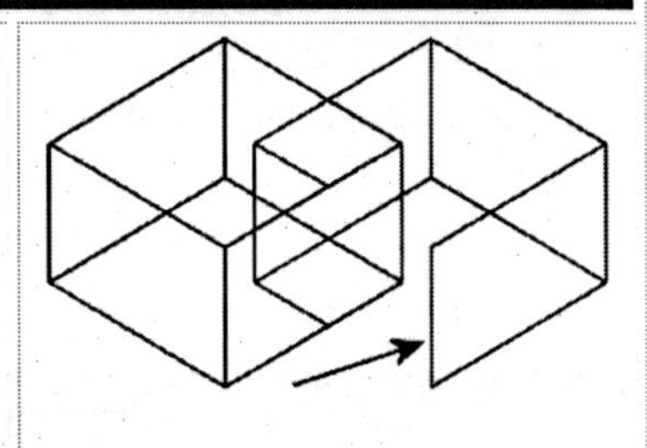

9. Change the **Projmode** [Project] to the **View** setting [View] as indicated. (Note: you can use cursor or dynamic input menus if you prefer.)

   Select object to trim or shift-select to extend or [Fence/Crossing/Project/Edge/eRase/Undo]: ***p***

   Enter a projection option [None/Ucs/View] <Ucs>: ***v***

10. Now trim the line that wouldn't trim previously.

   Select object to trim or shift-select to extend or [Fence/Crossing/Project/Edge/eRase/Undo]:

It trims now for two reasons: (1) AutoCAD has projected the **cutting edge** and **object to trim** against your screen (the current view) as if your screen defined the XY plane, and (2) AutoCAD has extended the cutting edge according to the **Edgemode** setting.

11. Complete the command.

   Select object to trim or shift-select to extend or [Fence/Crossing/Project/Edge/eRase/Undo]: ***[ENTER]***

Your drawing looks like the figure at right.

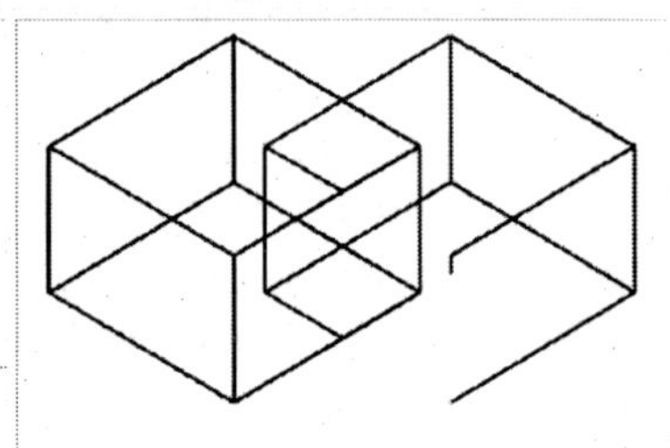

12. Exit the drawing without saving.

   Command: ***quit***

Now let's look at the ***Extend*** command.

Do This: 6.1B	Extending in Z-Space

I. Open the *ext* file in the C:\Steps3D\Lesson06 folder. The drawing looks like the figure at right.

II. Be sure the **Edgemode** system variable is set to **1** and the **Projmode** system variable is set to **0**.

III. Follow these steps.

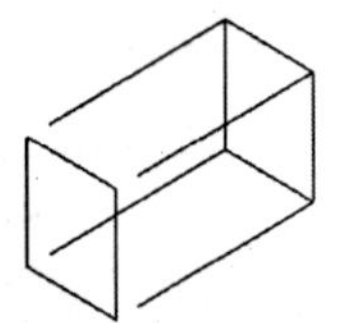

6.1B: EXTENDING IN Z-SPACE

1. Begin the ***Extend*** command and select the line indicated as your boundary edge.

   Command: ***ex***

   Current settings: Projection=None, Edge=Extend

   Select boundary edges ...

   Select objects or <select all>:

   Select objects: ***[ENTER]***

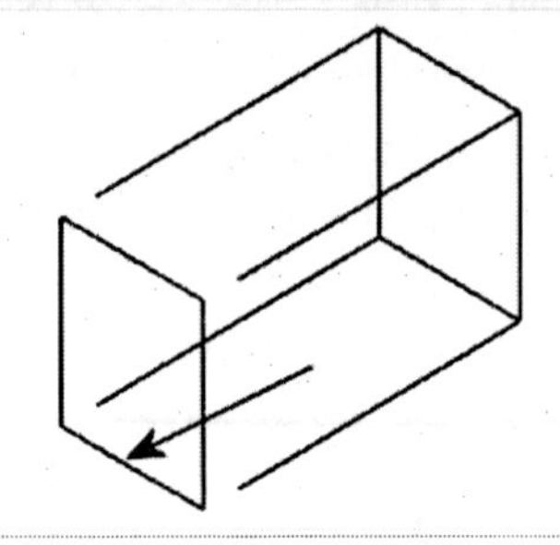

6.1B: Extending in Z-Space

2. Select the four open-ended lines to extend. Select the bottom lines first.

    Select object to extend or shift-select to trim or [Fence/Crossing/Project/Edge/Undo]:

Notice that the lower lines extend but not the upper lines. As in the ***Trim*** command, with the **Projmode** set to **0**, the lines must be in the same XY plane. The upper lines don't share the boundary edge's plane, so they didn't extend.

3. Change the **Projmode** [Project] to the **UCS** setting [Ucs].

    Select object to extend or shift-select to trim or [Fence/Crossing/Project/Edge/Undo]: *p*

    Enter a projection option [None/Ucs/View] <None>: *u*

4. Now extend the lines that wouldn't extend previously.

    Select object to extend or shift-select to trim or [Fence/Crossing/Project/Edge/Undo]:

They extend now because AutoCAD has projected them against the XY plane of the current UCS.

5. Complete the command.

    Select object to extend or shift-select to trim or [Fence/Crossing/Project/Edge/Undo]: ***[ENTER]***

Your drawing looks like the figure at right.

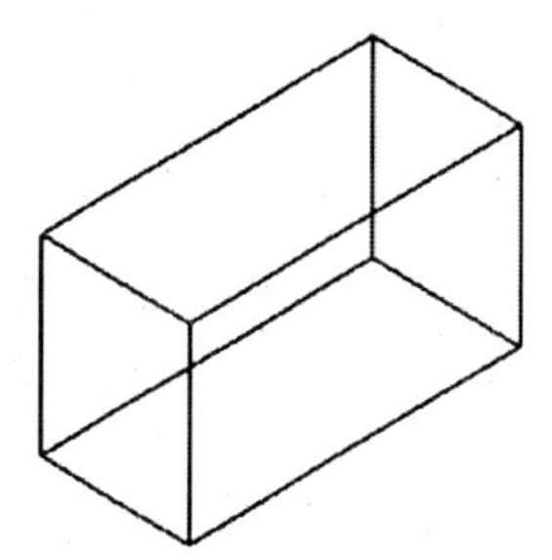

6. Repeat Step 1, but select the boundary edge indicated at right.

    Command: *ex*

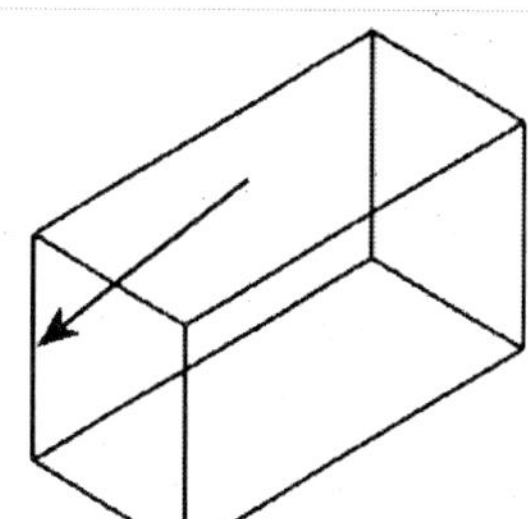

7. Select the line indicated here to extend.

    Select object to extend or shift-select to trim or [Fence/Crossing/Project/Edge/Undo]:

Notice that the line won't extend. With **Projmode** set to **1**, the lines must intersect in a two-dimensional projection of the current UCS. The lines don't, so the selected **object to extend** didn't extend.

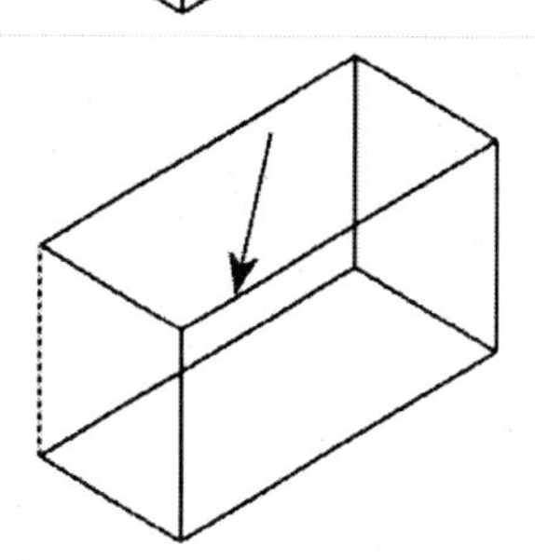

8. Change the **Projmode** [Project] to the **View** settings [View].

    Select object to extend or shift-select to trim or [Fence/Crossing/Project/Edge/Undo]: *p*

    Enter a projection option [None/Ucs/View] <Ucs>: *v*

6.1B: Extending in Z-Space

9. Now extend the line that wouldn't extend previously.

   Select object to extend or shift-select to trim or [Fence/Crossing/Project/Edge/Undo]:

As with the ***Trim*** command, it extends now because AutoCAD is looking at the lines projected against your screen (the current view).

10. Complete the command.

    Select object to extend or shift-select to trim or [Fence/Crossing/Project/Edge/Undo]: ***[ENTER]***

Your drawing looks like the figure at right.

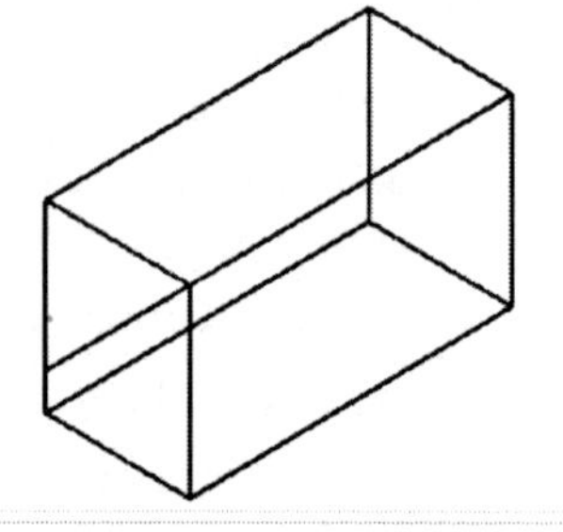

11. Exit the drawing without saving.

    Command: ***quit***

You've seen that, although still fairly simple to use, the ***Trim*** and ***Extend*** commands have some different options with which you must become familiar if you're to use them to full advantage in Z-Space.

But there are some additional things that I should highlight before continuing.

- These commands work on the same objects for which you used them in 2D space – you can't trim/extend a 3D face, mesh, region or solid.
- You can't use a 3D face, mesh, or solid as a cutting edge or boundary edge when trimming/extending.
- You *can*, however, use a region as a cutting edge or boundary edge when trimming or extending.

6.2 Moving in Z-Space

You've seen how AutoCAD has adapted several modification tools you already knew to help you in Z-Space and with three-dimensional objects. It's good to maintain some familiarity between the 2D and 3D worlds.

Where to Find It:	
Command Line:	***3DMove***
Hotkey(s):	***3m***
Ribbon (Tab/Panel):	Home – Modify – **3D Move**
Menu:	Modify – 3D Operations – **3D Move**
Toolbar:	Modeling – **3D Move**

You might think that moving objects in Z-space would be simple enough using Cartesian coordinates – just add the Z! Okay, it really is. But AutoCAD does include a command that makes moving to a general location fairly simple. What's more, it really doesn't require a lot of adjustment to use.

The sequence is identical to the ***Move*** command. Then why use it at all? Notice the tool for the ***3DMove*** command. This handy gizmo – oddly called a gizmo – appears after you've identified your base point or accepted the **Displacement** option. It will help you relocate objects in Z-space.

3D Move Gizmo

How (you ask)?

Well, when you place your cursor on one of the axes, AutoCAD will move the object along that axis. It even provides a ghost line to let you know which axis it's using. But perhaps more importantly, you can select a single face (tessellation division) on a mesh – by holding down the CTRL key while you pick – and use the gizmo to move just that face. Further, when you move the gizmo over a 3D Face or the face of a solid object with the dynamic UCS toggled on, AutoCAD will automatically change the UCS to assist your move. (Now that's a powerful tool!)

But wait! If you right click on the gizmo, AutoCAD will display the cursor menu shown in Figure 6.001. Use the top frame of this menu to toggle between the gizmos for **3D Move**, **3D Rotate** (Section 6.3, p.132), and **3D Scale** (Section 6.4, p.134). This menu also provides the following options.

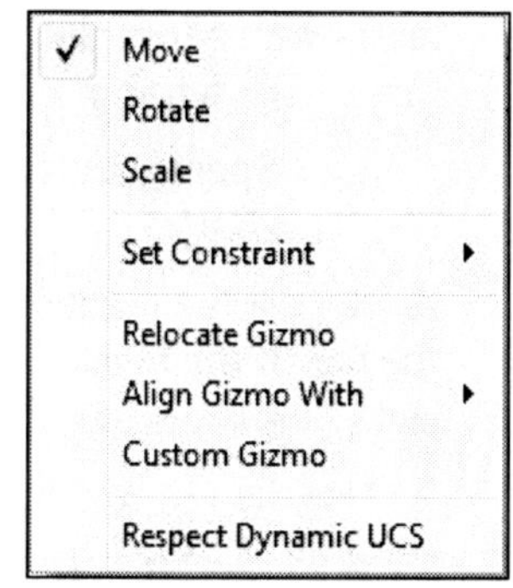

Figure 6.001

- **Set Constraint** (to **X**, **Y**, or **Z**, **XY**, **YZ**, or **ZX**) – use this to constrain (restrict) the operation to a specific axis.
- **Relocate Gizmo** – use this to move the gizmo to a more convenient location. The new location becomes your base point.
- **Align Gizmo With (World UCS, Current UCS, Face)** – use this to manually align the gizmo before proceeding with the operation.
- **Custom Gizmo** – this option allows you to redefine the gizmo by identifying one, two, or three points, or even an object.
- **Respect Dynamic UCS** – this is just another Dynamic UCS toggle like the one on the status bar.

Oh, the ***3DMove*** command also allows you to enter X,Y,Z coordinates in a Cartesian format, but you can do that with the standard ***Move*** command.

Try this.

Do This: 6.2A	Moving Objects in Z-Space

I. Open the *move* file in the C:\Steps3D\Lesson06 folder. It looks like the figure at right.

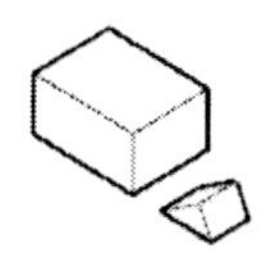

II. Set the **3D Wireframe** visual style current.

III. Follow these steps.

6.2A: MOVING OBJECTS IN Z-SPACE
1. We want to move the wedge to "about" the center of the top of the block. You can do it with the ***Move*** command – use OSNAPs to move it adjacent to one of the sides, then eyeball a second move to about the center of the block. But let's see how we can do it with the ***3DMove*** command. Enter the command. **Command: *3dmove***
2. Select the wedge. **Select objects:**
3. Notice that the **3D Move** gizmo appears when AutoCAD asks for a base point. **Specify base point or [Displacement] <Displacement>:**
4. Place your cursor over the X-axis as shown. When you see the ghost line, pick and move the object to the center of the bottom of the block. AutoCAD relocates it. (Notice that AutoCAD shifts to grip prompts.) **** MOVE **** **Specify move point or [Base point/Copy/Undo/eXit]:** 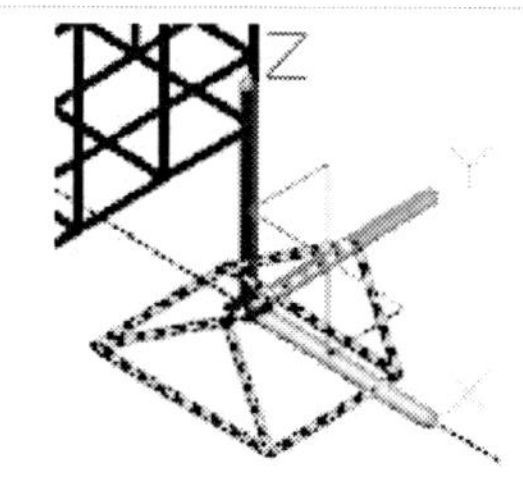

6.2A: Moving Objects in Z-Space

5. This time, place your cursor over the blue Z-axis grip as shown. Move the wedge to the top of the block.

    **** MOVE ****

    Specify move point or [Base point/Copy/Undo/eXit]:

6. Repeat the procedure to roughly center the wedge on the top of the block. Your drawing looks like the figure at right (shown in **3D Hidden** visual style).

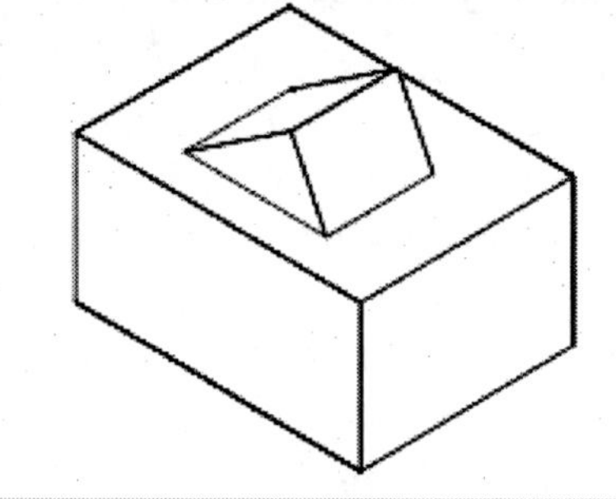

7. Use the ESC key to clear the selection.

8. With the **3D Wireframe** visual style set, hold down the CTRL key and pick the centermost tessellation division on the right face of the block (as shown).

    Notice that the CTRL key allows you to select one (or several) individual faces and that the 3D Move Gizmo automatically displays.

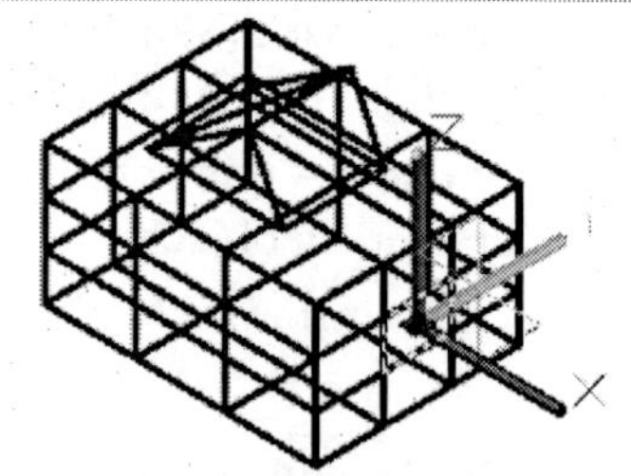

9. Use the gizmo to move the tessellation division to the right. Notice the grip procedure you're using!

    **** STRETCH ****

    Specify stretch point or [Base point/Undo/eXit]:

    (Shown here in **Realistic** visual style.)

10. Exit the drawing without saving.

Now do this, repeat the last exercise without entering the ***3DMove*** command at all. As my son would say, "well, duh!" The gizmo appears* when you select a three-dimensional object in much the same way grips appear … and you've already seen that you use grips procedures when you use a gizmo.

Kinda makes you wonder why they even have a ***3DMove*** command, doesn't it? (Well, in AutoCAD's defense, the command preceded the gizmo by a few releases.)

Let's look at how the gizmo might help us rotate an object.

6.3 Rotating About an Axis

Rotating about a 2-dimensional base point was easy – you simply selected what to rotate and a base point. Then you told AutoCAD what angle you wanted.

Rotating about an axis is slightly more complex. But if you ever need to rotate an object in Z-Space, you'll find the ***3DRotate*** command quite handy. It works like this:

Command: *3drotate*

Where to Find It:	
Command Line:	***3DRotate***
Hotkey(s):	***3r***
Ribbon (Tab/Panel):	Home – Modify – **3D Rotate**
Menu:	Modify – 3D Operations – **3D Rotate**
Toolbar:	Modeling – **3D Rotate**

* You can change which gizmo appears by default using the **Move Gizmo** flyout on the ribbon's **Home** tab – **Subobject** panel. You can also toggle between the active gizmos here as well as on the cursor menu shown in Figure. 6.001, p.131. We'll spend more time with the gizmos in Lesson 7.

Current positive angle in UCS: ANGDIR=counterclockwise ANGBASE=0
Select objects: ***[Select the object(s) you want to rotate.]***
Select objects: ***[Confirm completion of the selection set.]***

At this point, the prompts change from the familiar 2D approach and AutoCAD presents the **3D Rotate** gizmo. This tool works in much the same way the **3D Move** gizmo worked – place your cursor over one of the axes handles and AutoCAD will rotate the selected objects about that axis.

3D Rotate Gizmo

Here are the remaining prompts:

Specify base point: ***[Put the grip in the center of the gizmo over the point about which you want to rotate.]***
Pick a rotation axis: ***[Select one of the axes handles – just as you selected an axis with the 3D Move gizmo.]***
Specify angle start point or type an angle: ***[Pick a starting point or enter an angle – if you opt to enter an angle, AutoCAD rotates the object by that angle and doesn't prompt for an angle end point.]***
Specify angle end point: ***[The object rotates dynamically, so rotate until you're happy and pick the ending point.]***

Now, let's give this command a try.

Do This: 6.3A	Rotating Objects in Z-Space

I. Open the *ro3d* file in the C:\Steps3D\Lesson06 folder. The drawing looks like the figure at right.
II. Follow these steps.

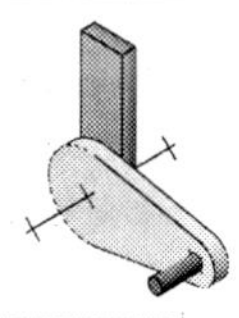

6.3A: ROTATING OBJECTS IN Z-SPACE	
1. Enter the ***3DRotate*** command. **Command:** ***3r***	
2. Select the handle. **Select objects:** **Select objects:** ***[ENTER]*** The **3D Rotate** gizmo appears.	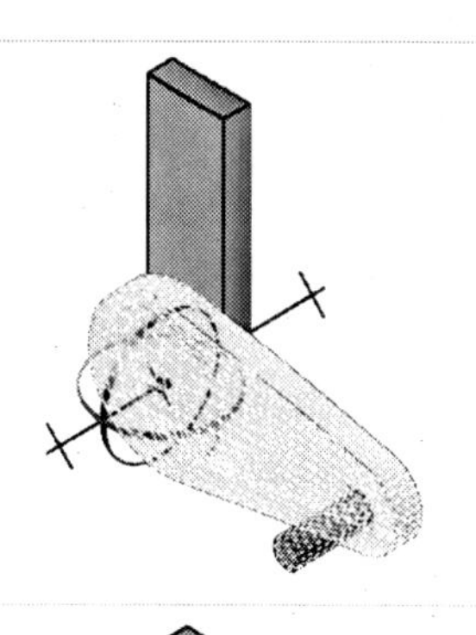
3. Pick the node in the center of the handle. **Specify base point:** The gizmo moves to the base point.	
4. Pick the (green) Y-axis. AutoCAD presents a ghost line (right). **Pick a rotation axis:** Notice that the selected axis changes colors to help identify which one you've selected.	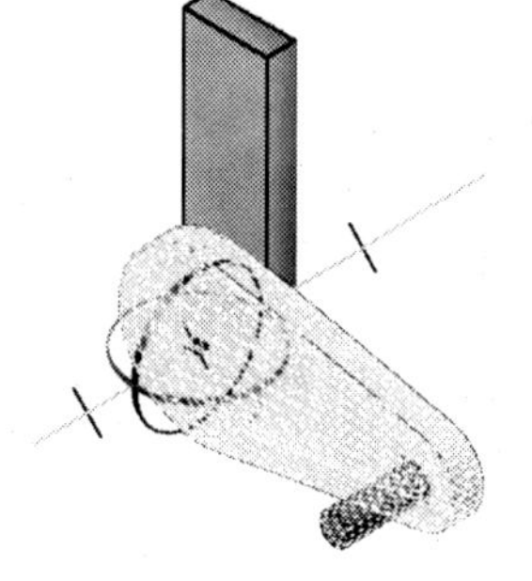
5. Pick the node in the center of the small end of the handle. **Specify angle start point or type an angle: _cen of**	

6.3A: Rotating Objects in Z-Space

6. Move your cursor up and down. Notice that the handle moves dynamically. Enter an angle or pick a rotation you like.

Specify angle end point:

AutoCAD rotates the handle.

7. Now try this, select the handle without entering a command. Notice that AutoCAD presents the **3D Move** gizmo.

8. Right click on the gizmo and select **Rotate** [Rotate] from the menu. Notice that the gizmo changes.

9. Right click again and select **Relocate Gizmo** [Relocate Gizmo] from the menu. Place it at the node in the center of the handle.

10. Now repeat Steps 4 – 6. Notice the prompt.

**** ROTATE ****

Specify rotation angle or [Base point/Copy/Undo/Reference/eXit]:

As with the ***3DMove*** command, you really didn't need to use the ***3DRotate*** command, did you?*

7. Exit the drawing without saving it.

Command: ***quit***

6.4 Mirroring Three-Dimensional Objects

The differences between the ***Mirror3d*** and ***Mirror*** commands are really quite similar to the differences between ***Rotate3d*** and ***Rotate*** the commands. Rather than selecting a point around which to rotate an object in 2D space, you had to pick an axis to satisfy the ***Rotate3d*** command. Rather than picking two points on a mirror *line* as you did in 2D space, you must pick three points to identify a mirror *plane* (the actual face of the mirror) when you use the ***Mirror3d*** command.

Where to Find It:	
Command Line:	*Mirror3D*
Hotkey(s):	*3dmirror*
Ribbon (Tab/Panel):	Home – Modify – **3D Mirror**
Menu:	Modify – 3D Operations – **3D Mirror**

The options offered by the ***Mirror3d*** command follow.

Command: ***mirror3d***

Select objects: ***[Select the object(s) you want to mirror.]***

Select objects: ***[Confirm the selection set.]***

Specify first point of mirror plane (3 points) or [Object/Last/Zaxis/View/XY/YZ/ZX/3points] <3points>: ***[Use this and the next two options to identify the mirror plane. Notice that AutoCAD has no gizmo for mirroring!]***

Specify second point on mirror plane:

Specify third point on mirror plane:

Delete source objects? [Yes/No] <N>: ***[This option is the same as the 2D*** **Mirror** ***command – hit*** ENTER ***to keep the source objects or enter*** **Y** ***to remove them.]***

* You didn't need the ***3DRotate*** command, but it does provide a bit more precision than the gizmo alone.

It might look frightening compared with the ***Mirror*** command, but once you've used it, you'll find it fairly simple and straightforward. Let's consider each of the plane-defining options.

- The default option is to specify **3 points** on the plane. AutoCAD needs you to define the mirror plane by picking any three points on it. Once you've done that, AutoCAD will prompt you to

  Delete source objects? [Yes/No] <N>:

  This tells AutoCAD whether you wish to mirror the source or make a mirrored copy.
- The **Object** option is probably the easiest. If you have an object drawn that can serve as an mirror plane, all you have to do is select it. When you choose this option, AutoCAD prompts

  Select a circle, arc, or 2D-polyline segment:

  - If you select a circle or arc, AutoCAD rotates the objects parallel to the plane of the circle or arc and about an imaginary axis drawn through the center of it.
  - AutoCAD treats a 2D-polyline arc as an arc.
- The **Last** option refers to the last axis you used in the ***Mirror3d*** command.
- When you use the **View** option, AutoCAD mirrors the objects about the current viewport's plane.
- The **Zaxis** option aligns asks for a point on the mirror plane and then a point on the Z-axis. AutoCAD prompts:

  Specify point on mirror plane:

  Specify point on Z-axis (normal) of mirror plane:
- **XY/YZ/ZX** tells AutoCAD to align the mirror-plane with one of the standard planes associated with a selected point.

Let's get right to an exercise.

Do This: 6.4A	Mirroring Objects in Z-Space

I. Open the *Star* file in the C:\Steps3D\Lesson06 folder. The drawing looks like the figure at right.
II. Set the **Endpoint** running OSNAP.
III. Follow these steps.

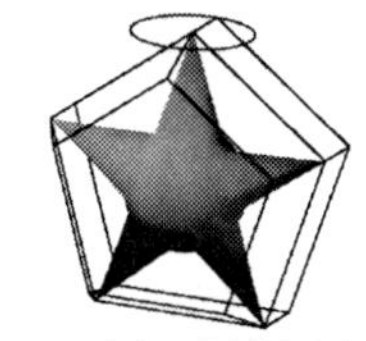

6.4A: MIRRORING OBJECTS IN Z-SPACE

1. Enter the ***Mirror3d*** command.

   Command: ***mirror3d***

2. Select the star.

   Select objects:

   Select objects: ***[ENTER]***

6.4A: Mirroring Objects in Z-Space

3. We'll use the default **3points** approach first. Pick the points indicated.

    Specify first point of mirror plane (3 points) or [Object/Last/Zaxis/View/XY/YZ/ZX/ 3points] <3points>:

    Specify second point on mirror plane:

    Specify third point on mirror plane:

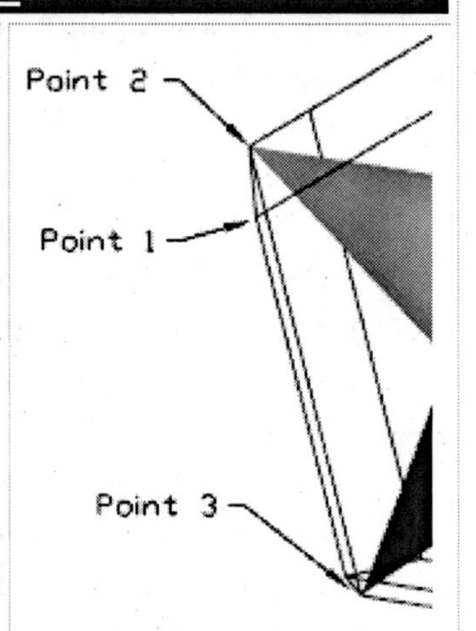

4. Don't delete the source objects.

    Delete source objects? [Yes/No] <N>: ***[ENTER]***

Your drawing looks like the following left figure. The star has been mirrored along the plan you identified. Rotate the view to see it from above (following right figure) for a better understanding of the angles.

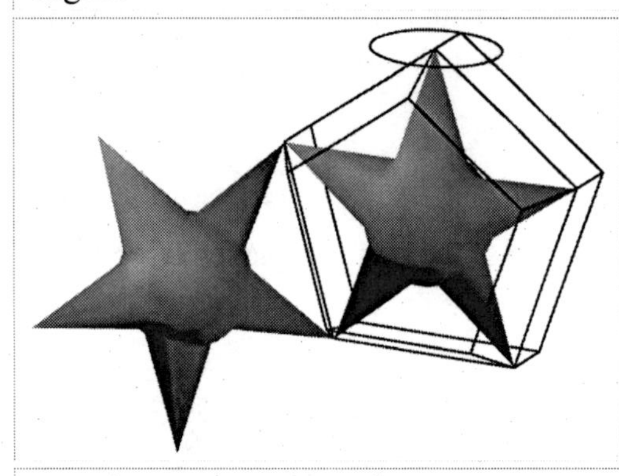

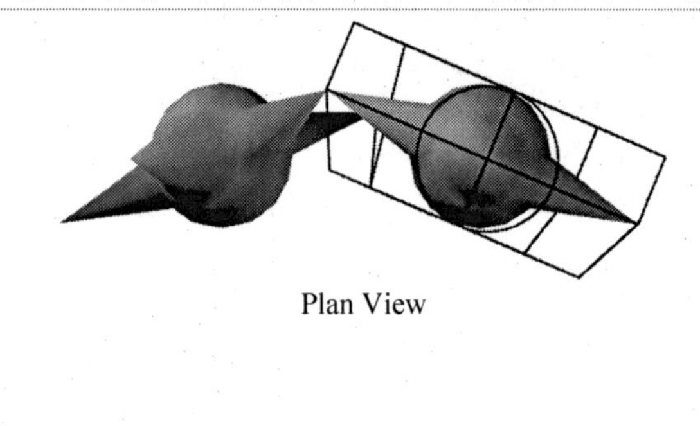

Plan View

5. Erase the new star.

    Command: ***e***

6. Let's use the **Object** option to stand the star on its head. Repeat Steps 1 and 2.

    Command: ***mirror3d***

7. Choose the **Object** option Object.

    Specify first point of mirror plane (3 points) or [Object/Last/Zaxis/View/XY/ YZ/ZX/3points] <3points>: ***o***

8. Select the star's halo (the circle).

    Select a circle, arc, or 2D-polyline segment:

9. This time, delete the source objects.

    Delete source objects? [Yes/No] <N>: y

Your drawing looks like the figure at right. The star has been mirrored using the plane in which the circle was drawn.

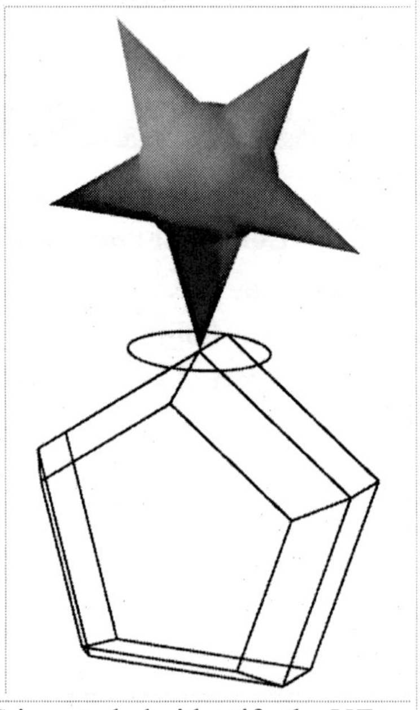

10. Now we'll mirror the star using the YZ plane. (Turn on the UCS icon to help identify the YZ plane.) Repeat Steps 1 and 2.

    Command: ***[ENTER]***

6.4A: Mirroring Objects in Z-Space

11. Choose the **YZ** option [YZ] and select the leftmost point of the star.

Specify first point of mirror plane (3 points) or [Object/Last/Zaxis/View/XY/ YZ/ZX/3points] <3points>: *yz*

Specify point on YZ plane <0,0,0>:

12. Don't delete the source objects.

Delete source objects? [Yes/No] <N>:

Your drawing looks like the figure at right.

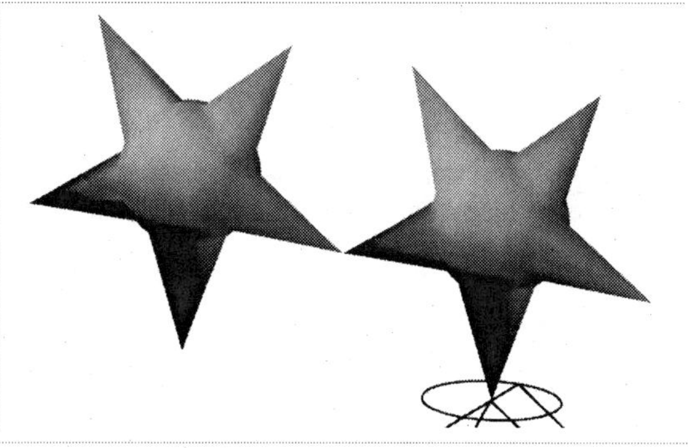

13. Exit the drawing without saving your changes.

Command: *quit*

If there were only one suggestion I could make about both the ***Rotate3d*** and ***Mirror3d*** commands, it would be to always check your image from more than one viewpoint (preferably three or four). Remember that, in Z-Space, object positions seen from one angle are not necessarily true three-dimensional positions.

6.5 Scaling Three-Dimensional Objects

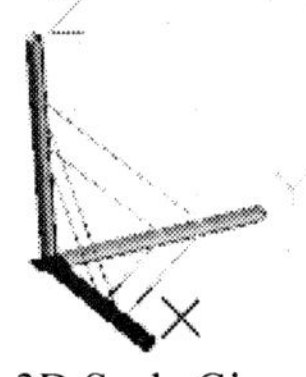

3D Scale Gizmo

The ***3DScale*** command resembles its 2D counterpart with a tweak or two. This one requires the use of the 3D Scale gizmo!

Where to Find It:	
Command Line:	***3DScale***
Hotkey(s):	***3s***
Ribbon (Tab/Panel):	Home – Modify – **3D Scale**

Let's look at the gizmo first. Notice that it looks like the 3D Move gizmo (p.130) with spider webs. You'll use these "web lines" to control what you scale. Refer to the following table.

Select	To:
	Select the area closest to the XYZ vertex on the gizmo to scale the selected object(s) uniformly.
	Select between the parallel lines of the web to scale along a specific plane.
	Select an axis to scale along that axis.

The command sequence looks like this.

Command: *3dscale*

Select objects: ***[Select the object you wish to scale.]***

Select objects: ***[Confirm the selection, the 3D Scale gizmo appears.]***

Specify base point: ***[Select a point, AutoCAD will place the node at the center of the gizmo on the base point.]***

Pick a scale axis or plane: ***[Refer to the table and select the part of the gizmo that corresponds to what you wish to scale.]***

Specify scale factor or [Copy/Reference] <1.0000>: ***[The rest of the sequence is identical to the 2d command.]***

Let's get right to an exercise.

Do This: 6.5A	Scaling Objects in Three Dimensions

I. Open the *3dScale* file in the C:\Steps3D\Lesson06 folder. The drawing looks like the figure at right.

II. Turn the UCS Icon off.

III. Follow these steps.

6.5A: Scaling Objects in Three Dimensions	
1. Enter the ***3DScale*** command. **Command:** ***3s***	
2. Select the cylinder. **Select objects:** Notice that the 3D Scale gizmo appears in the center of the selection.	
3. Use 0,0,0 as your base point. **Specify base point:** ***0,0,0*** Notice that the gizmo relocates.	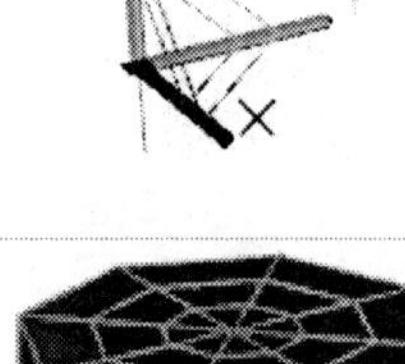
4. Select the gizmo to scale the object along the Z-axis. **Pick a scale axis or plane:**	
5. Scale the cylinder down about .75 times. **Specify scale factor or [Copy/Reference] <1.0000>:** ***.75*** The cylinder looks like the figure at right.	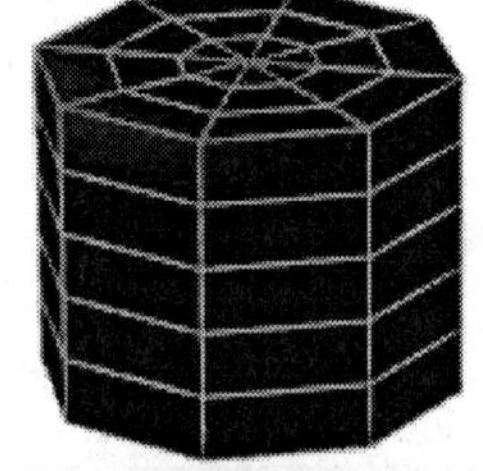
6. Undo the change. Let's try something a bit more challenging.	
7. Pick the **Front** face on the cube to change the view.	
8. Hold down the CTRL key and select the middle tessellation divisions with a crossing window (to select the ones in the back as well as the front. AutoCAD selects the divisions.	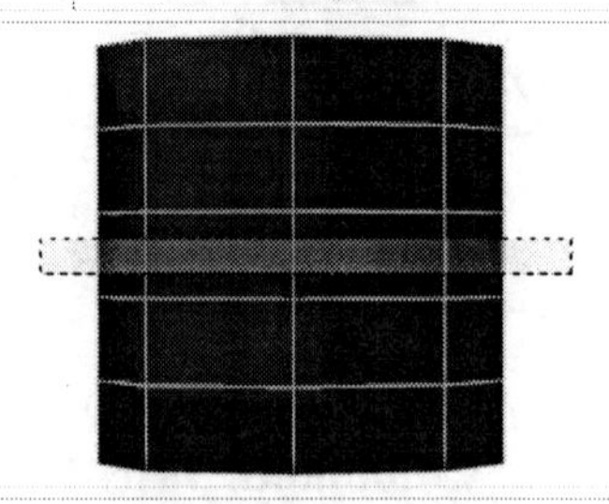
9. Use the cube to change to a top view.	
10. Use the **Relocate Gizmo** option Relocate Gizmo on the cursor menu to locate the gizmo in the center of the cylinder in this view. Use the cursor menu to select the **Scale** gizmo Scale, as well. (Notice that we still haven't begun the ***3DScale*** command?)	

6.5A: SCALING OBJECTS IN THREE DIMENSIONS

11. Pick the area between the two web lines and drag inward until the gizmo is about ½ size.

12. Clear the grips (and gizmo) with the ESC key [Esc].

13. Pick the top/right corner of the cube to restore the view. (Adjust the view as needed to see the results of your scale.)

14. Okay, now for grins, change the **Smoothness** property to **Level 2**. (I know, you've heard it before, but com'n, this really is soo coool!)

15. Save the drawing and close it.

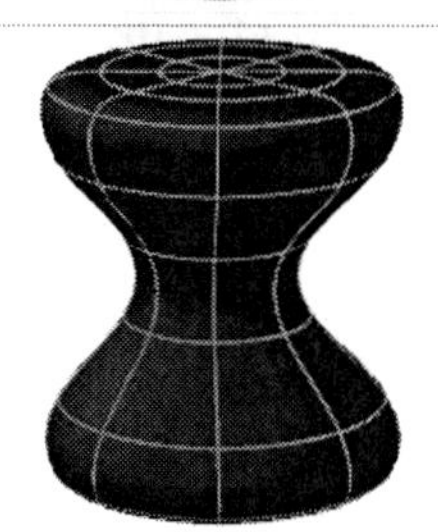

Are you beginning to see the potential of mesh objects?

6.6 Three Dimensional Arrays

Where to Find It:	
Command Line:	***3DArray***
Hotkey(s):	***3a***
Ribbon (Tab/Panel):	Home – Modify – **3D Array**
Menu:	Modify – 3D Operations – **3D Array**
Toolbar:	Modeling – **3D Array**

The most important difference between the two-dimensional rectangular array and the three-dimensional rectangular array is the addition of prompts for number and spacing of levels. The most important difference between the two-dimensional polar array and the three-dimensional polar array is that, rather than selecting a center point of the array, you must identify two points on an axis.

I should mention another important difference between the ***Array*** and ***3DArray*** commands. ***3DArray*** has no dialog box with which to work. But if you're comfortable with the Array dialog box, the command line prompts and options will be familiar to you.

Let's array some objects in Z-Space.

Do This: 6.6A	**Arraying Objects in Z-Space – Rectangular Arrays**

I. We'll create a three-dimensional piperack. Open the *3darray-rec* file in the C:\Steps3D\Lesson06 folder. The drawing looks like the figure at right.

II. Follow these steps.

6.6A: Arraying Objects in Z-Space – Rectangular Array

1. Enter the ***3DArray*** command.

   Command: ***3a***

2. Select the vertical 10' I-Beam.

   Select objects:

   Select objects: ***[ENTER]***

3. Accept the default **Rectangular** type of array.

   Enter the type of array [Rectangular/Polar] <R>: ***[ENTER]***

4. Tell AutoCAD you want two rows, three columns, and two levels.

   Enter the number of rows (---) <1>: ***2***

   Enter the number of columns (|||) <1>: ***3***

   Enter the number of levels (...) <1>: ***2***

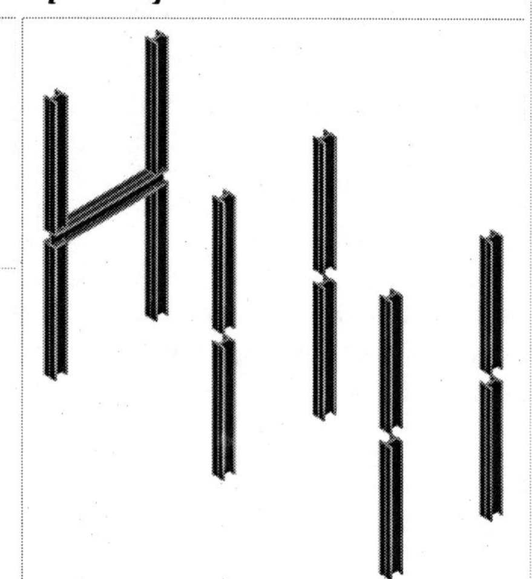

5. Specify the distances as shown.

   Specify the distance between rows (---): ***9'***

   Specify the distance between columns (|||): ***15'***

   Specify the distance between levels (...): ***11'***

Your drawing looks like the figure at right. (Adjust your view as required to see the entire drawing.)

6. Now we'll array the horizontal support. Repeat the ***3DArray*** command.

   Command: ***[ENTER]***

7. Select the horizontal support.

   Select objects:

8. Accept the default **Rectangular** type of array.

   Enter the type of array [Rectangular/Polar] <R>: ***[ENTER]***

9. You'll want to create one row, three columns, and two levels …

   Enter the number of rows (---) <1>: ***[ENTER]***

   Enter the number of columns (|||) <1>: ***3***

   Enter the number of levels (...) <1>: ***2***

10. …at the spacing indicated.

    Specify the distance between columns (|||): ***15'***

    Specify the distance between levels (...): ***11'***

Your drawing looks like the figure at right.

11. Save the drawing as *MyPiperack* in the C:\Steps3D\Lesson06 folder, and then exit.

    Command: ***saveas***

Do This: 6.6B	Arraying Objects in Z-Space – Polar Arrays

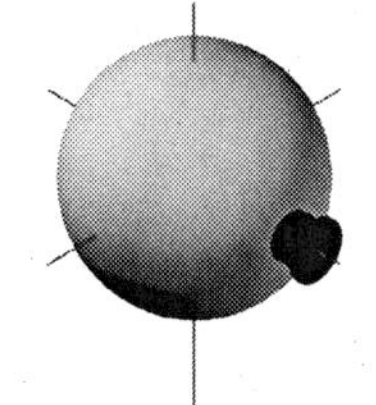

I. Now we'll use the **Polar** option of the ***3DArray*** command. Open the *3darray-polar* file in the C:\Steps3D\Lesson06 folder. The drawing looks like the figure at right.

II. Set the **Intersection** and **Endpoint** running OSNAPs. Clear all other settings.

III. Freeze the **obj2** layer to temporarily remove the sphere.

IV. Follow these steps.

6.6B: Arraying Objects in Z-Space – Polar Arrays

1. Enter the ***3DArray*** command.

   Command: ***3a***

2. Select the nozzle.

   Select objects:

   Select objects: ***[ENTER]***

3. Tell AutoCAD you wish to create a **Polar** array Polar.

   Enter the type of array [Rectangular/Polar] <R>: ***p***

4. We'll create four copies of the nozzle and fill a full circle.

   Enter the number of items in the array: ***4***

   Specify the angle to fill (+=ccw, -=cw) <360>: ***[ENTER]***

5. We do want to rotate the nozzles as they're copied.

   Rotate arrayed objects? [Yes/No] <Y>: ***[ENTER]***

6. Select the intersection of the guidelines as the **center point of array**.

   Specify center point of array:

7. Pick the rightmost endpoint of the FRONT-BACK horizontal line (the one running from front to back).

   Specify second point on axis of rotation:

Your drawing looks like the figure at right.

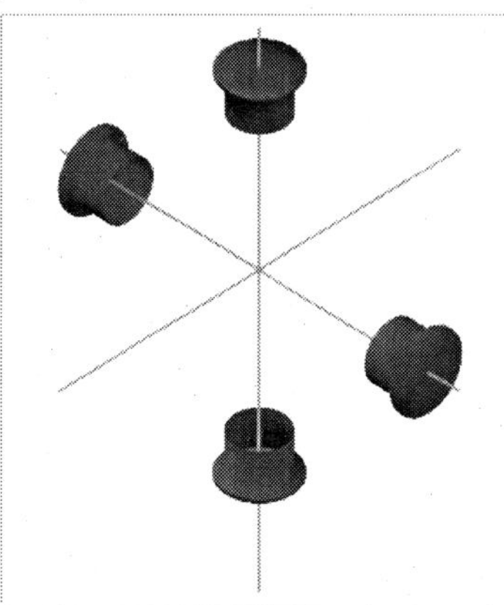

8. Repeat Steps 1 through 7, but this time select an endpoint on the vertical line in Step 7.

Your drawing looks the figure at right.

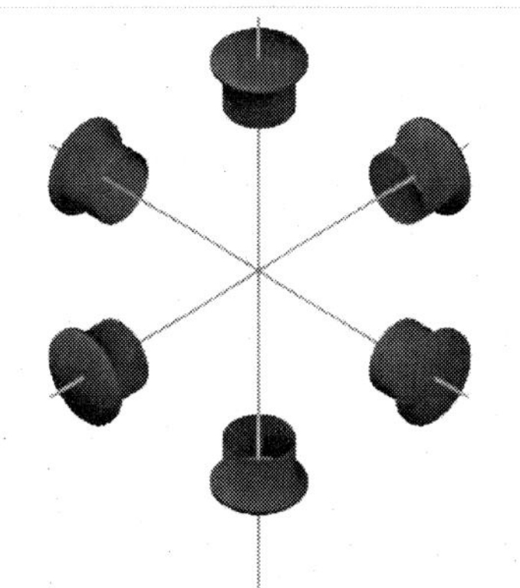

6.6B: Arraying Objects in Z-Space – Polar Arrays

9. Thaw the **obj2** layer and freeze the **marker** layer. Your drawing looks like the figure at right.

10. Save the drawing as *Weird Vessel* in the C:\Steps3D\Lesson06 folder, and then exit.

    Command: ***saveas***

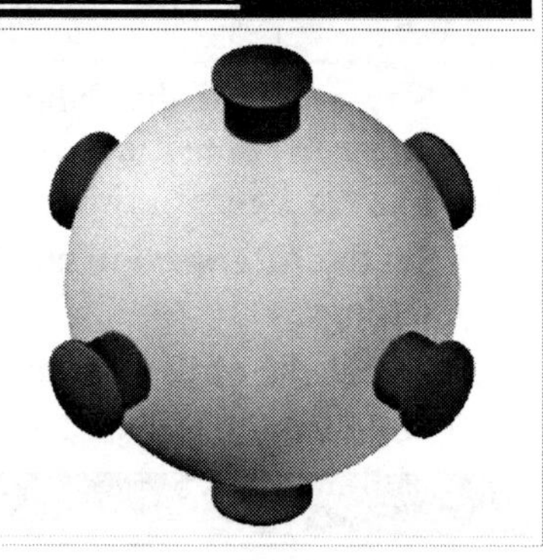

6.7 Aligning Three-Dimensional Objects

When you aligned objects in the basic text, you used two source points and two destination points. The main difference between that two-dimensional exercise and aligning three-dimensional objects is that AutoCAD requires three points of alignment for the three-dimensional object.

Another difference lies in your ability to scale the aligned objects as you did in the 2D exercise. You won't have that option with three-dimensional objects. But you can always use the ***Scale*** command once the objects are aligned.

Where to Find It:	
Command Line:	***3DAlign***
Hotkey(s):	***3al***
Ribbon (Tab/Panel):	Home – Modify – **3D Align**
Menu:	Modify – 3D Operations – **3D Align**
Toolbar:	Modeling – **3D Align**

You can use the ***Align*** command you already know, or you can use the newer ***3DAlign*** command. It looks like this:

Command: ***3dalign***

Select objects: ***[Select the object you wish to align to another object.]***

Select objects: ***[ENTER]***

Specify source plane and orientation ...

Specify base point or [Copy]: ***[You can select up to three points on the selected object to match three points on the target object.]***

Specify second point or [Continue] <C>:

Specify third point or [Continue] <C>:

Specify destination plane and orientation ...

Specify first destination point: ***[Select the points you wish to match to the points you selected in the last steps.]***

Specify second destination point or [eXit] <X>:

Specify third destination point or [eXit] <X>:

Let's perform a three-dimensional alignment using AutoCAD's ***3DAlign*** command.

Do This: 6.7A	Aligning Three-Dimensional Objects

I. Open the *align* file in the C:\Steps3D\Lesson06 folder. We'll align the eastern face of the wedge with the western face of the box.

II. Follow these steps. (Note: If you have trouble, change to the **2D Wireframe** visual style.)

6.7A: Aligning Three-Dimensional Objects

1. Enter the ***3DAlign*** command.
   Command: ***3al***
2. Select the wedge.
   Select objects:
   Select objects: ***[ENTER]***
3. Select the alignment points as indicated.
   Specify source plane and orientation ...
   Specify base point or [Copy]: ***[Point 1a]***
   Specify second point or [Continue] <C>: ***[Point 2a]***
   Specify third point or [Continue] <C>: ***[Point 3a]***

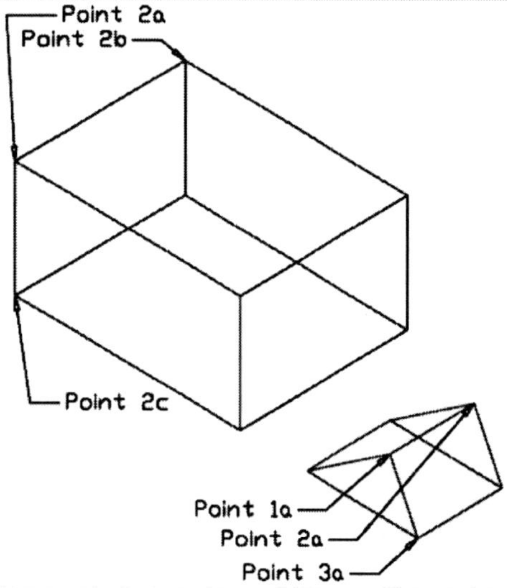

4. Now match the points with the destination points.
   Specify destination plane and orientation ...
   Specify first destination point: ***[Point 2a]***
   Specify second destination point or [eXit] <X>: ***[Point 2b]***
   Specify third destination point or [eXit] <X>: ***[Point 3c]***

   Your drawing looks like the figure at right. 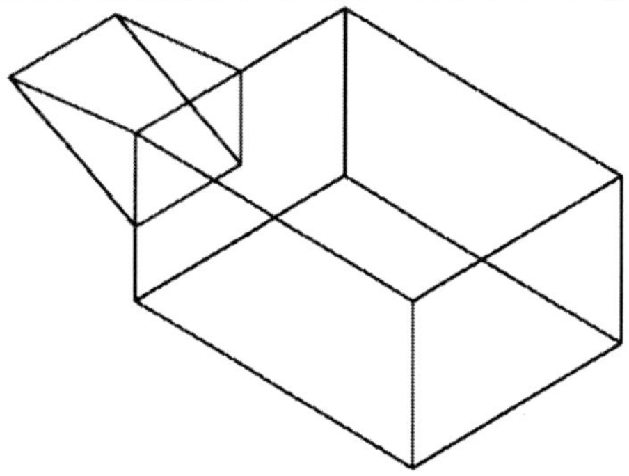
5. Exit the drawing without saving it.
   Command: ***quit***

I wish they were all that easy!

The really bright side to the ***Align*** command is that it doesn't care what types of objects are being aligned. You can align the objects you aligned in 2D space or you can align 3D faces, 3D meshes, regions, or solids!

6.8 Extra Steps

Much of the material you've covered in this lesson and will cover in Lesson 7 involves either new or improved tools introduced in R2010. It would be a good idea to investigate the older tools we used to edit meshes and 3D faces.

Open the LegacyMeshTools.pdf file found at our Files site (http://www.uneedcad.com/Files/LegacyMeshTools.pdf) and go through the lesson exercises found there.

6.9 What Have We Learned?

Items covered in this lesson include:

- *Modification tools used in Z-Space*
  - ***Trim***
  - ***Extend***
  - ***3DRotate***
  - ***Mirror3D***
  - ***3DMove***
  - ***3DArray***
  - ***3DAlign***
  - ***3DScale***

This has been a busy (and full) lesson, but you've learned so much!

When combined with your knowledge of wireframe and mesh modeling (and the tools in Lesson 7), these tools will enable you to create almost any structure you wish to draw. With some creative use of visual styles, you can produce professional-quality, colorful drawings of almost anything for any industry!

But what must you have that I can't provide?

PRACTICE ... PRACTICE ... PRACTICE!

Remember: Only through practice does training become experience. And it's experience that creates successful, efficient, economical, and sound designs; and it's experience that earns top dollar!

So repeat any lesson as needed for the proper training, and then work through the exercises at the end of the lesson for experience.

Our next lesson concludes the wonderful world of Mesh Modeling.

6.10 Exercises

1. Using the *Star-Root* file in the C:\Steps3D\Lesson06 folder, create the star drawing we used in Exercise 6.4A, p.135. (Hint: Grips make this exercise much easier.)
2. Create the *Three-Dimensional Chess Board.* Follow these guidelines:
   2.1. Each square is 1½".
   2.2. The boards are rotated at 15° increments.
   2.3. The post is 1" diameter.
   2.4. The frames are ½" wide x ¾" deep.
   2.5. The boards are 8" apart.
   2.6. Save the drawing as *My3DChess* in the C:\Steps3D\Lesson06 folder.

3D Chess Board

3. Create the *Sailboat* drawing shown at right. Follow these guidelines:
   3.1. This is a toy sailboat. The boat itself is 6 "x 2" x ¾".
   3.2. The keel is 3" below the bottom of the boat.
   3.3. The mast is 7" long x 1/8" diameter.
   3.4. The boom is 5" long x 1/8" diameter.
   3.5. Save the drawing as *MySBoat* in the C:\Steps3D\Lesson06 folder.

4. Create the *Conveyor Belt* drawing shown below. Follow these guidelines:
   4.1. Draw only one shaft and one roller (use the ***Revsurf*** command for best results).
   4.2. Use the 3D commands from this lesson to arrange the guides and the rollers on the guides.
   4.3. Use splines and the ***Rulesurf*** command to create the belt.
   4.4. I used a **Surftab1** setting of **18** or the shaft and roller, and **Surftab1** setting of **100** and **Surftab2** setting of **36** when I created the belt.
   4.5. Save the drawing as *MyBelt* in the C:\Steps3D\Lesson06 folder.

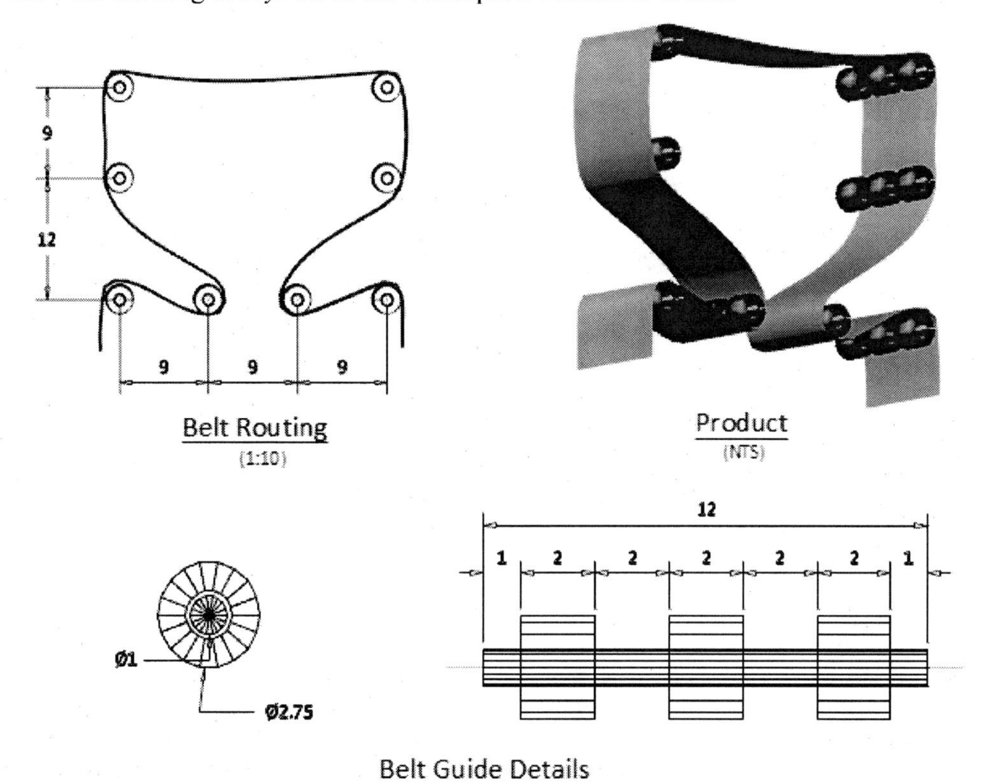

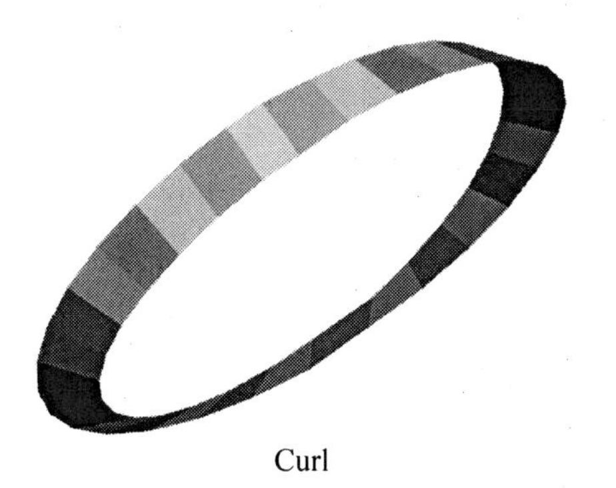
Curl

5. Here's another challenge! Create the *Curl* drawing shown here. Follow these guidelines:
   5.1. I started with a 1" line.
   5.2. There are 30 faces in all.
   5.3. The ring is ~4¼" diameter (I started with a line at 1,1 and arrayed it about point 4,4).
   5.4. I used three layers and two colors.
   5.5. The faces rotate 180°.
   5.6. Save the drawing as *MyCurl* in the C:\Steps3D\Lesson06 folder.

6. Starting with the *MyPiperack* file you created in Exercise 6.6A, p.139 (or the *Piperack* file if that one isn't available), create the piping drawing. Follow these guidelines:
   6.1. The tank has a 15' diameter and a 10' height. The top has a 3' pointed cone.
   6.2. Pipe is 12" diameter (12.75" ID, or outer diameter).
   6.3. Elbows are 18" from open face to centerline of bend.
   6.4. There's a 1/8" gasket between the flange and the nozzle at the tank.
   6.5. **Surftab1** and **Surftab2** values are 16.
   6.6. The dike wall around the tank is 2' high. The top of the wall is one mesh grid wide.
   6.7. I used ***Revsurf*** to create the elbows and a mesh cylinder to create the pipe.
   6.8. Save the drawing as *MyPipingPlan* in the C:\Steps3D\Lesson06 folder.

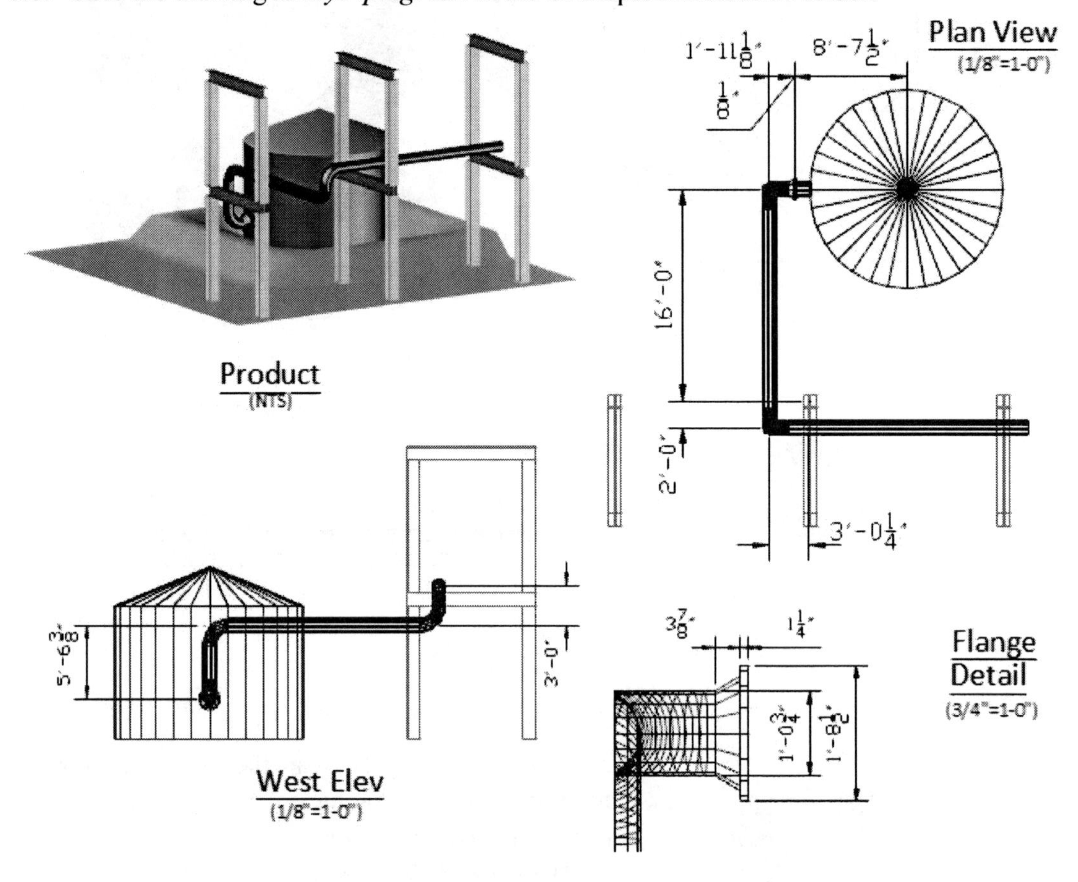

7. Create the *Propeller* drawing shown below. Follow these guidelines:
   7.1. The blade is a three-dimensional curve – use a spline as the arc and rise to the end of the upper line as shown in the *top blade detail*. Use as many vertices a you need – but I wouldn't use less than five.
   7.2. I used a **Surftabl** setting of **16** and **Surftab2** setting of **18** to create the blade.
   7.3. Once you've drawn the blade, turn it into a block. Insert the block into its proper place on the hub, but then explode it.
   7.4. Use the ***Rotate3d*** command to rotate the blade 105° on the hub.
   7.5. Use the 3D mesh editing tools you learned in this lesson to attach the ends of the blade to the hub.
   7.6. (Hint: The ***Stretch*** command works as well in Z-Space as it did in 3D space.)
   7.7. Save the drawing as *MyProp* in the C:\Steps3D\Lesson06 folder.

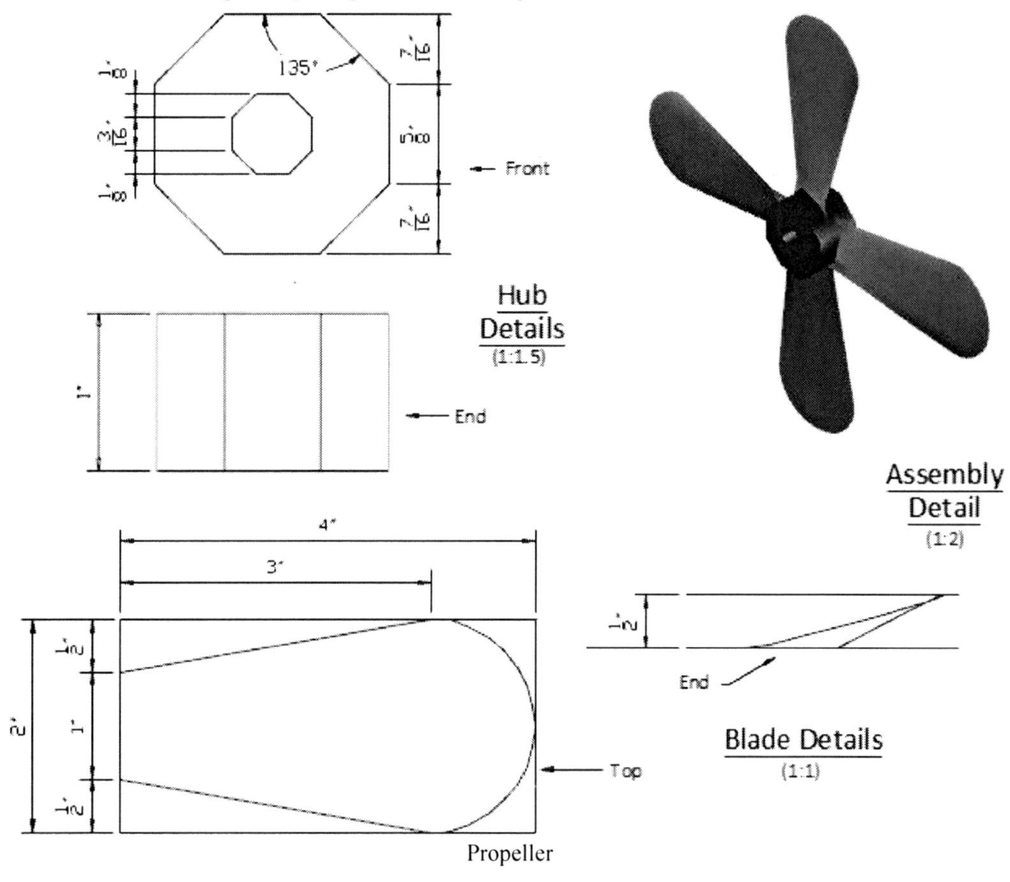

Propeller

6.11 For Web-Based Review Questions, visit: http://foragerpub.com/AcadFiles/2010/2010.htm

Lesson

Following this lesson, you will:

✓ *Know how to perform pre-smoothness mesh editing*

✓ *Know how to use subobject selection filters*

- ***Face***
- ***Edge***
- ***Vertex***

✓ *Know how to convert from one model type to another*

- ***ConvToSurface***
- ***ConvToSolid***
- ***MeshSmooth***

✓ *Know how to edit smooth meshes*

Mesh Editing

In today's world, you'll find quite a market for good mesh-creation and editing skills. Although CAD applications tout their solid modeling abilities over meshes, most 3D animation and games packages work exclusively with meshes. I've found, however, that creating a lot of my support objects in AutoCAD and exporting them into Poser, 3DS Max, or Lightwave saves a lot of time. Unfortunately, exporting a mesh or a solid is not always a flawless procedure.

It's a good idea, then, to get familiar with creating and editing mesh files here in AutoCAD. We'll even look at converting the files within AutoCAD before exporting them into one of the other packages.

7.1 Meshes, Surfaces, Solids – Which Do I Use?

By the time you finish this text, you'll have no problem answering this question. But that's getting ahead of ourselves. Let's look briefly at each before we continue.

- Collectively, surfaces (3dfaces) make up a mesh. You may have noticed a lack of page space spent in this text to editing surfaces; that's because there'll almost always be a better way to produce what you want without having to explode a mesh or, heaven forbid, create a 3d face! (If you just have to know how to edit a 3d face, review the archive document: http://www.uneedcad.com/Files/LegacyMeshTools.pdf, Section 6.1.2.)
- You probably found creating mesh primitives a very easy, almost fun task. Most people do. Creating the more complex meshes wasn't much more difficult. But to turn these primitives into something more elaborate can be tedious and time consuming. (And let's face it, some animation meshes can be *very* elaborate!) For the most part, you'll have to edit meshes on a face-by-face, edge-by-edge, or even vertex-by-vertex basis! (It's actually as tedious as it sounds!) Mesh editing, however, provides some smoothing tools that you just can't get anywhere else. (You can smooth a solid, but AutoCAD will turn it into a mesh first.) You'll spend this lesson studying methods to modify meshes.
- Creating solid primitives matches mesh primitives for simplicity (and fun). But you'll find editing a solid a real pleasure after your efforts toward mesh editing in this lesson.

Luckily, mesh editing shares some (well, a couple) of the solid editing tools. But more importantly, AutoCAD recognized the need to use solid editing tools sometimes and mesh editing tools other times – often on the same model! So, they make it possible to convert quickly and easily from one to the other. We'll look at those tools in Section 7.4 (p.164).

Let's get started.

7.2 Pre-Smoothness Editing

7.2.1 Subobject Selection & 3D Editing Tools

Editing a mesh means editing one or more parts of a web of edges, faces, and vertices. You'll move, scale, rotate, stretch, extrude … or more simply put, you'll modify one of these objects using one of the many modification tools you already know.

> To select an individual edge, face, or vertex, hold down the CTRL key on your keyboard as you select the object. Continue holding the CTRL key until you've selected all the edges, faces, or vertices you wish to modify.

You have a few ways to control, or filter, what you modify. These include a button on the **Subobject** panel of the ribbon's **Mesh Modeling** tab, a cursor menu selection, and a system variable setting (**SubObjSelectionMode**). Refer to the following chart.

FILTER	SUBOBJECT PANEL	CURSOR MENU	SUBOBJSELECTIONMODE SETTING
Edge		**Subobject Selection filter – Edge**	2
Face		**Subobject Selection filter – Face**	3
Vertex		**Subobject Selection filter – Vertex**	1

It'll be easier if we look at each piece of a mesh separately.

Modifying Mesh Faces

Modifying faces offers the fastest and most visible method for producing changes to your mesh. Use the 3D gizmos and/or commands to move, rotate, and scale a mesh face (or faces) for macro effects.

> AutoCAD allows you to change the gizmo without issuing a command. Select the gizmo of choice on the **Subobject** panel of the ribbon's **Mesh Modeling** tab or on the cursor menu when you right click on a gizmo. (See the cursor menu explanation in Lesson 6, Section 6.2, p.130.)

Let's jump right in!

Do This: 7.2.1A	**Modifying Mesh Faces**

I. Open the *Mesh1* file in the C:\Steps3D\Lesson07 folder. The drawing looks like the figure at right. (It's a simple mesh box with the default setup.)

II. Follow these steps.

7.2.1A: MODIFYING MESH FACES

1. On the **Subobject** panel of the ribbon's **Mesh Modeling** tab, select the **Face** tool (**Filter** flyout). Be sure the **Move Gizmo** appears in the same panel.

2. Hold down the CTRL key and select the top left faces indicated at right. AutoCAD indicates the selected faces with a grip, and it displays the **Move Gizmo**.

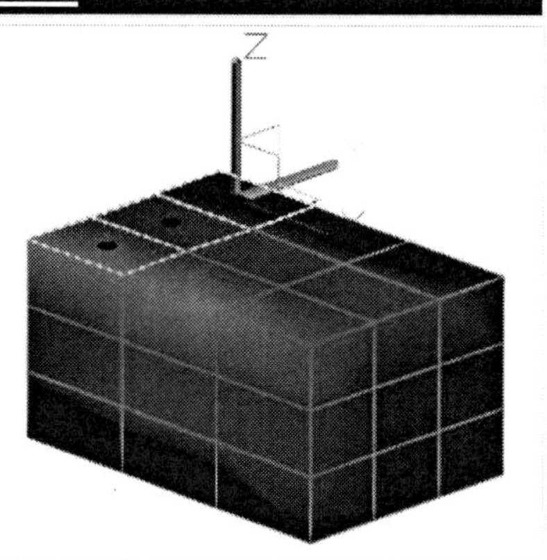

3. Pick the Z axis. [Note: You can pick and drag on a gizmo for an "about here" modification, or you can pick on the gizmo and release the mouse button. If you release the button, AutoCAD will begin a more precise grips procedure.]

4. AutoCAD begins the grips **Stretch** procedure (we'll use this to move the faces). Move the faces upward *1"*.

   **** STRETCH ****

   Specify stretch point or [Base point/Undo/eXit]: ***1***

Your drawing looks like the figure at right.

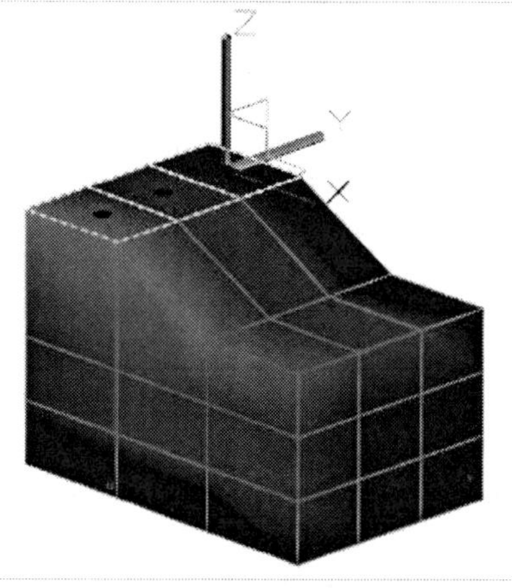

7.2.1A: Modifying Mesh Faces

5. Change to the **Scale Gizmo**. (Use either the cursor menu or the **Subobject** panel's gizmo button.)

6. Right click on the gizmo and select **Relocate Gizmo** [Relocate Gizmo] on the cursor menu.

7. Notice that the gizmo now follows your cursor. Pick the grip in the center face.
AutoCAD locates the gizmo here.

8. Pick between the web lines for the XY plane.

9. AutoCAD repeats the grips **Scale** procedure. We'll scale the faces by ½.

**** SCALE ****

Specify scale factor or [Base point/Undo/Reference/eXit]: ***.5***

10. Clear the gizmo [Esc]. Your drawing looks like the figure at right.

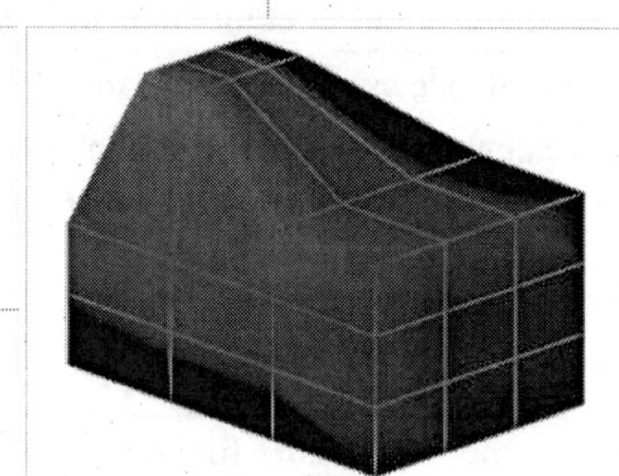

11. Change to the **Rotate Gizmo**.

12. Hold down the CTRL key and select the lower row of faces along the right side of the mesh.

13. Relocate the gizmo to the FRONT-RIGHT-BOTTOM endpoint of the mesh as indicated at right.

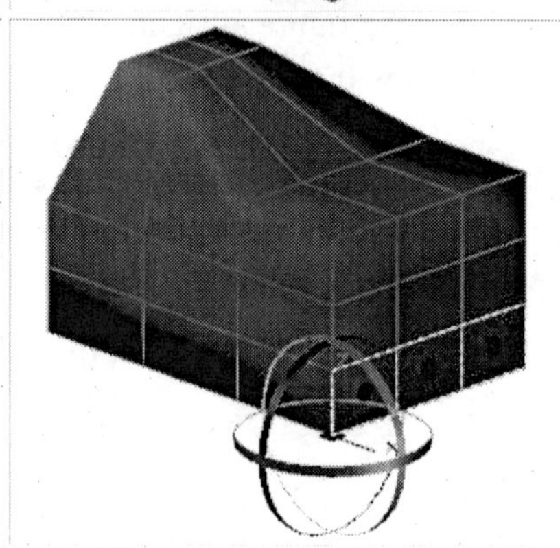

14. Pick the green, Y axis.

15. AutoCAD begins the grips **Rotate** procedure. Rotate the faces 30°.

**** ROTATE ****

Specify rotation angle or [Base point/Undo/Reference/eXit]: ***30***

16. Clear the gizmo [Esc]. Your drawing looks like the figure at right.

17. Save and close the drawing.

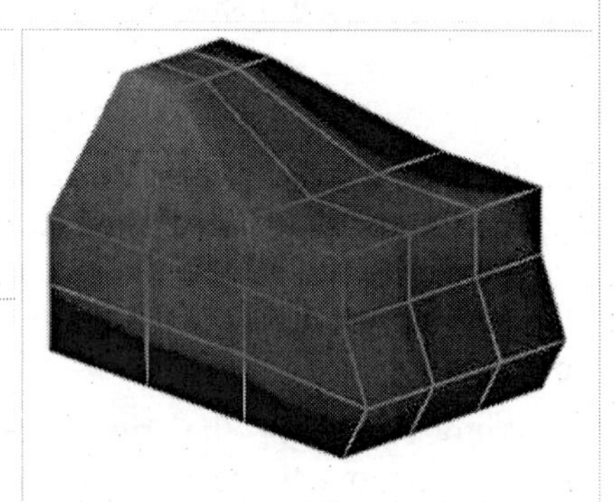

Modifying Mesh Edges

Edges, as you might expect, edge, or frame a face. As with faces, you can move, rotate, or scale an edge using 3D commands or gizmos. After faces, editing edges produces the fastest and most profound effect on your mesh. But modifying edges takes a bit more work than modifying faces (after all, each face has four separate edges … and many faces share edges!). Modify edges for more subtle changes.

Let's take a look.

Do This: 7.2.1B	Modifying Mesh Edges

III. Open the *Star* file in the C:\Steps3D\Lesson07 folder. The drawing looks like the figure at right. (It's a simple mesh cylinder with the following setup: Axis 16, Height 1, Base 2.)

IV. Let's use the mesh's edges to turn it into a three-dimensional star. Follow these steps.

7.2.1B: MODIFYING MESH EDGES

1. On the **Subobject** panel of the ribbon's **Mesh Modeling** tab, select the **Edge** tool (**Filter** flyout). Be sure the **Move Gizmo** appears in the same panel.

2. View the cylinder from the top.

3. Hold down the CTRL key while placing a selection window around the area indicated. You're trying to select only the vertical edge within the window.
AutoCAD selects the edge and displays the 3D Move Gizmo.

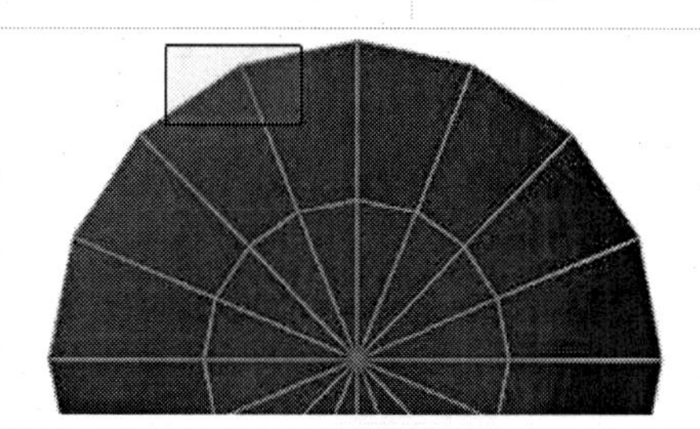

4. Place your cursor within the XY plane of the gizmo, pick and drag the edge inward to the intersection of the tessellations indicated at right.

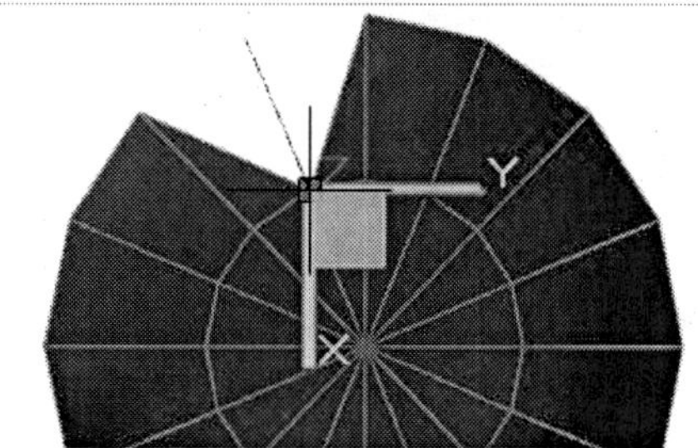

5. Repeat Steps 3 and 4 until half the cylinder looks like the figure at right.

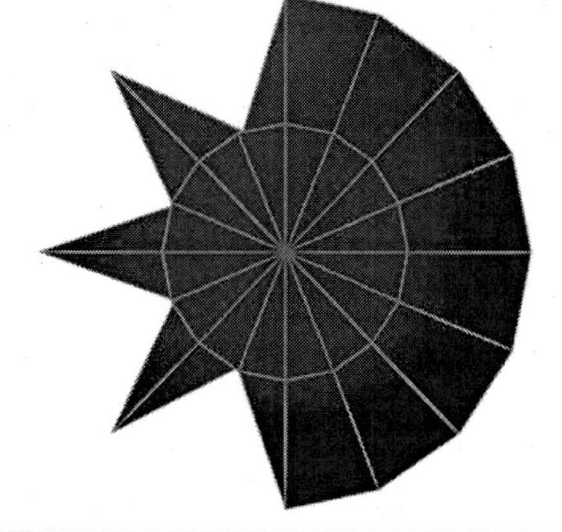

6. These steps should have given you some feel for moving edges. But there's a better way to achieve our star. Undo the changes, but stay in the top view.

7. Hold down the CTRL key while selecting every other edge on the perimeter of the cylinder. (Use windows as we've been doing.)

8. Use either the **Subobject** panel buttons or the cursor menu to change to the **Scale Gizmo**.

7.2.1B: Modifying Mesh Edges

9. Relocate the gizmo to the center of the cylinder – point 0,0,.0625. (To precisely relocate the gizmo, use the cursor menu and select **Relocate Gizmo** Relocate Gizmo. Once you've selected this option, just enter the coordinates at the command prompt.)

Command: *0,0,.0625*

10. Pick between the outer web lines on the gizmo. AutoCAD begins the grips **Scale** procedure. Set the scale to ***.5***.

**** SCALE ****

Specify scale factor or [Base point/Undo/Reference/eXit]: *.5*

11. Clear the gizmo [Esc].

Your star looks like the figure at right.

12. Now let's sharpen the tips of the star. Hold down the CTRL key and select each of the tips with a window.

13. Move the gizmo to the center of the cylinder as you did in Step 9.

14. With the edges still selected, change to the front view.

15. Pick the Z axis on the Scale Gizmo. AutoCAD again begins the grips **Scale** procedure. This time, set the scale to ***.1***.

**** SCALE ****

Specify scale factor or [Base point/Undo/Reference/eXit]: *.1*

16. Clear the gizmo and set the view to TOP-FRONT-RIGHT (use the view cube). (Adjust for a better view if you wish.)

Your star looks like the figure at right.

17. Let's add a bit of razzle-dazzle. Restore the top view and again select the edges at the tips of the star's arms.

18. Select the **Rotate Gizmo** either from the **Subobject** panel or the cursor menu.

19. Relocate the gizmo to the center of the star as you did in Step 9.

20. Pick the Z axis (the blue, outer ring of the gizmo).

AutoCAD begins the grips Rotate procedure.

21. Rotate the points of the star ***22.5°***.

**** ROTATE ****

Specify rotation angle or [Base point/Undo/Reference/eXit]: *22.5*

7.2.1B: MODIFYING MESH EDGES

22. Clear the gizmo and set the view to TOP-FRONT-RIGHT. (Adjust for a better view if you wish.) Your star looks like the figure at right.

23. Save, but don't close it yet.

How's that for a star? (I modeled it after the star on the tip of an old pair of spurs.)
Let's see what we can do with vertices.

Modifying Mesh Vertices

I'd save the vertex work for the really fine work on your model. Use them as final locators for specific points or to do some last minute touch up work.

> Obviously, rotating or scaling a specific vertex won't have any discernable effect on your model, but you can rotate and scale the space between multiple selected vertices.

Let's indent the center of our star.

Do This: 7.2.1C	Modifying Mesh Vertices

I. Be sure you're still in the *Star* file in the AutoCAD\Steps3D\Lesson07 folder. If not, please open it now.
II. Follow these steps.

7.2.1C: MODIFYING MESH VERTICES

1. On the **Subobject** panel of the ribbon's **Mesh Modeling** tab, select the **Vertex** tool (**Filter** flyout). Be sure the **Move Gizmo** appears in the same panel.
2. Hold down the CTRL key while select the center of the top of the star. AutoCAD displays the **Move Gizmo**.
3. Pick on the Z axis of the gizmo and move the vertex down .06".
4. Adjust your view and repeat Step 3 to move the bottom gizmo upward the same distance.
5. Restore the TOP-FRONT-RIGHT view. Your star looks like the figure at right.
6. Save the drawing.

If you have a couple minutes, undo the changes you made in the last exercise and try to accomplish the same task by scaling the distance between the vertices rather than moving them. Which was faster? Which was more precise?

7.2.2 Splitting an Face

You'll often find yourself coming up a bit short on faces; that is, occasionally, a single face won't provide the amount of detail you need for a project. You have a few things you can do to remedy this situation.

- You can start over – recreate the original mesh with more tessellation divisions. This may bring on groans and a desire to break something or say things your mother would frown upon. The biggest problem with this approach lies in the number of faces that inevitably result in places where you really don't need them. This can increase the amount of work required to modify an otherwise simple part of the mesh. In other words, you can't localize the increase in faces to a certain area.
- You can refine the mesh – as you'll see in Section 7.3.2 (p.161), you can localize the number of new faces to a selected area with the refine procedure, but you only have limited control over how many new faces this may produce (more broken objects and your mom washes your mouth out with soap!). Additionally, to refine a mesh, it must have a **Smoothness** property greater than zero (more on that in Section 7.3, p.160). You're generally better off not smoothing a mesh until you have the basic work completed.
- You can split an existing face – as Goldilocks said, "this one is juuust right" (for the simple fix).

Where to Find It:	
Command Line:	***MeshSplit***
Hotkey(s):	***split***
Ribbon (Tab/Panel):	Mesh Modeling – Mesh Edit – **Split Mesh Face**
Menu:	Modify – Mesh Editing – **Split Face**

Once you split a face, you can do anything to the new faces, edges, and vertices that you can do to the original mesh's faces, edges, and vertices. But more faces (and subsequent edges and vertices) provide opportunity for a more detailed model.

The procedure is simple enough:

> **Command:** ***meshsplit***
>
> **Select a mesh face to split:** ***[Be sure you filter for Face. You can only split one face at a time.]***
>
> **Specify first split point:** ***[Specify the points on one of the selected face's edge where you wish AutoCAD to locate your new edge. AutoCAD presents this icon when your cursor is located in a proper selection area (you can't select within the face itself). I suggest you use OSNAPs for more precision in locating your split points.]***
>
> **Specify second split point:** ***[Locate the other endpoint of the new edge.]***

You may have some difficulty using OSNAPs in this procedure (AutoCAD finds all the faces and edges regardless of whether or not you can see them), but practice will help.

Let's split a face or two.

Do This: 7.2.2A	Splitting a Mesh Face

I. Reopen the *Mesh1* file in the C:\Steps3D\Lesson07 folder.

II. Select the **Face** filter and the **Move** gizmo on the **Subobject** panel.

III. Follow these steps.

7.2.2A: Splitting a Mesh Face

1. Begin the ***MeshSplit*** command.

   Command: ***split***

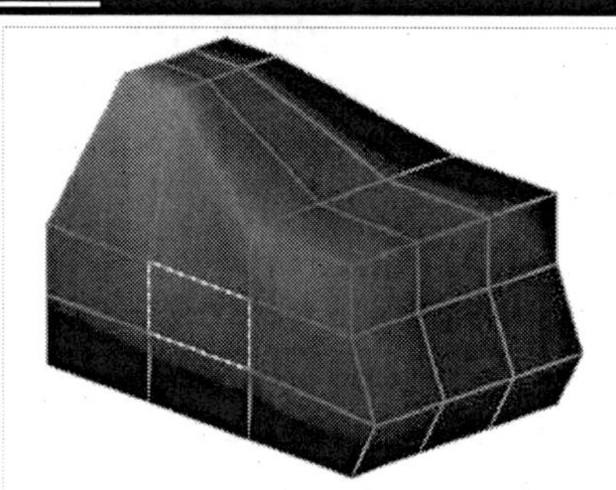

2. Select the center face on the front of the mesh as indicated.

   Select a mesh face to split:

3. Select the midpoint of the left and right vertical lines (this may be easier in a 3D Wireframe visual style).

   Specify first split point:

   Specify second split point:

   Your mesh looks like the figure at right. Notice that you now have two faces where you had only one previously.

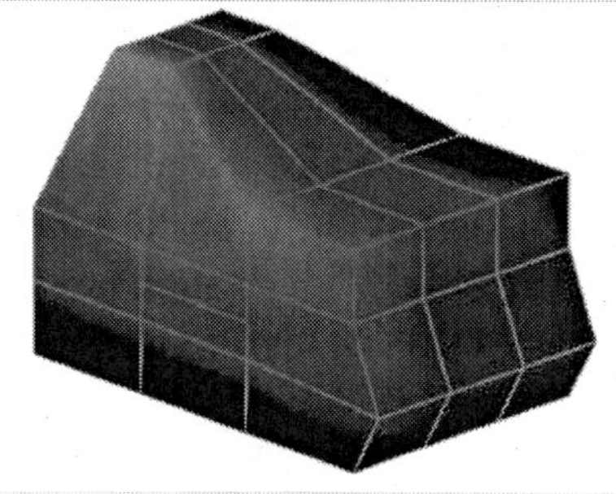

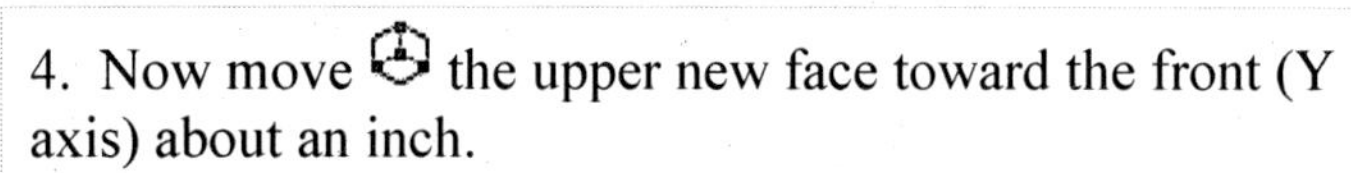

4. Now move the upper new face toward the front (Y axis) about an inch.

   The mesh looks like the figure at right.

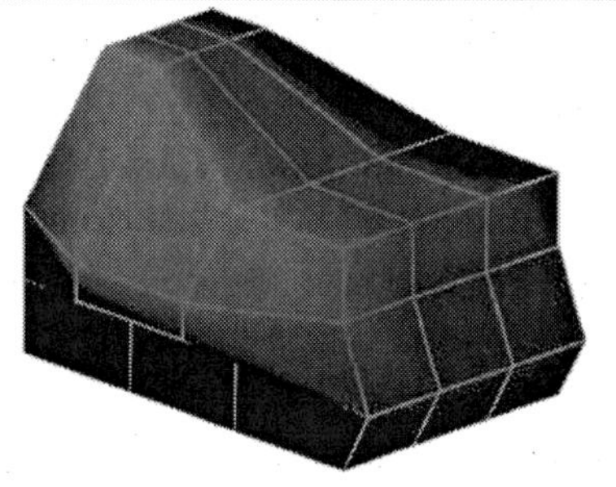

5. That's not very attractive. Isn't there a way, you might ask, to move just the face without having to distort the rest of the front?

   But of course! Undo the move, save the other changes, and let's look at *extruding* faces.

7.2.3 Extruding a Face

Meshes share the ***Extrude*** command with solids, which we begin studying in our next lesson. In fact, I robbed most of this section from the first solid lesson of the *One Step at a Time* text, '09 edition! So study it well now, it's a wonderful tool and one of the few that span the 3D modeling world's various incarnations.

One of the easiest ways to move a mesh face without distorting adjacent faces is simply to *extrude* the face. AutoCAD will move the face while adding new faces between the moved face and its adjacent faces.

The command sequence looks like this:

Command: ***extrude***

Current wire frame density: ISOLINES=4

Select objects to extrude: ***[Select the face(s) you want to extrude.]***

Where to Find It:	
Command Line:	***Extrude***
Hotkey(s):	***ext***
Ribbon (Tab/Panel):	Mesh Modeling – Mesh Edit – **Extrude Face**
Menu:	Draw – Modeling – **Extrude**
Toolbar:	Modeling – **Extrude**

Select objects to extrude: ***[Confirm the selection set]***

Specify height of extrusion or [Direction/Path/Taper angle] <5.0000>: ***[Tell AutoCAD how far you want to extrude the face or select one of the other options.]***

There aren't many options to confuse you, but what they can accomplish will astound you. Let's look at each line.

- AutoCAD first lets you know how many *isolines* it'll use to display the object. Isolines will be important when we work with solids; you can ignore them for now.
- The first option occurs right after the **Select objects** prompts. With the **Direction** option, you can pick two points to identify direction and length of the extrusion.
- Next, you can select an object that'll define the extrusion **Path**. The path object can be a line, arc, or 3DPoly. The results can be quite elaborate.
- Another option that can produce elaborate results is the **taper for extrusion** option. The default (**0**) produces a nice straight extrusion. An angle entry, however, can turn a box into a pyramid!

You'll find extrusions to be the handiest of tools when you work with solids, and they're handy for meshes as well. With meshes, however, you can get some startling results – especially when you smooth an extruded object. Keep that in mind when you prepare your mesh! (I'll show you what happens in our next exercise.)

We'll spend some time with extrude in this exercise.

Do This: 7.2.3A	Extruding a Mesh Face

I. Open the *Mesh2* file in the C:\Steps3D\Lesson07 folder. It's looks like the *Mesh1* file we completed in our last exercise, but it has some additional objects.

II. Follow these steps.

7.2.3A: EXTRUDING A MESH FACE

1. We'll start by creating a simple extrusion. Begin the ***Extrude*** command.

   Command: ***ext***

2. Hold down the CTRL key and select the upper face (half-face?) on the front of the mesh.

   Select objects to extrude:

   Select objects to extrude: ***[ENTER]***

3. Give the extrusion a height of 2".

   Specify height of extrusion or [Direction/Path/Taper angle] <1.0000>: ***2***

   Your mesh looks like the figure at right. Notice that AutoCAD has moved the selected face and added faces between it and the adjoining faces. (Treat these new faces the same as any other mesh faces.)

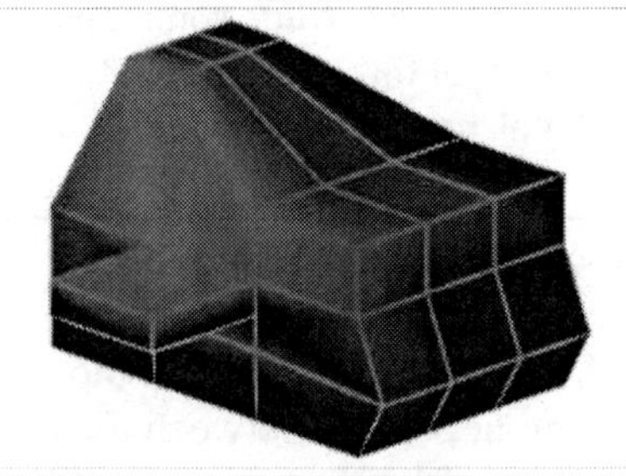

4. Thaw the **marker** layer and set the **obj2** layer current. (Notice the arc that appears in the front of the mesh.)

5. Begin an extrusion and select the face from which the arc appears.

   Command: ***ext***

   Current wire frame density: ISOLINES=4

   Select objects to extrude:

   Select objects to extrude: ***[ENTER]***

7.2.3A: Extruding a Mesh Face

6. This time, select the **Path** option …
    Specify height of extrusion or [Direction/Path/Taper angle] <2.0000>: ***p***

7. … and select the arc.
    Select extrusion path or [Taper angle]:

The extrusion looks like the figure at right. Notice the number of new faces AutoCAD has created; as with the simple extrusion, you can treat these new faces the same as any other mesh faces. Notice also that AutoCAD has extruded the new faces onto the original face's layer – not the current layer.

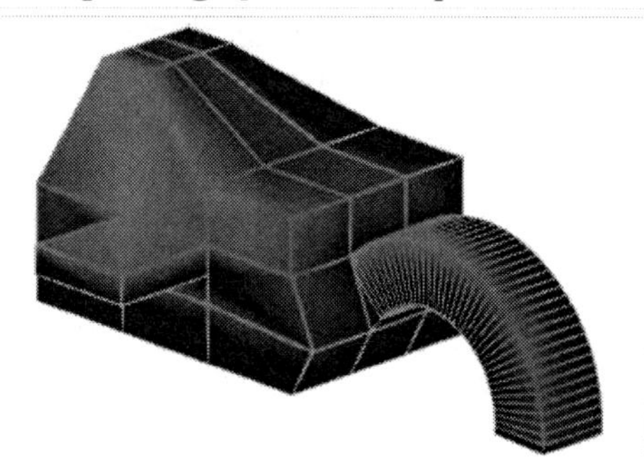

8. Repeat the ***Extrude*** command and select the center face on the raised top.
    Command: ***[ENTER]***

9. This time, use the **Taper angle** option.
    Specify height of extrusion or [Direction/Path/Taper angle] <2.0000>: ***t***

10. Give the extrusion a taper angle of 15° …
    Specify angle of taper for extrusion <0>: ***15***

11. … and make it 1".
    Specify height of extrusion or [Direction/Path/Taper angle] <2.0000>: ***1***

The extrusion looks like the figure at right.

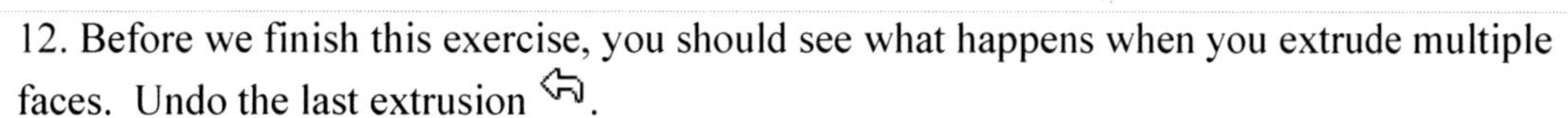

12. Before we finish this exercise, you should see what happens when you extrude multiple faces. Undo the last extrusion.
    Command: ***u***

15. Extrude the three adjoining faces along the raised top. Use a simple extrusion and make it 1".
    Command: ***ext***

Notice that AutoCAD appears to have created three different extrusions. In fact, it has! *This is a key difference between mesh and solid extrusions: extruding adjoining* solid *faces combine into a single extrusion!*

16. If you really want to see what's happened in the last extrusion, set the **Smoothness** property to **Level1**.

Notice the difference? Be aware of this when you extrude a face which you plan to eventually smooth! Examine the results on the other extrusions as well.

17. Close the drawing without saving it.

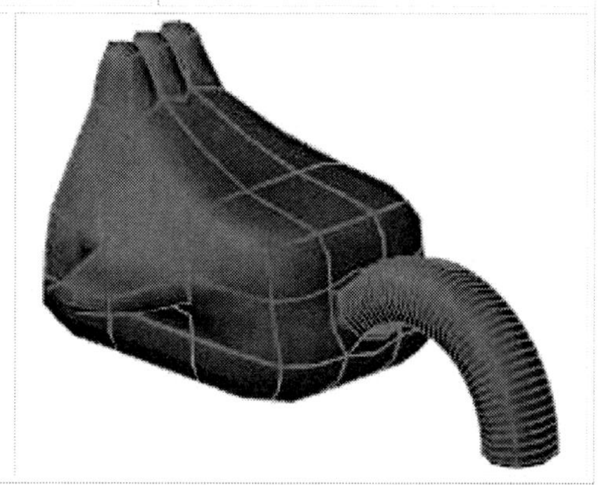

Most solid modelers consider ***Extrude*** a "can't-do-without-it" tool. (You will, too, once you get into solids!) We'll look at extrude again in Lesson 8, p.177.

7.3 The Smoothness Property

We've already experimented a bit with the **Smoothness** property – it's the property that turns a faceted (hard-edged) object into a smooth "free-form" (aka. *organic*) design. Some tools, such as ***MeshRefine***, won't work unless you have a smoothness level greater than zero. But unless you have to use one of these tools, it's best to hold setting a smoothness level until you've about finished modifying your mesh. Of course, you can set the smoothness level temporarily if you just have to see how your model is developing. (In fact, ***MeshSmoothMore*** and ***MeshSmoothLess*** discussed in Section 7.3.1 are both transparent commands!)

7.3.1 More or Less – The Smoothness Level

AutoCAD provides a couple approaches for setting a smoothness level: the Properties palette, which you've seen, and the ***MeshSmoothMore*** or ***MeshSmoothLess*** commands. Each prompts for a **mesh object to increase** [or **decrease**] **the smoothness level**.

AutoCAD allows four smoothness levels. See their effects on mesh primitives in Figure 7.002. As you can see, the roundness of the mesh increases with the level.

Where to Find It:	
Command Line:	***MeshSmoothMore*** or ***MeshSmoothLess***
Hotkey(s):	***more*** or ***less***
Ribbon (Tab/Panel):	Home – Mesh – **Smooth More** or **Smooth Less**
	Mesh Modeling – Mesh– **Smooth More** or **Smooth Less**
Menu:	Modify – Mesh Editing – **Smooth More** or **Smooth Less**
Toolbar:	Smooth Mesh – **Smooth More** or **Smooth Less**

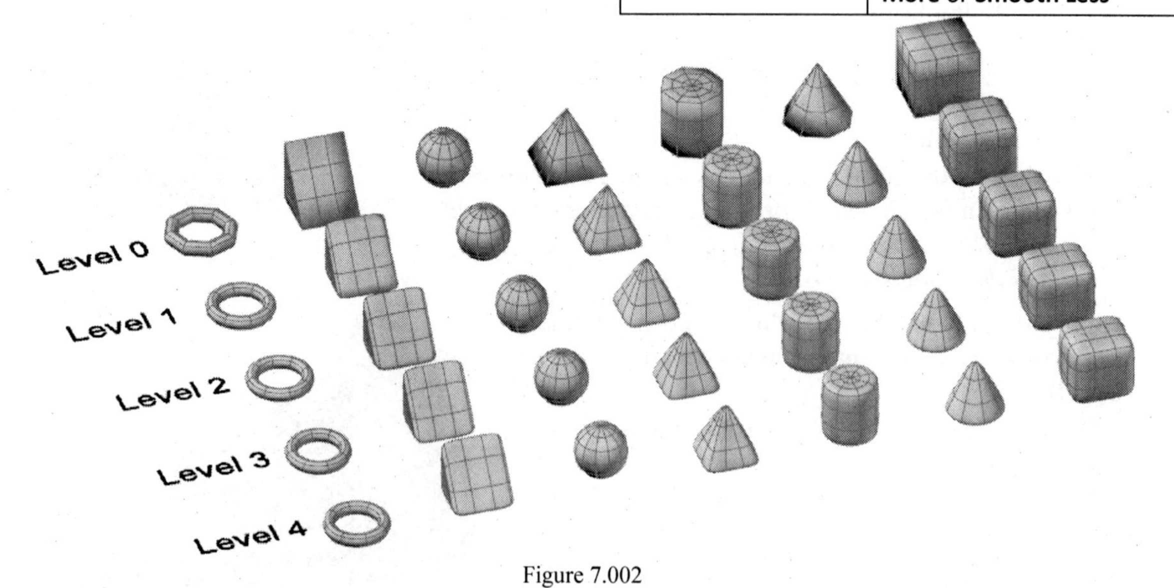

Figure 7.002

Smoothness works best on mesh primitives, but can have an odd effect even on them depending upon the level of smoothness you apply and the amount of modification they've undergone. But you can also crease a selected edge – selectively removing the smoothness in areas of the mesh.

7.3.2 Creasing a Mesh

Creasing a mesh removes the smoothness along selected edges. I've creased the top edges on the mesh box in Figure 7.003. The command prompts:

Where to Find It:	
Command Line:	***MeshCrease*** or ***MeshUnCrease***
Hotkey(s):	***crease*** or ***uncrease***
Ribbon (Tab/Panel):	Mesh Modeling – Mesh– **Add Crease** or **Remove Crease**
Menu:	Modify – Mesh Editing – **Crease** or **Uncrease**
Toolbar:	Smooth Mesh – **Mesh Crease** or **Mesh Uncrease**

Command: ***MeshCrease***

Select mesh subobjects to crease: ***[Select an edge, face, or vertex to crease. When you select an edge, AutoCAD creases just that edge. When you select a face or vertex, AutoCAD creases all four edges that form it.]***

Select mesh subobjects to crease: ***[Confirm the selection.]***

Specify crease value [Always] <Always>: ***[Accepting the default –*** **Always** ***– means that AutoCAD will crease the selection regardless of the smoothness level. Enter a value – 1 through 4 – means that AutoCAD will crease the selection only up to that level. In other words, if you enter a value of 2, AutoCAD will smooth the selection for levels 3 and 4 but not for levels 0, 1, and 2.]***

Figure 7.003

Removing a crease only involves selecting the crease to remove.

Command: ***MeshUncrease***

Select crease to remove: ***[Select the crease.]***

Select crease to remove: ***[Confirm the selection.]***

Let's smooth and crease an object.

Do This: 7.3.2A	**Smoothing and Creasing a Mesh**

I. Start a new drawing using the *Acad3D* template file.

II. Set the view to **Parallel**, create a 3x3x3 mesh box (size isn't important), and remove the grid.

III. Follow these steps.

7.3.2A: SMOOTHING AN D CREASING A MESH

1. Start the ***MeshSmoothMore*** command and select the box.

Command: ***more***

Select mesh objects to increase the smoothness level:

Your box looks like the figure at right. This is **Smoothness Level1**.

7.3.2A: Smoothing an d Creasing a Mesh

2. Increase the mesh to **Smoothness Level2**. You can repeat Step 1 or select the mesh and use the Properties palette.
Notice the difference?

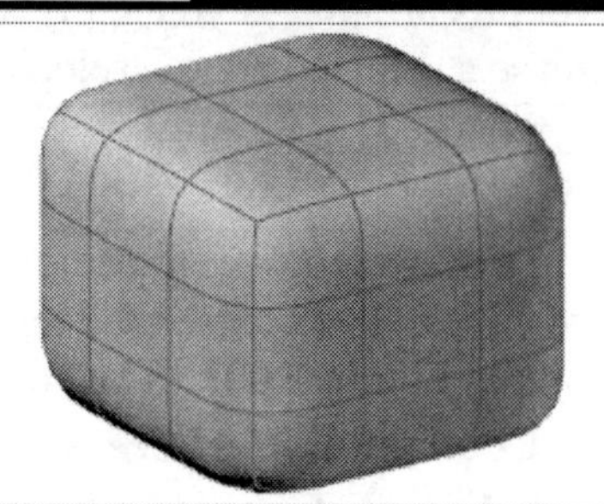

3. Set the **Subobject** filter to **Edge**.

4. Begin the ***MeshCrease*** command.
   Command: *crease*

5. Select the edges around the top of the mesh.
   Select mesh subobjects to crease:

6. Tell AutoCAD to crease the mesh only up to Level3.
   Specify crease value [Always] <Always>: *3*

The mesh looks like the figure at right.

7. AutoCAD creased the mesh because it has a smoothness property of **Level2**. Change it to **Level4**.
Notice the difference? AutoCAD smoothed the edges (although less than the rest of the box) because you told it to only crease up to **Level3**.

8. Reset the mesh to **Level2**.

9. Extrude the center face of the top of the mesh.
Notice (right) that the smoothness level effects the results of the extrusion.

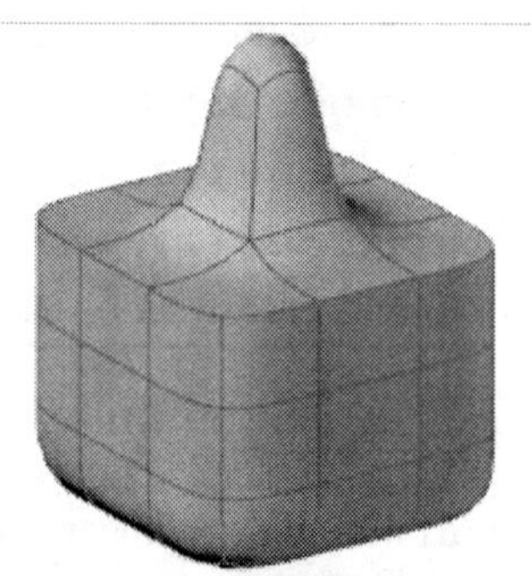

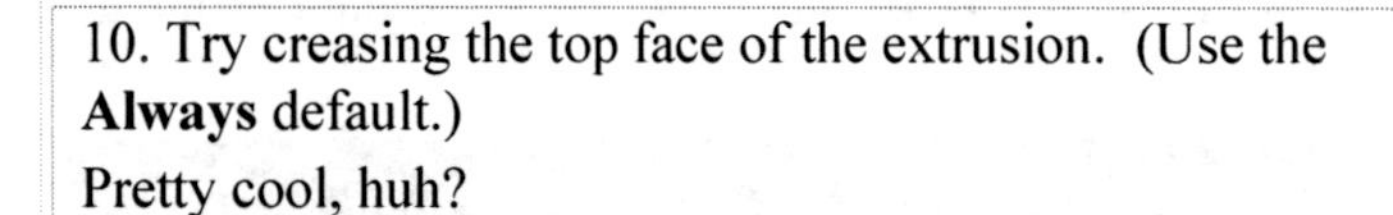

10. Try creasing the top face of the extrusion. (Use the **Always** default.)
Pretty cool, huh?

11. Save the drawing as *I'm So Smooth* in the C:\Steps3D\Lesson07 folder.

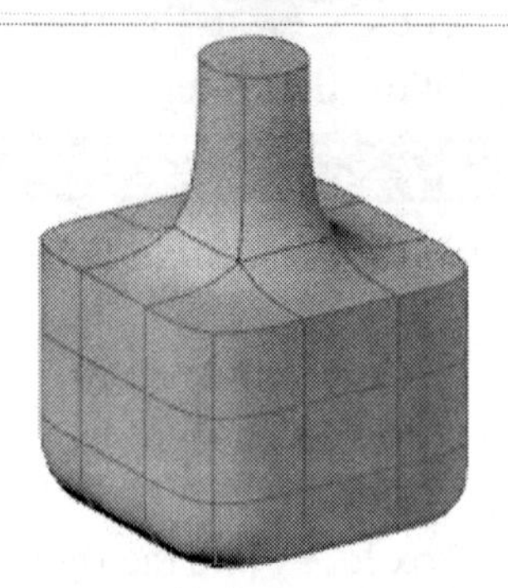

I'm sure you can see the potential of these cool tools! But there're still more to come!

7.3.3 Refining a Mesh

Refining a mesh adds faces. This can lead to a much finer model, but it can also greatly increase the complexity of the model … and the amount of work required to modify it. So here's a good rule to follow: *refine and add faces as needed, but make sure you need them before you add them.* A corollary to that rule should be: *only refine the part of the mesh that needs refining.* In other words, don't refine the entire mesh when refining a face will do.

Where to Find It:	
Command Line:	***MeshRefine***
Hotkey(s):	***refine***
Ribbon (Tab/Panel):	Home – Mesh – **Mesh Refine**
	Mesh Modeling – Mesh– **Refine Mesh**
Menu:	Modify – Mesh Editing – **Refine Mesh**
Toolbar:	Smooth Mesh – **Refine Mesh**

The ***MeshRefine*** command looks like this:

Command: *MeshRefine*

Select mesh object or face subobjects to refine: *[Select what you wish to refine.]*

Select mesh object or face subobjects to refine: *[Confirm the selection.]*

The number of faces AutoCAD adds to the mesh depends upon its level of smoothness. Use the chart to determine how many faces will appear in the place of a single face selection.

LEVEL	# OF FACES	LEVEL	# OF FACES
0	*	3	64
1	4	4	256
2	16		

As you can see, you really won't want to refine a face more than is absolutely necessary. But if you have to have a **Level4 Smoothness** property and you don't need 256 faces, change to a **Level2 Smoothness** property, refine the necessary face(s), and then reset the **Smoothness** property to **Level4**!

Take a look at how this works.

Do This: 7.3.2A	Refining a Mesh

I. Start a new drawing using the *Acad3D* template file.
II. Set the view to **Parallel**, create a 3x3x3 mesh box (size isn't important), and remove the grid.
III. Set the **Smoothness** property of the mesh to **Level1**.
IV. Follow these steps.

7.3.2A: Refining a Mesh

1. Begin the ***MeshRefine*** command.

Command: *refine*

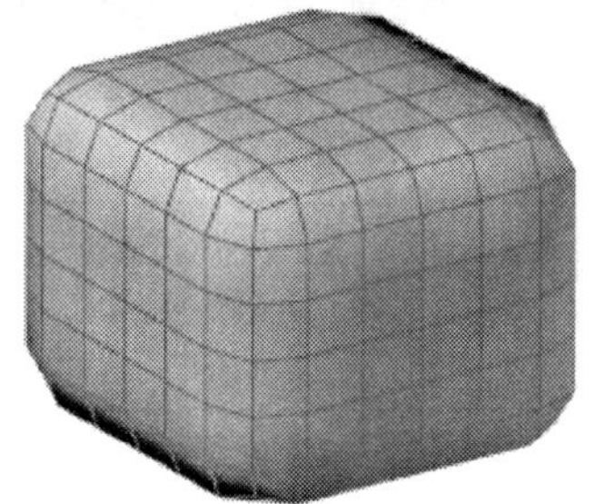

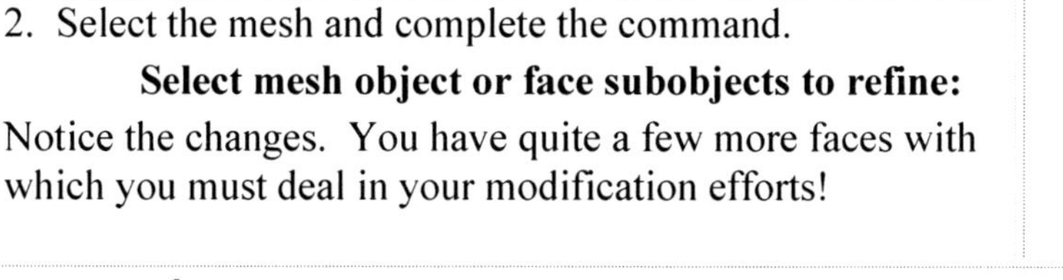

2. Select the mesh and complete the command.

Select mesh object or face subobjects to refine:

Notice the changes. You have quite a few more faces with which you must deal in your modification efforts!

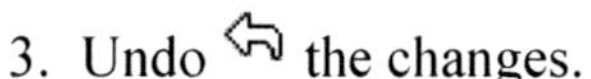

3. Undo the changes.

* You can't refine a **Level0** mesh.

7.3.2A: Refining a Mesh	
4. Repeat Steps 1 and 2 on a single face. Does that make your modifications look a bit easier? Notice that the edges of the faces appear distorted. You can set the **Smoothness** property to **Level0** to fix this while you work on the faces.	
5. Undo the changes. 6. Set the **Smoothness** property to **Level4** and repeat Step 4. Doesn't that make you glad you can refine just a single face rather than the entire mesh?! 7. Close the drawing without saving the changes.	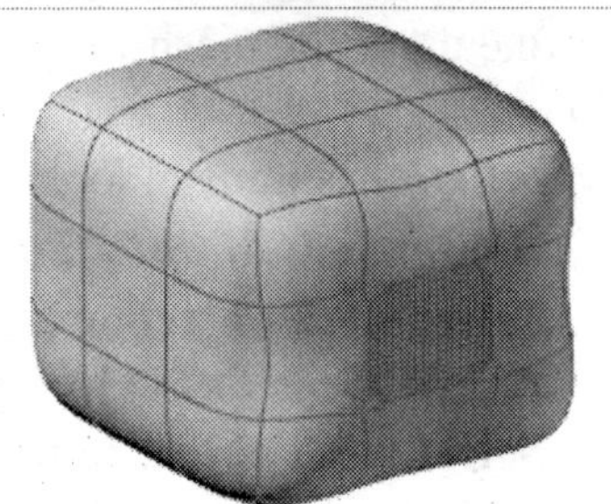

Remember, you can treat each new face just as you treat any other face on the mesh!

7.4 From One to Another – Model Conversion

After all this effort to conquer meshes, have you noticed anything lacking? Well, how about putting a hole in one? How would you do it?

The truth is, there isn't an easy way to put a hole in a mesh. You can, however, easily put a hole in a solid object. You may find it difficult, on the other hand, to extrude a single tessellated face on a solid.

So many model types … how does one choose?

Oh, it's not that bad. AutoCAD understands that you can do some things with a solid which you can do with a mesh … and vice versa. So it provides tools for converting your model from one to another.

We'll look (briefly) at four conversion tools, but the first two have a system variable that effects the results. We'll look at the results of the **SmoothMeshConvert** system variable setting as we proceed, but the following table of the tools you'll find on the ribbon's **Mesh Modeling** tab, **Convert Mesh** panel, will give you some idea of what to expect.

Ribbon Icon	Setting	Value	What It Creates
	Smooth, Optimized	0	This setting creates a smooth (rounded) model with as few faces as possible.
	Smooth, Not Optimized	1	This setting creates a smooth (rounded) model with as many faces as the original mesh possessed.
	Faceted, Optimized	2	This setting creates a sharp (non-rounded) model with as few faces as possible.
	Faceted, Not Optimized	3	This setting creates a sharp (non-rounded) model with as many faces as the original mesh possessed.

7.4.1 Converting Meshes and Solids to Surfaces

Use the ***ConvToSurface*** command to convert meshes, 2D solids (*not* 3D solids), regions, planar 3D faces, and open polylines, lines, and arcs with thickness to surfaces. You can then use surface tools to modify the surfaces.

You'll find that ***ConvToSurface*** presents some different results as you adjust the **SmoothMeshConvert** system variable.

Where to Find It:	
Command Line:	***ConvToSurface***
Ribbon (Tab/Panel):	Mesh Modeling – Convert Mesh– **Convert to Surface**
Menu:	Modify – 3D Operations – **Convert to Surface**

Samples of a mesh converted to a surface appear in the following table.

SMOOTH, OPTIMIZED	SMOOTH, NOT OPTIMIZED	FACETED, OPTIMIZED	FACETED, NOT OPTIMIZED

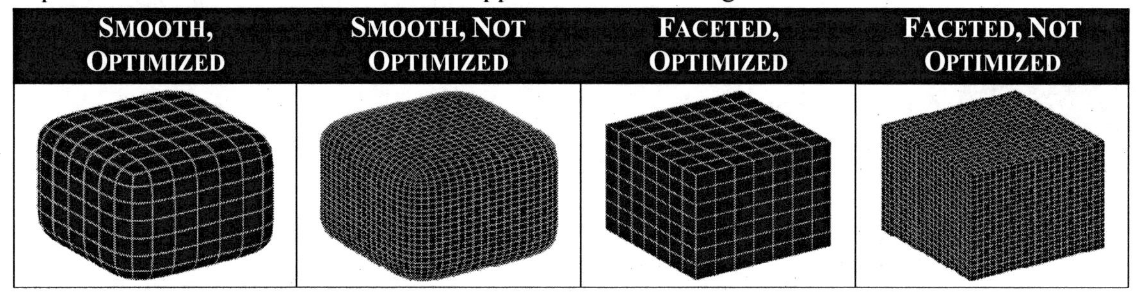

We haven't discussed surface modification tools in this text because they just don't hold a candle to meshes. You can, however, get familiar with them by studying the supplement: http://www.uneedcad.com/Files/LegacyMeshTools.pdf. Remember, a surface model is a collection of 3D faces.

7.4.2 Convert to Solid

The procedure for converting to a solid resembles that of converting to a surface. Use the ***ConvToSolid*** command to convert meshes, closed polylines (with thickness), and closed surfaces. Note that you can't convert a polyline with non-uniform or zero width to a solid.

Samples of a mesh converted to a solid appear in the following table.

Where to Find It:	
Command Line:	***ConvToSolid***
Ribbon (Tab/Panel):	Mesh Modeling – Convert Mesh– **Convert to Solid**
Menu:	Modify – 3D Operations – **Convert to Solid**

SMOOTH, OPTIMIZED	SMOOTH, NOT OPTIMIZED	FACETED, OPTIMIZED	FACETED, NOT OPTIMIZED

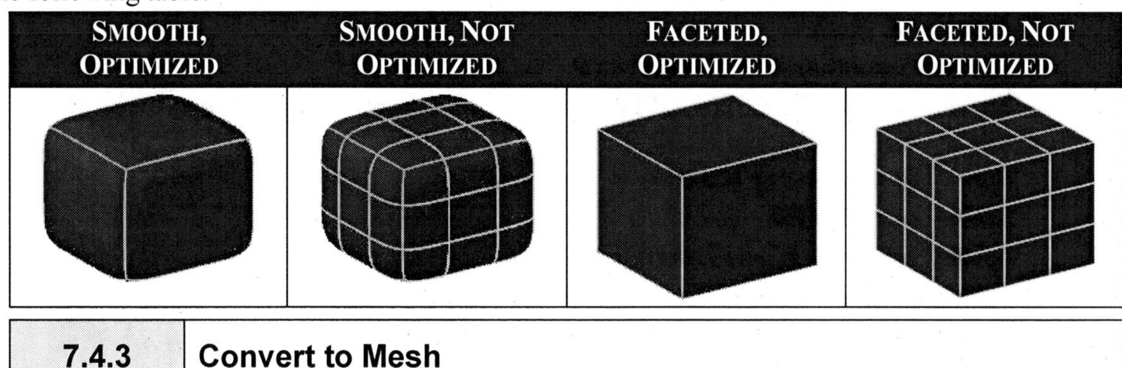

7.4.3 Convert to Mesh

Converting a solid to a mesh involves a different procedure from the previous two we've discussed. (Unfortunately, AutoCAD hasn't graced us with a ***ConvToMesh*** command.) Further, the command – ***MeshSmooth*** – isn't affected by the **SmoothMeshConvert** system variable!

The ***MeshSmooth*** command proceeds like this:

Command: *MeshSmooth*

Select objects to convert: *[Select the objects you want to convert to a mesh.]*

Select objects to convert: *[Confirm the selection.]*

You can convert polygon meshes, surfaces, and solids to meshes. Once AutoCAD has completed the conversion, treat the new mesh as you would any other mesh.

Where to Find It:	
Command Line:	***MeshSmooth***
Ribbon (Tab/Panel):	Home – Mesh – **Smooth Object**
	Mesh Modeling –Mesh– **Smooth Object**
Menu:	Modify – 3D Operations – **Convert to Solid**

Some of the mesh editing commands (such as ***MeshRefine*** and ***MeshSmoothMore/MeshSmoothLess***) will offer to convert a selected object should you select something that isn't already a mesh.

7.5 A Mesh Project

We've just about covered meshes, but let's try a putting-it-all-together exercise before we move on to solids.

We're going to create a timing light. (For those of you who've never worked on older cars, it's a tool used to set the timing for when your engine's spark plugs fire.) It looks a bit like a ray gun you might see on an old Buck Rogers flick. (Good golly! This guy must be older than space dust!)

Do This: 7.5A **A Mesh Project**

I. Open the *Timing Light Setup* file in the C:\Steps3D\Lesson07 folder.

II. Set the **Mesh Box Tessellation Divisions** to: **Length** = 3, **Width** = 3, and **Height** = 4.

III. Turn Ortho on.

IV. Follow these steps.

7.5A: A MESH PROJECT

1. Turn the UCS **90°** on the Y axis.

Command: *ucs*

2. Create a mesh box centered at 0,0,0. Give it a length of 2", width of 1.75", and height of 9.25".

Command: *mesh*

Current smoothness level is set to : 0

Enter an option [Box/Cone/CYlinder/Pyramid/Sphere/Wedge/Torus/SEttings] <Box>: *b*

Specify first corner or [Center]: *c*

Specify center: *0,0,0*

Specify corner or [Cube/Length]: *l*

Specify length <1.0000>: *2*

Specify width <1.0000>: *1.75*

Specify height or [2Point] <1.000>: *9.25*

3. Reset the UCS to **World** and adjust your view so you can better see the mesh. (I like to turn off the UCS Icon, but do what makes you comfortable.)

Command: *ucs*

7.5A: A Mesh Project

4. Set the front view current.

5. Set the **Subobject** filters to select a **Face** and use the **Rotate Gizmo** .

6. Hold down the CTRL key and put a window around the right end of the mesh. AutoCAD displays the Rotate Gizmo, but it's in the wrong place.

7. Use the endpoint OSNAP to move the gizmo Relocate Gizmo to the upper right (visible) endpoint of the mesh.

8. Pick on the gizmo's (green) Y axis and tell AutoCAD to rotate the selected faces 15°.

   **** ROTATE ****

   Specify rotation angle or [Base point/Undo/Reference/eXit]: ***15***

9. Repeat Steps 5 through 8 on the other end of the mesh.

   Your mesh looks like the figure at right.

10. Set the **Subobject** filters to select an **Edge** and use the **Move Gizmo** .

11. Use the same windowing technique you used so far to select and move the edges of the mesh to meet the dimensional requirements shown at right.

3.75 2 2.25 1.25

12. Adjust to a TOP-FRONT-RIGHT view.

13. Crease the edges indicated here. (Be sure to crease the RIGHT-BOTTOM edges, too.)

    Command: ***crease***

    Select mesh subobjects to crease:

    12 objects found.

    Specify crease value [Always] <Always>: ***[ENTER]***

14. Adjust to a BOTTOM-FRONT-RIGHT view and set the **Subobject** filters to select a **Face** .

7.5A: A Mesh Project

15. ***Extrude*** the face indicated by 1/16".

    Command: _.EXTRUDE

    Current wire frame density: ISOLINES=41 found

    Specify height of extrusion or [Direction/Path/Taper angle] <9.2500>: *.0625*

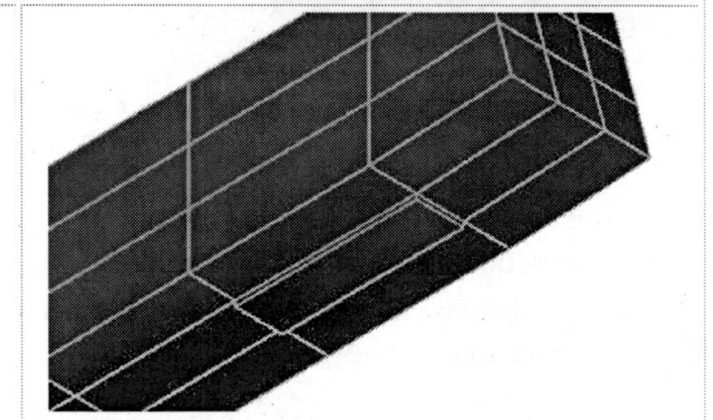

16. Now rotate the same face 15° – use the right edge as the pivot point.

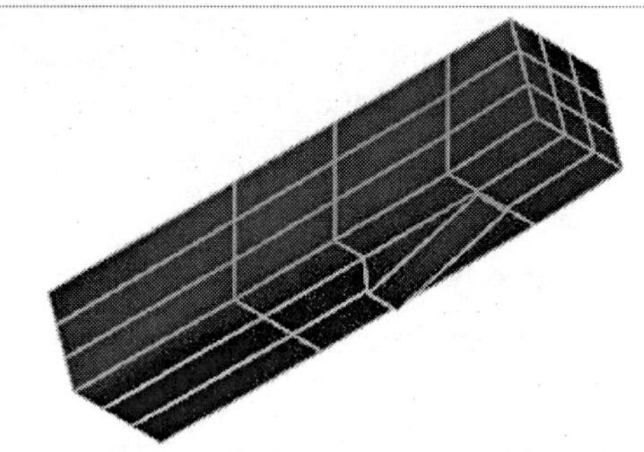

17. ***Extrude*** the same face 1.375" and again 2.5".

18. ***Crease*** **(Always)** the bottom face on the handle. Your mesh looks like the figure at right.

19. Don't forget to save occasionally.

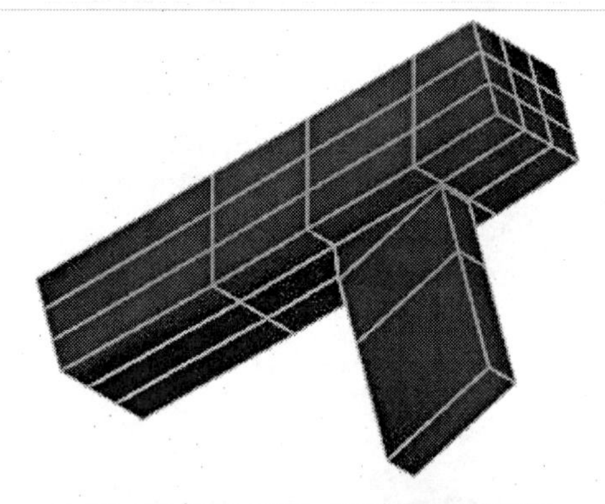

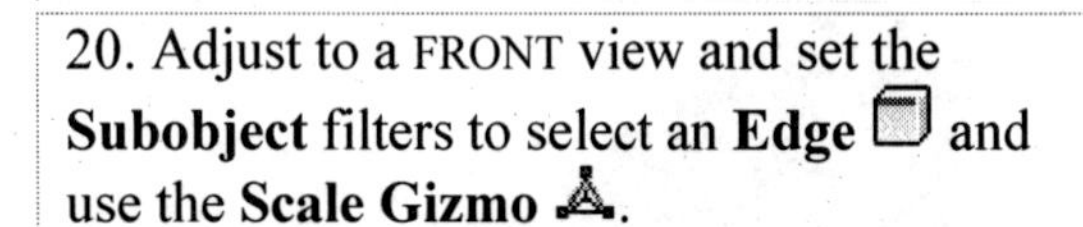

20. Adjust to a FRONT view and set the **Subobject** filters to select an **Edge** and use the **Scale Gizmo**.

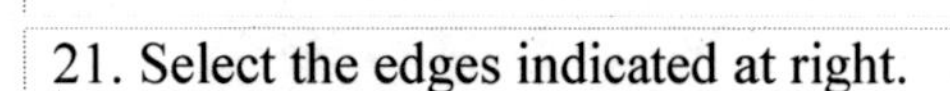

21. Select the edges indicated at right.

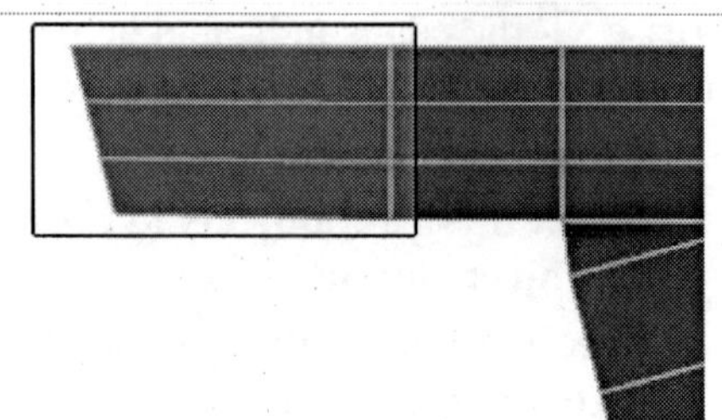

22. Adjust to a LEFT view and move the gizmo Relocate Gizmo to ***0,0,0***.

23. Scale the edges in the YZ plane to 80%.

    **** SCALE ****

    Specify scale factor or [Base point/Undo/Reference/eXit]: *.8*

24. Adjust to the TOP-LEFT-FRONT view.

24. Add creases to the perimeter edges of the front, and then set the **Smoothness** property to **Level4** for the mesh.

Your mesh looks like the figure at right. This is the general shape of your timing light, but we can still tweak it a bit. Let's add a trigger and shape the light lens.

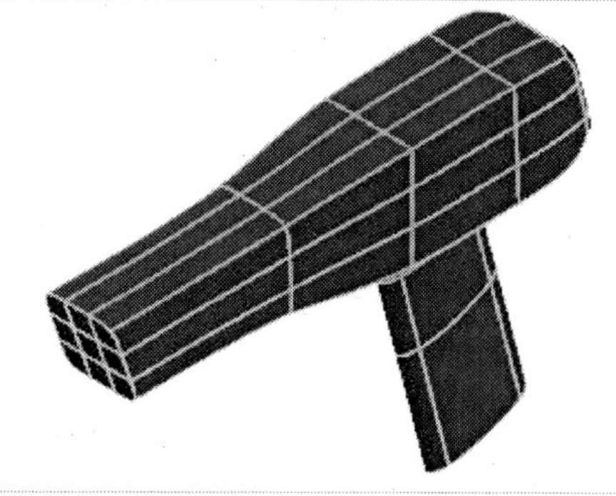

25. Reset the **Smoothness** property to **None**.

26. Adjust to the BOTTOM-LEFT-FRONT view and set the **Subobject** filters to select a **Face** and use the **Rotate Gizmo**.

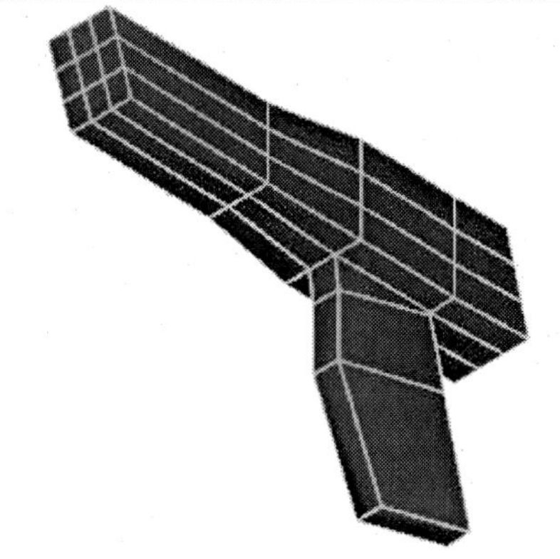

27. Rotate the middle front face of the handle by 15°.

**** ROTATE ****

Specify rotation angle or [Base point/Undo/Reference/eXit]: ***15***

28. Set the LEFT view current and adjust your view so you can see this face more closely.

29. Set the **UCS** to **View**.

Command: ***ucs***

30. On layer **obj2**, trace the tessellation division lines forming the front of the handle, then divide these lines into three parts. (Set the **PDMode** to ***3*** or a setting that will enable you to better see the nodes.) Erase the lines and leave the nodes.

31. Using the nodes you created in Step 30, split the mesh face into three faces, as indicated.

Command: ***split***

Your mesh looks like the figure at right. (Freeze the **obj2** layer and set **obj3** current.)

32. Restore the World UCS, adjust to the BOTTOM-LEFT-FRONT view, and set the **Subobject** filters to select a **Face**.

7.5A: A MESH PROJECT

33. Extrude the new center face ½", and ***Crease*** **(Always)** the edges along its base.

34. Split the front of the trigger along its midpoints, from front to back.

35. Move the new edge to the right ¼".

36. Now for some razzle dazzle; select all the faces on the trigger (use a crossing window in the FRONT view). Then change the **Color** property to **Black**.

The mesh looks like the figure at right. (Just the lens to fix now!)

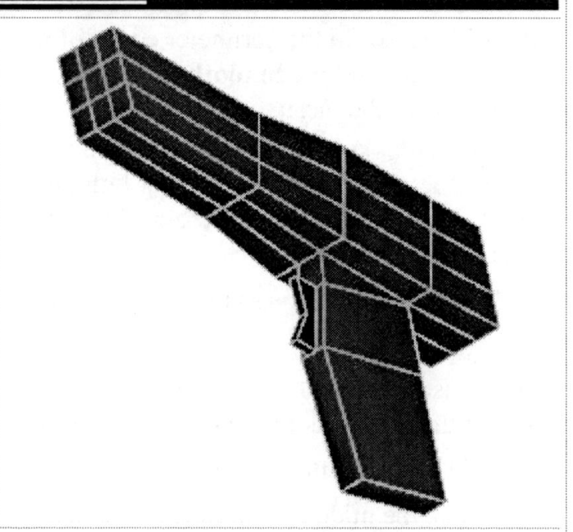

37. Adjust to the LEFT view, and set the **Subobject** filters to select a **Face**. Set the **Smoothness** property to **Level1**.

38. Scale the middle face by a factor of ***1.5***. (Be sure your gizmo is located in the center of the face and you scale along the YZ plane.)

39. Select all the outer faces within the "nozzle" of the light.

40. Refine the selected faces.

Command: *refine*

41. Select the faces that encircle the center face and move them to the right (along the X axis) ½".

42. Now move the center face to the left (along the X axis) 3/8".

43. Extrude the center face another 1/8".

44. Finally, scale the center face outward about 1.25x.

45. Now select all the faces involved in the lens (the center face and those formed by the last extrusion.), and change their color to **Cyan**.

46. Change the **Smoothness** property to **Level4**.

Your timing light looks like the figure at right.

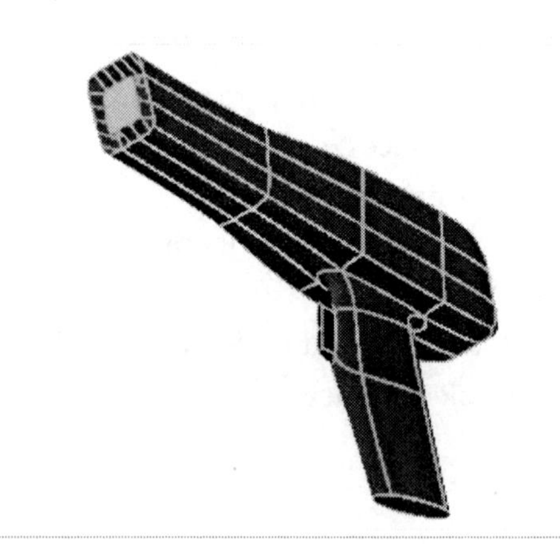

47. Save and close the drawing.

As my son would say, "Dude! That was so cool!"

7.6	Extra Steps (The XEdges Command)

This command doesn't actually convert objects to a wireframe, but it does create a wireframe defined by the edges of a 3D model (mesh, solid, region, surface, or even subobjects).

The sequence is one of the easy ones.

> **Command:** ***xedges***
>
> **Select objects:** ***[Select a mesh, solid, region, surface or subobject]***
>
> **Select objects:** ***[Confirm the selection.]***

Let's see this one in action.

Where to Find It:	
Command Line:	***XEdges***
Ribbon (Tab/Panel):	Home – Solid Editing – (Extract Edges flyout) **Extract Edges**
	Mesh Modeling –Section (subpanel) – **Extract Edges**
Menu:	Modify – 3D Operations – **Extract Edges**

Do This: 7.4.4A	Creating a Wireframe from a Mesh

I. Reopen the *Mesh1* file in the C:\Steps3D\Lesson07 folder.

II. Set the **obj1** layer current.

III. Follow these steps.

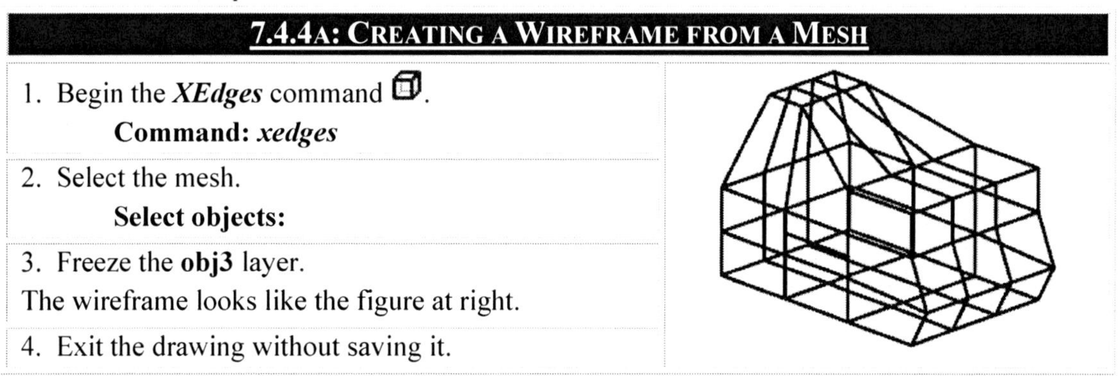

7.4.4A: Creating a Wireframe from a Mesh	
1. Begin the ***XEdges*** command. **Command:** ***xedges***	
2. Select the mesh. **Select objects:**	
3. Freeze the **obj3** layer. The wireframe looks like the figure at right.	
4. Exit the drawing without saving it.	

Try doing a list on one of the wireframe lines. Notice that they're all polylines (not lwpolylines, but the real McCoy!).

7.7	What Have We Learned?

Items covered in this lesson include:

- *Pre-Smoothness Editing Tools*
- *Mesh Editing*
- *Subobject Selection Filters*
  - ***Face***
  - ***Edge***
  - ***Vertex***
- *Commands and System Variables*
  - ***SubObjSelectionMode***
  - ***MeshSplit***
  - ***Extrude***
  - ***MeshSmoothMore***
  - ***MeshSmoothLess***
  - ***MeshCrease***
  - ***MeshUnCrease***
  - ***MeshRefine***
  - ***SmoothMeshConvert***
  - ***ConvToSurface***
  - ***ConvToSolid***
  - ***MeshSmooth***
  - ***Xedges***
- *Model Conversion*

Okay, we've wrapped up our studies of meshes. What do you think?

You'll get a chance to practice what you know about meshes in the Exercises section that follows – and practice is a good thing (and a requirement for true mastery of these complex models). But let me leave meshes with a couple observations and maybe a hint or two.

- A great many objects in the modeling world utilize meshes as their construction base. You'll find this truer in the game and animation worlds than in construction, but how many of us don't have a cartoon or two in our hidden dreams?
- With the introduction of "organic" or "free-form" mesh modeling, AutoCAD has created a bridge between the construction, animation, and game worlds. It's a good idea to get comfortable with the tools.
- You'll soon discover that solid modeling provides many tools that you don't have in mesh modeling. (The opposite is also true!) In most respects, frankly, solid modeling is easier. I know, that's an opinion, but I've used both. Still, you can't "smooth" a solid the way you can smooth a mesh, so remember those conversion tools!
- If you have the opportunity, open up a native file in 3DS Max, Poser, Maya, or Lightwave. Notice that the models in each are meshes (*not* solids).

Get some practice, and then move on to solids!

7.8 Exercises

1. Using the *MeshIsoSetup* template in the C:\Steps3D\Lesson07 folder, create the *Battery* drawing shown at right. Follow these guidelines:
   1.1. I started with a mesh cylinder: Axis = 8, Height = 2, Base = 5.
   1.2. Use the dimensions as a guide.
   1.3. Crease the outer, upper and lower edges.
   1.4. Crease the first circular edge in from the outside (upper and lower).
   1.5. Save the drawing as *MyBattery* in the C:\Steps3D\Lesson07 folder.

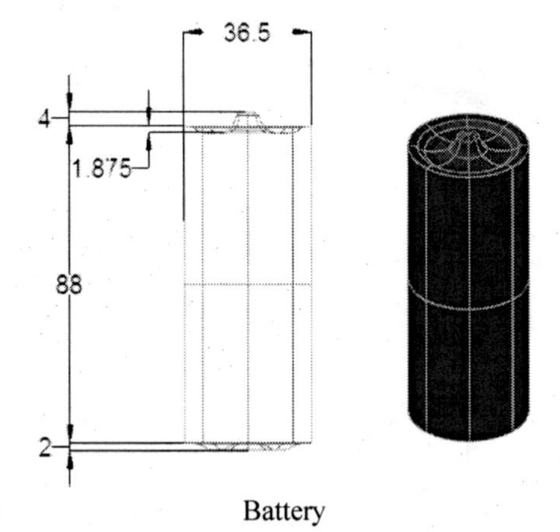

Battery

2. Using the *MeshIsoSetup* template in the C:\Steps3D\Lesson07 folder, create the *flash drive* shown at right. Follow these guidelines:
   2.1. Start with a mesh box, with the following tessellation divisions: Length = 3, Width = 3, Height = 3. Give it these dimensions: 50x18x7.
   2.2. Split the two outer front/back, center top/bottom faces on the right end of the mesh to provide the faces for the plug-in part of the drive.
   2.3. To create the plug-in, extrude the end faces 12mm. Then crease all the faces on the extrusion so the plug-in won't smooth.
   2.4. Next, extrude the center face on the top inward 1mm (for the manufacturers label). Crease the five faces involved in this intention.
   2.5. Finally, smooth the mesh to Level2.

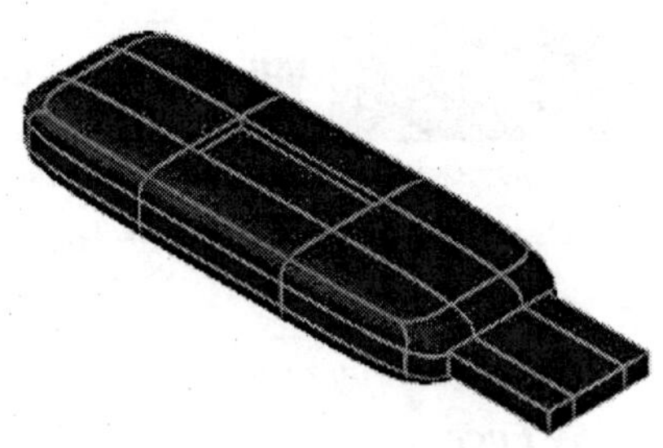
Flash Drive

3. Using the *MeshSetup* template in the C:\Steps3D\Lesson07 folder, create the *horn* shown at right. Follow these guidelines:
   3.1. Start with a mesh cylinder with the following tessellation divisions: Axis = 8, Height = 7, Base = 3. Make the cylinder 11.5" long x 6"Ø.
   3.2. Using 0,0,0 as the base point, scale each of the tessellation edges along the XY plane by the following amounts (starting at the bottom edge): 100%, 65%, 35%, 5%, 35%, 8%, 10%, and 15%.
   3.3. Move each level of edges along the Z axis as follows (starting at the top): 0", 1", 1.5", 1.5", 1.5", -1", and 0".
   3.4. Move the top's center faces down along the Z axis by ½".
   3.5. Move the bottom's center faces up along the Z axis by 1½".
   3.6. Smooth the mesh to Level4.
   3.7. Save the mesh as *MyHorn* in the C:\Steps3D\Lesson07 folder.

Horn

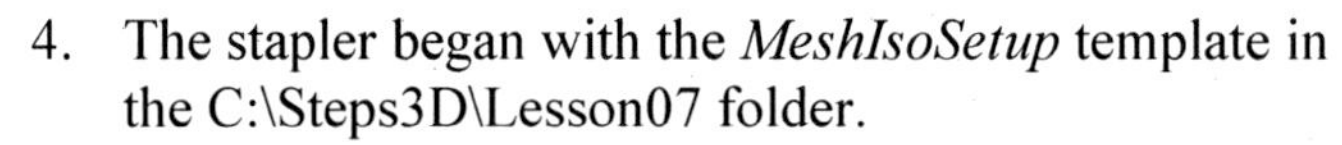

4. The stapler began with the *MeshIsoSetup* template in the C:\Steps3D\Lesson07 folder.
   4.1. Start with a mesh box with the following tessellation divisions: Length = 2, Width = 1, Height = 4.
   4.2. The overall measurement of the base is 55x30x55.
   4.3. The top and bottom are 105mm extrusions; the staple holder is a 95mm extrusion.
   4.4. Smooth the mesh until you're happy with the outcome. (I removed the isolines for the picture.)
   4.5. Save the mesh as *MyStapler* in the C:\Steps3D\Lesson07 folder.

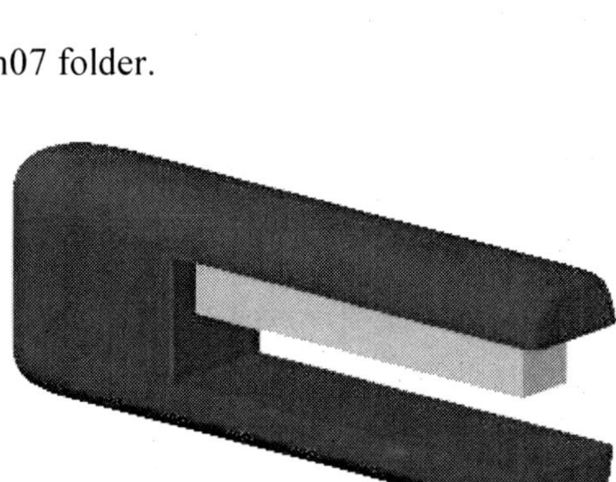

Stapler

5. This last exercise has no dimensions, it's more an idea of things you can create when you just sit down and play for 45 minutes or an hour. See what you can do with the idea!

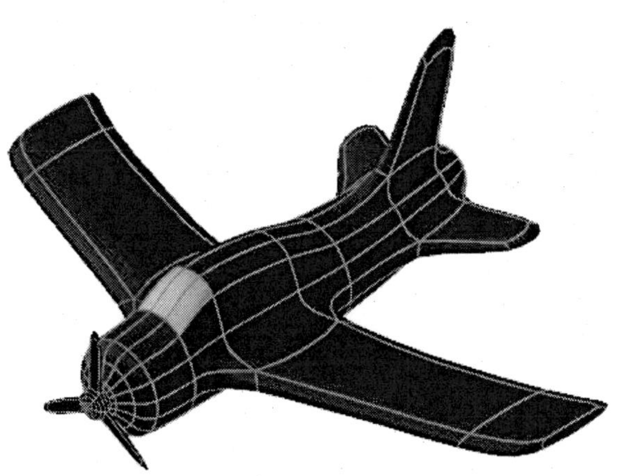

Toy Plane

7.9	**For Web-Based Review Questions, visit: http://foragerpub.com/AcadFiles/2010/2010.htm**

Lesson

Following this lesson, you will:

✓ *Know how to create AutoCAD's Solid Modeling Building Blocks*

- ***Box***
- ***Wedge***
- ***Cone***
- ***Sphere***
- ***Cylinder***
- ***Torus***
- ***Pyramid***
- ***Helix***
- ***PolySolid***

✓ *Know how to convert objects – 2D to 3D*

- ***Extrude***
- ***Thicken***
- ***PressPull***
- ***Revolve***
- ***Sweep***

✓ *Know how and why to use AutoCAD's Isolines system variable*

✓ *Creating 3D Object from 2D Objects*

Solid Modeling Creation Tools – The Basics

Understanding some of the history of three-dimensional AutoCAD might help you prepare for this lesson.

AutoCAD began its trek into Z-Space by creating the Z-axis. The Z-axis gave us the ability to create three-dimensional lines and circles for the first time. AutoCAD called this development Wireframe Modeling.

But although the creation of a Z-axis was no small feat for programmers, Wireframe Modeling came up short in its usefulness to draftsmen. After all, a skeleton without skin is a fairly transparent accomplishment.

AutoCAD "covered" the need by developing Surface (and later Mesh) Modeling. Here, we gained the 3DFace (and related) commands that could be used to "stretch a blanket" over the wireframe. This appeared to solidify AutoCAD's three-dimensional experiment. But the success, like its models, was hollow.

AutoCAD programmers still dreamed of a model that would be "just like the real thing." That is, they wanted the computer to be able to reflect mass properties – solids where the object was solid, and spaces where the solid was empty. They wanted a solid object to be *a solid object – not a loose conglomeration of circles and lines. So AutoCAD developed Solid Modeling.*

Obviously, I couldn't give you the full history in these few paragraphs. The reason for these paragraphs, then, is to let you know that developers of Solid Modeling had Wireframe and Surface Modeling on which to build. (Interestingly, Mesh Modeling, in its current incarnation, also had Solid Modeling to build upon!)

What does that mean to you now? Simply that having studied the intricacies of the more primitive modeling techniques, you're well prepared (better, perhaps, than you might think) for tackling this most remarkable of AutoCAD's modeling tools.

8.1 What Are Solid Modeling Building Blocks?

Most people refer to Solid Modeling building blocks as primitive solids. But frankly, that term isn't as descriptive as it might be. Building blocks are toys with which we all played as children. You're already familiar with their basic shapes – box, wedge, cone, cylinder, sphere, and torus. You studied all these as part of your mesh primitive models.

Additionally, we'll include homemade shapes as part of our building blocks (didn't you wish you could do that when you were a child?). To create these, we'll use the solids equivalent of the ***Revsurf*** command – ***Revolve*** – and we'll revisit the ***Extrude*** command.

Therefore, to answer the question, "What are Solid Modeling building blocks?" (for the test), let me give you a quick definition. Solid Modeling building blocks are predefined and user-defined solid shapes with which you build your model.

Let's look at each of them.

8.2 3D Solids from 2D Regions and Solids

8.2.1 *Thicken*

You have a few ways to convert existing structures – surfaces, faces, and closed objects – into solid objects. These commands – ***Thicken***, ***Extrude***, and ***PressPull*** – work on different types of existing objects and knowing which is which will save a lot of frustration.

Thicken works on surfaces only. That sounds simple enough, but be aware that it doesn't work on 3D Faces or meshes (the source of potential aggravation). You must have created the surface using one of the surface tools – ***ConvToSurface*** (good on 2D solids, regions, lines/arcs/polylines with thickness, and *planar* faces) or ***Planesurf***. (Remember that you can also convert many polylines and circles with thickness to solids using the ***ConvToSolid*** command !)

Where to Find It:	
Command Line:	***Thicken***
Ribbon (Tab/Panel):	Home – Solid Editing – **Thicken**
Menu:	Modify – 3D Operations – **Thicken**

The command goes like this:

Command: *thicken*

Select surfaces to thicken: *[Select the surface(s) to thicken.]*

Select surfaces to thicken: *[Confirm the selection set.]*

Specify thickness <0.0000>: *[Tell AutoCAD how thick to make it.]*

That's simple enough! But be aware that thicken doesn't allow for anything other than a straight path along the current UCS.

8.2.2 *Extrude*

One of the easiest ways to create a three-dimensional solid is simply to *extrude* a two-dimensional object. This means that AutoCAD will take the two-dimensional object and "stretch" it or "pull" it into Z-Space. The objects on which AutoCAD can perform this engineering marvel (and produce solids rather than extruded mesh faces) are 3D faces, closed polylines, circles, ellipses, closed splines, donuts, regions, and 2D solids. The results may surprise you!

We saw the command sequence in Lesson 7, p.157, but let's look at the solid aspects of the ***Extrude*** command.

Where to Find It:	
Command Line:	***Extrude***
Hotkey(s):	***ext***
Ribbon (Tab/Panel):	Home – Modeling – **Extrude**
	Mesh Modeling – Mesh Edit – **Extrude Faces**
Menu:	Draw – Modeling – **Extrude**
Toolbar:	Modeling – **Extrude**

AutoCAD first lets you know how many *isolines* it'll use to display the object. We ignored isolines when we discussed meshes, but let me explain them now.

Remember when you drew complex mesh models? You had to identify the number of faces to use by answering some prompts or adjusting the values of the **Surftab1** and **Surftab2** system variables.

When drawing a solid object, the shape is unaffected by the surftab settings. A round solid object is round regardless of the number of lines AutoCAD uses to *show* that it's round. But using a large number of lines (*isolines*) to show something is round takes a bit more memory and regeneration time, so AutoCAD allows you to control the number.

You'll control the number of isolines used to show a rounded solid object with the **Isolines** system variable. (You can toggle isolines on or off with the **Isolines** button in the **Edge Effects** panel of the ribbon's **Render** tab, or by setting the **VSEdges** system variable to one or zero.)

This will become clearer in our next exercise.

8.2.3 *PressPull*

If you thought ***Thicken*** and ***Extrude*** were easy, take a look at ***PressPull***. Here's the sequence:

Where to Find It:	
Command Line:	***PressPull***
Ribbon (Tab/Panel):	Home –Modeling – **Press/Pull**
Toolbar:	Modeling – **PressPull**

Command: *presspull*

Click inside bounded areas to press or pull.

All you do with ***PressPull*** once you've picked

inside a bounded area is to push or pull until the 3D object is as big as you want it! It works on polylines, regions, 3D faces, and 2D solids to create a 3D solid. What's more – you can use it to press or pull any face of an existing 3D solid! (You can even create extensions or tunnels with it!) Sooo cool!

Let's make some 3D solid objects.

Do This: 8.2A	From 2D to 3D – Without Kicking and Screaming!

I. Open the *regions & solids* file in the C:\Steps3D\Lesson08 folder. The drawing looks like the figure at right. (The I-Beam is a polyline, the square is a solid, and the circle is a circle. The oddball shape is a region and the rectangle is a polyline drawn with thickness.)

II. Set the **obj1** layer current and the **Isolines** system variable is set to **4**.

III. Follow these steps.

8.2A: From 2D to 3D

1. Enter the ***Extrude*** command.

   Command: ***ext***

2. Select the I-Beam.

   Current wire frame density: ISOLINES=4

   Select objects to extrude:

   Select objects to extrude: ***[ENTER]***

3. Enter a **height of extrusion** of **5.**

   Specify height of extrusion or [Direction/Path/Taper angle]: ***5***

   The I-Beam looks like the figure at right. Notice that the original object disappears and that AutoCAD creates the new 3D solid object on the current layer.

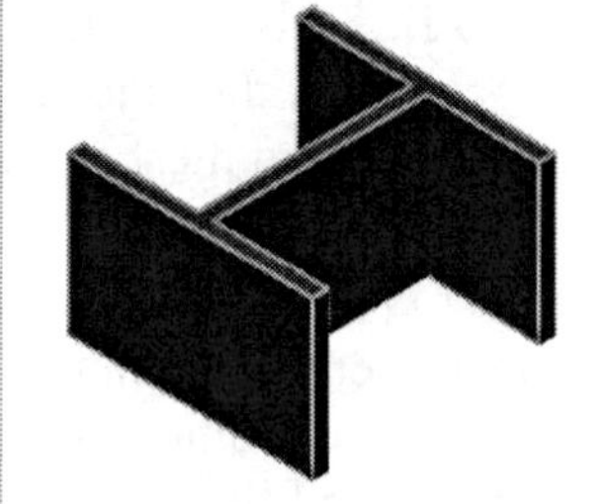

4. Set the **3D Wireframe** visual style current.

5. Repeat the ***Extrude*** command.

6. Select the circle.

   Select objects to extrude:

7. Tell AutoCAD to use a **Path** Path to guide the extrusion …

   Specify height of extrusion or [Direction/Path/Taper angle] <5.0000>: ***p***

8. … and select the line in the center of the circle.

   Select extrusion path or [Taper angle]:

   The extrusion looks like the figure at right. (Pretty bad, huh?)

8.2A: FROM 2D TO 3D

9. ***Undo*** the last command.

10. Let's see how the **Isolines** system variable affects the extrusion. Change it to 16.

Command: ***isolines***

Enter new value for ISOLINES <4>: ***16***

11. Repeat Steps 5 through 8.

The extension now looks like the figure at right. (Better?)

12. Reset the **Realistic** visual style Realistic.

13. Now let's look at the last prompt. Repeat the ***Extrude*** command and select the solid (the square).

14. Let's taper Taper angle this one …

Specify height of extrusion or [Direction/Path/Taper angle] <-0.9132>: ***t***

15. …at a 30° angle.

Specify angle of taper for extrusion <0>: ***30***

16. Make it 5 units tall.

Specify height of extrusion or [Direction/Path/Taper angle] <-0.9132>: ***5***

How's this one? (That's one way to build a quick, solid pyramid! We'll see another later in this lesson.)

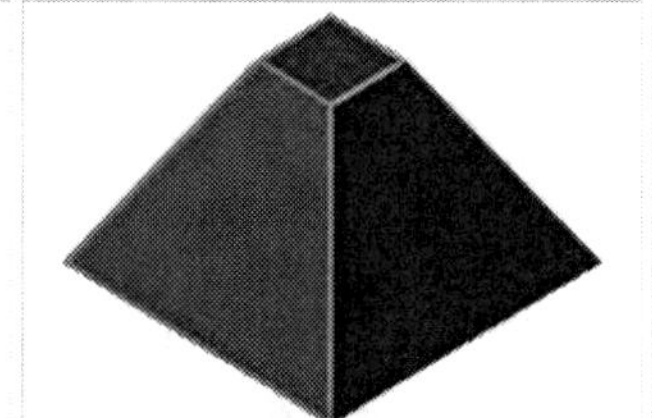

17. Let's try something cool – enter the ***PressPull*** command.

Command: ***presspull***

18. Try it on some of the faces of the box in the back. Try pulling the extrusion. Now try pushing (pressing) the extrusion back through the box.

Click inside bounded areas to press or pull.

(This is when we hear the cheerleaders yell, "Yeah! AutoCAD!")

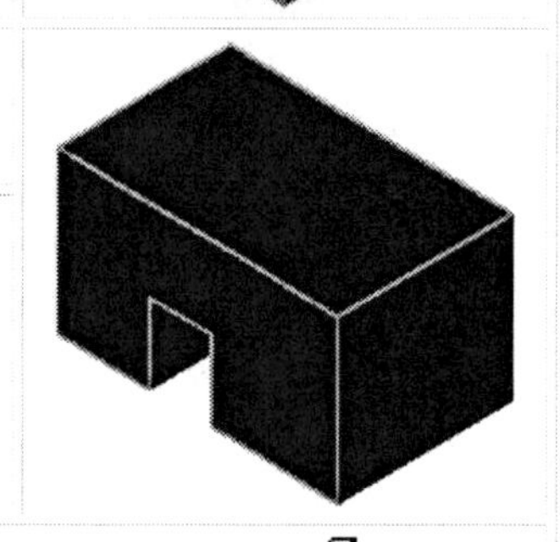

19. Okay, let's look at a couple more commands. Enter the ***ConvToSolid*** command. (We'll use it to convert the polyline rectangle to a 3D solid.)

Command: ***convtosolid***

20. Select the polyline rectangle.

Select objects:

Select objects: ***[ENTER]***

Not much to that one, is there?

21. Only one command left – but we have to convert the oddball region to a surface first. Use the ***ConvToSurface*** command and select it now.

Command: ***convtosurface***

8.2A: FROM 2D TO 3D

22. Now thicken it.

Command: ***thicken***

Select surfaces to thicken:

Select surfaces to thicken: ***[ENTER]***

Give it a thickness of about one inch.

Specify thickness <0.0000>: ***1***

23. Save the drawing.

Command: ***qsave***

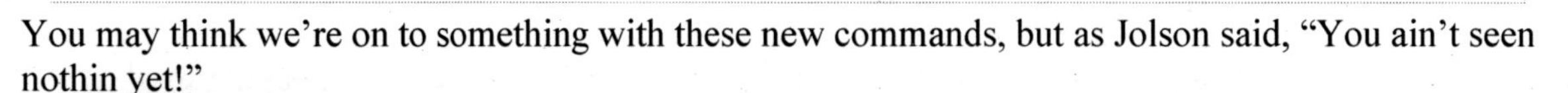

You may think we're on to something with these new commands, but as Jolson said, "You ain't seen nothin yet!"

8.3 Drawing the Solid Modeling Building Blocks

Extruding two-dimensional objects into Z-Space is handy. But solids offer many of the same predefined shapes that you used in Mesh Modeling – plus some additional shapes. We'll look at these now, and then we'll look at some other cool 3D solid creation tools.

Let's look at the predefined Solid Modeling shapes and, when applicable, their similarities to (and differences from) their Mesh Modeling counterparts.

8.3.1 Box

Use the ***Box*** command to draw any size box whose sides are parallel or perpendicular to the current UCS. This is the command sequence:

Command: ***box***

Specify first corner or [Center]: ***[Identify the first corner of the box.]***

Specify corner or [Cube/Length]: ***[Identify the opposite corner of the box.]***

Specify height or [2Point]<0.0000>: ***[Tell AutoCAD how tall to make the box.]***

Where to Find It:	
Command Line:	***Box***
Ribbon (Tab/Panel):	Home – Modeling – **Box**
Menu:	Draw – Modeling – **Box**
Toolbar:	Modeling – **Box**

The first thing you probably noticed is that the command sequence is nearly identical to the ***Mesh*** **Box** option (except that the solid sequence starts with the ***Box*** *command* rather than a **Box** *option*). This'll make the ***Box*** command all the easier to master.

We'll draw some boxes in our exercise to see the similarities to meshes.

Do This: 8.3.1A — Drawing Solid Boxes

I. Start a new drawing using the *3D Objects* template in the C:\Steps\Lesson07 folder.

II. Be sure polar snap is off.

III. Follow these steps.

8.3.1A: DRAWING SOLID BOXES

1. Enter the ***Box*** command.

Command: ***box***

8.3.1A: Drawing Solid Boxes

2. Specify the corners of the box as shown.

    Specify first corner or [Center]: ***4,4***

    Specify other corner or [Cube/Length]: ***@4,2***

3. Give it a **height** of **1.5** units as indicated.

    Specify height or [2Point] <0.0000>: ***1.5***

That was simple, wasn't it?

4. This time, let's draw a cube. Repeat the command.

    Command: ***[ENTER]***

5. We'll use the **Center** option Center.

    Specify first corner or [Center]: ***C***

6. And place the center as indicated.

    Specify center: ***5,5,2.5***

7. Tell AutoCAD to draw a **Cube** Cube.

    Specify corner or [Cube/Length]: ***c***

8. Make the sides of the cube **2** units.

    Specify length: ***2***

Your drawing looks like this. Notice that the center we indicated is the center of the box along all three axes – X, Y, and Z.

9. We'll draw one more to see the **Length** option. Repeat the command.

    Command: ***[ENTER]***

10. Pick the FRONT-BOTTOM-RIGHT corner of the upper box (at coordinates 6,4,1.5) as the first **corner**.

    Specify first corner or [Center]:

11. Use the **Length** option Length.

    Specify corner or [Cube/Length]: ***l***

12. Move your cursor toward the BACK-RIGHT and turn ortho on. Finally, specify the length, width, and height as shown.

    Specify length <2.0000>: 2

    Specify width: 2

    Specify height or [2Point] <2.000>: 2

Your drawing looks like the figure at right.

13. Save the drawing as *MyBoxes* in the C:\Steps3D\Lesson08 folder.

    Command: ***save***

It's really no different from drawing mesh boxes, is it? Does it remind you of playing with blocks when you were a child? Well, now you can make a living playing with those blocks!

8.3.2 Wedge

The similarities between the ***3D*** command's **Box** and **Wedge** options hold true for the ***Box*** and ***Wedge*** commands as well. The command sequence for the ***Wedge*** command is nearly identical to the **Wedge** option of the ***Mesh*** command.

We'll draw a couple of wedges for practice.

Where to Find It:	
Command Line:	***Wedge***
Hotkey(s):	***we***
Ribbon (Tab/Panel):	Home –Modeling – (Box flyout) **Wedge**
Menu:	Draw – Modeling – **Wedge**
Toolbar:	Modeling – **Wedge**

Do This: 8.3.2A Drawing Solid Wedges

I. Start a new drawing using the *3D Objects* template in the C:\Steps\Lesson07 folder.

II. Follow these steps.

8.3.2A: DRAWING SOLID WEDGES

1. Enter the ***Wedge*** command.

   Command: *we*

2. Start the wedge as shown.

   Specify first corner or [Center]: *1,1*

3. Point the wedge away from the screen by using a negative X value, as shown, and give it a height of 1.5 units.

   Specify other corner or [Cube/Length]: *@-4,2*

   Specify height or [2Point]<2.000>: *1.5*

   Your wedge looks like the figure at right.

4. Let's try the **Cube** option. Repeat the command.

   Command: *[ENTER]*

5. Pick the bottom corner (at coordinate 1,1) as the **first corner**.

   Specify first corner or [Center]:

6. Use the **Cube** option Cube...

   Specify corner or [Cube/Length]: *c*

7. ... and give it a length of **2**.

   Specify length <2.000>: *[ENTER]*

   Your drawing looks like the figure at right.

   Obviously, you haven't drawn a cube but a wedge (the diagonal half of a cube) based on the cube you specified.

8. Save the drawing as *MyWedges* in the C:\Steps3D\Lesson08 folder.

   Command: *save*

You probably noticed that the ***Wedge*** command doesn't have a rotation angle prompt. You can orient the solid wedge as you draw it (using positive or negative numbers), or you can rotate it to point it in the desired direction.

8.3.3	Cones and Cylinders

Where to Find It:	
Command Line:	***Cone***
Ribbon (Tab/Panel):	Home – 3D Modeling – (Box flyout) **Cone**
Menu:	Draw – Modeling – **Cone**
Toolbar:	Modeling – **Cone**

Where to Find It:	
Command Line:	***Cylinder***
Hotkey(s):	***cyl***
Ribbon (Tab/Panel):	Home – 3D Modeling – (Box flyout) **Cylinder**
Menu:	Draw – Modeling – **Cylinder**
Toolbar:	Modeling – **Cylinder**

The command sequence for solid cones (or cylinders) is identical to their mesh counterparts, and almost identical to each other!

Let's draw each and compare the results to the results of the same exercise using the mesh tools.

Do This: 8.3.3A	Drawing Solid Cones and Cylinders

I. Open the *Cones&Cylinders* file in the C:\Steps\Lesson07 folder. It looks like the figure at right. These are meshes; we'll recreate each as a solid.

II. Set **Isolines** to *24* and the **Realistic** visual style current.

III. Follow these steps.

8.3.3A: Drawing Solid Cones and Cylinders

1. Enter the ***Cone*** command.

   Command: ***cone***

2. We'll use default options on the first cone/cylinder. Place the first cone at coordinate **2,8** (the first cylinder at coordinate **8,8**).

   Specify center point of base or [3P/2P/Ttr/Elliptical]: ***2,8***

3. Give the cone/cylinder a base radius of **2** and a height of **4**.

   Specify base radius or [Diameter]: ***2***

   Specify height or [2Point/Axis endpoint/Top radius]<2.000>: ***4***

4. Repeat Steps 1 through 3 using the ***Cylinder*** command.

   Command: ***cyl***

   Your solids looks like the figure at right. Compare them with the adjacent meshes.

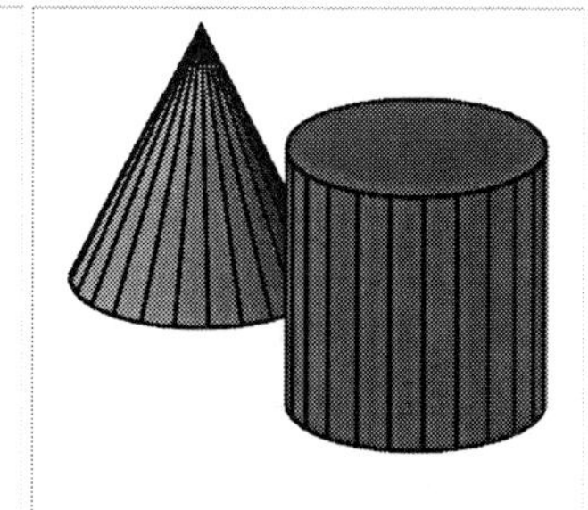

5. Let's use our new commands to draw an elliptical cone and then an elliptical cylinder. Repeat the ***Cone*** command.

   Command: ***cone***

6. Choose the **Elliptical** option [Elliptical].

   Specify center point of base or [3P/2P/Ttr/Elliptical]: ***e***

7. Place the **endpoint of first axis** at coordinate **0,4**. (Place the **endpoint of first axis** of the cylinder at **6,4**.)

   Specify endpoint of first axis or [Center]: ***0,4***

8.3.3A: DRAWING SOLID CONES AND CYLINDERS

8. Place the **other endpoint of first axis** at coordinate **4,4**. (Place the **other endpoint of first axis** for the cylinder at **10,4**.)

   Specify other endpoint of first axis: ***4,4***

9. Place the **endpoint of second axis** 1" to the north.

   Specify endpoint of second axis: @1<90

10. Finally, make the cone (cylinder) 4" tall.

    Specify height or [2Point/Axis endpoint/Top radius] <4.0000>: 4

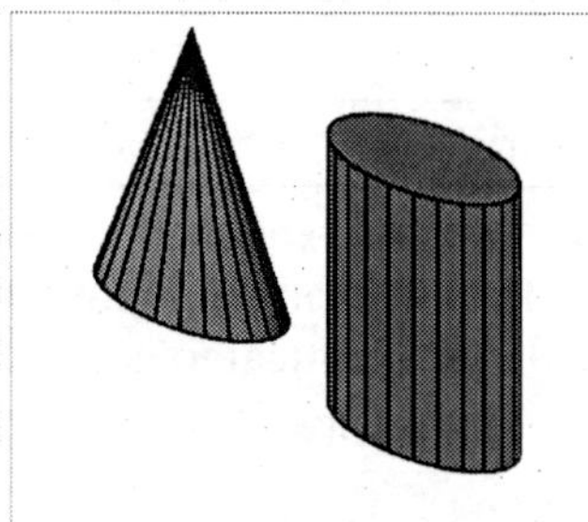

11. Repeat Steps 5 through 10 using the ***Cylinder*** command.

    Command: ***cyl***

The solid ellipse cone/cylinder looks like the figure at right. Compare them with the adjacent meshes.

12. Now use the **Top radius** option to change the top. Repeat the ***Cone*** command.

    Command: ***cone***

13. Place the base point of the cone at **2,0**.

    Specify center point of base or [3P/2P/Ttr/Elliptical]: ***2,0***

14. Use the **Diameter** option [Diameter] to give the cone a diameter of 4".

    Specify base radius or [Diameter] <2.0000>: ***d***

    Specify diameter <4.0000>: ***4***

15. Now use the **Top radius** option [Top radius] to define the top radius at 1".

    Specify height or [2Point/Axis endpoint/Top radius] <4.0000>: ***t***

    Specify top radius <0.0000>: ***1***

16. Finally, give it a height of 4".

    Specify height or [2Point/Axis endpoint] <4.0000>: ***4***

The cone now looks like the figure at right.

17. Save the drawing.

    Command: ***save***

Of course, you can also use the other options made familiar by your 2-dimensional use of the ***Circle*** and ***Ellipse*** commands. It's interesting that these commands use the same procedures to produce such similar objects. Perhaps, in the future, AutoCAD will reduce them to one command with a **Cone/Cylinder** option.

Did you notice the differences in the faces/isolines of the objects? You'd find it much easier to edit the face of a mesh than a solid. On the other hand, you'll find editing the solid much easier in other ways. Aren't you glad AutoCAD comes with the conversion tools (***ConvToSolid*** and ***MeshSmooth***)? (If you're not now, you *will be*!)

8.3.4	Sphere

You're already familiar with the sphere's options from your use of the ***Mesh*** command.

Although AutoCAD doesn't prompt for the number of longitudinal or latitudinal segments, it's a good idea to set the **Isolines** system variable to a large enough number for proper viewing (unless you're using the **Realistic** or **Conceptual** visual style). But that's something you should do early in the drawing session. It doesn't have to be repeated for each command.

Where to Find It:	
Command Line:	***Sphere***
Ribbon (Tab/Panel):	Home – Modeling – (Box flyout) **Sphere**
Menu:	Draw – Modeling – **Sphere**
Toolbar:	Modeling – **Sphere**

Draw a sphere.

Do This: 8.3.4A	Drawing a Solid Sphere

I. Start a new drawing using the *3D Objects* template in the C:\Steps\Lesson07 folder.

II. Set the visual style to **Realistic** and remove isolines.

III. Follow these steps.

8.3.4A: DRAWING A SOLID SPHERE

1. Enter the ***Sphere*** command.

    Command: *sphere*

2. Locate the **center of sphere** as indicated …

    Specify center point or [3P/2P/Ttr]: *4,4,4*

3. … and give it a **radius** of **2**.

    Specify radius or [Diameter] <2.0000>: *2*

Your drawing looks like the figure at right.

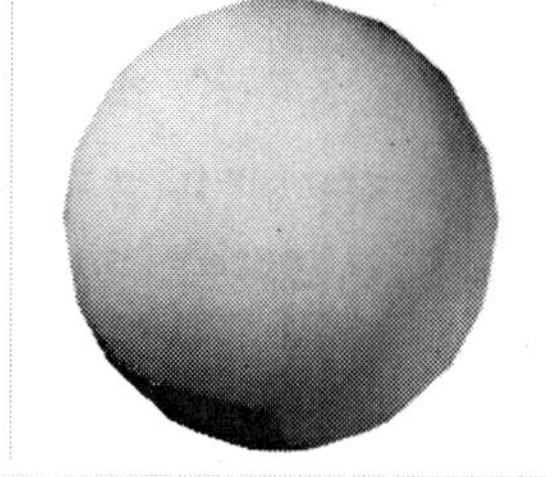

4. Save the drawing as *MySphere* in the C:\Steps3D\Lesson08 folder.

    Command: *save*

There's nothing else to show you about spheres. AutoCAD simplicity – what a marvel!

8.3.5	Torus

The command sequence for a solid torus (like the other solid sequences) is nearly identical to its mesh model counterpart.

As with the ***Sphere*** command, the options reflect those of the ***Circle*** command.

Draw a torus.

Where to Find It:	
Command Line:	***Torus***
Hotkey(s):	***tor***
Ribbon (Tab/Panel):	Home –Modeling – (Box flyout) **Torus**
Menu:	Draw – Modeling – **Torus**
Toolbar:	Modeling – **Torus**

Do This: 8.3.5A	Drawing a Solid Torus

I. Start a new drawing using the *3D Objects* template in the C:\Steps\Lesson07 folder.

II. Set the visual style to **Conceptual** and remove isolines.

III. Follow these steps.

8.3.5A: Drawing a Solid Torus

1. Enter the ***Torus*** command.

   Command: ***tor***

2. Locate the torus as indicated.

   Specify center point or [3P/2P/Ttr]: ***4,4,1***

3. Size the torus and the tube as indicated.

   Specify radius or [Diameter] <2.0000>: ***3***

   Specify tube radius or [2Point/Diameter]: ***.5***

   Your drawing looks like the figure at right.

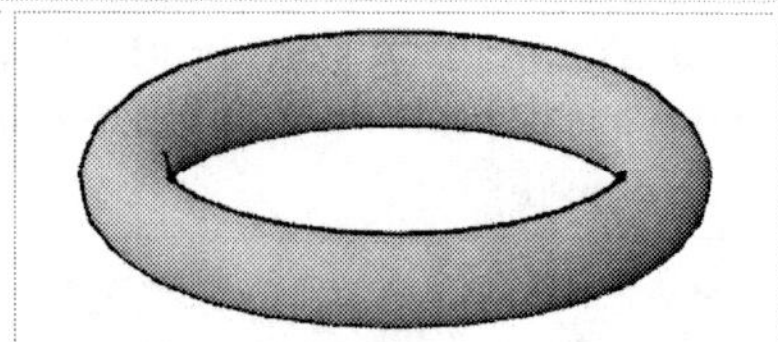

4. The ***Torus*** command cries for experimentation. Let's play a little. What happens when the tube diameter is larger than the torus diameter? (Let's find out).

   Repeat the ***Torus*** command.

   Command: ***[ENTER]***

5. Let's put this torus in the center of the first one.

   Specify center point or [3P/2P/Ttr]: ***4,4,1***

6. Just for fun, let's give the radius of the torus a negative number …

   Specify radius or [Diameter] <3.0000>: ***-3***

7. … and the tube a larger (absolute) number.

   Specify tube radius or [2Point/Diameter] <0.5000>: ***6***

   Your drawing looks like this. (Whoa, cool! See what you can discover with a bit of experimentation!)

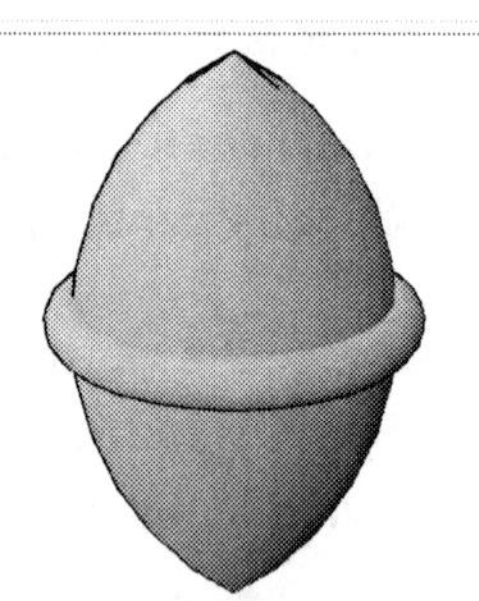

8. Remember the ***Subtract*** command (Lesson 3, p.71)? Subtract the outer torus from the inner torus.

   Command: ***su***

   Your solid looks like the figure at right. Try that with a mesh!

9. Save the drawing as *MyTorus* in the C:\Steps3D\Lesson08 folder.

   Command: ***save***

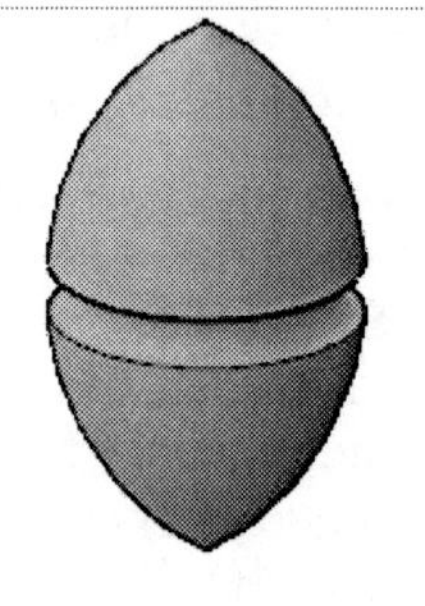

Oh, the fun you can have with AutoCAD, time, and a little imagination!

8.3.6 PolySolid

Here's a handy command that doesn't use the circle prompts and isn't available to mesh enthusiasts. Use ***PolySolid*** to create three-dimensional walls or other solids on the run.

It looks like this:

Command: ***polysolid***

Height = 4.0000, Width = 0.2500, Justification = Center ***[AutoCAD presents its initial PolySolid settings.]***
Specify start point or [Object/Height/Width/Justify] <Object>: ***[Pick a start point.]***
Specify next point or [Arc/Undo]: ***[Continue picking points as you would with the*** **Line** ***or*** **PLine** ***commands.]***
Specify next point or [Arc/Undo]:
Specify next point or [Arc/Close/Undo]:

Where to Find It:	
Command Line:	*PolySolid*
Hotkey(s):	*psolid*
Ribbon (Tab/Panel):	Home – Modeling – **Polysolid**
Menu:	Draw – Modeling – **Polysolid**
Toolbar:	Modeling – **Polysolid**

PolySolid looks and acts a lot like the ***PLine*** command, but the differences are remarkable.

- The **Object** option provides a quick and easy way to convert existing objects to polysolids. Objects available for conversion include: lines, arcs, 2D polylines, and circles. (Note that multilines are *not* included on the list.) Objects can be open or closed.
- **Width** provides the same opportunity as the ***PLine*** **width** option, but **Height** allows the same opportunity for Z-space.
- **Justify** works much like the same option in the ***Mline*** command, although the options it offers go by different names (**Left**, **Center**, **Right** – justification based on the direction of the first line segment).

Height, Width, and **Justify** options all affect how AutoCAD draws the polysolid even when you use the **Object** option.

Do This: 8.3.6A	Drawing a PolySolid

I. Open the *2DFlrPln* file in the C:\Steps\Lesson07 folder. (Users of *AutoCAD: One Step at a Time* may recognize the floor plan as the one they drew in the basic book. I've converted the outer walls of the multiline to polylines for this exercise.)

II. Follow these steps.

8.3.6A: DRAWING A POLYSOLID

1. Enter the ***Polysolid*** command.

    Command: ***psolid***

2. Adjust the **Height** [Height] to 8'…

    Specify start point or [Object/Height/Width/Justify] <Object>: ***h***
    Specify height <0'-4">: ***8'***

3. … the **Width** [Width] to 5½" …

    Specify start point or [Object/Height/Width/Justify] <Object>: ***w***
    Specify width <0'-0 1/4">: ***5.5***

4. … and the **Justification** [Justify] to left.

    Specify start point or [Object/Height/Width/Justify] <Object>: ***j***
    Enter justification [Left/Center/Right] <Right>: ***l***

8.3.6A: Drawing a PolySolid

5. Now use the **Object** [Object] option …

 Specify start point or [Object/Height/Width/Justify] <Object>: ***[ENTER]***

… to select the outer polyline along the left side of the floor plan.

6. Repeat the **Object** option [Object] to select the outer polyline along the right side of the floor plan.

 Command: ***[ENTER]***

Your drawing looks like the following figure (shown with a **Realistic** visual style).

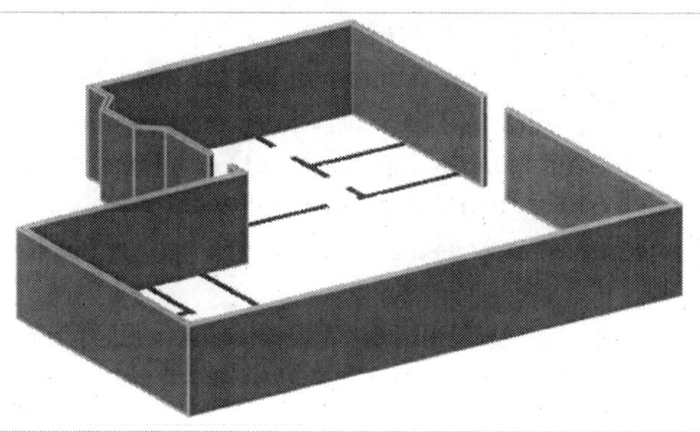

7. Save and exit the drawing.

 Command: ***qsave***

You could have drawn the outline from scratch, but I find it easier to do the layout in 2D and then make the conversion. You can now do the inner walls and use the commands you'll find in Lesson 9 to add openings for doors and windows. Use the information in Lesson 11 to add the actual 3-dimensional door and window blocks.

There are some things to remember about this procedure:

- AutoCAD drew the wall on the object's original layer – *not* the current layer.
- AutoCAD deleted the original object after it created the polysolid.
- AutoCAD will convert arcs as well as lines.

8.3.7 The Solid *Pyramid*

I suppose it's only fair; you can draw surface pyramids so why not solid ones?

The options should be familiar from both the ***Polygon*** command and the mesh ***Pyramid*** procedure.

Let's draw some pyramids and see if this approach is any more fun than the surface approach.

Where to Find It:	
Command Line:	***Pyramid***
Hotkey(s):	***pyr***
Ribbon (Tab/Panel):	Home – Modeling – (Box flyout) **Pyramid**
Menu:	Draw – Modeling – **Pyramid**
Toolbar:	Modeling – **Pyramid**

Do This: 8.3.7A Drawing a Solid Pyramid

I. Start a new drawing using the *3D Objects* template in the C:\Steps\Lesson07 folder.

II. Set **Realistic** as the current visual style.

III. Follow these steps.

8.3.7A: Drawing a Solid Pyramid

1. Enter the ***Pyramid*** command.

   Command: ***pyr***

2. AutoCAD lets you know what you're working with and asks you to locate the pyramid. Put it at the location indicated.

   4 sides Circumscribed

   Specify center point of base or [Edge/Sides]: ***0,0***

3. Use polar entry to tell AutoCAD that you want a 10" radius Polar: 10.0000 < 0°.

   Specify base radius or [Inscribed] <14.1421>:

4. Tell AutoCAD that you want a height of 15".

   Specify height or [2Point/Axis endpoint/Top radius]: ***15***

   Your pyramid looks like the figure at right.

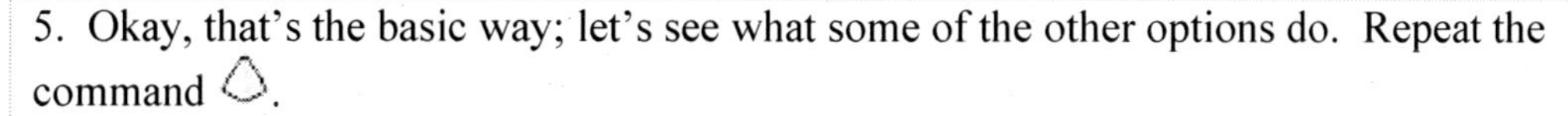

5. Okay, that's the basic way; let's see what some of the other options do. Repeat the command.

   Command: ***[ENTER]***

6. Let's do something the ancients never did, let's give our pyramid twelve sides Sides.

   Specify center point of base or [Edge/Sides]: ***s***

   Enter number of sides <4>: ***12***

7. Better put it a good distance from the first one.

   Specify center point of base or [Edge/Sides]: ***30,0***

8. And we'll inscribe Inscribed this one in a 10" radial circle Polar: 10.0000 < 0°.

   Specify base radius or [Inscribed] <14.1421>: ***I***

   Specify base radius or [Circumscribed] <14.1421>: ***10***

9. Let's give this one a flat top Top radius in a radius of 4".

   Specify height or [2Point/Axis endpoint/Top radius] <15.0000>: ***t***

   Specify top radius <0.0000>: ***4***

10. Finally, give it an **Axis endpoint** Axis endpoint as indicated.

    Specify height or [2Point/Axis endpoint] <15.0000>: ***a***

    Specify axis endpoint: ***@30,0,15***

    Your drawing looks like the figure at right.

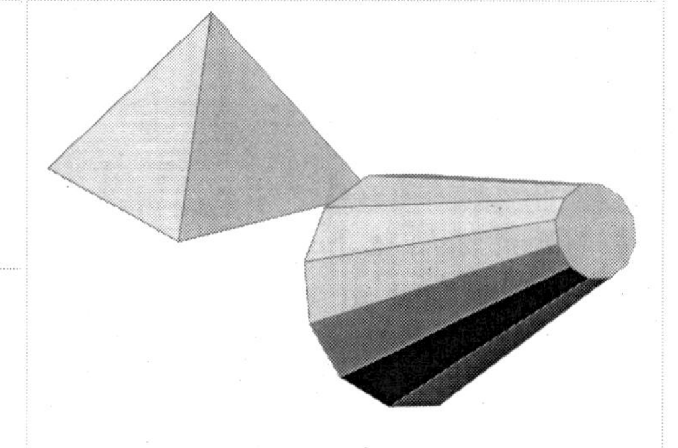

11. Save the drawing with a convenient name, and close it.

    Command: ***save***

One more basic command – but this one's a head-turner!

8.3.8	Creating a *Helix*

This one was a long time in the making; and boy, was it anxiously anticipated!

Ever try to draw a spring or a coil? Prior to the '07 release, it was a nightmare for those few courageous souls who dared the attempt. But those were the olde days.

Here's how we do it now:

Command: ***helix***

Number of turns = 3.0000 Twist=CCW *[some basic setup information]*

Specify center point of base: ***[Where do you want it?]***

Specify base radius or [Diameter] <1.0000>: ***[How big do you want the base?]***

Where to Find It:	
Command Line:	***Helix***
Ribbon (Tab/Panel):	Home – Draw (subpanel) – **Helix**
Menu:	Draw – **Helix**
Toolbar:	Modeling – **Helix**

Specify top radius or [Diameter] <1.0000>: ***[How big do you want the top?]***

Specify helix height or [Axis endpoint/Turns/turn Height/tWist] <1.0000>: ***[How tall do you want it?]***

Interestingly, most of the options appear at the end of the command. The first three prompts contain no surprises. Let's look at that last one.

- **Specify helix height** is pretty straightforward. Tell AutoCAD how tall to make the helix.
- **Axis endpoint** – no real difficulty here either. As you did with other commands, pick or identify an axis endpoint for the end opposite the one you identified in the first prompt. If necessary, AutoCAD will lean the helix over to reach the point you identify.
- **Turns** is a unique option. Use this to redefine the number of turns you want in your helix. (The default is 3 as indicated in the informational line that began the command.)
- **turn Height** refers to the distance covered by a single turn. If you use this option, AutoCAD will automatically adjust the number of turns to reach the indicated height. If you've entered a desired number of turns (the **Turn** option), you can't also enter a **turn Height**.
- **tWist** just wants to know which way to go with the helix – clockwise (CW) or counterclockwise (CCW). It prompts:

  Enter twist direction of helix [CW/CCW] <CCW>:

Let's give it a try.

Do This: 8.3.8A	Drawing a Helix (Spring or Coil)

I. Open *Helix* in the C:\Steps\Lesson07 folder.

II. Follow these steps.

8.3.8A: Drawing a Helix
1. Enter the ***Helix*** command. **Command:** ***helix***
2. AutoCAD gives you some starting information, then asks where to put the helix. Locate it as indicated. **Number of turns = 3.0000 Twist=CCW** **Specify center point of base:** ***0,0***

8.3.8A: Drawing a Helix

3. Give it a base radius of 2 …

    Specify base radius or [Diameter] <1.0000>: ***2***

… and a top radius to match.

    Specify top radius or [Diameter] <2.0000>: ***[ENTER]***

4. Make it 6" tall.

    Specify helix height or [Axis endpoint/Turns/turn Height/ tWist] <1.0000>: ***6***

The helix looks like the figure at right.

5. Okay, let's try some options. Repeat the command.

    Command: ***[ENTER]***

6. Put this one to the left of the first.

    Number of turns = 3.0000 Twist=CCW
    Specify center point of base: ***-10,0***

7. Use the same base diameter, but give the top a radius of 1.

    Specify base radius or [Diameter] <2.0000>: ***[ENTER]***
    Specify top radius or [Diameter] <2.0000>: ***1***

8. We'll turn [tWist] this one clockwise [CW].

    Specify helix height or [Axis endpoint/Turns/turn Height/tWist] <6.0000>: ***w***
    Enter twist direction of helix [CW/CCW] <CCW>: ***CW***

9. Give it six turns [Turns].

    Specify helix height or [Axis endpoint/Turns/turn Height/tWist] <6.0000>: ***t***
    Enter number of turns <3.0000>: ***6***

10. Now let's make this one 6" tall, too, for comparison.

    Specify helix height or [Axis endpoint/ Turns/turn Height/tWist] <6.0000>: ***[ENTER]***

Your helix looks like the figure at right.

11. Save the drawing, and close it.

    Command: ***qsave***

A helix, coil, or spring drawn like this doesn't bring much to a drawing. After all, it's still a two-dimensional object (a line) even if it does twist through Z-space. We'll reopen this drawing in our next section and I'll show you how to give it some true 3-dimensional qualities.

8.4 Creating More Complex Solids

8.4.1 Using the *Revolve* Command

After we studied the mesh primitive model objects in Lesson 4, we spent another lesson studying several commands that helped us create more complex mesh models.

In our solids studies, we've replaced ***Tabsurf*** easily with the ***Extrude*** command that we saw at the beginning of this lesson (p.177). But it may please (or frighten) you know that most of the mesh-specific commands have no direct equivalents in the solids world. Of course, that being said, now I'll tell about the exceptions to that statement!

The first exception involves your favorite command (and mine) – ***Revsurf***. Remember the nifty shapes we created on our train back in Lesson 5 (Ex. 5.2.3A, p.109 – the top of the smokestack, the wheels, and the bell)? It'd be a shame not to be able to create such objects as solids.

For that reason, AutoCAD has provided the ***Revolve*** command. But unlike the other solid commands and their mesh counterparts, ***Revolve*** is just a bit more difficult to use than ***Revsurf***. But this is mostly because of the additional options involved. Here's the command sequence:

Where to Find It:	
Command Line:	***Revolve***
Hotkey(s):	***rev***
Ribbon (Tab/Panel):	Home – Modeling – (Extrude flyout) **Revolve**
Menu:	Draw – Modeling – **Revolve**
Toolbar:	Modeling – **Revolve**

Command: *revolve*

Current wire frame density: ISOLINES=64

Select objects to revolve: *[Select the object that defines the basic shape of the object you wish to create; you can select multiple objects.]*

Select objects to revolve: *[Confirm the selection set.]*

Specify axis start point or define axis by [Object/X/Y/Z] <Object>: *[Select a point on the axis of revolution.]*

Specify axis endpoint: *[Select another point to define the axis.]*

Specify angle of revolution or [STart angle] <360>: *[Tell AutoCAD how much of a revolution you want.]*

> If your selection of objects to revolve involves closed objects, ***Revolve*** will create a 3D solid. If, however, your selection involves open objects, ***Revolve*** will actually create a surface object! (Look out, ***Revsurf***! You may be replaced!)
>
> Most objects to be revolved are 2D objects. You can, however, also revolve the face of a 3D solid!

The first options don't occur until AutoCAD prompts for an **axis** (of revolution). Then you have four!

- The default option requires that you specify a point on the axis. AutoCAD will then ask you to specify another point to define the axis.
- You can also define the axis by **Object**. When you choose this option, AutoCAD asks you to **Select an object**. Select an object that exists in the current XY plane and AutoCAD will do the rest.
- The **X/Y/Z (axis)** options will revolve the object about the selected axis using coordinate 0,0 as the center of the revolution.
- AutoCAD presents the last prompt after you've made the revolution **axis** decision. This prompt allows you to control the **angle of revolution** (how much of a revolution do you want?). Simply enter an angle in degrees. You can also define the **STart angle** of your revolution.

Let's see the ***Revolve*** command in action.

Do This: 8.4.1A	Drawing a 3D Solid with the *Revolve* Command

I. Open the *finial* file in the C:\Steps3D\Lesson08 folder. The drawing looks like the figure at right.

II. Notice the orientation of the UCS and where it's centered; then turn off the UCS icon.

III. Be sure the **obj1** layer is current.

IV. Follow these steps.

8.4.1A: *REVOLVE*

1. Enter the ***Revolve*** command.

   Command: *rev*

2. Select the shape.

   Select objects to revolve:

   Select objects to revolve: *[ENTER]*

3. Using OSNAPs, pick the endpoints of the vertical line to define your **axis** (of revolution). (Pick the bottom endpoint first.)

   Specify axis start point or define axis by [Object/X/Y/Z] <Object>:

4. Revolve the object **270°**.

   Specify angle of revolution or [STart angle] <360>: *270*

Your drawing looks like the figure at right. Notice that AutoCAD creates the solid on the current layer.

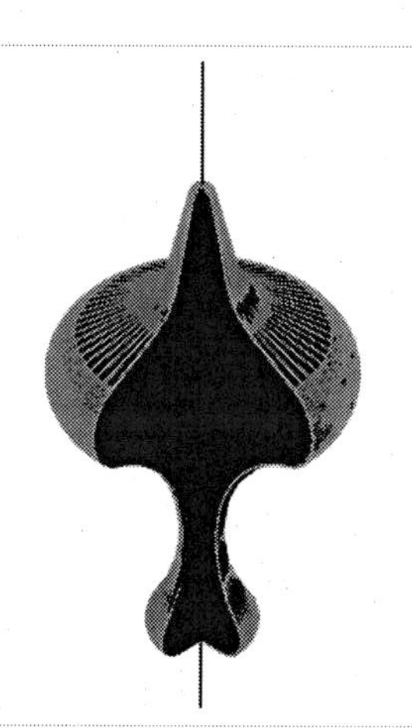

5. Undo the change.

   Command: *u*

6. Let's use an object to define our axis. Repeat Steps 1 and 2.

   Command: *rev*

7. Tell AutoCAD you'll use an **Object** [Object] to define the **axis.**

   Specify axis start point or define axis by [Object/X/Y/Z] <Object>: *o*

8. Then select the vertical line.

   Select an object:

9. Accept the default **360°** this time.

   Specify angle of revolution or [STart angle] <360>: *[ENTER]*

Remove the isolines, and your drawing looks like the figure at right.

10. Undo the change.

    Command: *u*

8.4.1A: *REVOLVE*

11. Let's use the X-axis to define our axis of revolution. Repeat Steps 1 and 2.
    Command: ***rev***

12. Tell AutoCAD to revolve the objects about the X-axis [X].
    Specify axis start point or define axis by [Object/X/Y/Z] <Object>: ***x***

13. Accept the **360°** default rotation.
    Your drawing again looks like the figure at right. (Okay. See if you can rotate it to look like that. Hint: Use the view cube – RIGHT-BOTTOM edge.) Your finial has become an ashtray! Doesn't that deserve another "Whoa, cool"?.

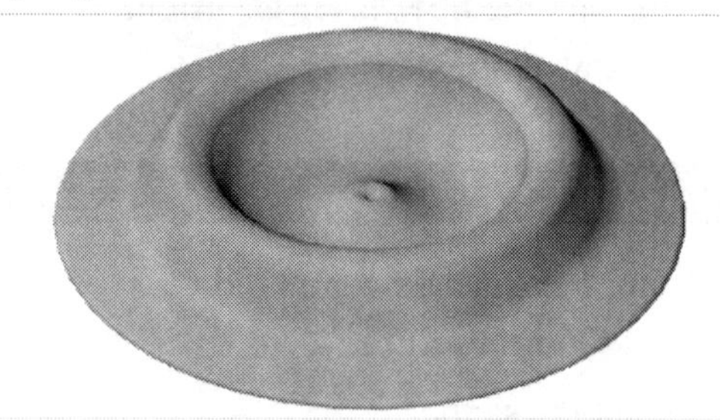

To see how ***Revolve*** works with open objects (it does the same things as it did in our last exercise – but it creates surfaces rather than solids), follow these steps while you're still in the *finial* file:

1. Freeze all layers that don't end in -sur. Conversely, thaw all layers that do end in -sur.
2. Repeat the last exercise.

The object that you'll revolve now is an open object.

8.4.2 Using the *Sweep* Command

Where to Find It:	
Command Line:	***Sweep***
Ribbon (Tab/Panel):	Home – Modeling – (Extrude flyout) **Sweep**
Menu:	Draw – Modeling – **Sweep**
Toolbar:	Modeling – **Sweep**

Sweep works something like ***Revolve*** except that you're not restricted to moving around an axis. With Sweep, you can "sweep" an object along a path to create a solid. You can sweep open or closed objects, including: lines, arcs (straight and elliptical), 2D polylines and splines, circles, ellipses, 2D solids, regions, and planar surfaces, faces and solids. Sweep the objects along a path made of: lines, arcs (straight and elliptical), 2D polylines and splines, 3D polylines and splines, helices, and edges of solids or surfaces.

Wow, that's quite a versatile tool! It works like this:

Command: ***sweep***

Current wire frame density: ISOLINES=4

Select objects to sweep: ***[Select the object you want swept; this is the object that will define the shape of your object.]***

Select objects to sweep: ***[Confirm the selection.]***

Select sweep path or [Alignment/Base point/Scale/Twist]: ***[Select the object that will define the path of your sweep.]***

Options include:

- **Alignment** allows you to adjust the object to be tangent to the path (default) or not.
- The **Base point** option allows you to define the base point of the object as it is swept along the path.
- **Scale**, of course, allows you to adjust the size of the object being swept. AutoCAD will graduate the scale from one end of the path to the other.

- With the **Twist** option, you can control the angle of the object being swept or, by default, allow banking of the object as it follows the path.

Let's take a look.

Do This: 8.4.2A	Drawing a 3D Solid with the *Sweep* Command

I. Reopen the *helix* file in the C:\Steps3D\Lesson08 folder.
II. Use the **Conceptual** visual style with no isolines.
III. Follow these steps.

8.4.2A: *SWEEP*

1. Enter the ***Sweep*** command.
   Command: *sweep*

2. Select the tiny circle in front of the helix on the right.
   Select objects to sweep:
   Select objects to sweep: *[ENTER]*

3. Select the helix on the right.
   Select sweep path or [Alignment/Base point/Scale/Twist]:

   Your swept spring looks like the figure at right. Notice that AutoCAD removed the object you selected to sweep.

4. Undo the changes.
   Command: *u*

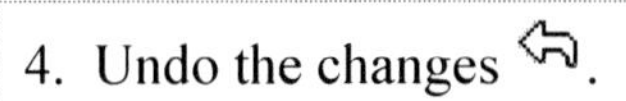

5. Let's look at the **Scale** option. Repeat Steps 1 and 2.
   Command: *[ENTER]*

6. Select the **Scale** option.
   Select sweep path or [Alignment/Base point/Scale/Twist]: *s*

7. Scale the object by ½, and select the helix on the left as your path.
   Enter scale factor or [Reference]<1.0000>: .5
   Select sweep path or [Alignment/Base point/Scale/ Twist]:

   The results look like the figure at right.

8. Save the drawing.
   Command: *qsave*

Experiment with the other shape if you wish.

A few things to remember about the sweep command include:

- AutoCAD places the new object on the current layer.
- AutoCAD removes the original of the object being swept.
- AutoCAD doesn't remove the object that forms the path of your sweep.

8.5 Extra Steps

- Read through the *3D Solids – Creating* section of the *User's Guide*. (Follow this path: Help – Help; then on the **Index** tab, enter **3D Solids** and double click on some of the topics.) You can pick up some tips here on each of the tools we've covered in this lesson … and some that are yet to come. Don't forget to try some of the procedures outlined on the **Procedures** tabs.
- Experiment with the ***Loft*** command (Section 5.3.2, p.113) with solids.

8.6 What Have We Learned?

Items covered in this lesson include:

- *AutoCAD's Solid Modeling building blocks*
  - ***Extrude***
  - ***Box***
  - ***Wedge***
  - ***Cone***
  - ***Sphere***
  - ***Cylinder***
  - ***Torus***
  - ***Revolve***
  - ***Polysolid***
  - ***Helix***
  - ***Pyramid***
  - ***Thicken***
  - ***Sweep***
  - ***PressPull***
- *Support for the building blocks*
  - ***Isolines***
  - ***VSEdges***

This has been another fun lesson! (We need these occasionally.) The commands have been simple and straightforward.

In Lesson 8, you saw how to draw familiar shapes as solids rather than meshes. You also had the opportunity to do some 2D-to-3D conversions. Did you feel like a kid again – opening a new box of blocks for the first time and exploring each wooden shape? Did your mind slip ever so slightly into that thin mist that inevitably precedes any great discovery? Did you start to create vague mental objects using the shapes as building blocks? How many times did you begin a thought with the words, "I can use this for …" or "This is a lot easier than …"?

The Solid Modeling bug has bitten you!

Actually, you may not be quite bit … yet. But wait until you finish Lessons 9 and 10! There you'll get to play with your new building blocks in ways you never dreamed possible back in your nursery. You'll see ways to combine blocks … ways that were simply not possible until the advent of the computer. Wait until you see …

But I'm getting ahead of myself. First, we must finish this lesson. Do the exercises. Get some practice. Answer the review questions. And then we can proceed!

8.7 Exercises

To get the opportunity to compare Solid Modeling with Mesh Modeling, repeat the exercises you did in Lessons 3, 4, 5, 6, and 7. Use solids whenever possible.

1. through 6. Create the drawings in Lesson 4 Exercises 9 through 14 using solids. Save the drawings to the C:\Steps3D\Lesson08 folder.

7. through 10. Create the drawings in Lesson 5 Exercises 2 and 6 through 8 using solids instead of surface models. Save the drawings to the C:\Steps3D\Lesson08 folder.

11. Create the *Double Helix* drawing. Follow these guidelines.
    11.1. The balls are 1" diameter.
    11.2. The rods are 1/8" diameter x 1" long.
    11.3. Each pairing rotates 15°.
    11.4. Save the drawing as *MyGenes* in the C:\Steps3D\Lesson08 folder.

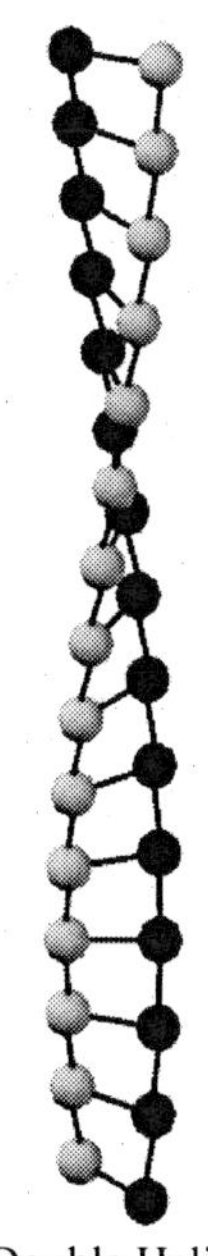

Double Helix

12. Create the *Paper Clip* drawing shown in the following figure.
    12.1. The drawing is set up and dimensioned using metrics.
    12.2. The paper clip is a single solid object.
    12.3. Save the drawing as *MyClip* in the C:\Steps3D\Lesson08 folder.

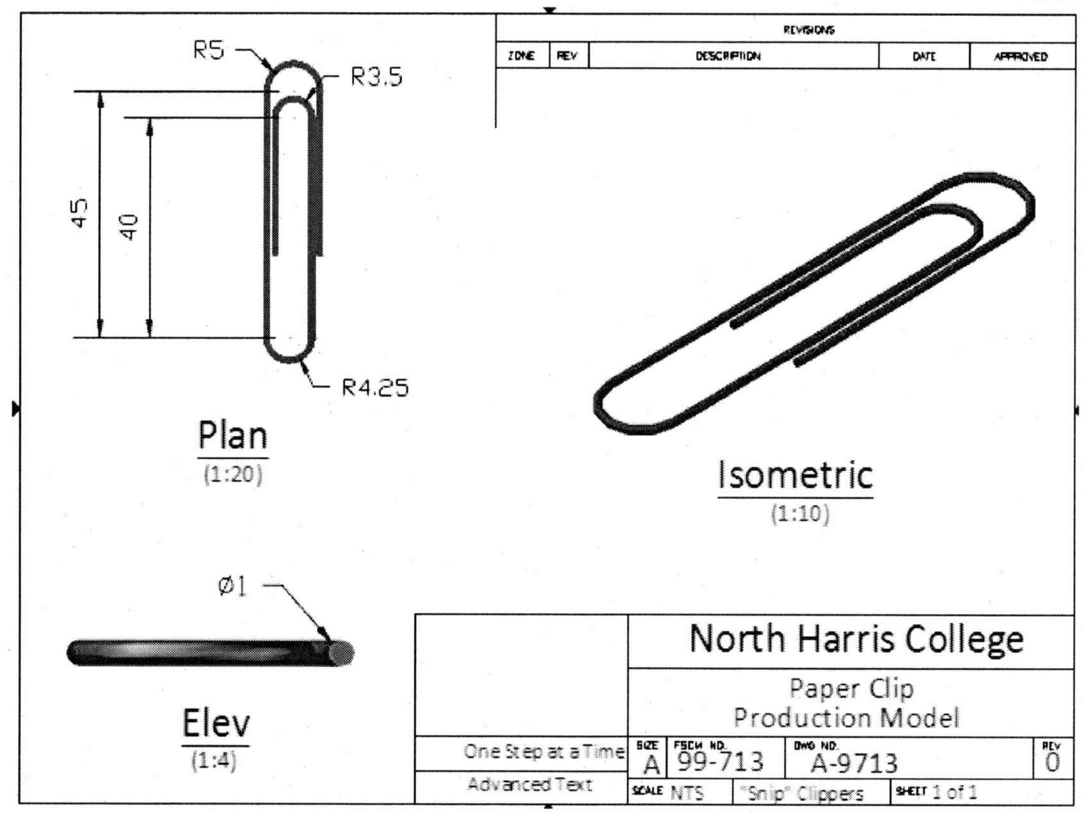

13. Create the *Fence* drawing. Follow these guidelines.
    13.1. The posts are 4 x 4s (3½" x 3½").
    13.2. The slats are 1 x 4s (¾" x 3½"). (I started with splines and turned them into regions.)
    13.3. The fence is 6" above the ground.
    13.4. The center rail is a 1 x 2 (¾" x 1½").
    13.5. Save the drawing as *MyFence* in the C:\Steps3D\Lesson08 folder.

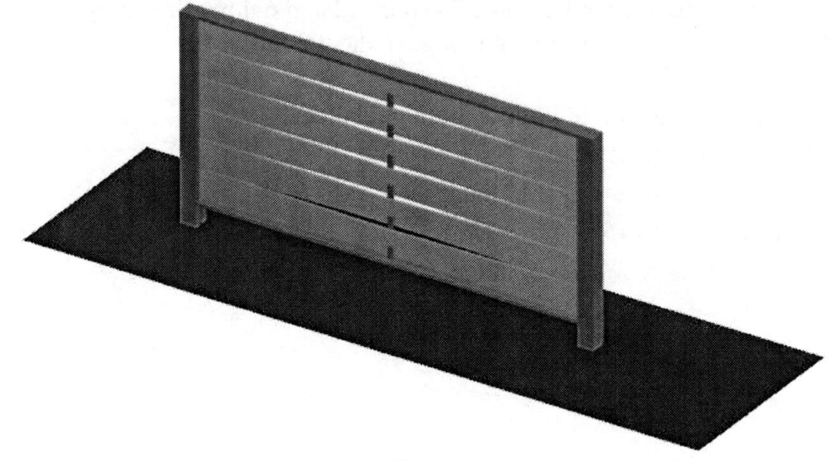

Fence

14. Create the *Fan Cover* drawing in the following figure.
    14.1. Each of the wires is 1/16" in diameter (including the torus around the frame).
    14.2. The center plate is 1/8" thick.
    14.3. Save the drawing as *MyFanCover* in the C:\Steps3D\Lesson08 folder.

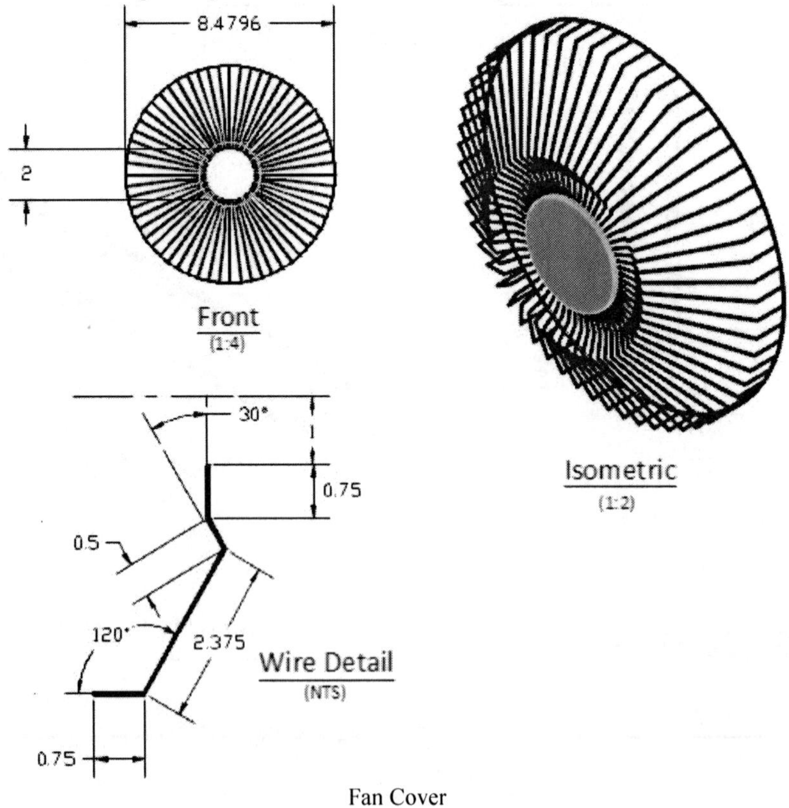

Fan Cover

15. Create the *Round Planter* drawing. Follow these guidelines.
    15.1. Use the *ANSI A Title Block* in AutoCAD's template folder.
    15.2. Text size is 3/16" and 1/8".
    15.3. Save the drawing as *MyPlanter* in the C:\Steps3D\Lesson08 folder.

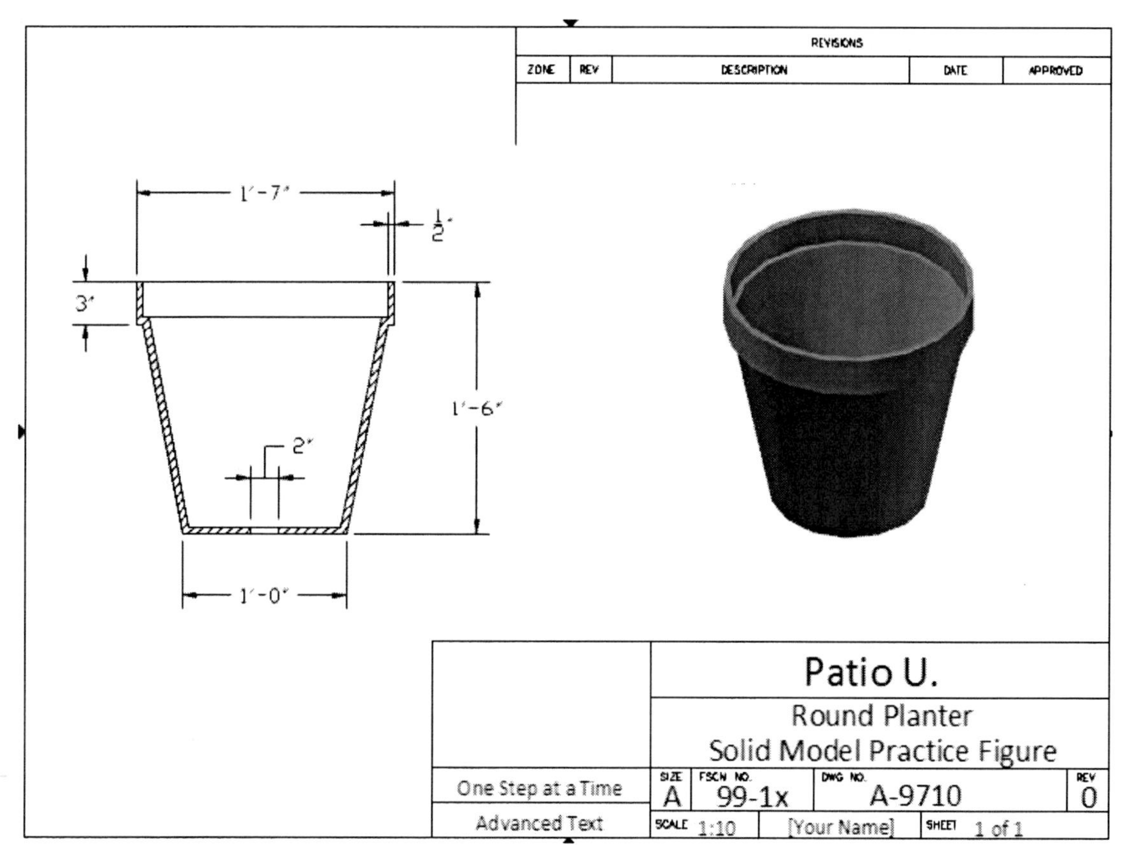

Planter

16. Create the *Patio Planter Box* drawing shown here. Follow these guidelines.
    16.1. Use the *ANSI A Title Block* found in AutoCAD's template folder.
    16.2. Text size is 3/16" and 1/8".
    16.3. Save the drawing as *MyPlanterBox* in the C:\Steps3D\Lesson08 folder.

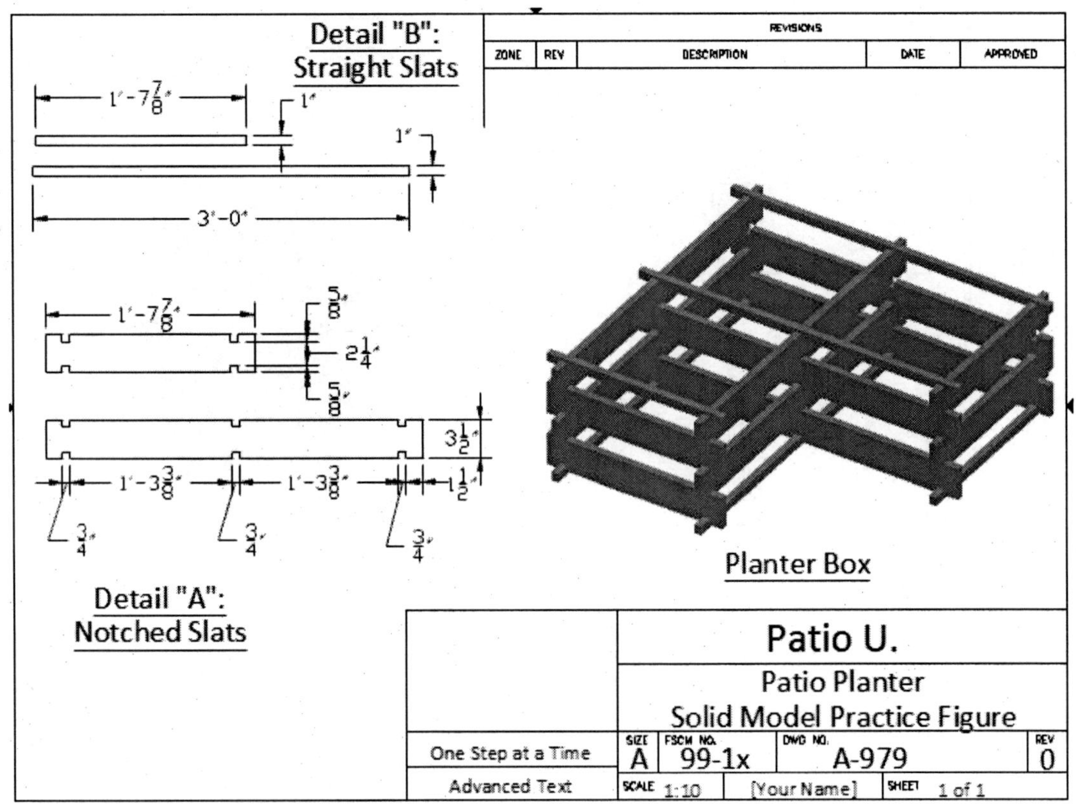

Pation Planter

8.8	**For Web-Based Review Questions, visit: http://foragerpub.com/AcadFiles/2010/2010.htm**

Lesson 9

Following this lesson, you will:

✓ *Know how to create composite solids from AutoCAD's solid building blocks using these commands:*

- ***Union***
- ***Subtract***
- ***Intersect***
- ***Slice***
- ***Interfere***

✓ *Know how to create a cross section using the **Section** and **SectionPlane** commands*

✓ *Know how to shape solids using these commands:*

- ***Fillet***
- ***Chamfer***

Composite Solids

I think I can … I think I can … I think I can …

Watty Piper's *The Little Engine That Could*

You've made it through the beginnings of Solid Modeling. You've experienced successes and near misses throughout your study, but the semester is half over. You may be tired and thinking more about Christmas or Easter or Labor Day than AutoCAD. You may be wondering, "Why AutoCAD ... why school ... why spend all this time and money to educate (or reeducate) myself?"

Let's pause for a paragraph or two for some words of encouragement. Let me tell you where you are in the overall scheme of (AutoCAD) things.

In all professions, there's a turning point – the point where the draftsman becomes the engineer, where the painter becomes the artist, the idea becomes the design. In every life there is (hopefully) a time where adolescence gives way to adulthood. In the world of computer drafting, you're at that turning point. You're about to leave the CAD Draftsman designation behind and become a true CAD Operator.

In the next few lessons, you'll discover how to take the building blocks – 3D solids and all of the basic and advanced modifying tools you've learned (and some you'll learn now) – and create *objects*. (Notice I didn't say, create "*drawings*.") You'll show the objects you create *in* drawings, but be assured that you'll be creating objects.

Once you've accomplished that very doable goal, I'll show you how to apply materials to those objects – how to show the wood grain on a table or make glass transparent. We'll cover this in Lesson 12, p.284.

But for now, let me offer these words of encouragement: Approach these lessons with the confidence of a graduate moving into graduate school. You've many successes under your belt, but the best is yet to come!

---

*I'll group the remaining solid construction tools into loose categories to help your understanding. These include Construction/Shaping Tools (**Slice**, **Section**, **Interfere**, **Union**, **Subtract**, **Intersect**, **Fillet**, and **Chamfer**), Solid Editing Tools (the multifaceted **SolidEdit** command, grips and the Properties palette), and some Print Setup commands exclusively for use with solid objects (**Solprof**, **Soldraw**, **Solview**, **Flatshot**, and **3DPrint**).*

*We'll consider the Construction/Shaping tools in Lesson 9, and the many editing tools – including those hidden within the **SolidEdit** command in Lesson 10, p.223. We'll save the Print Setup commands for Lesson 11, p.251.*

9.1 Solid Construction Tools

When you were a child, did you ever wish you could fuse your playing blocks together in order to preserve a particularly clever building effort? You wanted to keep that castle or tower forever to demonstrate your prowess with the tools of your trade. You wanted your family and friends to be able to see – years from now – how you built the perfect model!

And then your sister rode through on her tricycle and your dreams were shattered.

Well, the AutoCAD programmers had sisters, too. So they created a way to permanently fuse their computer blocks so that no one could ever disassemble them. Then they followed their childhood fantasies and created ways to remove parts of their blocks by using different shapes to define the carving. And they invented ways to find interferences between their blocks and to create cross sections …

… and their childhood dreams became reality in Solid Modeling Construction Tools.

Let's see how they work.

9.1.1 Union

The ***Union*** command behaves very much like a computer-controlled welding rod. It combines two solid objects into one. But unlike the welding rod, it leaves no seams that can break!

It's one of the simplest tools you'll ever hope to find. The command sequence is

Command: *union*

Select objects: *[Select the solids you wish to weld.]*

Select objects: *[Confirm the selection.]*

Where to Find It:	
Command Line:	***Union***
Hotkey(s):	***uni***
Ribbon (Tab/Panel):	Home – Solid Editing – **Union**
Menu:	Modify – Solid Editing – **Union**
Toolbar:	Modeling – **Union**

They just don't come any easier. But the value of the ***Union*** command can't be overstated. It turns simple objects like cylinders and boxes into production models like flanges, tools, doorstops, doorknobs, and much, much more.

We have to try this one. We'll create a flange over the next few exercises by using construction tools and several cylinders.

Do This: 9.1.1A — Welding Solid Objects with the *Union* Command

I. Open the *flange* file in the C:\Steps3D\Lesson09 folder. The drawing looks like the figure at right.

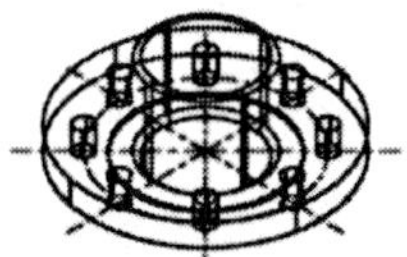

II. Follow these steps.

9.1.1A: WELDING SOLID OBJECTS WITH THE UNION COMMAND

1. Enter the ***Union*** command.

   Command: *uni*

2. Select the three cylinders indicated (bottom, large base, and outer neck).

   Select objects:

3. Confirm the selection and switch to the **Conceptual** visual style.

   Select objects: *[ENTER]*

   Your drawing looks like the figure at right. The selected cylinders have become a single unit. (Try to erase one of them to verify this.)

4. Restore the previous view.

5. Save the drawing, but don't exit.

   Command: *qsave*

Of course, the ***Union*** command is only one side of the coin. If you can add objects to each other, you should be able to remove one object from another as well.

Let's look at the ***Subtract*** command.

9.1.2 Subtract

We created a solid flange in our last exercise, but a flange has little use if nothing can flow through it. Enter the ***Subtract*** command.

You're already familiar with the ***Subtract*** command from Lesson 3. We used it in Section 3.3.3, p.71, to put windows in our walls. Subtract works the same on 3D solids as it did on regions.

We'll use it to remove the bolt holes and the core of our flange.

Do This: 9.1.2A	Removing One Solid from Another

I. Be sure you're still in the *flange* file in the C:\Steps3D\Lesson09 folder. If not, please open it now.

II. Set the 3D Wireframe visual style current.

III. Follow these steps.

9.1.2A: REMOVING ONE SOLID FROM ANOTHER

1. Enter the ***Subtract*** command.

   Command: *su*

2. AutoCAD asks you to select the object from which you wish to remove something. Select the solid you created in our last exercise (select the outermost cylinder). Confirm the selection.

   Select solids, surfaces, and regions to subtract from ..
   Select objects:
   Select objects: *[ENTER]*

3. AutoCAD now wants to know what to remove. Select the innermost cylinder and each of the smaller cylinders arrayed about the flange (the bolt holes).

   Select solids, surfaces, and regions to subtract
   Select objects:
   Select objects: *[ENTER]*

4. Reset the **Conceptual** [Conceptual] visual style.
   Your drawing looks like the figure at right.

5. Save and close the drawing.

   Command: *qsave*

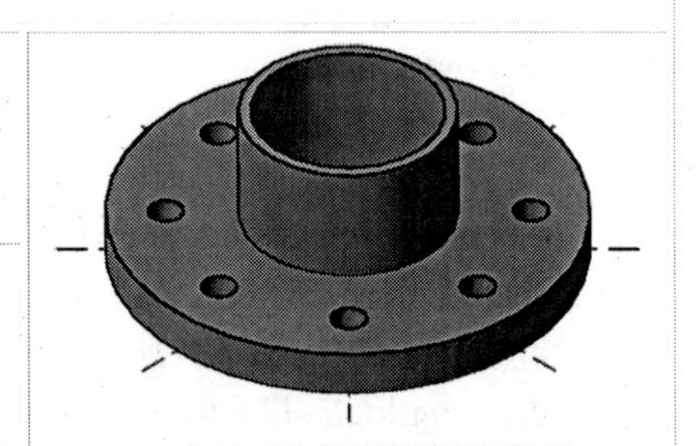

These first two exercises have been fairly easy and straightforward. But take a moment and fill in the blanks in the following sentences with as many answers as you can. (Time yourself and see how many answers you can find in 60 seconds.)

I can use the ***Union*** command to create __________ out of __________.

I can use the ***Subtract*** command to create __________ out of __________.

Here are some hints:

- Look about the room and consider objects on the desk, floor, walls, and shelves.
- Imagine that you have the blocks with which you played as a child, but now you have a bottle of glue, a drill, and a chisel, as well.

Are you beginning to see the possibilities?

9.1.3	Intersect

The ***Intersect*** command comes in handy when creating intricate multisided figures. It works by removing everything that doesn't intersect something else. This will become clearer with an exercise, but first look at the command sequence.

Where to Find It:	
Command Line:	*Intersect*
Hotkey(s):	*in*
Ribbon (Tab/Panel):	Home – Solid Editing – **Intersect**
Menu:	Modify – Solid Editing – **Intersect**
Toolbar:	Modeling – **Intersect**

Command: ***intersect***

Select objects: ***[Select objects that intersect each other.]***

Select objects: ***[Confirm the selection set.]***

If only they all accomplished as much – as easily! Let's see what we can do with the ***Intersect*** command.

Do This: 9.1.3A	Creating Objects at Intersections

I. Open the *emerald* file in the C:\Steps3D\Lesson09 folder. The drawing looks like the figure at right. (It's two octogons extruded at 30° to become solid objects. We mirrored the objects to have the four solids you see now.)

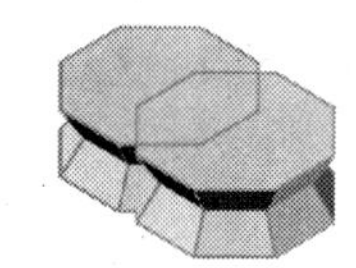

II. Follow these steps.

9.1.3A: CREATING OBJECTS AT INTERSECTIONS

1. Enter the ***Intersect*** command.

 Command: ***in***

2. Select the four octagon solids.

 Select objects:

 Select objects: ***[ENTER]***

 Your drawing looks like the figure at right.

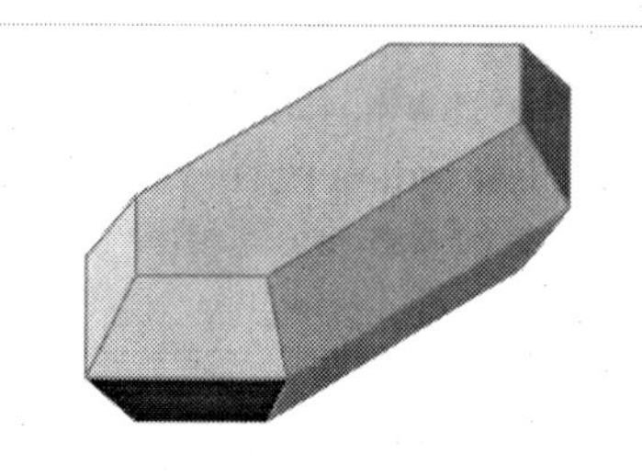

3. Save the drawing, but don't exit.

 Command: ***qsave***

Wow! What else can we do?!

9.1.4	Slice

Did you know that the value of precious stones often increases when they're cut just right? Let's cut our emerald.

Where to Find It:	
Command Line:	*Slice*
Hotkey(s):	*sl*
Ribbon (Tab/Panel):	Home – Solid Editing – **Slice**
Menu:	Modify – 3D Operations – **Slice**

The command we'll use is called ***Slice***. Use the ***Slice*** command to make a straight cut or remove a piece of an object as if cutting it away with a knife. The sequence offers more options than the others we've seen in this lesson, but that means more opportunities to cut it the way you want it cut.

It looks like this:

Command: ***slice***

Select objects to slice: ***[Select the solid object to cut.]***

Select objects to slice: ***[Confirm the selection set.]***

Specify start point of slicing plane or [planar Object/Surface/Zaxis/View/XY/YZ/ZX/ 3points] <3points>: ***[Pick two points on the object to define the slicing plane.]***

Specify second point on plane:

Specify a point on desired side or [keep Both sides] <Both>: ***[Pick a point on the side of the object you wish to keep.]***

The options should be familiar from your study of the ***Rotate3d*** and ***Mirror3d*** commands, but let's go over them again to be sure.

- The default option requires that you select two points (a **start point** and a **second point** on the plane) to define a slicing plane. This is like drawing the knife blade that'll be slicing through the object.
- The **3points** option (enter *3* or just hit ENTER to accept the default option) requires that you select three points to define the slicing plane. It allows a bit more precision in your slicing.
- The **planar Object** option is still the easiest. If you have an object (circle, ellipse, arc, spline, or polyline) drawn through the object you want to slice, you can select it as your slicing plane.
- The **Surface** option works in the same way as the **planar Object** option except that you can use a surface as a cutting plane. The results may surprise you!
- The **Zaxis** option is difficult to follow. It prompts like this:

  Specify a point on the section plane:

  Specify a point on the Z-axis (normal) of the plane:

  - The first prompt is asking for a point on the slicing plane.
  - The next prompt is asking for a point on the Z-axis of the slicing plane. Use this point to orient the slicing plane – essentially by picking a point to define which way is "up" if you're standing on the slicing plane.
- The **XY/YZ/ZX** options allow you to define the slicing plane by identifying a single point on the chosen plane. AutoCAD defines these planes according to the current UCS.
- The last option occurs at the final prompt. It allows you to keep either a selected piece or both pieces of the object after you've sliced it.

Let's cut our gemstone.

Do This: 9.1.4A	**Slicing 3D Solid Objects**

I. Be sure you're still in the *emerald* file in the C:\Steps3D\Lesson09 folder. If not, please open it now.

II. Follow these steps.

9.1.4A: SLICING 3D SOLID OBJECTS
1. Enter the ***Slice*** command.   **Command:** ***sl***
2. Select the emerald.   **Select objects to slice:**   **Select objects to slice:** ***[ENTER]***

9.1.4A: Slicing 3D Solid Objects

3. Accept the **3Point** option and then select the points indicated.

    Specify start point of slicing plane or [planar Object/Surface/Zaxis/View/XY/YZ/ZX/ 3points] <3points>: ***[ENTER]***

    Specify first point on plane:

    Specify second point on plane:

    Specify third point on plane:

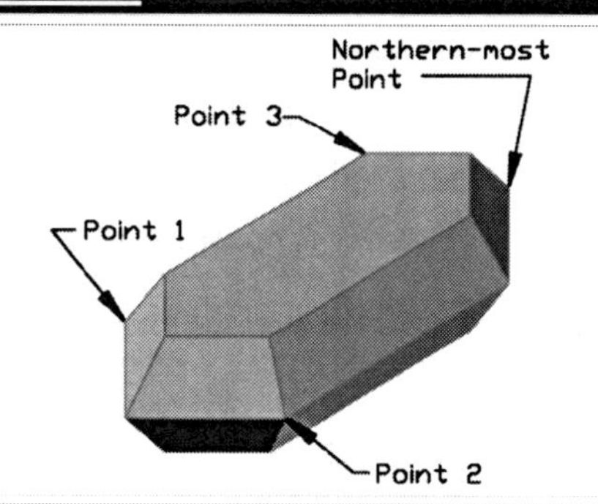

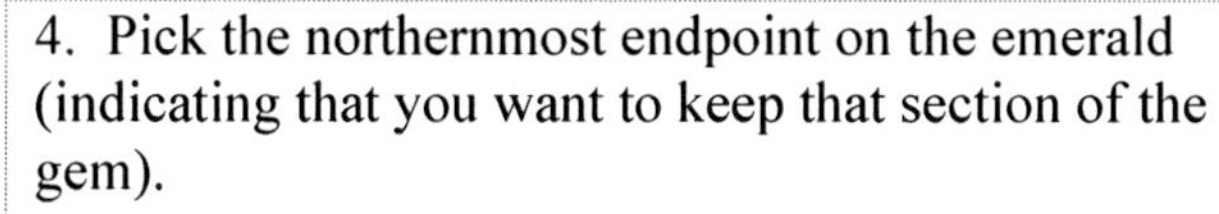

4. Pick the northernmost endpoint on the emerald (indicating that you want to keep that section of the gem).

    Specify a point on desired side or [keep Both sides] <Both>:

Your drawing looks like the figure at right.

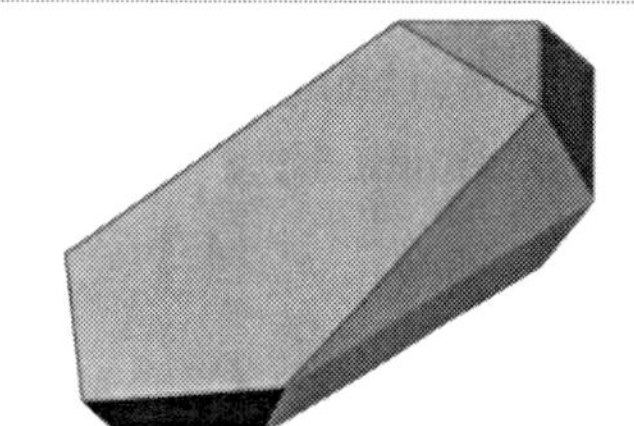

5. Undo the changes.

6. Let's try the **planar Object** option. First, thaw the **obj1** layer. Notice the circle that intersects the emerald.

    Command: ***sl***

7. Repeat Steps 1 and 2.

    Command: ***sl***

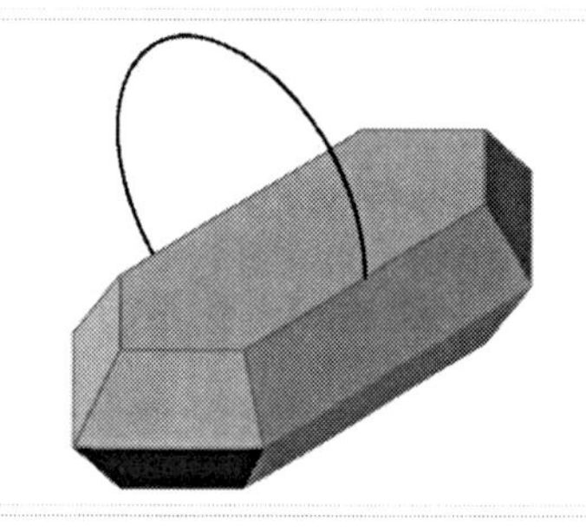

8. Choose the **planar Object** option planar Object.

    Specify start point of slicing plane or [planar Object/Surface/Zaxis/View/XY/ YZ/ZX/3points] <3points>: ***O***

9. Select the circle …

    Select a circle, ellipse, arc, 2D-spline, 2D-polyline to define the slicing plane:

10. … and pick the northernmost point on the emerald.

    Specify a point on desired side or [keep Both sides] <Both>:

Your drawing looks like the figure at right.

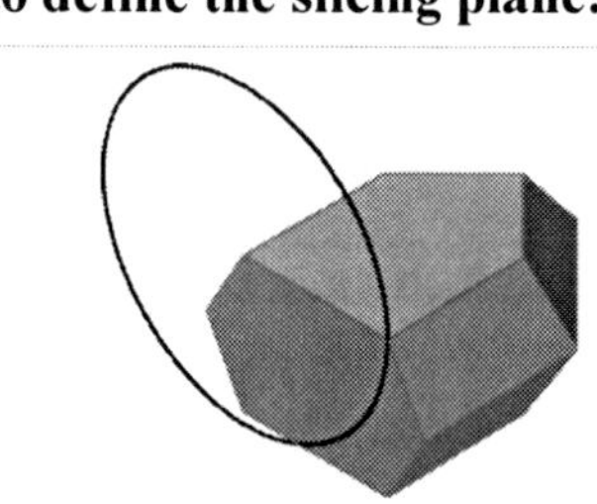

11. Undo the changes.

    Command: ***u***

12. Let's try the **Surface** option. First, thaw the **obj4** layer. Notice the surface that intersects the emerald.

13. Repeat Steps 1 and 2.

    Command: ***sl***

14. Choose the **Surface** option Surface.

    Specify start point of slicing plane or [planar Object/Surface/Zaxis/View/XY/YZ/ZX/ 3points] <3points>: ***s***

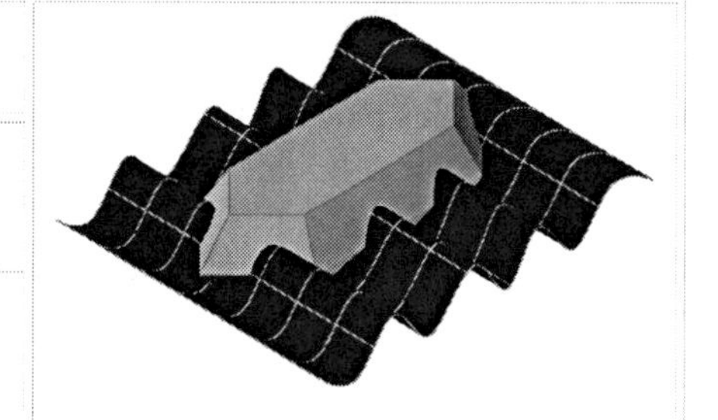

9.1.4A: Slicing 3D Solid Objects

15. Select the surface …

    Select a surface:

16. … and keep both sides of the emerald.

    Specify a point on desired side or [keep Both sides] <Both>: ***[ENTER]***

17. Erase the surface and the top of the emerald and freeze the **obj4** layer.
    Your drawing looks like the figure at right. (Pretty cool? Imagine what you can do with a really involved surface!)

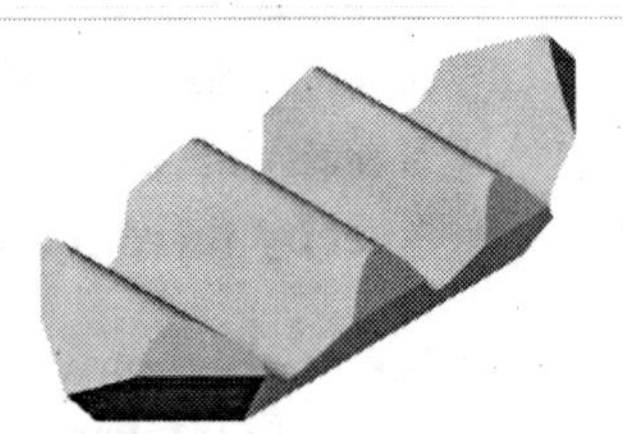

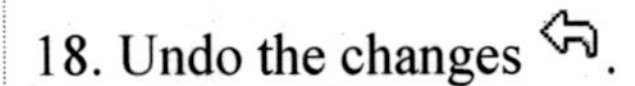

18. Undo the changes.

    Command: ***u***

19. Save the drawing.

    Command: ***qsave***

We've seen the most common uses of the ***Slice*** command, but take some time and experiment with the other options. As I said, the ***Slice*** command is slightly more involved than the commands we learned earlier, but you can see why.

Most people find one method of doing things easier than other methods. With the ***Slice*** command, AutoCAD gives you plenty of methods from which to choose. Knowing all the ways to accomplish an intended goal, however, may save you some time and hassle later when the preferred method refuses to work.

Where to Find It:	
Command Line:	***MassProp***
Menu:	Tools – Inquiry – **Region/Mass Properties**
Toolbar:	Inquiry – **Mass Properties**

Suppose you had to determine the volume of the object we created in our last exercise (the cut emerald). Can you think of an easy way? AutoCAD provides a tool to make the calculations downright easy – ***MassProp***. The ***MassProp*** command will compute not only the volume of the selected object but also the mass, bounding box, centroid, moments and products of inertia, radii of gyration, and principle moments and directions about the centroid. All you have to do is enter the command and select the object! (And your boss spent all those years in engineering school learning how to do this on a slide rule!)

9.1.5 Interferences

Where to Find It:	
Command Line:	***Interfere***
Hotkey(s):	***inf***
Ribbon (Tab/Panel):	Home – Solid Editing – **Interfere**
Menu:	Modify – 3D Operations – **Interference Checking**

Using the ***Interfere*** command, you can identify problems cheaply and easily *before* construction finds them.

Interfere identifies places in a drawing where one solid interferes with another. You'll use it more as a checking tool once you've completed the drawing than as a drawing tool itself – although you can use it to draw an interference.

The command sequence looks like this:

Command: ***interfere***

Select first set of objects or [Nested selection/Settings]: ***[Identify the solids you want to check.]***

Select first set of objects or [Nested selection/Settings]: ***[Confirm the selection.]***

Select second set of objects or [Nested selection/checK first set] <checK>: ***[Identify a second set of solids if you wish to check one against the other (otherwise, AutoCAD will check all the objects in the first set against each other).]***

At this point, AutoCAD presents the Interference Checking dialog box (Figure 9.002).

Let's take a look at our options first; then we'll look at the dialog box.

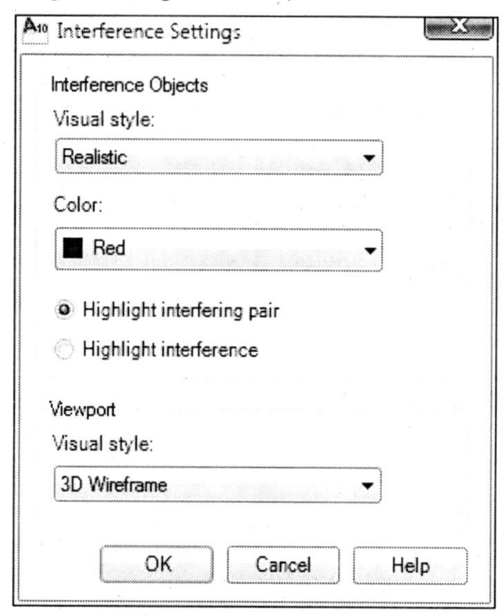

Figure 9.001

- **Select first set of objects** is fairly straightforward. Pick the object you wish to check for interferences. But what about those other options?
  - **Nested selection** allows you to select objects that reside in blocks or Xrefs for checking.
  - **Settings** calls the Interference Settings dialog box (Figure 9.001). Here you can set which visual style and color you'd like AutoCAD to use when showing interferences. You can also have AutoCAD highlight the entire interfering pair or just the interference (using the radio buttons). In the **Viewport** frame, you can tell AutoCAD which visual style to use for the viewport where you're checking. It's a good idea to use different visual styles in these frames to make it easier to see your interferences.
- **Select second set of objects** allows you to select a second set of objects to check against the set you selected at the first prompt. It might be easier just to select all the objects you want to check against each other during the first prompt, and then use the **checK first set** option. This way, AutoCAD checks everything in the selection set against everything else in the selection set. It might bog down in a very large drawing; otherwise, it's a bit easier on you.

Once AutoCAD has completed its check, it presents the Interference Checking dialog box (Figure 9.002). This one is easier than it looks.

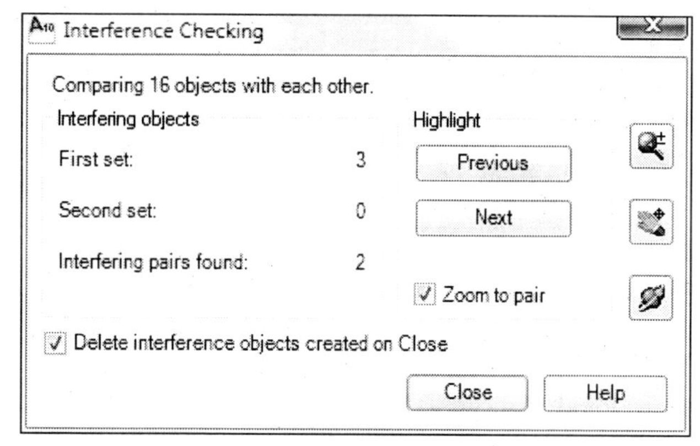

Figure 9.002

- The **Interfering objects** frame tells you how many interfering objects it found in your selection sets. Then it tells you how many interfering pairs it found.
- The **Highlight** frame makes it easier for you to see the interferences – even allowing you to move between them with the **Previous** and **Next** buttons. A check next to **Zoom to pair** makes your efforts more rewarding. You can also use the three buttons to the left of the Highlight frame – **Zoom dynamic**, **Pan**, and **3D Orbit** – to assist your viewing.
- Finally, when you check for interferences, AutoCAD will automatically create an object where the interferences occur. (This is something like using the ***Intersect*** command, but AutoCAD doesn't delete anything.) You can keep this object to further study the interference, or you can **Delete interference objects on close** with a check in the box.

Let's see this one in action.

Do This: 9.1.5A	Finding Interferences

I. Open the *pipe09* file in the C:\Steps3D\Lesson09 folder. The drawing looks like the figure at right. (It's a simple piping plan with a two-level piperack. Can you see any interferences?)

II. Follow these steps.

9.1.5A: FINDING INTERFERENCES

1. Begin the ***Interfere*** command.
   Command: *inf*
2. We'll check the entire drawing for interferences. At the prompt, type ***all***.
   Select first set of objects or [Nested selection/Settings]: *all*
   Select first set of objects or [Nested selection/Settings]: *[ENTER]*
3. We'll check all of the solids against each other, so hit ENTER at the prompt.
   Select second set of objects or [Nested selection/checK first set] <checK>: *[ENTER]*
4. (Refer to Figure 9.002, p.209.) AutoCAD found three solids hitting each other in two instances of interference.
5. Use the **Previous** and **Next** buttons to examine each of the interferences. Interference **2 of 2** is shown at right. (Notice that AutoCAD shows the interference with a Red Realistic visual style as was indicated in the settings dialog box – Figure 9.001, p.209.)

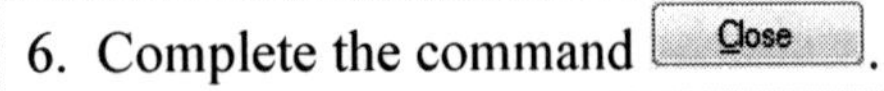

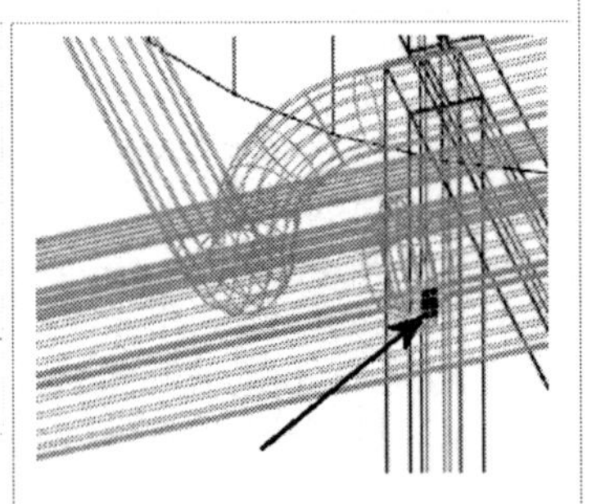

6. Complete the command Close.
7. Exit the drawing without saving it.
   Command: *quit*

In a few simple steps, you've located a problem that might have cost tons of money in redesign and construction costs.

We have another timesaver to see, but first, let's visit some old friends.

9.2 Using Some Old Friends on Solids – *Fillet* and *Chamfer*

Remember how much fun you had drawing the drill jig back in Lesson 8 of the basic text? You used the ***Fillet*** and ***Chamfer*** commands to make the corners. Where would you be if you couldn't use those convenient tools on 3D solids?

Luckily, AutoCAD saw the need to round and mitre corners on solid objects and made the tools available – with a few necessary adjustments. Look at the command sequences when these two are used on solids (we'll begin with the ***Fillet*** command):

Where to Find It:	
Command Line:	*Fillet*
Hotkey(s):	*f*
Ribbon (Tab/Panel):	Home – Modify – **Fillet**
Menu:	Modify – **Fillet**
Toolbar:	Modify – **Fillet**
Tool Palette:	Modify – **Fillet**

Command: *fillet*
Current settings: Mode = TRIM, Radius = 0.5000

Select first object or [Undo/Polyline/Radius/Trim/Multiple]: ***[Select a solid.]***

Enter fillet radius: ***[Enter the desired radius.]***

Select an edge or [Chain/Radius]: ***[Select the edge to fillet.]***

The command begins just as it did when you studied it in the basic text. But when you select a solid object at the **Select first object** prompt, AutoCAD recognizes the solid and asks for some different input.

- It immediately prompts for a **radius** – something it didn't do for a two-dimensional object.
- You can hit ENTER at the **Select an edge** prompt and AutoCAD will assume that you intend to fillet the edge you selected at the **Select first object** prompt. It'll then proceed to fillet that edge. Alternately, you can choose one of the other options:
  - When you pick a single edge on the surface of a solid while using the **Chain** option, AutoCAD should pick the other lines on that surface that are sequential and tangential to the one you selected. [Frankly, I've never been impressed by the way this works (or doesn't work).]
  - The **edge** option (the default) simply allows you to select the edges to fillet one at a time.
  - The **Radius** option allows you to change the radius of each edge you select.

The command sequence for the ***Chamfer*** command is

Where to Find It:	
Command Line:	*Chamfer*
Hotkey(s):	*cha*
Ribbon (Tab/Panel):	Home – Modify – (Fillet flyout) **Chamfer**
Menu:	Modify – **Chamfer**
Toolbar:	Modify – **Chamfer**
Tool Palette:	Modify – **Chamfer**

Command: ***chamfer***

(TRIM mode) Current chamfer Dist1 = 0.5000, Dist2 = 0.5000

Select first line or [Undo/Polyline/ Distance/Angle/Trim/mEthod/Multiple]: ***[Select a solid.]***

Base surface selection...

Enter surface selection option [Next/OK (current)] <OK>: ***[Each edge naturally has two surfaces that are next to it; hit ENTER if the correct surface is highlighted or type* N *to toggle between the surfaces until the appropriate one highlights.]***

Specify base surface chamfer distance <0.5000>: ***[Enter the chamfer distances.]***

Specify other surface chamfer distance <0.5000>:

Select an edge or [Loop]: ***[Select the edge to chamfer.]***

- The first option, as explained, allows you to select the correct surface to chamfer.
- The next two options – **Specify base surface chamfer distance** and **specify other surface chamfer distance** – allow you to accept or change the chamfer distances (notice that there's no **Angle** option here as there is for two-dimensional objects).
- The last option – **Select an edge or [Loop]** – allows you to pick the edges to chamfer individually or, when you use the **Loop** option, collectively around the entire surface.

Let's try the ***Fillet*** and ***Chamfer*** commands on a solid.

Do This: 9.2A	Solid Fillets and Chamfers

I. Open the *flange* file in the C:\Steps3D\Lesson09 folder. If you haven't completed it yet, open the *Flange-done* file instead. We'll countersink the boltholes, mitre the weld neck (the upper cylinder), and fillet the edge.

II. Follow these steps.

9.2A: SOLID FILLETS AND CHAMFERS

1. Let's begin with a simple fillet. Enter the ***Fillet*** command.

    Command: *f*

2. Select the top surface of the flange's base (pick the outer edge).

    Current settings: Mode = TRIM, Radius = 0.5000

    Select first object or [Undo/Polyline/Radius/Trim/Multiple]:

3. Set the radius to ¼ unit.

    Enter fillet radius <0.5000>: .25

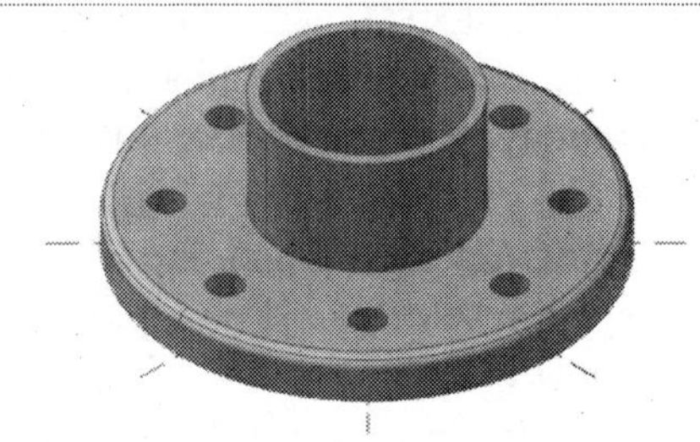

4. Complete the command.

    Select an edge or [Chain/Radius]: *[ENTER]*

Your drawing looks like the figure at right.

5. Now we'll mitre the weld neck. Enter the ***Chamfer*** command.

    Command: *cha*

6. Select the outer edge of the upper cylinder …

    (TRIM mode) Current chamfer Dist1 = 0.5000, Dist2 = 0.5000

    Select first line or [Undo/Polyline/Distance/Angle/Trim/mEthod/Multiple]:

7. … and adjust the surface selected until just the outer circle of the cylinder is highlighted (right).

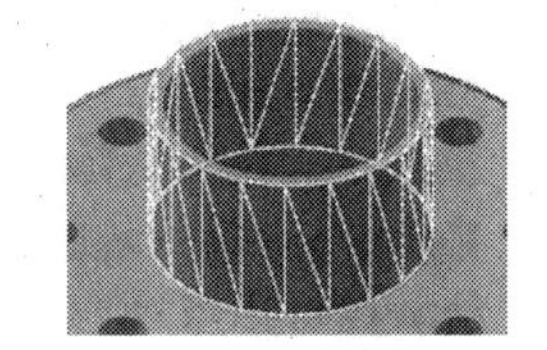

    Base surface selection...

    Enter surface selection option [Next/OK (current)] <OK>: *n*

    Enter surface selection option [Next/OK (current)] <OK>: *[ENTER]*

8. Set the chamfer distances to 1/8 unit.

    Specify base surface chamfer distance <0.5000>: *.125*

    Specify other surface chamfer distance <0.5000>: *.125*

9. Select the outer edge of the upper cylinder and then confirm the selection.

    Select an edge or [Loop]:

    Select an edge or [Loop]: *[ENTER]*

Your drawing looks like the figure at right.

10. Now countersink the bolt holes. (This'll be easier with a **3D Wireframe** visual style.) Repeat the ***Chamfer*** command.

    Command: *cha*

11. Select the upper surface of the base of the flange.

    (TRIM mode) Current chamfer Dist1 = 0.1250, Dist2 = 0.1250

    Select first line or [Undo/Polyline/Distance/Angle/Trim/mEthod/Multiple]:

    Base surface selection...

    Enter surface selection option [Next/OK (current)] <OK>: *[ENTER]*

9.2A: Solid Fillets and Chamfers

12. Set the chamfer distances to ¼ unit.

    Specify base surface chamfer distance <0.1250>: ***.25***

    Specify other surface chamfer distance <0.1250>: ***.25***

13. Select the upper circle around each of the bolt holes.

    Select an edge or [Loop]:

14. Complete the command, reset the **Realistic** visual style, and remove the isolines.

    Select an edge or [Loop]: ***[ENTER]***

Your drawing looks like the figure at right.

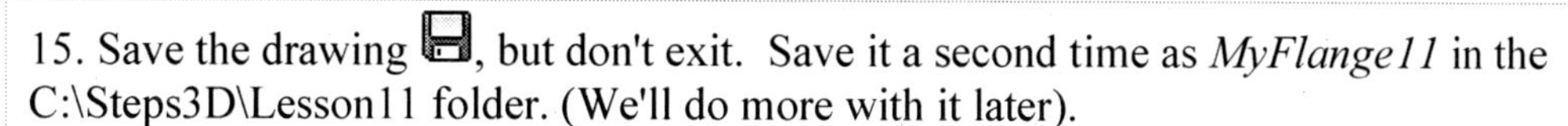

15. Save the drawing, but don't exit. Save it a second time as *MyFlange11* in the C:\Steps3D\Lesson11 folder. (We'll do more with it later).

    Command: ***qsave***

    Command: ***saveas***

Close the drawing.

Now let's look at that other timesaver – the ***Section*** command.

9.3 Creating Cross Sections with *SectionPlane*

Where to Find It:	
Command Line:	***SectionPlane***
Hotkey(s):	***splane***
Ribbon (Tab/Panel):	Home – Section – **Section Plane**
	Mesh Modeling – Section – **Section Plane**
Menu:	Draw – Modeling – **Section Plane**

Have you ever completed the tedious cross section of an object only to discover that you missed something (perhaps a line or an arc that was difficult to see)? Well, AutoCAD has just the tool for you!

The ***SectionPlane*** command creates cross sections of solid objects.

Let's see.

Command: ***sectionplane***

Select face or any point to locate section line or [Draw section/Orthographic]: ***[Locate the section plane.]***

Specify through point: ***[Pick a second point on the plane.]***

That looks deceptively simple. What about those options?

- **Draw section** allows you to define your section with multiple points – producing something of a jogged section. It prompts:

  Specify start point:

  Specify next point:

  Specify next point or ENTER to complete:

  Specify point in direction of section view: ***[Pick a point on the cutting plane.]***

- **Orthographic** lets you pick from some preset settings to make a straight section plane. It prompts:

  Align section to: [Front/bAck/Top/Bottom/Left/Right] <Front>:

But the really cool part of the ***SectionPlane*** command occurs after you've created the section plane. You can move it around adjusting the section. You can even use grips to adjust what you see in terms of **Section Plane**, **Section Boundary**, and **Section Volume**.

- **Section Plane** is the plane of the section.
- A **Section Boundary** shows the XY extent of the section (with a boundary box). The Z-extent is infinite.
- **Section Volume** uses a 3D box to show the extent of the section in all directions.

The ***SectionPlane*** command doesn't look like much when you use it. You need to "activate" the section first with the ***LiveSection*** command. You'll see this in action in our exercise.

Where to Find It:	
Command Line:	***LiveSection***
Ribbon (Tab/Panel):	Home – Section (subpanel) – **Live Section**
	Mesh Modeling – Section – **Live Section**

This command will become a lot clearer with an example.

Do This: 9.3A	**Creating Cross Sections with *SectionPlane***

I. Reopen the *flange* file (or the *flange-done* file) in the C:\Steps3D\Lesson09 folder.

II. Follow these steps.

9.3A: CREATING CROSS SECTIONS WITH SECTIONPLANE

1. Enter the ***SectionPlane*** command.
   Command: *splane*

2. Let's draw a jogged section. Use the **Draw Section** option Draw section.
   Select face or any point to locate section line or [Draw section/Orthographic]: *d*

3. Start left of the flange, move to the center, then move to the back.
   Specify start point: *[Start left of the flange.]*
   Specify next point: _cen of *[Use an OSNAP to get the center of the flange.]*
   Specify next point or ENTER to complete: *[Move to the back.]*
   Specify next point or ENTER to complete: *[ENTER]*

4. Now tell AutoCAD which part of the flange you want to see. Pick somewhere in the BACK-LEFT area.
   Specify point in direction of section view:

   The results look something like the figure at right. (Doesn't look like much yet, does it?)

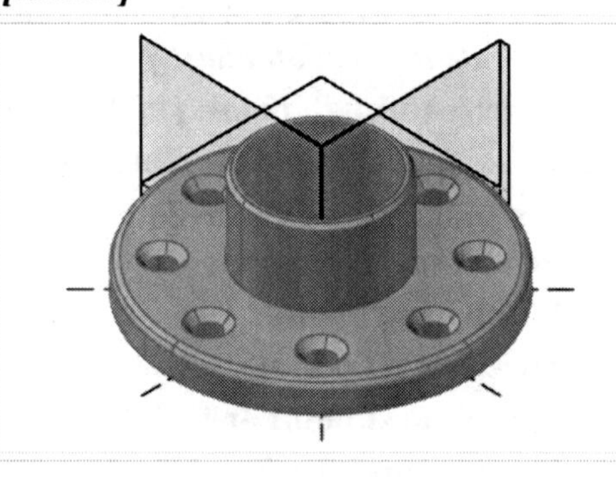

5. Enter the ***LiveSection*** command and select the section you just drew. (Select the piece on the "ground.")

Command: *livesection*

Select section object:

Now you can see the section (right).

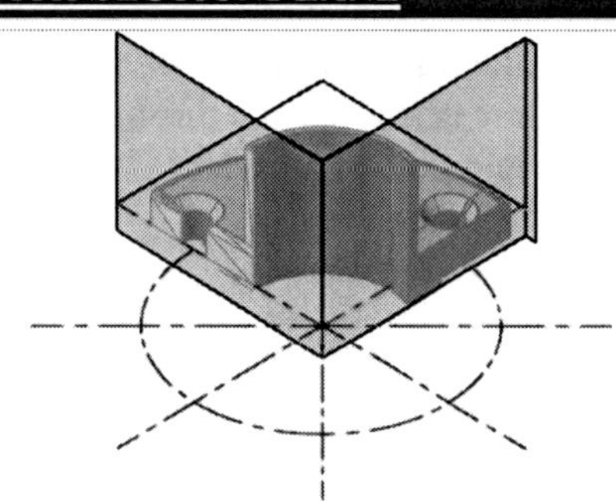

6. Undo the changes and we'll create a section plane we can do a little more with.

Command: *undo*

7. Enter the ***SectionPlane*** command .

Command: *sectionplane*

8. Create a **Front** **Orthographic**

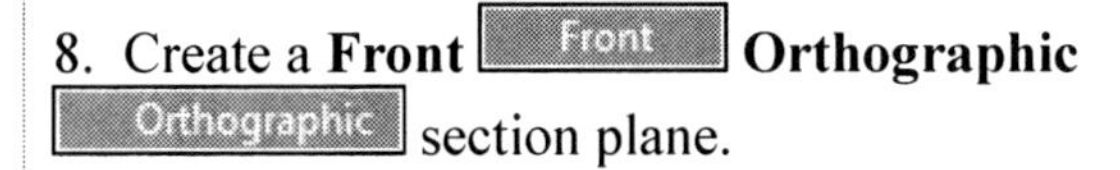

section plane.

Select face or any point to locate section line or [Draw section/ Orthographic]: *o*

Align section to: [Front/bAck/Top/Bottom/Left/ Right] <Front>: *f*

Notice that the ***LiveSection*** works automatically here.

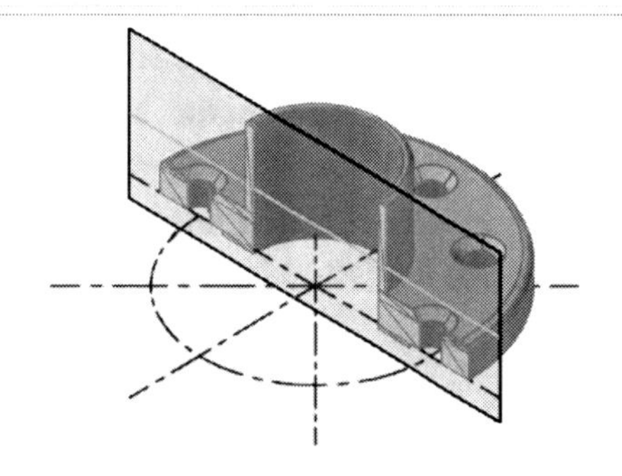

9. Now we can have some fun. Pick on the plane and notice the gizmo and grips. Use the gizmo to move the plane back and forth. Notice that the section changes dynamically relative to the position of the plane. (This works easier if you turn off your OSNAPs.)

10. Use the center (flip) arrow to flip the section. (Flip it back after you've experimented.)

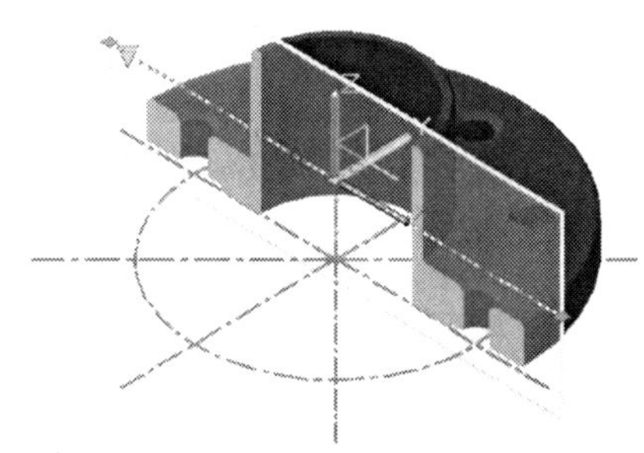

11. Use the ***Rotate*** gizmo to rotate the section about 45° around the center of the flange.

command: *ro*

Notice that the section itself rotates – not the object being sectioned.

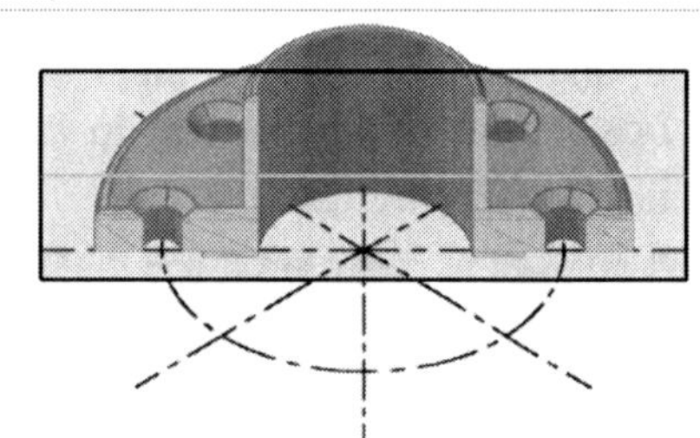

9.3A: Creating Cross Sections with SectionPlane

12. We like this section, so let's keep it. Select the section line (the gray line through the center of the plane), right click, and select **Generate 2D/3D section** from the menu

Generate 2D/3D section... .

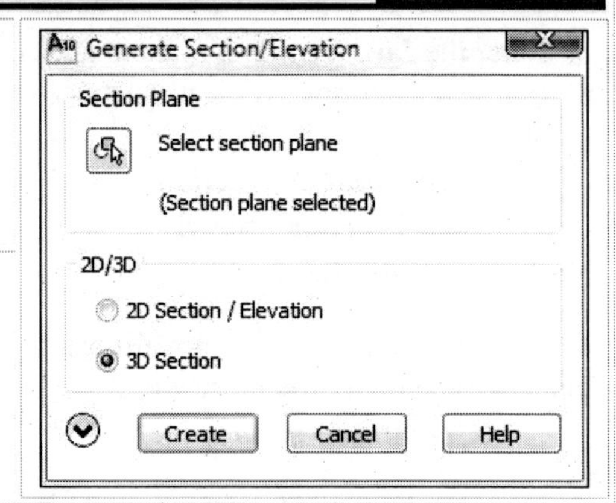

13. AutoCAD helps you out with a dialog box (right). Take a minute to explore the possibilities here, then create a 3D section.

14. Place the section next to the flange for comparison.

Units: Inches Conversion: 1.0000

Specify insertion point or [Basepoint/Scale/X/Y/Z/Rotate]:

Enter X scale factor, specify opposite corner, or [Corner/XYZ] <1>: ***[ENTER]***

Enter Y scale factor <use X scale factor>: ***[ENTER]***

Specify rotation angle <0>: ***[ENTER]***

It looks like the following figure.

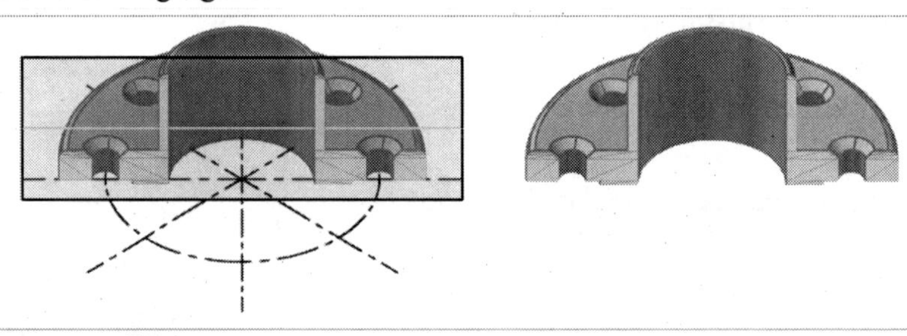

15. Save the drawing.

Command: ***qsave***

9.4 Extra Steps

Go back to the list of items you created at the end of Section 9.1.2, p.204. Take a few hours (or an afternoon) and see how many of them you can create in AutoCAD using the tools you learned in this lesson. I can't think of a better way to gain experience (or identify questions).

Some hints for this exercise:

- Don't attempt to draw anything that won't fit in a shoebox.
- One of those flexible, 6" rulers with inches on one side and millimeters on the other will serve you well (now and in the future).
- Try (at first) to limit yourself to objects that'll require no more than three of the basic shapes you've learned.
- You'll find other terrific objects to draw in garages and kitchens.

9.5 What Have We Learned?

Items covered in this lesson include:

- *Tools used to create composite solids*
  - ***Union***
  - ***Subtract***
  - ***Intersect***
  - ***Slice***
  - ***Interfere***
- *Tools used to shape solids*
  - ***Fillet***
  - ***Chamfer***
- *Other Solid Modeling tools*
  - ***SectionPlane***
  - ***Massprop***
  - ***LiveSection***

Well, what do you think? Wouldn't it have been fun to have these tools when you were playing with blocks as a child?

As I promised, you've stopped drawing pictures of things and have actually begun creating objects using the tools in AutoCAD's "shop." I hope you can sense the potential of these tools from what you've seen here.

> When I was in junior high school, I read a book by Jack London called *Call of the Wild.* It was about a dog that was taken from an easy life in the Northwest and forced to pull a sled in the Klondike during the Gold Rush. Buck (the dog) had many adventures (learning experiences) as he adapted to the wild frontier life of the arctic. But all the while – with increasing intensity – he felt a call from the wild to move out on his own. He experimented with the urging – often leaving camp for days at a time to explore the wilderness. In the end, after learning all that he could in the safety and comfort of the camps, he answered the call and moved out to live with the other wild creatures.
>
> At this point in your AutoCAD training, you should be experimenting on your own, just as Buck did. You'll find thrills – and chills – as you discover things about the software that even the masters don't know. You'll make some mistakes, but the adventure lies in overcoming the mistakes (that's what makes learning fun!).
>
> In a few short chapters, you'll be on your own (with the other wild creatures in the design world). Learn all that you can now!

9.6 Exercises

1. through 8. Create the "su" drawings in Appendix B using solids. Use solid primitives and the composite solid creation tools you learned in this lesson to make each drawing a single object. Save the drawings in the C:\Steps3D\Lesson09 folder.

9. Create the hinge. Refer to the following guidelines.
    9.1. The hinge is a single solid object.
    9.2. Fully dimension the hinge as shown.
    9.3. Place it with the title block of your choice on an 11" x 8½" sheet of paper.
    9.4. Save the drawing as *MyHinge* in the C:\Steps3D\Lesson09 folder.

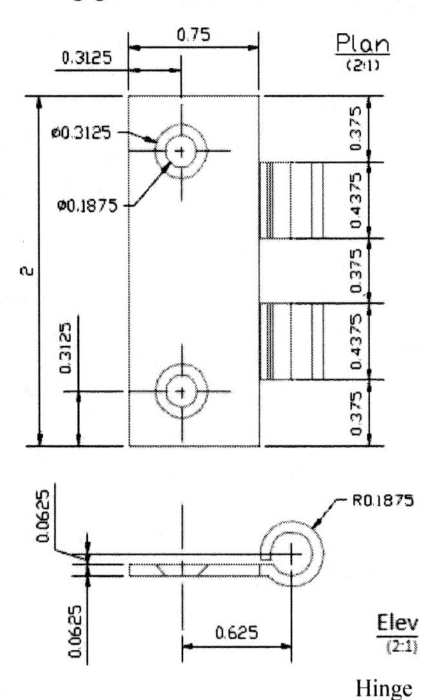

Hinge

10. Create the *Flange* (p.219). Refer to the following guidelines.
    10.1. The flange is a single solid object.
    10.2. Place it with the title block of your choice on an 11" x 8½" sheet of paper.
    10.3. Bold holes are 3/8" diameter.
    10.4. The center hole is 2¼" diameter.
    10.5. Save the drawing as *MyFlange* in the C:\Steps3D\Lesson09 folder.

11. Create the *Flange Gear* (p.219). Refer to the following guidelines.
    11.1. The flange gear is a single solid object.
    11.2. Fully dimension the object as shown.
    11.3. Place it with the title block of your choice on an 11" x 8½" sheet of paper.
    11.4. Save the drawing as *MyAnchor* in the C:\Steps3D\Lesson09 folder.

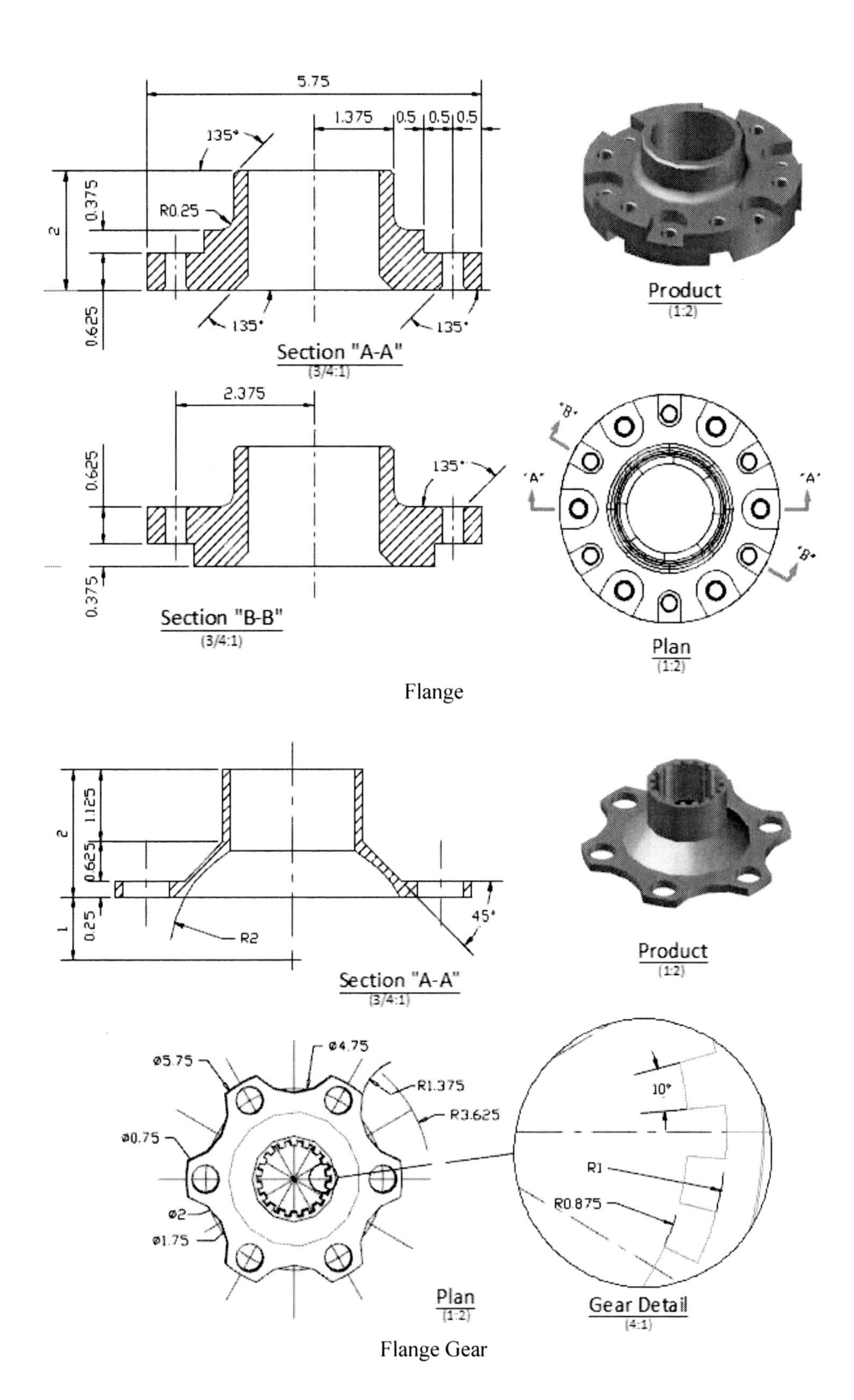

Flange

Flange Gear

12. Create the *Floating Support and Anchor*. Refer to the following guidelines. (Hint: I created the top of the brace with wedges whose length and height were 4".)
    12.1. Each piece is a single solid object.
    12.2. Fully dimension the object as shown.
    12.3. Place it with the title block of your choice on an 11" x 8½" sheet of paper.
    12.4. Save the drawing as *MyFlangeGear* in the C:\Steps3D\Lesson09 folder.

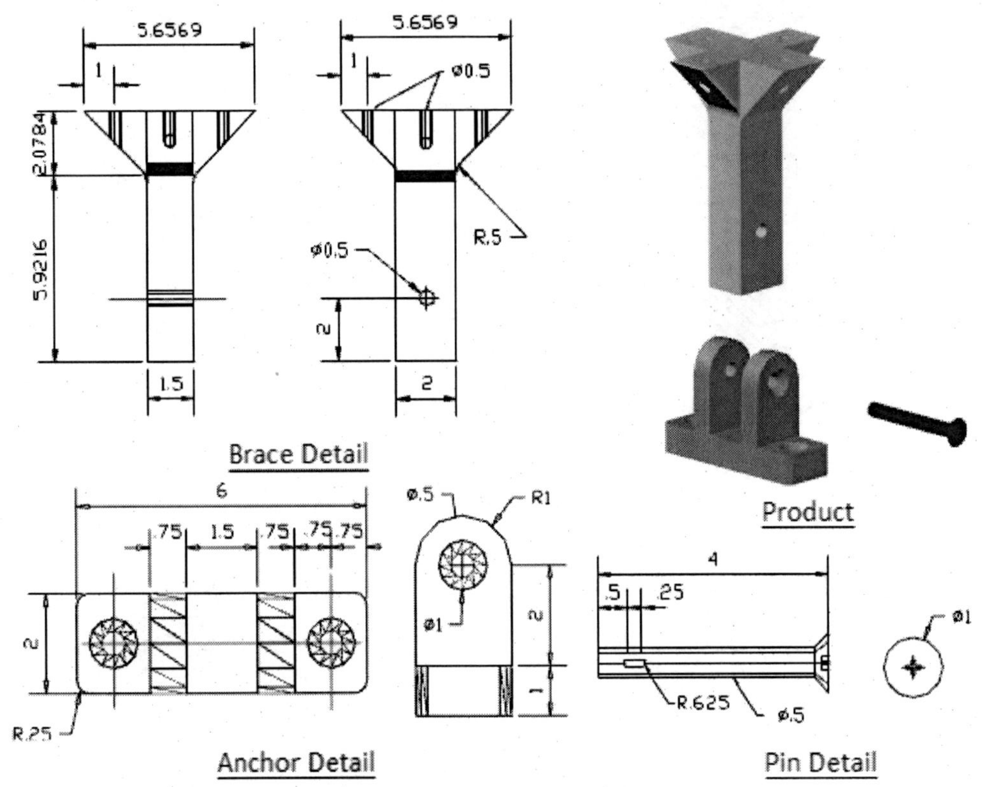

Floating Support and Anchor

13. Create the *Table Lamp* (p.221). Refer to the following guidelines.
    13.1. The base is a solid object.
    13.2. The top is 1/8" thick hollow glass ball.
    13.3. Fully dimension the object as shown.
    13.4. Place it with the title block of your choice on an 11" x 8½" sheet of paper.
    13.5. Save the drawing as *MyTableLamp* in the C:\Steps3D\Lesson09 folder.

14. Create the *Dining Chair* (p.221). Refer to the following guidelines.
    14.1. Each leg begins at 1" diameter, but balloons to 1½" in the middle.
    14.2. The leg bracing is ¾" diameter; the back dowels are ½" diameter.
    14.3. The back support is 1" squared.
    14.4. Fully dimension the object as shown.
    14.5. Place it with the title block of your choice on an 11" x 8½" sheet of paper.
    14.6. Save the drawing as *MyDiningChair* in the C:\Steps3D\Lesson09 folder.

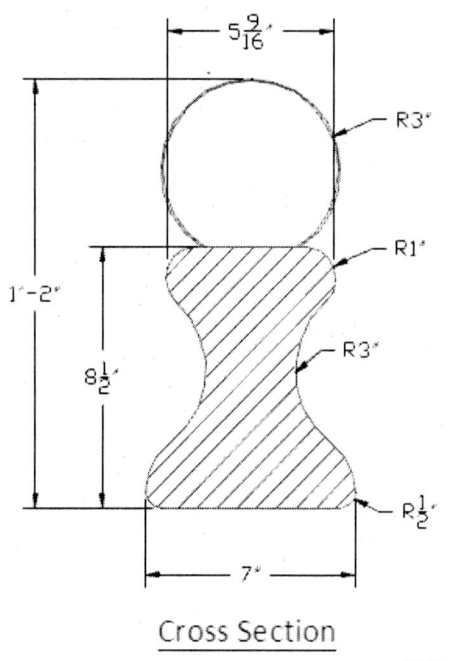

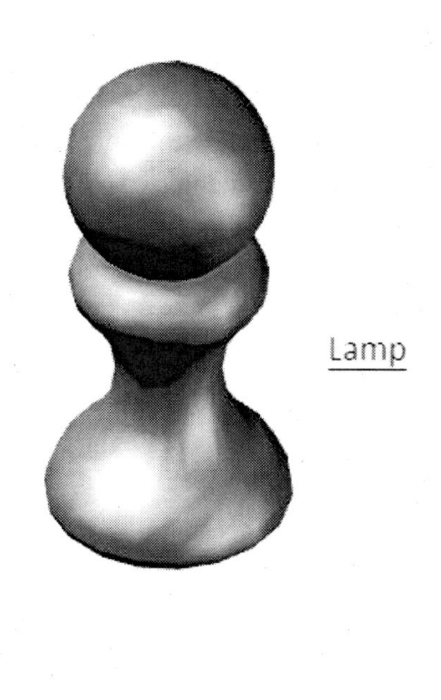

Table Lamp

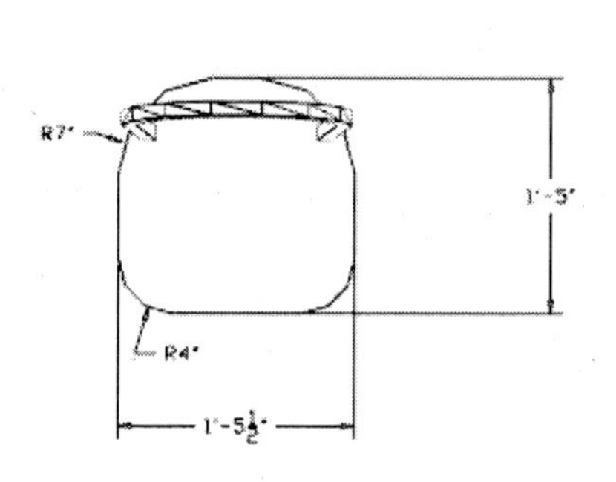

Plan

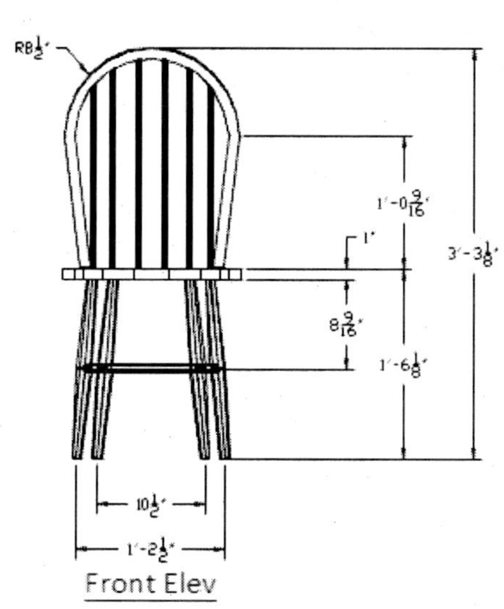

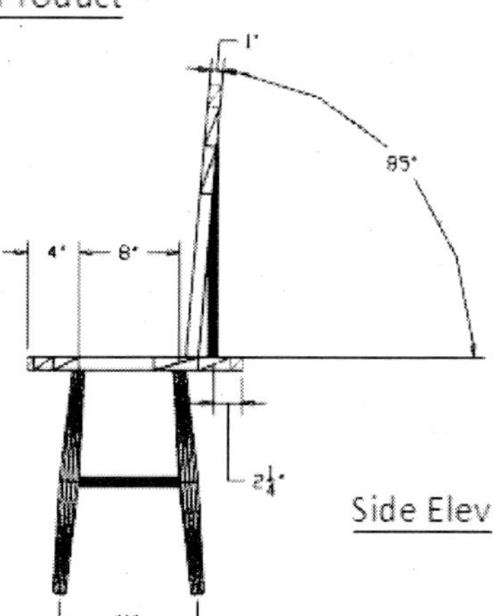

Dining Chair

15. Create the *Fountain*. Refer to the following guidelines.
    15.1. Each piece (including the water) is a separate solid object.
    15.2. Fully dimension the objects as shown.
    15.3. Save the drawing as *MyFountain* in the C:\Steps3D\Lesson09 folder.

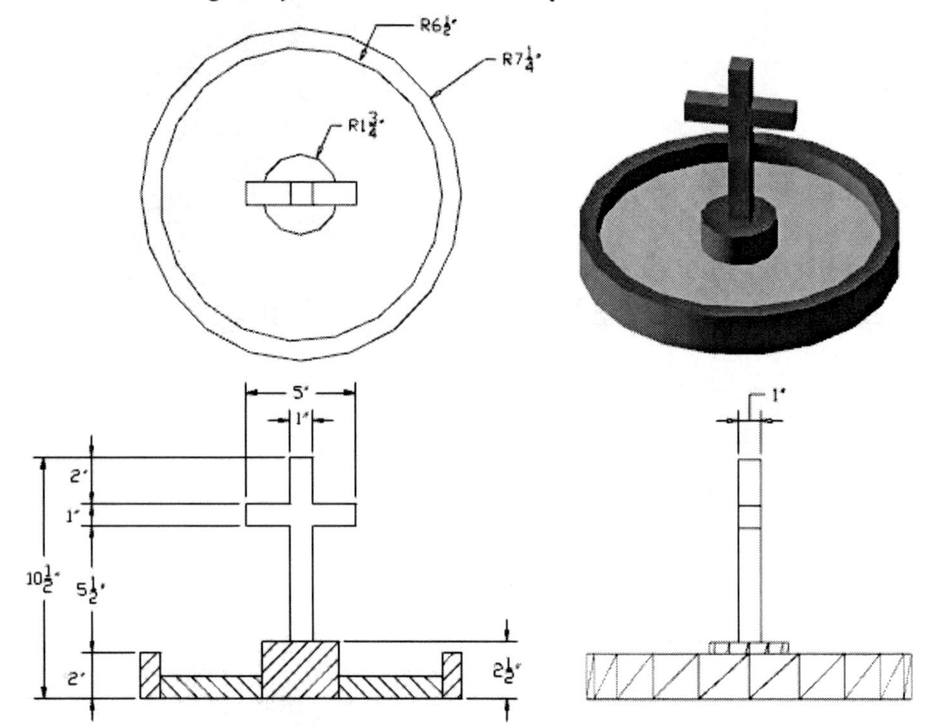

Thanks to Casey Peel for permission to use this drawing.

9.7	**For Web-Based Review Questions, visit: http://foragerpub.com/AcadFiles/2010/2010.htm**

Lesson

Editing 3D Solids

Following this lesson, you will:

- ✓ *Know how to edit 3D Solids with grips and gizmos*
- ✓ *Know how to edit 3D Solids with some basic tools*
  - ***Move***
  - ***Rotate***
  - ***Extrude***
  - ***XEdges***
- ✓ *Know how to use the **SolidEdit** command's various tools:*
  - ***Face***
    - ***Extrude***
    - ***Move***
    - ***Rotate***
    - ***Offset***
    - ***Taper***
    - ***Delete***
    - ***Copy***
    - ***color***
  - ***Edge***
    - ***Copy***
    - ***coLor***
  - ***Body***
    - ***Imprint***
    - ***seParate solids***
    - ***Clean***
    - ***Shell***
- ✓ *Know how to use the **SolidCheck** system variable*

*Over the course of your studies, I've pointed out many of AutoCAD's redundant features. (Indeed, by now you know me to be a great advocate of AutoCAD redundancy.) In this lesson, we'll present a feature that duplicates some other features of 3D solid editing. But this command – **SolidEdit** – has some routines and twists that make it a favorite to solid modelers everywhere.*

*Unfortunately, **SolidEdit** is not a simple command. In fact, it's a command in the tradition of **PEdit** or **Splinedit**. In other words, expect a multi-tiered command with multiple options per tier.*

On the fortunate side, however, AutoCAD has greatly increased the involvement of grips and gizmos in its solid editing arsenal. These might make you a bit more comfortable with an otherwise daunting task. Additionally, AutoCAD makes it easy to select a single face as an object for some of its easier modifying tools!

*I'll show you all three approaches; I'll begin with the **SolidEdit** command approach, show you what grips and gizmos can do, and where possible, show you the basic tools you can use. (The trick to editing solids, as you'll soon see, lies in knowing which tool to use for which job!)*

Let's get started.

10.1 A Single Command, But It Does So Much - *SolidEdit*

Actually, as a command by itself, ***SolidEdit*** doesn't accomplish a thing. The ***SolidEdit*** command should be considered a ticket into a realm where many new commands dwell – each capable of something beneficial to the solid modeler.

If we tried to study ***SolidEdit*** as a single command, we might find it somewhat overwhelming. But luckily, AutoCAD divided the command options into three categories – **Face**, **Edge**, and **Body**. In fact, AutoCAD's ***SolidEdit*** command prompt looks like this:

Enter a solids editing option [Face/Edge/Body/Undo/eXit] <eXit>:

Each of the options (categories) presents a separate tier of choices designed to help modify a solid object. (The other two options – **Undo** and **eXit** – are the standard options for most commands.) We'll use these natural divisions to study each option as a category, or grouping of several routines.

AutoCAD also provides a **Solid Editing** ribbon panel (**Home** tab) and toolbar, each with buttons that quickly access the options in the categories. We'll use these buttons throughout our lesson.

10.2 Changing Faces – The Face Category

The Face category contains the bulk of ***SolidEdit***'s commands. This category includes routines (or command options) designed to alter the faces of a solid.

> Don't confuse the face on a 3D solid with a 3D face. Remember, you'll find a 3D face on a *surface model*. A solid model has a 3D *solid* face.

To get to the Face category's options, follow this sequence:

Command: ***solidedit***

Solids editing automatic checking: SOLIDCHECK=1

Enter a solids editing option [Face/Edge/Body/Undo/eXit] <eXit>: ***f***

Enter a face editing option

[Extrude/Move/Rotate/Offset/Taper/Delete/Copy/coLor/mAterial/Undo/eXit] <eXit>:

You can then select the routine you wish to use.

Of course, an easier way to access a specific ***SolidEdit*** routine would be simply to pick the desired choice on the ribbon's Solid Editing panel or the Solid Editing toolbar.

Let's look at each of the routines as though they were individual commands.

10.2.1 Extruding the Thickness of a 3D Solid Face

How do you change the individual faces of a 3D solid?

Well, one way is by using the **Extrude** option of the Face category. The option works something like the ***Extrude*** command – but only on selected faces.

Where to Find It:

Command Line:	***Solidedit – Face – Extrude***
Ribbon (Tab/Panel):	Home – Solid Editing – (Faces flyout) **Extrude faces**
Menu:	Modify – Solid Editing – **Extrude faces**
Toolbar:	Solid Editing – **Extrude faces**

Here's the sequence:

Command: ***solidedit***

Solids editing automatic checking: SOLIDCHECK=1

Enter a solids editing option [Face/Edge/Body/Undo/eXit] <eXit>: ***f***

Enter a face editing option [Extrude/Move/Rotate/Offset/Taper/Delete/Copy/coLor/mAterial/Undo/eXit] <eXit>: ***e [Select the Extrude option.]***

Select faces or [Undo/Remove]: ***[Pick a face or an edge(s) of the face you wish to extrude – AutoCAD will highlight the two faces that form that edge.]***

Select faces or [Undo/Remove/ALL]: ***[Confirm the selection set.]***

Specify height of extrusion or [Path]: ***[How far do you want to extrude?]***

Specify angle of taper for extrusion <0>: ***[How much of a taper do you want?]***

AutoCAD will extrude the selected face and return to the Face category of the ***SolidEdit*** command.

Let's give it a try.

Do This: 10.2.1A — Extruding a 3D Solid Face

I. Open the *SE-Box* file in the C:\Steps3D\Lesson10 folder. The drawing is a simple box. (The current viewpoint is 1,-2,1).

II. Follow these steps.

10.2.1A: EXTRUDING A 3D SOLID FACE

1. Enter the command sequence shown to access the **Extrude** option of the Face category.

Alternately, you can pick the **Extrude faces** button on the **Solid Editing** panel. (Note: Picking the **Extrude faces** button replaces the entire sequence shown in this step.)

Command: ***solidedit***

Solids editing automatic checking: SOLIDCHECK=1

Enter a solids editing option [Face/Edge/Body/Undo/eXit] <eXit>: ***f***

Enter a face editing option [Extrude/Move/Rotate/Offset/Taper/Delete/Copy/coLor/mAterial/Undo/eXit] <eXit>: ***e***

2. Select the upper face.

Select faces or [Undo/Remove]:

10.2.1A: EXTRUDING A 3D SOLID FACE

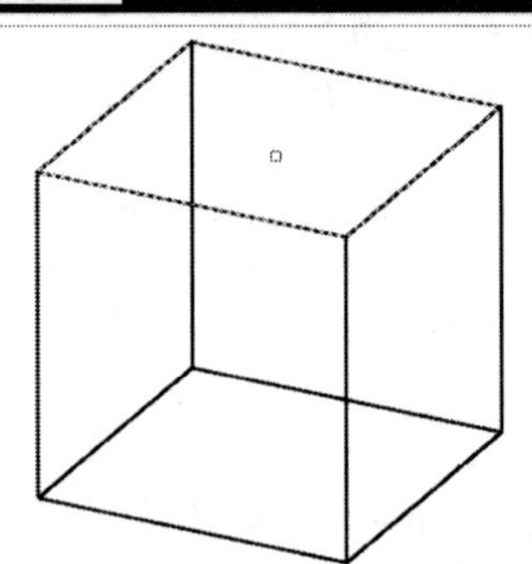

3. Confirm the selection set.

    Select faces or [Undo/Remove/ALL]:
    [ENTER]

4. Tell AutoCAD to use an extrusion height of **1** and a taper angle of **30°**.

    Specify height of extrusion or [Path]: ***1***
    Specify angle of taper for extrusion <0>: ***30***

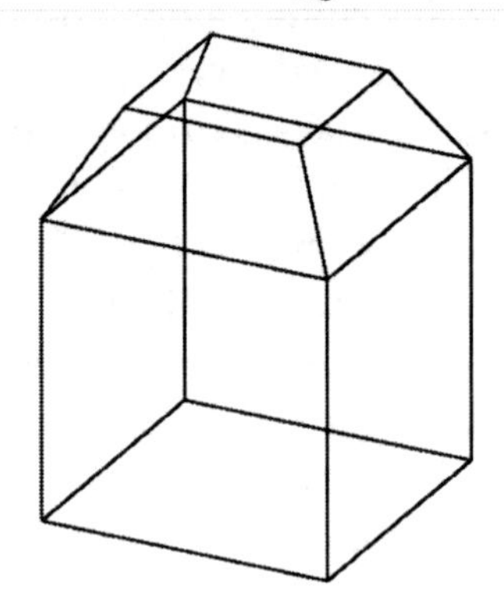

5. By default, AutoCAD validates that the task you've outlined is possible and then extrudes the object.
Hit ***enter*** twice to exit the command.

    Solid validation started.
    Solid validation completed.
    Enter a face editing option
    [Extrude/Move/Rotate/Offset/Taper/Delete/ Copy/coLor/mAterial/Undo/eXit] <eXit>:
    [ENTER]
    Solids editing automatic checking: SOLIDCHECK=1
    Enter a solids editing option
    [Face/Edge/Body/Undo/eXit] <eXit>:
    [ENTER]

Your drawing looks like the figure at right.

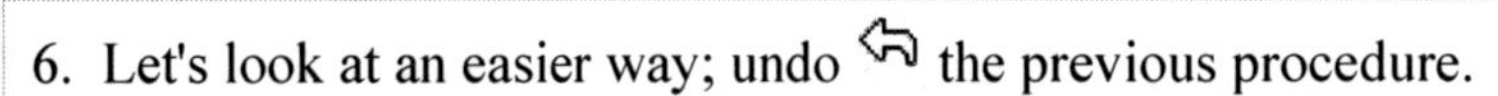

6. Let's look at an easier way; undo the previous procedure.

    Command: ***u***

7. We'll use AutoCAD's single-face selection ability to bypass this complicated procedure in favor of a simple command approach. Begin the ***Extrude*** command.

    Command: ***ext***

8. Hold down the CTRL key while selecting just the top face of the box.

    Select objects to extrude:
    Select objects to extrude: ***[ENTER]***

9. Use the same **Taper angle** and **height** as before.

    Specify height of extrusion or [Direction/Path/Taper angle]: ***t***
    Specify angle of taper for extrusion: ***30***
    Specify height of extrusion or [Direction/Path/Taper angle]: ***1***

10. Note the difference between this extrusion and the last – the extruded face (the second procedure) has produced a separate solid. But that's easily fixed. Enter the ***Union*** command.

    Command: ***uni***

11. Select the top and bottom solids.

    Select objects:

10.2.1A: EXTRUDING A 3D SOLID FACE

12. Save the drawing.

Command: ***qsave***

Which procedure did you find easier? Unfortunately, the ***SolidEdit*** command doesn't have simple command substitutes for all its tools!

Let's look at a similar option.

10.2.2 Moving a Face on a 3D Solid

The **Move** routine of the Face category proves to be quite handy when it becomes necessary to relocate part of a 3D solid – such as a bolt hole – that was improperly placed. (Okay, it's not as handy as other approaches – but let's not get ahead of ourselves!)

Where to Find It:	
Command Line:	***Solidedit – Face – Move***
Ribbon (Tab/Panel):	Home – Solid Editing – (Faces flyout) **Move faces**
Menu:	Modify – Solid Editing – **Move faces**
Toolbar:	Solid Editing – **Move faces**

The command sequence looks like this (beginning with the ***SolidEdit*** command's **Face** option):

Enter a face editing option [Extrude/Move/Rotate/Offset/Taper/Delete/Copy/coLor/mAterial/Undo/eXit] <eXit>: ***m [select the Move option]***

Select faces or [Undo/Remove]: ***[The face selection options are the same as those in the Extrude option.]***

Select faces or [Undo/Remove/ALL]:

Specify a base point or displacement: ***[The next options are identical to the basic two-dimensional*** **Move** ***command's options.]***

Specify a second point of displacement:

AutoCAD will then move the selected face and return to the Face category of the ***SolidEdit*** command.

Let's give it a try.

Do This: 10.2.2A	Moving a 3D Solid Face

I. Be sure you're still in the *SE-Box* file in the C:\Steps3D\Lesson10 folder. If not, please open it now.

II. Thaw the **obj2** layer. Notice the cylinder that appears inside the box.

III. Follow these steps.

10.2.2A: MOVING A 3D SOLID FACE

1. Use the ***Subtract*** command to remove the cylinder from the box.

Command: ***su***

Your drawing looks like the figure at right. (I've shown it here using the **3D Hidden** visual style.)

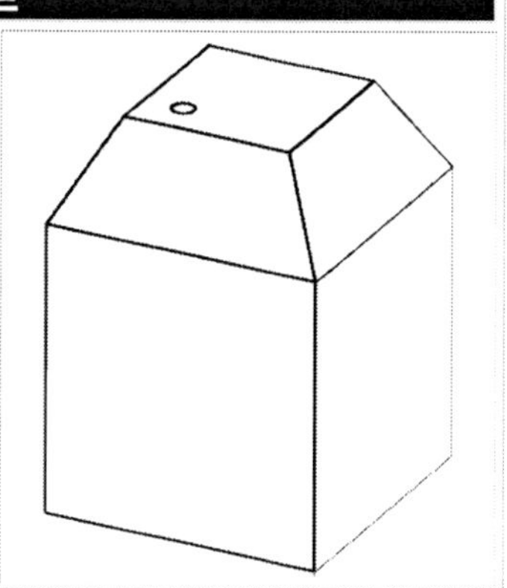

10.2.2A: Moving a 3D Solid Face

2. Enter the command sequence shown to access the **Move** option of the Face category. Alternately, you can pick the **Move Faces** button on the **Solid Editing** panel.

Command: ***solidedit***

Solids editing automatic checking: SOLIDCHECK=1

Enter a solids editing option [Face/Edge/Body/Undo/eXit] <eXit>: ***f***

Enter a face editing option [Extrude/Move/Rotate/Offset/Taper/Delete/Copy/coLor/mAterial/Undo/eXit] <eXit>: ***m***

3. Select the cylinder inside the box (restore the **3D Wireframe** visual style if necessary).

Select faces or [Undo/Remove]:

Select faces or [Undo/Remove/ALL]: ***[ENTER]***

4. I'll use the displacement method to move the hole **1** unit east and **1** unit north on the 3D solid.

Specify a base point or displacement: ***1,1***

Specify a second point of displacement: ***[ENTER]***

5. As with the **Extrude** routine, AutoCAD validates the procedure before actually moving the hole.
Hit ***enter*** twice to exit the command.

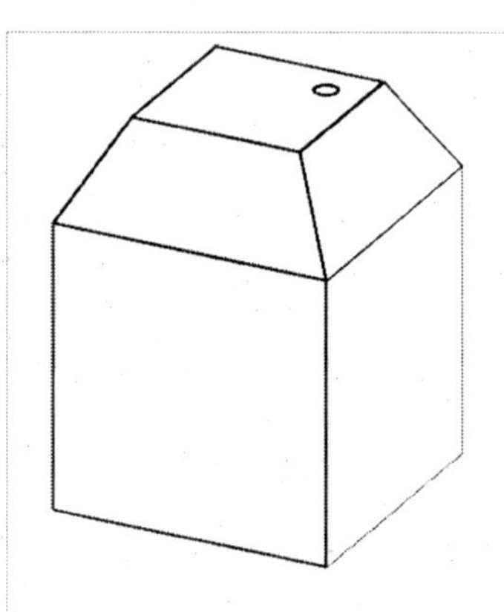

Solid validation started.

Solid validation completed.

Enter a face editing option [Extrude/Move/Rotate/Offset/Taper/Delete/Copy/coLor/mAterial/Undo/eXit] <eXit>: ***[ENTER]***

Solids editing automatic checking: SOLIDCHECK=1

Enter a solids editing option [Face/Edge/Body/Undo/eXit] <eXit>: ***[ENTER]***

Your drawing looks like this (**3D Hidden** visual style shown).

6. As we found in our last exercise – there's an easier approach. Undo the move.

Command: ***u***

7. Let's try the single-face selection approach with the simple ***Move*** command.

Command: ***m***

8. Hold down the CTRL key and select the cylinder.

Select objects:

Select objects: ***[ENTER]***

9. And use the displacement method to move it as before.

Specify base point or [Displacement] <Displacement>: ***1,1***

Specify second point or <use first point as displacement>: ***[ENTER]***

10. Save the drawing, but don't exit.

Command: ***qsave***

You can also use the **Move** gizmo to move the faces just as you did with meshes!

10.2.3 Rotating Faces on a 3D Solid

Where to Find It:	
Command Line:	***Solidedit – Face – Rotate***
Ribbon (Tab/Panel):	Home – Solid Editing – (Faces flyout) **Rotate faces**
Menu:	Modify – Solid Editing – **Rotate faces**
Toolbar:	Solid Editing – **Rotate faces**

Like the other Face category options, **Rotate** emulates another command. But in the case of the **Rotate** routine, it doesn't emulate the ***Rotate*** command but rather the ***Rotate3d*** command. In other words, you'll have the opportunity to rotate a 3D solid face about an axis (as opposed to a point).

The command sequence (from the **Face** option selection) looks like this:

Enter a face editing option [Extrude/Move/Rotate/Offset/Taper/Delete/Copy/coLor/mAterial/Undo/eXit] <eXit>: ***r***

Select faces or [Undo/Remove]: ***[The*** **Select faces** ***routine is identical to that of the*** **Extrude** ***and*** **Move** ***routines.]***

Select faces or [Undo/Remove/ALL]: ***[Confirm the selection set.]***

Specify an axis point or [Axis by object/View/Xaxis/Yaxis/Zaxis] <2points>: ***[Select a point on the axis about which you wish to rotate the objects]***

Specify second point on the rotation axis: ***[Select a second point to identify the axis.]***

Specify rotation angle or [Reference]: ***[Tell AutoCAD how much to rotate the object(s).]***

[The rest of the options are identical to the **Rotate3d** ***command's options.]***

Let's consider each of the axis-defining options.

- The default option is to specify **2points** on the axis. AutoCAD needs you to define the axis of rotation by picking any two points on it. Once you've done that, AutoCAD will prompt you to

  Specify rotation angle or [Reference]:

  Then tell AutoCAD how much to rotate the selected objects.
- The **Axis by object** option is probably the easiest. If you have an object drawn that can serve as an axis, all you have to do is select it. When you choose this option, AutoCAD prompts

  Select a curve to be used for the axis:

  This prompt is a bit deceptive; you can select a line, circle, arc, ellipse, 2D or 3D polyline or spline. AutoCAD will align its axis of rotation with the object.
- When you use the **View** option, AutoCAD rotates the objects about an imaginary axis drawn perpendicular to your monitor's screen (in the current viewport).
- The **Xaxis/Yaxis/Zaxis** options align the axis of rotation with the X-, Y-, or Z-axis that runs through a selected point. AutoCAD prompts for the origin and rotation angle. Enter the point's coordinates or pick it (with an OSNAP) on the screen.

Let's give it a try.

Do This: 10.2.3A — Rotating a 3D Solid Face

I. Be sure you're still in the *SE-Box* file in the C:\Steps3D\Lesson10 folder. If not, please open it now.

II. Thaw the **obj3** layer. Notice the object that appears inside the box. (This might be easier using the **3D Wireframe** visual style.)

III. Use the ***Subtract*** command to remove the new object from the box. (It'll leave a slot.)

IV. Follow these steps.

10.2.3A: Rotating a 3D Solid Face

1. Enter the command sequence shown to access the **Rotate** option of the Face category. Alternately, you can pick the **Rotate Faces** button on the **Solid Editing** panel.

   Command: ***solidedit***

   Solids editing automatic checking: SOLIDCHECK=1

   Enter a solids editing option [Face/Edge/Body/Undo/eXit] <eXit>: ***f***

   Enter a face editing option [Extrude/Move/Rotate/Offset/Taper/Delete/Copy/coLor/mAterial/Undo/eXit] <eXit>: ***r***

2. Select the faces forming the slot.

   Select faces or [Undo/Remove]:

   Select faces or [Undo/Remove/ALL]: ***[ENTER]***

3. Select the upper- and lower-center points indicated for your rotation axis. (Pick the lower-center point first to properly set the Z-axis.)

   Specify an axis point or [Axis by object/View/Xaxis/Yaxis/Zaxis] <2points>:

   Specify the second point on the rotation axis:

4. Rotate the slot **45°**.

   Specify a rotation angle or [Reference]: ***45***

   Solid validation started.

   Solid validation completed.

5. Complete the command.

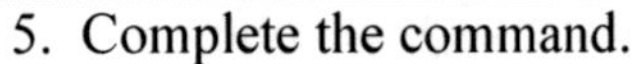

   Enter a face editing option [Extrude/Move/Rotate/Offset/Taper/Delete/Copy/coLor/mAterial/Undo/eXit] <eXit>: ***[ENTER]***

   Solids editing automatic checking: SOLIDCHECK=1

   Enter a solids editing option [Face/Edge/Body/Undo/eXit] <eXit>: ***[ENTER]***

   Your drawing looks like the figure at right.

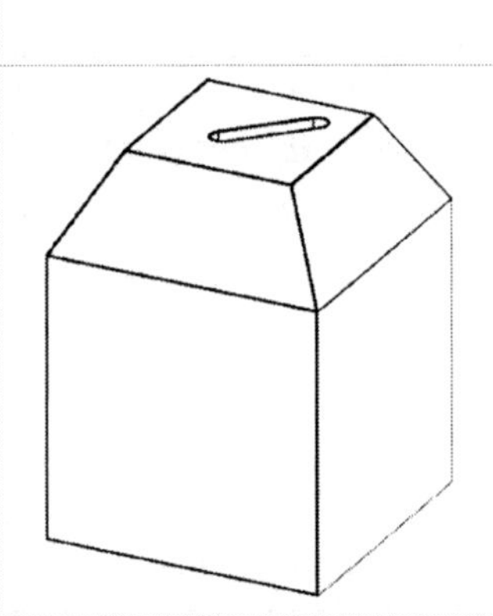

6. Once again, we can use the basic command to accomplish the same thing. Undo the rotation.

   Command: ***u***

7. Begin the ***Rotate*** command.

   Command: ***ro***

8. Hold down the CTRL key and select the faces that make up the slot.

   Select objects:

   Select objects: ***[ENTER]***

9. Use the upper center of the right-end arc as your base point and rotate the slot 45°.

   Specify base point: _cen of

   Specify rotation angle or [Copy/Reference] <0>: ***45***

10.2.3A: Rotating a 3D Solid Face

10. It just keeps getting easier – there's even a gizmo approach for this procedure. Undo the rotation.

    Command: ***u***

11. Without entering a command, hold down the CTRL key and select the faces that make up the slot.

12. Select the **Rotate** gizmo in the **Subobject** panel and move it to the same base point you used in Step 9.

13. Pick the Z axis on the gizmo.

14. Rotate the slot **45°**.

    **** ROTATE ****

    Specify rotation angle or [Base point/Undo/Reference/eXit]: ***45***

15. Save the drawing.

    Command: ***qsave***

Way cool! I wish all the ***SolidEdit*** tools had such redundancies!

10.2.4 Offsetting Faces on a 3D Solid

Where to Find It:	
Command Line:	***Solidedit – Face – Offset***
Ribbon (Tab/Panel):	Home – Solid Editing – (Faces flyout) **Offset faces**
Menu:	Modify – Solid Editing – **Offset faces**
Toolbar:	Solid Editing – **Offset faces**

The **Offset** routine of the Face category works very much like the two-dimensional command. There are, however, some quirks about it that you must know to avoid frustration.

The first of these quirks lies in the direction of the offset. When using the two-dimensional command, you pick the direction of the offset on the screen or by coordinate input. In Z-Space, this might present problems (since picking a point on the screen doesn't work well in Z-Space). Instead, you'll control the direction of the offset by using a positive or negative number to identify the distance of the offset. But here's the quirk: the positive number doesn't increase the size of the face being offset (it won't increase the size of the slot, as you'll see). Rather, it increases the *volume of the solid* (thus *decreasing* the size of the slot). A negative number, of course, has the opposite effect.

Another quirk is actually an omission on AutoCAD's part. Whereas you can offset an object through a point in two-dimensional space (using the ***Offset*** command), no such option is presented when using the ***SolidEdit*** command. You'll miss this convenience. (Hopefully, AutoCAD will remedy this oversight in the future.)

The command sequence for the Face category's **Offset** option is

> **Enter a face editing option**
>
> **[Extrude/Move/Rotate/Offset/Taper/Delete/Copy/coLor/mAterial/Undo/eXit] <eXit>:** ***o***
>
> **Select faces or [Undo/Remove]:** ***[The*** **Select faces** ***routine is identical to that of the SolidEdit routines you've already seen.]***
>
> **Select faces or [Undo/Remove/ALL]:**
>
> **Specify the offset distance:** ***[Enter the distance you wish to offset the selected faces.]***

Try enlarging the slot a bit.

Do This: 10.2.4A	Offsetting a 3D Solid Face

I. Be sure you're still in the *SE-Box* file in the C:\Steps3D\Lesson10 folder. If not, please open it now.

II. Follow these steps.

10.2.4A: OFFSETTING A 3D SOLID FACE

1. Enter the command sequence shown to access the **Offset** option of the Face category. Alternately, you can pick the **Offset Faces** button on the **Solid Editing** panel.

   Command: ***solidedit***

   Solids editing automatic checking: SOLIDCHECK=1

   Enter a solids editing option [Face/Edge/Body/Undo/eXit] <eXit>: ***f***

   Enter a face editing option [Extrude/Move/Rotate/Offset/Taper/Delete/Copy/coLor/mAterial/Undo/eXit] <eXit>: ***o***

2. Select the faces forming the slot.

   Select faces or [Undo/Remove]:

   Select faces or [Undo/Remove/ALL]: ***[ENTER]***

3. We want to increase the size of the slot (decreasing the volume of the solid). We'll enter a negative number. Offset the slot by 1/8" as indicated.

   Specify the offset distance: ***-.125***

4. Complete the command.
Your drawing looks like this (**3D Hidden** visual style shown).

5. Save the drawing, but don't exit.

   Command: ***qsave***

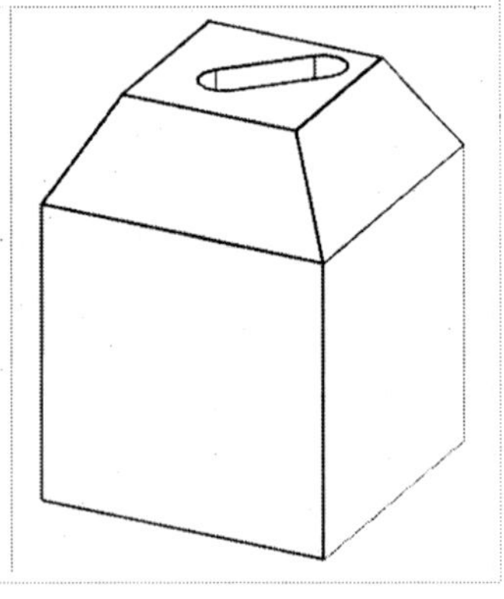

Sorry, no grips, gizmo, or simple command alternatives for this procedure – yet!

10.2.5	Tapering Faces on a 3D Solid

Tapering a face on a 3D solid isn't difficult, but you'll need to watch the positive and negative numbers just as you did when you offset a face. The command sequence is

Where to Find It:	
Command Line:	***Solidedit – Face – Taper***
Ribbon (Tab/Panel):	Home – Solid Editing – (Faces flyout) **Taper faces**
Menu:	Modify – Solid Editing – **Taper faces**
Toolbar:	Solid Editing – **Taper faces**

Command: ***solidedit***

Solids editing automatic checking: SOLIDCHECK=1

Enter a solids editing option [Face/Edge/Body/Undo/eXit] <eXit>: ***f***

Enter a face editing option [Extrude/Move/Rotate/Offset/Taper/Delete/Copy/coLor/mAterial/Undo/eXit] <eXit>: ***t***

Select faces or [Undo/Remove]: ***[The*** **Select faces** ***routine is identical to that of the SolidEdit routines you've already seen.]***

Select faces or [Undo/Remove/ALL]:

Specify the base point: ***[The base point doesn't have to be on the face itself; essentially, you're using the base point and second point to identify a direction, or axis, for the taper.]***

Specify another point along the axis of tapering: ***[Identify a second point along the axis.]***

Specify the taper angle: ***[Tell AutoCAD how much of an angle you wish to create.]***

Let's taper a hole in our solid.

Do This: 10.2.5A	Tapering a 3D Solid Face

I. Be sure you're still in the *SE-Box* file in the C:\Steps3D\Lesson10 folder. If not, please open it now.

II. Thaw the **obj4** layer. Notice the cylinder that appears inside the box.

III. Subtract the new cylinder from the box.

IV. Follow these steps.

10.2.5A: TAPERING A 3D SOLID FACE

1. Enter the command sequence shown to access the **Taper** routine of the Face category. Alternately, you can pick the **Taper Faces** button on the **Solid Editing** panel.

   Command: ***solidedit***

   Solids editing automatic checking: SOLIDCHECK=1

   Enter a solids editing option [Face/Edge/Body/Undo/eXit] <eXit>: ***f***

   Enter a face editing option [Extrude/Move/Rotate/Offset/Taper/Delete/Copy/coLor/mAterial/Undo/eXit] <eXit>: ***t***

2. Select one of the isolines defining the large hole. (This is easier with a 3D Wireframe visual style.)

   Select faces or [Undo/Remove]:

   Select faces or [Undo/Remove/ALL]: ***[ENTER]***

3. Use the center of the large right hole as the base point and the center of the other end as the other **point along the axis of tapering**.

   Specify the base point:

   Specify another point along the axis of tapering:

4. We'll reduce the size of the hole as it moves to the left. Enter a negative **taper angle** of **5°** as indicated.

   Specify the taper angle: ***-5***

5. Complete the command. Your drawing looks like the figure at right.

6. Save the drawing, but don't exit.

   Command: ***qsave***

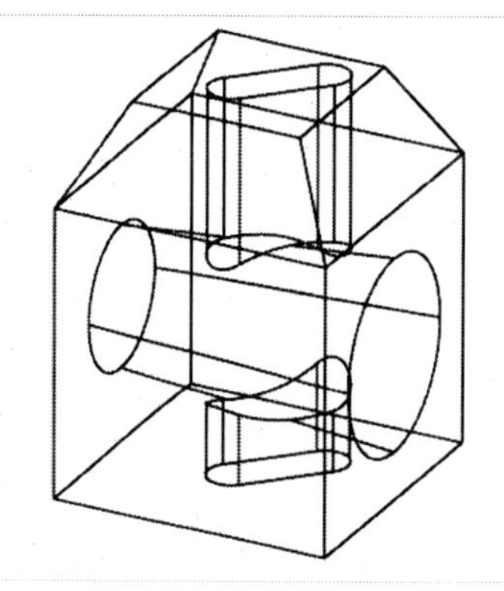

10.2.6	Deleting 3D Solid Faces

Our next routing provides a method for removing some faces (like fillets, chamfers, holes, etc.) from a 3D solid. The command sequence is one of the easiest (no points or axes to identify).

Command: ***solidedit***

Solids editing automatic checking: SOLIDCHECK=1

Enter a solids editing option [Face/Edge/Body/Undo/eXit] <eXit>: *f*

Enter a face editing option [Extrude/Move/Rotate/Offset/Taper/ Delete/Copy/coLor/mAterial/Undo/eXit] <eXit>: *d*

Select faces or [Undo/Remove]: *[The* **Select faces** *routine is identical to that of the SolidEdit routines you've already seen.]*

Select faces or [Undo/Remove/ALL]: *[ENTER]*

Where to Find It:	
Command Line:	***Solidedit – Face – Delete***
Ribbon (Tab/Panel):	Home – Solid Editing – (Faces flyout) **Delete faces**
Menu:	Modify – Solid Editing – **Delete faces**
Toolbar:	Solid Editing – **Delete faces**

Suppose we want to get rid of the large hole altogether. We'll use the **Delete** routine.

Do This: 10.2.6A	Deleting a 3D Solid Face

I. Be sure you're still in the *SE-Box* file in the C:\Steps3D\Lesson10 folder. If not, please open it now.

II. Follow these steps.

10.2.6A: DELETING A 3D SOLID FACE

1. Enter the command sequence shown to access the **Delete** routine of the Face category. Alternately, you can pick the **Delete Faces** button on the **Solid Editing** panel.

    Command: *solidedit*

    Solids editing automatic checking: SOLIDCHECK=1

    Enter a solids editing option [Face/Edge/Body/Undo/eXit] <eXit>: *f*

    Enter a face editing option [Extrude/Move/Rotate/Offset/Taper/Delete/Copy/ coLor/mAterial/Undo/eXit] <eXit>: *d*

2. Select the large, tapered hole (you'll find it easier to either select inside the hole or to select one of the internal isolines defining the inside face of the hole).

    Select faces or [Undo/Remove]:

3. Complete the command.

    Your drawing looks like the figure at right.

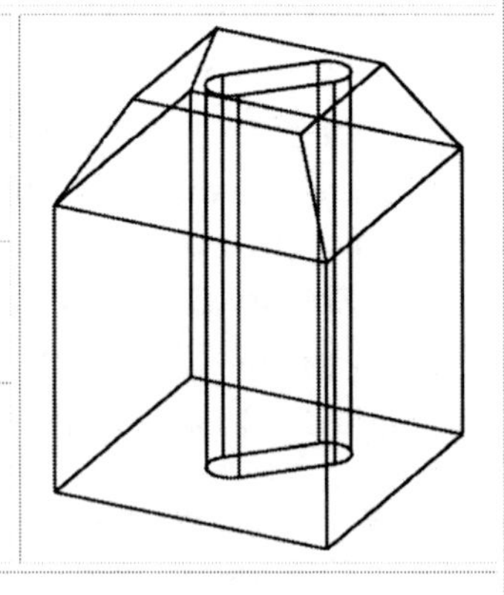

4. Okay, there's a much easier way. Undo the changes.

    Command: *u*

5. Hold down the CTRL key and select the hole.

6. Hit the DELETE key on your keyboard. (Sigh; why make it more difficult than that!)

7. Save the drawing, but don't exit.

    Command: *qsave*

10.2.7	Copying 3D Solid Faces as Regions or Bodies

Use the **Copy** routine of the Face category to create copies of one or more faces of a 3D solid. AutoCAD creates the copies as regions or bodies, which you can explode into individual lines, arcs, circles, and so forth, extrude into new 3D solid objects, or convert to surfaces (***ConvToSurface***). Unlike the results of the ***Copy*** command, however, the new objects exist on the layer that was current when the copies were created.

> A *body* is any structure that represents a solid or a Non-Uniform Rational B-Spline (NURBS) surface. NURB refers to a mathematical representation of a curved surface in a computer graphic.

Where to Find It:	
Command Line:	***Solidedit – Face – Copy***
Ribbon (Tab/Panel):	Home – Solid Editing – (Faces flyout) **Copy faces**
Menu:	Modify – Solid Editing – **Copy faces**
Toolbar:	Solid Editing – **Copy faces**

It's as easy to use as the ***Copy*** command. The Face category's command sequence is

Enter a face editing option [Extrude/Move/Rotate/Offset/Taper/Delete/Copy/coLor/mAterial/Undo/eXit] <eXit>: ***c***

Select faces or [Undo/Remove]: ***[The*** **Select faces** ***routine is identical to that of the SolidEdit routines you've already seen.]***

Select faces or [Undo/Remove/ALL]: ***[ENTER]***

Specify a base point or displacement: ***[The last prompts are the same as the*** **Copy** ***command's prompts.]***

Specify a second point of displacement:

We'll copy the top faces of our solid.

Do This: 10.2.7A	Copying a 3D Solid Face

I. Be sure you're still in the *SE-Box* file in the C:\Steps3D\Lesson10 folder. If not, please open it now.

II. Follow these steps.

10.2.7A: COPYING A 3D SOLID FACE

1. Enter the command sequence shown to access the **Copy** routine of the Face category. Alternately, you can pick the **Copy faces** button on the **Solid Editing** panel.

Command: ***solidedit***

Solids editing automatic checking: SOLIDCHECK=1

Enter a solids editing option [Face/Edge/Body/Undo/eXit] <eXit>: ***f***

Enter a face editing option [Extrude/Move/Rotate/Offset/Taper/Delete/Copy/coLor/mAterial/Undo/eXit] <eXit>: ***c***

2. Select the top face and the four angled faces around it.

Select faces or [Undo/Remove]:

Select faces or [Undo/Remove/ ALL]: ***[ENTER]***

3. Use the displacement method to copy the faces five units to the right as indicated.

Specify a base point or displacement: ***5,0***

Specify a second point of displacement: ***[ENTER]***

10.2.7A: COPYING A 3D SOLID FACE

4. Complete the command.

The copy looks like the figure at right.

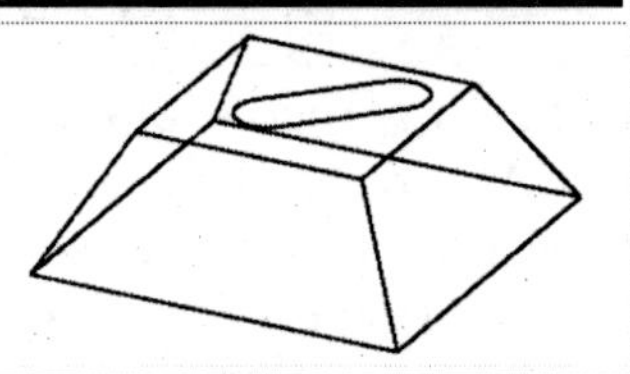

5. Convert the bodies to surfaces.

Command: ***convtosurface***

Select objects:

The surfaces look like the figure at right (shown with a **Realistic** visual style for clarity.)

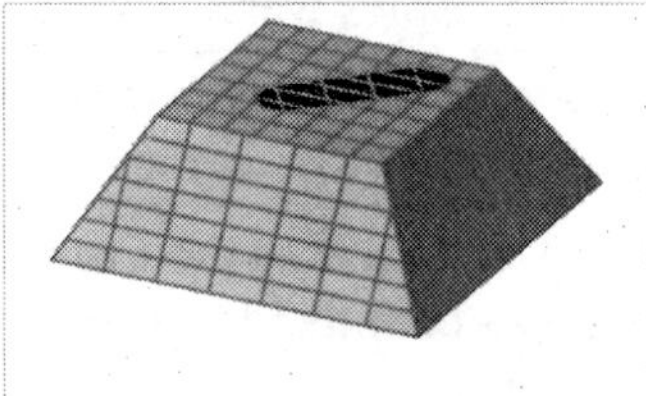

6. Save the drawing, but don't exit.

Command: ***qsave***

Cool!

10.2.8 Changing the Color of a Single Face

The **coLor** routine of the Face category is useful if you intend to shade or remove hidden lines from your drawing. It helps to distinguish between the different faces.

Where to Find It:	
Command Line:	***Solidedit – Face – coLor***
Ribbon (Tab/Panel):	Home – Solid Editing – (Faces flyout) **Color faces**
Menu:	Modify – Solid Editing – **Color faces**
Toolbar:	Solid Editing – **Color faces**

The Face category's command sequence is

Enter a face editing option [Extrude/Move/Rotate/Offset/Taper/ Delete/Copy/coLor/mAterial/Undo/eXit] <eXit>: ***l***

Select faces or [Undo/Remove]: ***[The*** **Select faces** ***routine is identical to that of the SolidEdit routines you've already seen.]***

Select faces or [Undo/Remove/ALL]: ***[ENTER]***

[AutoCAD presents the Color Selection dialog box; select the color you wish the face(s) to be.]

Let's make the slot a different color and view our 3D solid using the **Conceptual** visual style.

Do This: 10.2.8A — Changing the Color of a 3D Solid Face

I. Be sure you're still in the *SE-Box* file in the C:\Steps3D\Lesson10 folder. If not, please open it now.

II. Delete the 3D faces you created in the last exercise.

III. Follow these steps.

10.2.8A: CHANGING THE COLOR OF A 3D SOLID FACE

1. Enter the command sequence shown to access the **coLor** routine of the Face category.

Alternately, you can pick the **Color Faces** button on the **Solid Editing** panel.

Command: ***solidedit***

Solids editing automatic checking: SOLIDCHECK=1

Enter a solids editing option [Face/Edge/ Body/Undo/eXit] <eXit>: ***f***

10.2.8A: Changing the Color of a 3D Solid Face

Enter a face editing option [Extrude/Move/Rotate/Offset/Taper/Delete/Copy/coLor/mAterial/Undo/eXit] <eXit>: *c*

2. Select the faces that form the slot.

   Select faces or [Undo/Remove]:

   Select faces or [Undo/Remove/ALL]: *[ENTER]*

3. AutoCAD presents the Select Color dialog box. Select **Green.**

4. Complete the command.

5. Set the **Conceptual** visual style current [Conceptual]. Your drawing looks like the figure at right.

6. Save the drawing, but don't exit.

   Command: *qsave*

You'll notice a **Materials** option in the Face category. We'll look at materials when we get to Lesson 11.

10.3 Modifying Edges – The Edge Category

Use the ***XEdges*** command (Lesson 7, p.171) when you want to copy *all* the edges of the solid. The only real drawback is that it copies them in place (making it difficult to separate edges from the solid). On the bright side, it copies the edges onto the current layer so you can use layers to help you isolate the new objects.

The Edge category of the ***SolidEdit*** command contains only two real options (besides the **Undo/eXit** options). Both of these – **Copy** and **coLor** – repeat options found in the Face category. But here they're for use on single edges rather than entire faces.

The command sequence to access the **Edge** options of the ***SolidEdit*** command is

Command: *solidedit*

Solids editing automatic checking: SOLIDCHECK=1

Enter a solids editing option [Face/Edge/Body/Undo/eXit] <eXit>: *e*

Enter an edge editing option [Copy/coLor/Undo/eXit] <eXit>:

Prompts for both the **Copy** option and the **coLor** option are identical to their counterparts in the Face category.

Let's look at each in an exercise.

Where to Find It:	
Command Line:	***Solidedit – Edge – Copy***
Ribbon (Tab/Panel):	Home – Solid Editing – (Edges flyout) **Copy edges**
Menu:	Modify – Solid Editing – **Copy edges**
Toolbar:	Solid Editing – **Copy edges**

Where to Find It:	
Command Line:	***Solidedit – Edge – coLor***
Ribbon (Tab/Panel):	Home – Solid Editing – (Edges flyout) **Color edges**
Menu:	Modify – Solid Editing – **Color edges**
Toolbar:	Solid Editing – **Color edges**

Do This: 10.3A	Changing Edges on a 3D Solid

I. Be sure you're still in the *SE-Box* file in the C:\Steps3D\Lesson10 folder. If not, please open it now.

II. Change the visual style setting back to **3D Wireframe**.

III. Set **const** as the current layer.

IV. Follow these steps.

10.3A: Changing Edges on a 3D Solid

1. Enter the command sequence shown to access the **coLor** routine of the Edge category. Alternately, you can pick the **Color Edges** button on the **Solid Editing** panel.

    Command: ***solidedit***
    Solids editing automatic checking: SOLIDCHECK=1
    Enter a solids editing option [Face/Edge/Body/Undo/eXit] <eXit>: ***e***
    Enter an edge editing option [Copy/coLor/Undo/eXit] <eXit>: ***l***

2. Select the four edges that make up the right face of the box.

    Select edges or [Undo/Remove]:
    Select edges or [Undo/Remove]: ***[ENTER]***

3. AutoCAD presents the Select Color dialog box. Select **Red**.

4. AutoCAD changes the color of the selected lines (right) and then returns to the **edge editing option** prompt. Tell it you want to **Copy** Copy an edge as shown. (Alternately, you can pick the **Copy Edges** button on the **Solid Editing** panel).

    Enter an edge editing option [Copy/coLor/Undo/eXit] <eXit>: ***c***

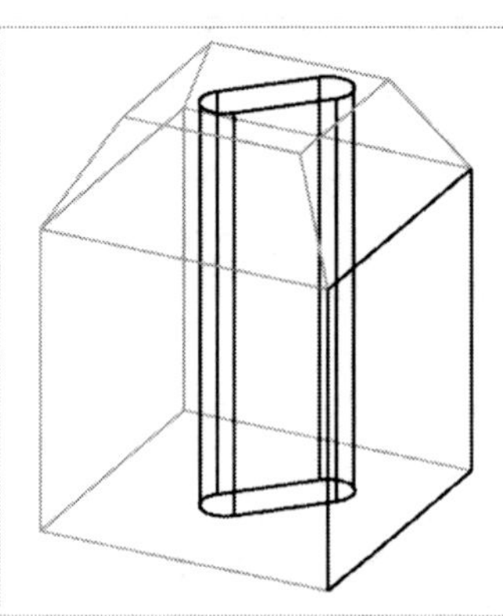

5. Select the same edges as in Step 2.

    Select edges or [Undo/Remove]:
    Select edges or [Undo/Remove]: ***[ENTER]***

6. Use the displacement method to copy the edges five units to the east, as indicated.

    Specify a base point or displacement: ***5,0***
    Specify a second point of displacement: ***[ENTER]***

7. Complete the command.

    Notice that the copies exist on the current layer and adopt the settings of that layer.

8. Save the drawing, but don't exit.

    Command: ***qsave***

How could you turn the new lines into a mesh?[*]

Not much difference occurs between these procedures and their counterparts in the Face category – except for the obvious effect on edges rather than faces.

Our next category, however, will be quite different!

[*] You could use the ***Edgesurf*** command

10.4 Changing the Whole 3D Solid – The Body Category

The Body category includes routines to modify a 3D solid as a whole. None of its five options (except the **Undo/eXit** options) has a counterpart in the 2D or 3D modification worlds. We'll look at each.

To access the Body category's options, follow this sequence:

> **Command:** ***solidedit***
> **Solids editing automatic checking: SOLIDCHECK=1**
> **Enter a solids editing option [Face/Edge/Body/Undo/eXit] <eXit>:** ***b***
> **Enter a body editing option**
> **[Imprint/seParate solids/Shell/cLean/Check/Undo/eXit] <eXit>:**

10.4.1 Imprinting an Image onto a 3D Solid

The **Imprint** routine of the Body category "imprints" (or draws) an image of a selected object – arc, circle, line, 2D or 3D polyline, ellipse, spline, region, or body – onto a 3D solid. Essentially, the imprint is a two-dimensional representation of the object on one of the faces of the 3D solid.

Once the impression has been made, *the line (or arc, spline, etc.) actually becomes a defining edge of a new face on the 3D solid.* So you can use this to help create new faces where they're needed.

> The buttons on the ribbon, toolbar, and menu all call the ***Imprint*** command, which uses the same prompts and does the same thing the **Imprint** option does. But you can use the command on a surface as well as a solid (and you don't have to wade through the ***SolidEdit*** command to get to the prompt!). We'll bypass the ***SolidEdit*** command at this point in favor of the more productive command.

The sequence to use the **Imprint** command is

> **Command:** ***Imprint***
> **Select a 3D solid or surface:** ***[Select the 3D solid or surface on which you wish to make the impression. The* Imprint *command begins with this prompt and allows you to select a surface as well as a solid.]***
> **Select an object to imprint:** ***[Select the object you wish to imprint.]***
> **Delete the source object [Yes/No] <N>:** ***[AutoCAD allows you the opportunity to keep or delete the object you're using to create your impression.]***
> **Select an object to imprint:** ***[You can continue to imprint objects if you wish.]***

Where to Find It:	
Command Line:	***Imprint***
Ribbon (Tab/Panel):	Home – Solid Editing – **Imprint edges**
Menu:	Modify – Solid Editing – **Imprint edges**
Toolbar:	Solid Editing – **Imprint**

This will become clearer with an exercise. Let's see what we can do with it.

Do This: 10.4.1A Imprinting Edges on a 3D Solid

I. Be sure you're still in the *SE-Box* file in the C:\Steps3D\Lesson10 folder. If not, please open it now.

II. Follow these steps.

10.4.1A: Imprinting Edges on a 3D Solid

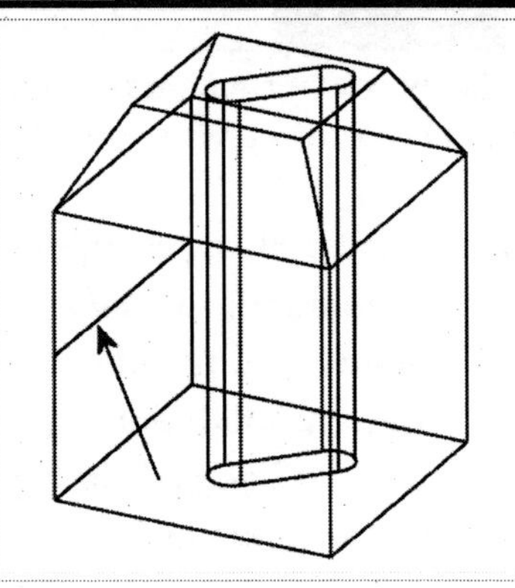

1. Draw a line between the midpoints of the vertical edges defining the left face of the 3D solid. We'll use this line to imprint a new edge.

   Command: ***l***

2. Enter the ***Imprint*** command. Alternately, you can pick the **Imprint** button on the **Solid Editing** panel.

   Command: ***imprint***

3. Select the 3D solid.

   Select a 3D solid or surface:

4. Select the line you created in Step 1.

   Select an object to imprint:

5. We won't need the line after we make the impression, so allow AutoCAD to delete it.

   Delete the source object [Yes/No] <N>: ***y***

6. We can continue to imprint objects, but we won't need to do so now. Complete the command.

   Select an object to imprint: ***[ENTER]***

7. Follow the procedure outlined in Exercise 10.2.3A, p.229, to rotate the new face 30°.

   Command: ***solidedit***

   Your drawing looks like the figure at right.

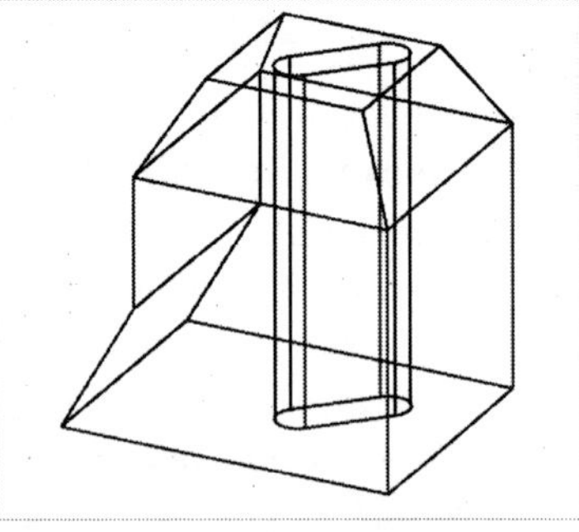

8. Save the drawing.

   Command: ***qsave***

You can probably see that **Imprint** will be one of the more useful of the ***SolidEdit*** command's options. You may find it easier to imprint and modify than to create a new 3D solid and join it (via the ***Union*** command) to an existing 3D solid.

10.4.2 Separating 3D Solids with the seParate Solids Routines

At first glance, the **seParate solids** option looked very promising – after all, it separates 3D solids into their constituencies. Simply put, this means that, if you created a 3D solid from a box and a cylinder, it would separate the 3D solid into the box and cylinder again. The way it works, however, *you can only separate the constituent objects when they don't actually touch.*

Still, there will be times when you find the **seParate solids** option quite handy. You'll see this in our exercise.

Where to Find It:	
Command Line:	***Solidedit – Body – SeParate***
Ribbon (Tab/Panel):	Home – Solid Editing – (Body flyout) **Separate**
Menu:	Modify – Solid Editing – **Separate**
Toolbar:	Solid Editing – **Separate**

The command sequence is one of the simplest:

Command: ***solidedit***
Solids editing automatic checking: SOLIDCHECK=1
Enter a solids editing option [Face/Edge/Body/Undo/eXit] <eXit>: ***b***
Enter a body editing option
[Imprint/seParate solids/Shell/cLean/Check/Undo/eXit] <eXit>: ***p***
Select a 3D solid: ***[Select the solid you wish to separate.]***

Let's take a look.

Do This: 10.4.2A	Separating Parts of a 3D Solid

I. Open the *SE-Box-2* file in the C:\Steps3D\Lesson10 folder. The drawing looks like the one we've been working, but has an additional object.
II. Use the ***List*** command to verify that all objects shown are part of a single 3D solid.
III. Follow these steps.

10.4.2A: Separating Parts of a 3D Solid
1. Enter the command sequence shown to access the ***seParate solids*** routine of the Body category. Alternately, you can pick the **Separate** button on the **Solid Editing** panel. **Command:** ***solidedit*** **Solids editing automatic checking: SOLIDCHECK=1** **Enter a solids editing option [Face/Edge/Body/Undo/eXit] <eXit>:** ***b*** **Enter a body editing option** **[Imprint/seParate solids/Shell/cLean/Check/Undo/eXit] <eXit>:** ***p***
2. Select either of the objects on the screen (as you've seen, although they don't touch, they're both part of a single 3D solid). **Select a 3D solid:**
3. Complete the command.
4. Erase the object on the left to verify that the 3D solids have separated. **Command:** ***e***
5. Save the drawing, but don't exit. **Command:** ***qsave***

That's easy enough – although it doesn't have a lot of use. It can, however, separate things that you joined inadvertently.

10.4.3	Clean

The **Clean** option of the Body category removes extra (redundant) edges and vertices – including imprinted and unused edges – from a 3D solid. Use it as a final cleanup tool once you've completed your 3D solid.

The command sequence is identical to that of the **seParate solids** option – and is, therefore, one of AutoCAD's simplest.

Where to Find It:	
Command Line:	***Solidedit – Body – Clean***
Ribbon (Tab/Panel):	Home – Solid Editing – (Body flyout) **Clean**
Menu:	Modify – Solid Editing – **Clean**
Toolbar:	Solid Editing – **Clean**

Did you notice the extra circular edge at the top of the new 3D solid? We imprinted it there by mistake and we don't need it. Let's use the **Clean** option to remove it.

Do This: 10.4.3A	Cleaning Up a 3D Solid

I. Be sure you're still in the *SE-Box-2* file in the C:\Steps3D\Lesson10 folder. If not, please open it now.

II. Follow these steps.

10.4.3A: CLEANING UP A 3D SOLID

1. Enter the command sequence shown to access the ***Clean*** routine of the Body category. Alternately, you can pick the **Clean** button on the **Solid Editing** panel.

Command: ***solidedit***
Solids editing automatic checking: SOLIDCHECK=1
Enter a solids editing option [Face/Edge/Body/Undo/eXit] <eXit>: ***b***
Enter a body editing option
[Imprint/seParate solids/Shell/cLean/Check/Undo/eXit] <eXit>: ***l***

2. Select the 3D solid.

Select a 3D solid:

3. Complete the command.

The object looks like the figure at right. Notice that AutoCAD has removed the extra circular edge on the top.

4. Save the drawing, but don't exit.

Command: ***qsave***

10.4.4	Shell

Where to Find It:	
Command Line:	***Solidedit – Body – Shell***
Ribbon (Tab/Panel):	Home – Solid Editing – (Body flyout) **Shell**
Menu:	Modify – Solid Editing – **Shell**
Toolbar:	Solid Editing – **Shell**

The **Shell** routine is one of the niftiest in the ***SolidEdit*** stable. With it, you can convert a 3D solid into a solid object similar to a surface or mesh model. In other words, you can convert a 3D solid into a hollow "shell" made up of a single 3D solid object.

To better understand this, consider your computer's monitor. Imagine the monitor with all the "guts" taken out – leaving just the plastic shell. The shell is a single object. The programmers at AutoCAD designed the **Shell** routine to create such objects!

The command sequence resembles those in the Face category:

Command: ***solidedit***
Solids editing automatic checking: SOLIDCHECK=1
Enter a solids editing option [Face/Edge/Body/Undo/eXit] <eXit>: ***b***
Enter a body editing option
[Imprint/seParate solids/Shell/cLean/Check/Undo/eXit] <eXit>: ***s***
Select a 3D solid: ***[Select the 3D solid you wish to shell.]***
Remove faces or [Undo/Add/ALL]: ***[Select an edge of the face(s) you wish to remove.]***

Remove faces or [Undo/Add/ALL]: ***[Complete the selection.]***

Enter the shell offset distance: ***[This figure defines the thickness of your shell.]***

Let's create a shell from our original object.

Do This: 10.4.4A	Shelling a 3D Solid

I. Be sure you're still in the *SE-Box-2* file in the C:\Steps3D\Lesson10 folder. If not, please open it now.

II. Change the viewpoint to 1,-2,-1.

III. Follow these steps.

10.4.4A: SHELLING A 3D SOLID	
1. Enter the command sequence shown to access the ***Shell*** routine of the Body category. Alternately, you can pick the **Shell** button on the **Solid Editing** panel. **Command:** ***solidedit*** **Solids editing automatic checking: SOLIDCHECK=1** **Enter a solids editing option [Face/Edge/Body/Undo/eXit] <eXit>:** ***b*** **Enter a body editing option** **[Imprint/seParate solids/Shell/cLean/Check/Undo/eXit] <eXit>:** ***s***	
2. Select the 3D solid. **Select a 3D solid:**	
3. Remove the bottommost face (pick in the middle of the face – not on an edge). **Remove faces or [Undo/Add/ALL]:** **Remove faces or [Undo/Add/ALL]:** ***[ENTER]***	
4. Make the shell thickness 1/8". **Enter the shell offset distance:** ***.125***	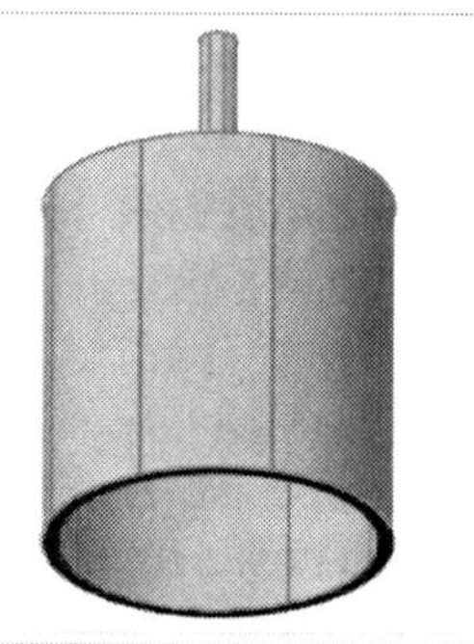
5. Complete the command. Your drawing looks like this (shown with a **Realistic** visual style).	
6. Save the drawing. **Command:** ***qsave***	

10.4.5	Checking to Be Certain You Have a ShapeManager® Solid

Where to Find It:	
Command Line:	***Solidedit – Body – Check***
Ribbon (Tab/Panel):	Home – Solid Editing – (Body flyout) **Check**
Menu:	Modify – Solid Editing – **Check**
Toolbar:	Solid Editing – **Check**

Have you noticed this note at the beginning of each of the options you've used in the ***SolidEdit*** command?

Solids editing automatic checking: SOLIDCHECK=1

This is telling you that the **SolidCheck** system variable has been set to **1** (that is, it's been activated). This means that AutoCAD will automatically check any 3D solid objects selected for editing to verify that they're valid *ShapeManager*® solids.

To keep it simple, this just means that other software than uses Autodesk's ShapeManager technology to produce 3D objects can use 3D solids created in AutoCAD. If the object fails the verification, it's a good idea to redraw it.

10.5 Extra Steps

Select a solid object and open the Properties palette. The information is fairly straightforward and about what you'd expect. Of course, you can change the settings that aren't grayed out. The Properties palette, then, provides another approach to editing 3D solids. But now, try to select just a single wall of the 3D solid. (Hold down the CTRL key as we did in our last exercise.)

What happens to the Properties palette? You can't do quite as much now. I suppose you'd better practice with the tools in this lesson then!

10.6 What Have We Learned?

Items covered in this lesson include

- *Grips and gizmos*
- *Tools used to edit 3D Solid Faces, including*
  - ***Extrude***
  - ***Move***
  - ***Rotate***
  - ***Offset***
  - ***Taper***
  - ***Delete***
  - ***Copy***
  - ***coLor***
- *Tools used to edit 3D Solid Edges, including*
  - ***Copy***
  - ***color***
  - ***XEdges***
- *Tools used to edit 3D Solid Bodies, including*
  - ***Imprint***
  - ***seParate solids***
  - ***Clean***
  - ***Shell***
- *The **SolidCheck** system variable*

Wow! What a lesson! Did you ever imagine a single command could have so many different options?!

You've learned many ways to create and modify 3D solids. As I promised, you're no longer simply a CAD draftsman. By experience and training, you've become a CAD operator. You no longer draw. Now you create actual objects in the computer. There's very little left in the three-dimensional world for you to learn! (Okay. Don't get too excited – we still have three more chapters!)

In our next lesson, I'll show you how to handle blocks in Z-Space. I'll also show you some tools to help you plot the objects you've learned to create. It'll be an easier and less involved lesson than this one, so you can relax a bit. After that, we'll look at rendering the objects you create (making them look "real" by assigning materials to them).

But first, as always, let's practice what we've learned.

10.7 Exercises

1. Create the *Level* drawing.
   1.1. Don't join the glass pieces to the body of the level.
   1.2. Add a 8½" x 11" title block.
   1.3. Save the drawing as *MyLevel* in the C:\Steps3D\Lesson10 folder.

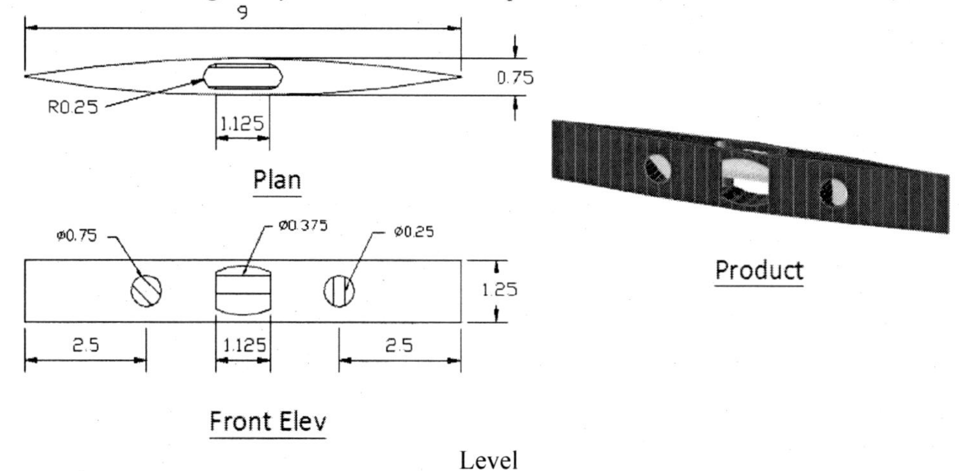

Level

2. Create the *Lid* drawing.
   2.1. Add a 8½" x 11" title block.
   2.2. Save the drawing as MyLid in the C:\Steps3D\Lesson10 folder.

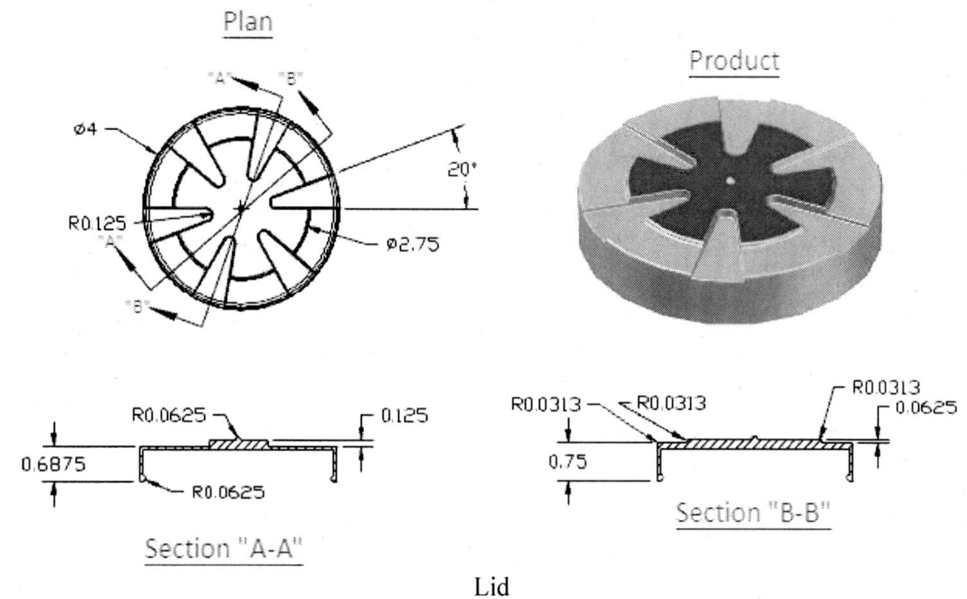

Lid

3. Open the *Slotted Guide #1* file in the C:\Steps3D\Lesson10 folder. Using only the procedures discussed in this lesson, create the drawing.
   3.1. Remember that a negative number entered as an extrusion height will extrude *into* the object.
   3.2. Remember that a negative number entered as a tapering angle will angle outward.
   3.3. Add a 8½" x 11" title block.
   3.4. Save the drawing as *MySG#1* in the C:\Steps3D\Lesson10 folder.

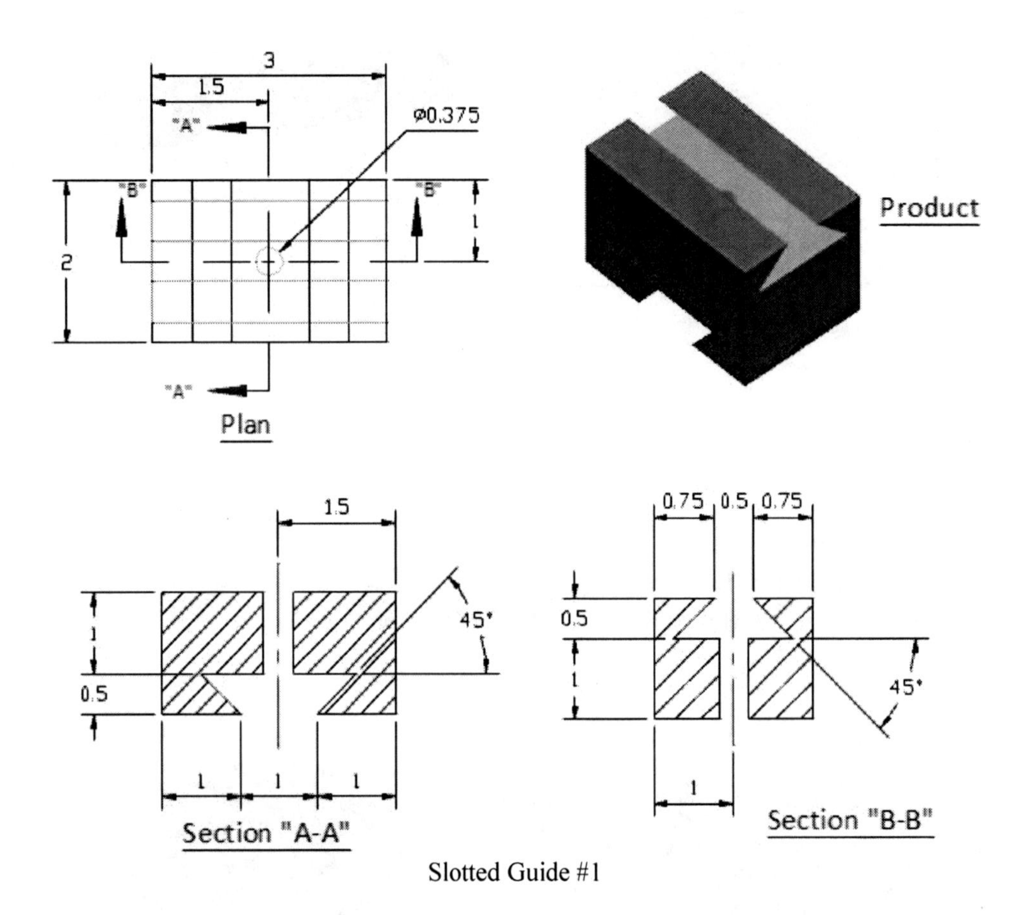

Slotted Guide #1

4. Open the *Slotted Guide #2* file in the C:\Steps3D\Lesson10 folder. Using only the procedures discussed in this lesson, create the drawing.
   4.1. Add a 8½" x 11" title block.
   4.2. Save the drawing as *MySG#2* in the C:\Steps3D\Lesson10 folder.

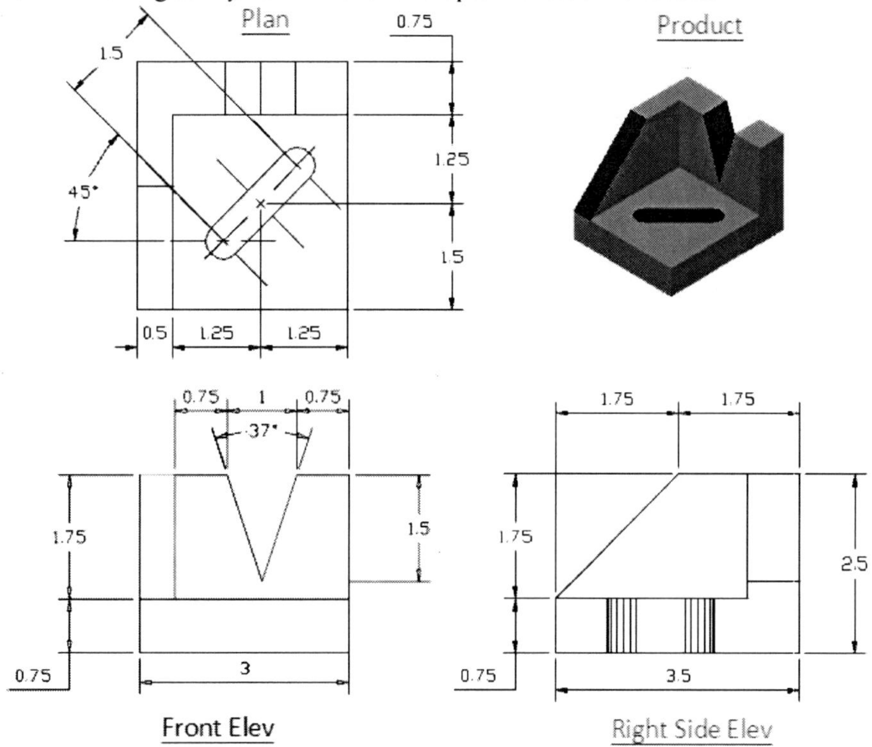

Slotted Guide #2

5. Create the *Plug* drawing.
   5.1. Add a 8½" x 11" title block.
   5.2. Save the drawing as *MyPlug* in the C:\Steps3D\Lesson10 folder.

Thanks to George Gilbert of G. Gilbert Engineering Services Ltd. for permission to use this drawing. For more on G. Gilbert Engineering Services Ltd., visit his web site at: http://ourworld.compuserve.com/homepages/george_gilbert/

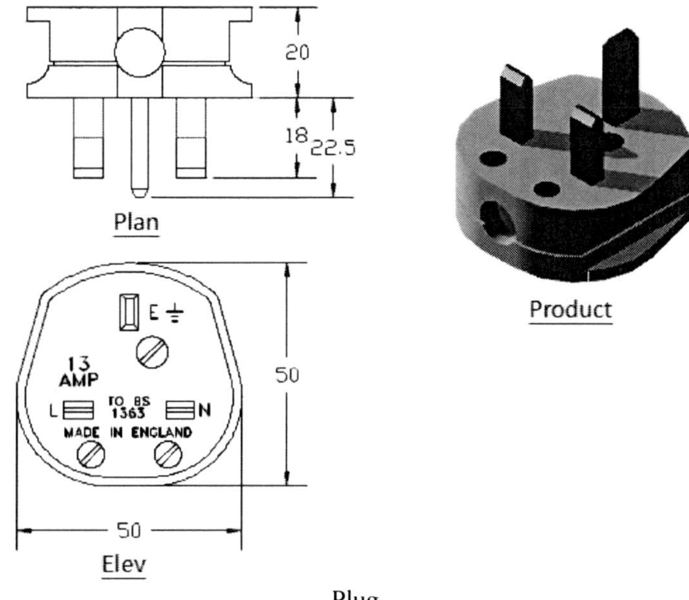

Plug

6. Create the *Corner Bracket* drawing.
    6.1. Use a 1/16" fillet along the edges.
    6.2. The rounded indentations are visible front and back.
    6.3. Add a 8½" x 11" title block.
    6.4. Save the drawing as *MyCB* in the C:\Steps3D\Lesson10 folder.

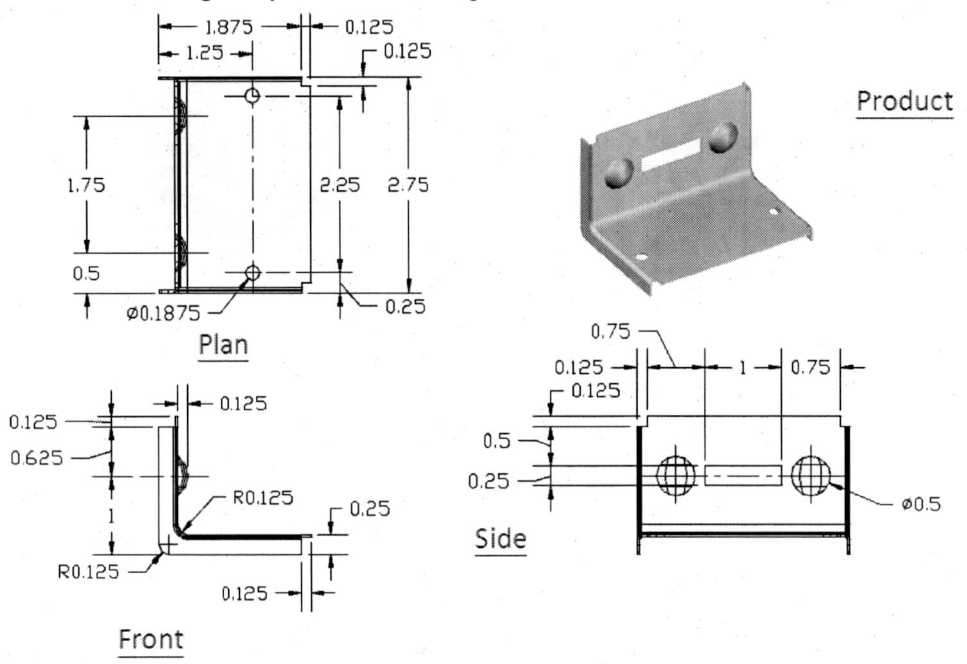

Corner Bracket

7. Create the *Power Switch* drawing.
    7.1. Add a 8½" x 11" title block.
    7.2. Save the drawing as *Switch* in the C:\Steps3D\Lesson10 folder.

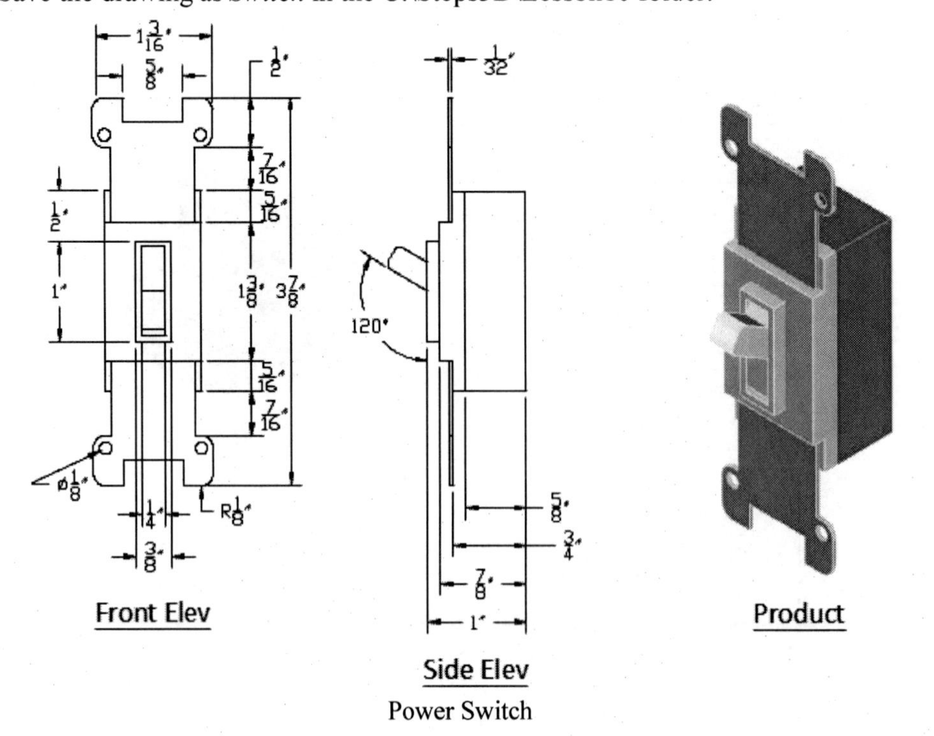

Power Switch

8. Create the *Wheel* drawing .
    8.1. Place this drawing on a C-size (22" x 17") sheet of paper. Use a title block of your choice.
    8.2. Fillet the rim and hub with a ¼" fillet.
    8.3. Save the drawing as *MyWheel* in the C:\Steps3D\Lesson10 folder.

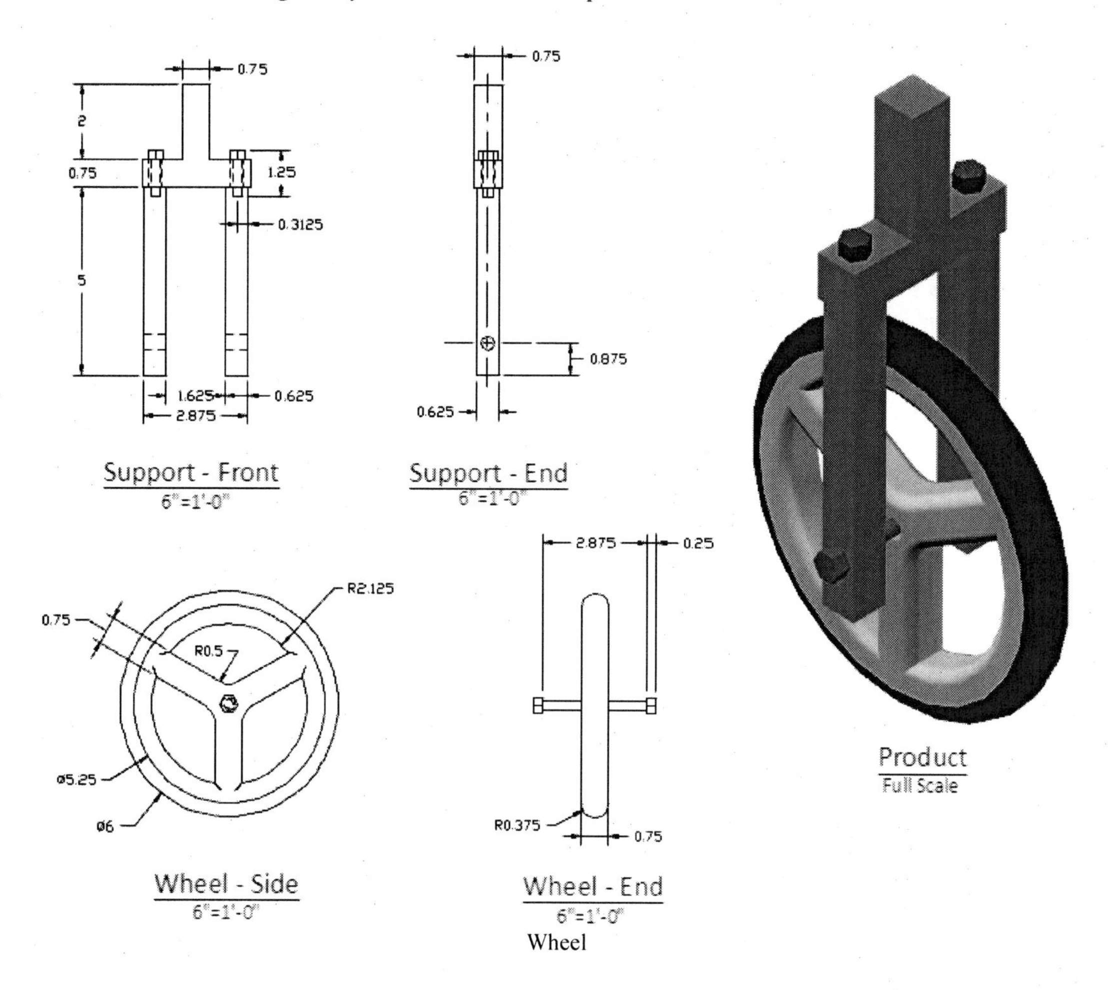

Wheel

10.8 **For Web-Based Review Questions, visit: http://foragerpub.com/AcadFiles/2010/2010.htm**

Lesson

Following this lesson, you will:

✓ *Know how to use blocks in Z-Space*

- *Creating three-dimensional blocks*
- *Inserting three-dimensional blocks*

✓ *Know how to use the Solid Plotting Tools*

- ***Solview***
- ***Soldraw***
- ***Solprof***
- ***Flatshot***
- ***3DPrint***

Three-Dimensional Blocks and Three-Dimensional Plotting Tools

As I mentioned when concluding the last lesson, there's very little left for you to learn with respect to creating three-dimensional objects in AutoCAD. This lesson, then, will serve as a wrap-up of Z-Space methods and techniques before we move on to our discussion of presentation tools.

In Lesson 11, we'll first consider the behavior of blocks in a three-dimensional world. While the differences between two-dimensional blocks and three-dimensional blocks can be dramatic, they don't necessarily have to be difficult. We'll consider the effects of Z-Space and working planes on creation and insertion of blocks as well as the use of attributes on a three-dimensional block.

Then we'll discuss some special tools designed to help you set up and plot 3D solids with considerably less difficulty than you might have had previously.

Let's begin.

11.1 Using Blocks in Z-Space

As with two-dimensional blocks, three-dimensional blocks can save you a tremendous amount of time and effort. But there are a few things the three-dimensional operator must consider.

11.1.1 Three-Dimensional Blocks and the UCS

First among these considerations is the working plane (the current UCS). *AutoCAD creates blocks against the plane of the current UCS* (Figure 11.001) – *not* the WCS. In other words, the XYZ values of the current UCS (the X-axis, Y-axis, and Z-axis values of the object) become part of the block definition.

Likewise, AutoCAD inserts bocks by matching the axis values of the block to the current UCS (Figure 11.002).

We'll see this in a series of exercises. First, we'll create two blocks – an elbow and some pipe. Our blocks will have two simple attributes defining what they are and their size. Second, we'll utilize our UCS and insertion scale factors to insert the blocks in a simple piping configuration. Third, we'll extract attribute data that'll give us a running total of the amount of pipe we've used.

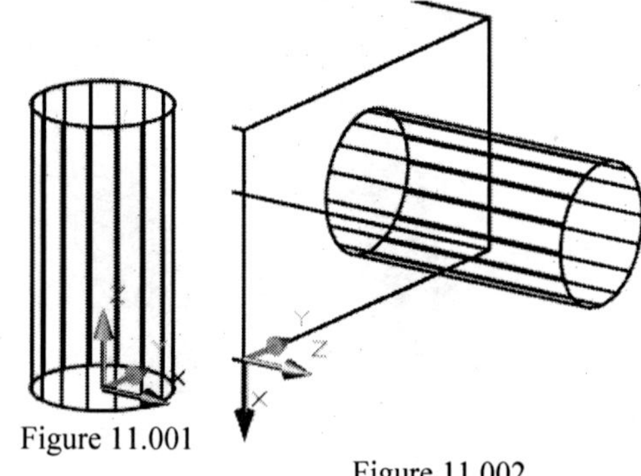

Figure 11.001

Figure 11.002

Let's begin.

Do This: 11.1.1A Creating 3-Dimensional Blocks

I. Open the *blocks* file in the C:\Steps3D\Lesson11 folder. The drawing looks like the figure at right.

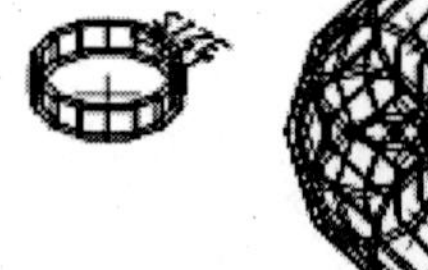

II. Notice that the elbow is open in the +X and +Y directions. Note also that the elbow and the pipe are drawn to scale as 4" fittings, and that the height of the pipe is 1".

III. Follow these steps.

11.1.1A: Creating 3-Dimensional Blocks

1. Make a note of the current UCS (it's equal to the WCS). Then make blocks (use the ***WBlock*** command) from the two objects you see.

Call the block on the left *Pipe*. Use the node as the insertion point (include the node and the two attributes in the block).

Call the block on the right *Ell*. Use the node inside the right opening of the elbow as the insertion point (include all three nodes and the two attributes in the block).

Be sure to write both blocks to the C:\Steps3D\Lesson11 folder.

11.1.1A: CREATING 3-DIMENSIONAL BLOCKS
2. Close the drawing.

Now we'll look at some new variations of the ***Insert*** command.

11.1.2 Inserting Three-Dimensional Blocks

The second three-dimensional block consideration involves the insertion scale of the block. In two-dimensional drafting, you could scale a block along the X- or Y-axis. Now you'll have an additional axis along which you can scale the block. You should give special attention to the use of a three-dimensional block because of the effect scale might have. Consider the figures below. We'll have an opportunity to use the scale options and see how the UCS affects the insertion when we insert our new blocks into a drawing.

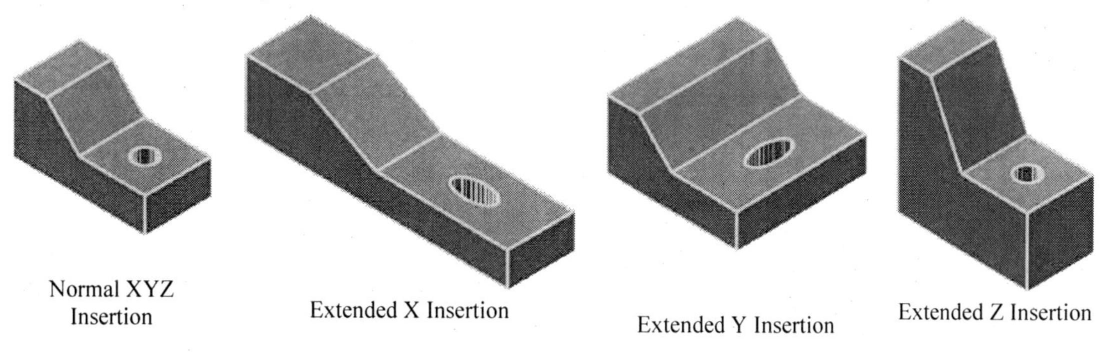

Normal XYZ Insertion — Extended X Insertion — Extended Y Insertion — Extended Z Insertion

Let's get started.

Do This: 11.1.2A — Inserting 3-Dimensional Blocks

I. Open the *Piping Configuration* file in the C:\Steps3D\Lesson11 folder. (I set the nodes to help guide you through the block insertions. The **UCSIcon** system variable has been set to **ORigin**.)

II. Set the running OSNAP to **Node**. Clear all other settings.

III. Set the current layer to **Pipe** and the current visual style to **Realistic**.

IV. Follow these steps.

11.1.2A: INSERTING 3-DIMENSIONAL BLOCKS	
1. Change the UCS. (Hint: Rotate the UCS 90°on the X-axis and then 270° on the Z-axis.) Relocate the UCS to the upper right node as shown. **Command: *ucs***	
2. Insert the *Ell* block at the 0,0,0 coordinate of the current UCS. (You created this block and placed it in the C:\Steps3D\Lesson11 folder in our last exercise.) Use a scale of 1 for each axis. Accept the default 4" size. The elbow looks like the figure at right.	

11.1.2A: INSERTING 3-DIMENSIONAL BLOCKS

3. Add the rest of the elbows. (Place the UCS at each node and insert the elbows at 0,0,0.)

Your drawing looks like the figure at right. (Your UCS icon may be in a different location.)

4. Reset the UCS to the WCS.

Command: *ucs*

5. Now we'll insert the pipe using a scale factor to make it fit between the elbows. Tell AutoCAD you wish to insert the *Pipe* block. Use the settings shown below. (The distance between the elbows we'll use is 8' – or 96".)

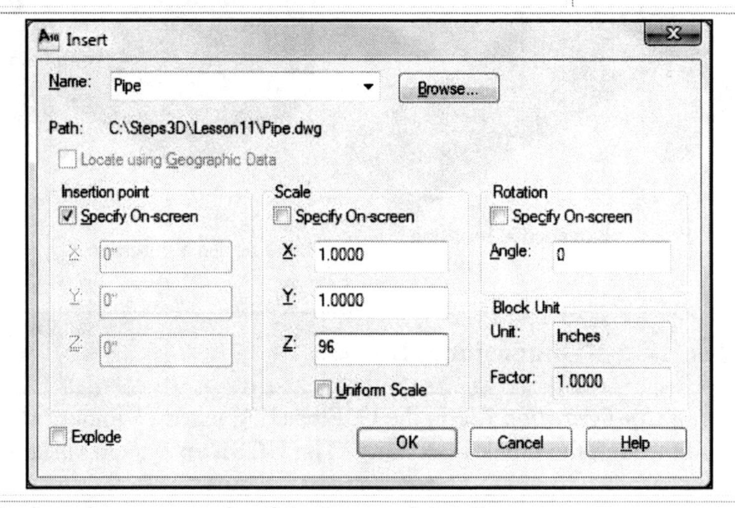

6. Insert the block at the upper node of the lower-right elbow.

Specify insertion point or [Basepoint/Scale/X/Y/Z/Rotate]:

7. Accept the default size of the pipe.

Enter attribute values

What is the size of the unit? <4">:

8. Repeat Steps 5 through 7 for the other vertical run of pipe.

Your drawing looks like the left figure following Step 9.

9. Adjust the UCS as needed to add the final run of pipe. The distance between the two elbows is 12".

Your drawing looks like the following figure (right).

11.1.2A: INSERTING 3-DIMENSIONAL BLOCKS

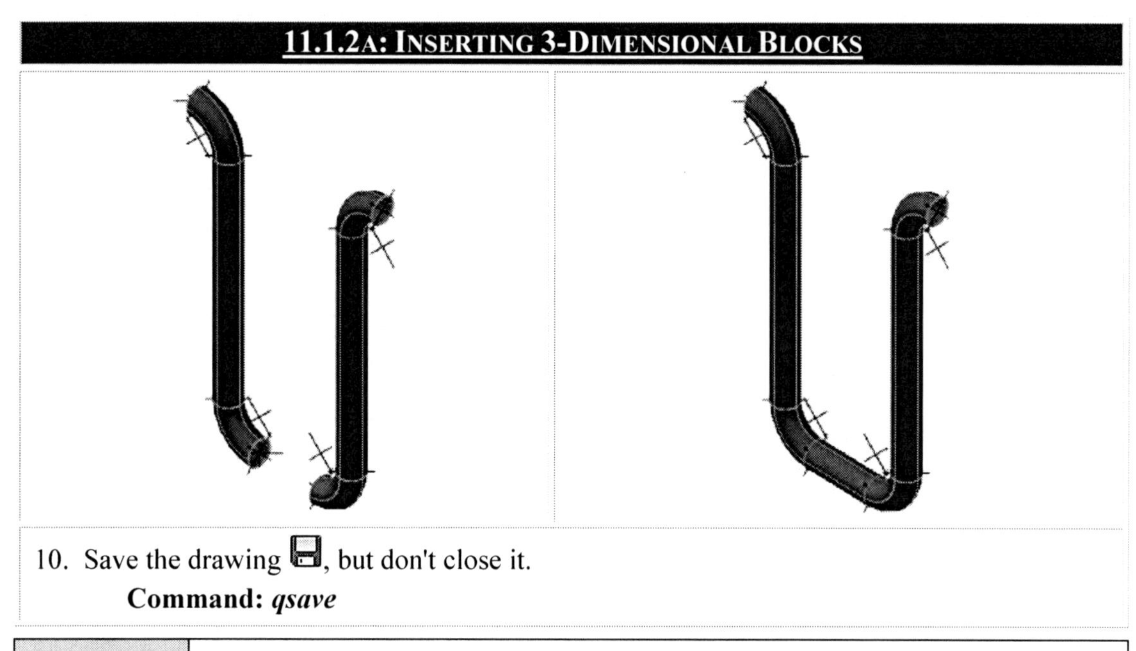

10. Save the drawing, but don't close it.
    Command: ***qsave***

11.1.3 Making Good Use of Attributes

The last (and possibly most rewarding) things we should consider when using three-dimensional blocks is the possible use of attributes. When combining the abilities you've already seen in this lesson with a clever use of attributes, you'll discover a remarkably useful way to create a bill of materials for most projects.

We'll extract the attribute data to our blocks to see that the *Pipe* blocks contain data about the length of each piece of pipe. This data is accurate enough to be used on a cutting list!

Sound nifty? Let's try it!

Do This: 11.1.3A **Extracting the Attributes**

I. Be sure you're still in the *Piping Configuration* file in the C:\Steps3D\Lesson11 folder. If you haven't finished this one, open *Piping Configuration - done*.
II. Set the UCS = World and restore the plan view (0,0,1).
III. Set the **Text** layer current and freeze the **Marker** layer.
IV. Follow these steps.

11.1.3A: EXTRACTING THE ATTRIBUTES

1. Begin the data extraction wizard (**Insert** tab – **Linking & Extraction** panel).
   Command: ***dataextraction***

2. **Create a new data extraction** [Create a new data extraction] and pick the **Next** button [Next >] to continue.

3. Call the data file something you can remember and save it [Save...] in the C:\Steps3D\Lesson11 folder.

4. We'll create our extraction using objects in the current drawing. [Select objects in the current drawing]
   Pick the **Select objects** button to return to the drawing.

11.1.3A: Extracting the Attributes

5. Select the piping configuration.

    Select objects:

    Select objects: ***[ENTER]***

6. Pick the **Next** button Next > to continue.

7. Select these objects (right) to **extract data from**.

8. Pick the **Next** button Next > to continue.

	Object	Display Name	Type
☐	3D Solid	3D Solid	Non-block
☐	Attribute Definition	Attribute Definition	Non-block
☑	Ell	Ell	Block
☑	Pipe	Pipe	Block
☐	Point	Point	Non-block

9. Extract the properties shown. You'll want the **Size** (from the **Attribute** category) and the **ScaleZ** from the **Geometry** category.

Properties

	Property	Display Name	Category
☐	Position X	Position X	Geometry
☐	Position Y	Position Y	Geometry
☐	Position Z	Position Z	Geometry
☐	Scale X	Scale X	Geometry
☐	Scale Y	Scale Y	Geometry
☑	Scale Z	Scale Z	Geometry
☑	SIZE	SIZE	Attribute

Category filter

- ☐ 3D Visualization
- ☑ Attribute
- ☐ Drawing
- ☐ General
- ☑ Geometry
- ☐ Misc

10. Pick the **Next** button Next > to continue.

11. AutoCAD shows you (following figure) that it has created a cutting list. Pick the **Next** button Next > to continue.

Count	Name	Scale Z	SIZE
1	Pipe	12.0000	4"
2	Pipe	96.0000	4"
4	Ell	1.0000	4"

12. Tell AutoCAD to insert data extraction table into drawing.

    ☑ Insert data extraction table into drawing

13. Pick the **Next** button Next > to continue.

14. Accept the default table style and pick the **Next** button Next > to continue.

15. Finish Finish the procedure and insert the table (shown) into a convenient place in the drawing.

16. Save the drawing.

    Command: ***qsave***

Count	Name	Scale Z	SIZE
1	Pipe	12.0000	4"
2	Pipe	96.0000	4"
4	Ell	1.0000	4"

How's that for a quick way to create a cutting list!?

Although we used pipe in our exercise, this method works as well when tracking board feet or lengths of steel. How might you use it in your profession?

11.2 Plotting a 3D Solid

We'll conclude our study of 3D solids with a group of special commands that AutoCAD designed to make plotting solid models easier.

First, we'll look at ***Flatshot***. ***Flatshot*** works something like taking a 2-dimensional photograph of a 3D solid object. You may find it handy for illustrations.

Next, we'll conquer the "Sol Group" – a group of paper space plotting tools that consists of three commands: ***Solview***, ***Soldraw***, and ***Solprof***. (You won't find a finer example of teamwork in the CAD world.) You'll use the first command – ***Solview*** – to set up the layout (the Paper Space viewports). Then you'll use the other two commands – ***Soldraw*** and ***Solprof*** – to create the actual drawings that go into the viewports.

Finally, we'll see the new kid in town; we'll explore ***3DPrint***. ***3DPrint*** prepares your model for the new 3D printers on the market.

Let's get started.

11.2.1 The *Flatshot* Command

Where to Find It:	
Command Line:	***Flatshot***
Hotkey(s):	***fshot***
Ribbon (Tab/Panel):	Home – Section (subpanel) – **Flatshot**

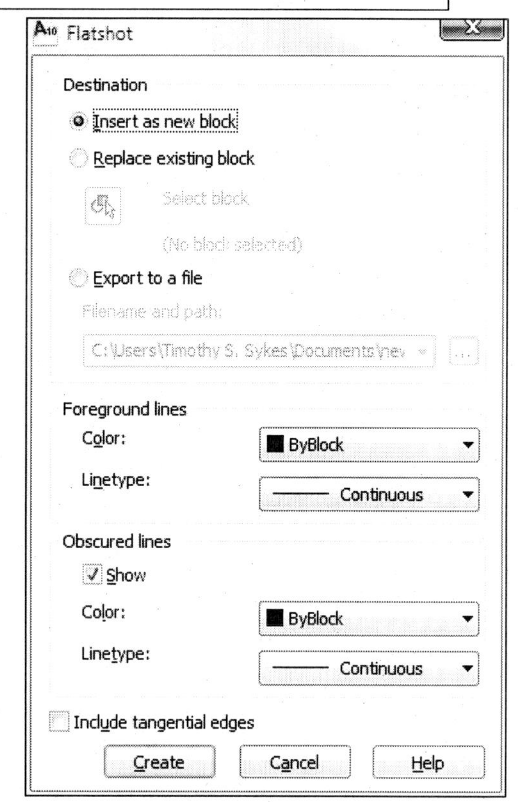

Figure 11.003

I think Autodesk designed ***Flatshot*** primarily as a detail-creation tool. It works well for just such a goal. I found, however, that you can also use it to convert 3-dimensional drawings to 2D drawings for plotting in Model Space. It takes a two-dimensional "photo" (image) of all three-dimensional objects *in the current viewport*.

This oddball command really presents nothing in the way of a command sequence – enter ***Flatshot*** at the command prompt, set up the Flatshot dialog box (Figure 11.003) as desired, and follow the command line sequence for insertion of a block. (AutoCAD inserts the flatshot as a block for easy manipulation.)

So let's look at the Flatshot dialog box.

- The **Destination** frame offers three radio options.
  - **Insert as new block** means that AutoCAD will create and insert the detail as a new block, which it gives a default name. Use the ***BEdit*** command to change the name if you wish.
  - You can also use the Flatshot to **Replace** [an] **existing block**. When you place the bullet next to this option, AutoCAD makes the **Select block** button available for you to choose the block you'll replace.
  - Using the **Export to a file** option works much like the ***WBlock*** command. Use the **Filename and path** box to name and locate the new block. You can edit the new block as a drawing of its own, or you can insert it into another file.

- Use the next two frames – **Foreground lines** and **Obscured lines** – to set the **Color** and **Linetype** of lines within the new block. You don't have to show the **Obscured lines** (lines behind objects), but if you do, use the options in this frame to make the lines appear hidden (use the **Hidden** line type).
- Put a check next to **Include tangential edges** for AutoCAD to create silhouette edges for curved surfaces.

> ***Flatshot*** actually projects the *current* view of the 3D objects in the *current* viewport against the XY-plane of the *current* UCS. That's a lot of "currents"; and that makes for a very dynamic command. This can cause a bit of confusion at first; but there's no substitute for experience!
>
> ***Practice Practice Practice***

Let's create a flatshot of our flange.

Do This: 11.2.1A	Creating a Flatshot Detail

I. Open the *MyFlange11a* file in the C:\Steps3D\Lesson11 folder. If that one isn't available, open *flange11*.

II. Set the **obj1** layer current.

III. Follow these steps.

11.2.1A: Creating a Flatshot Detail

1. Set the current **UCS** to **View**.

   Command: *ucs*

Note: This is an important step to help you align the created objects properly.

2. Enter the ***Flatshot*** command.

   Command: *fshot*

3. We'll accept the **Destination** defaults, but:
   - Set the **Foreground lines Color** to **ByLayer**
   - Accept the default **Foreground lines Linetype**
   - **Show** the **Obscured lines**, but change to **Color** to **ByLayer**
   - Use **Hidden** lines for the **Obscured lines Linetype** (you may have to load this linetype)

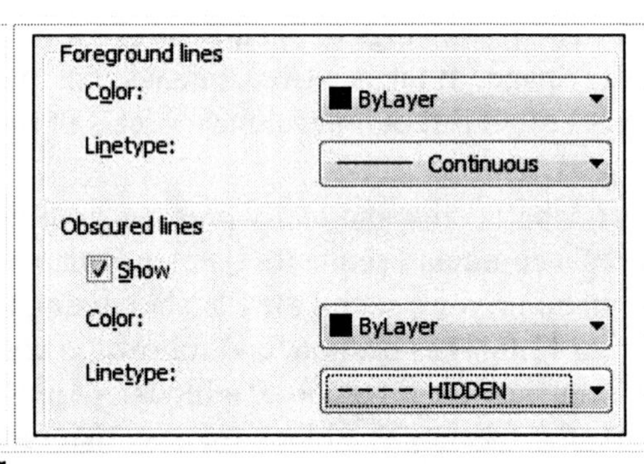

4. Insert the flatshot block to the upper right of the flange …

   Units: Inches Conversion: 1.00

   Specify insertion point or [Basepoint/Scale/X/Y/Z/Rotate]:

5. … at half scale …

   Enter X scale factor, specify opposite corner, or [Corner/XYZ] <1>: *.5*

   Enter Y scale factor <use X scale factor>: *[ENTER]*

6. … and at a 0° angle.

   Specify rotation angle <0>: *[ENTER]*

Your drawing looks like the figure at right.

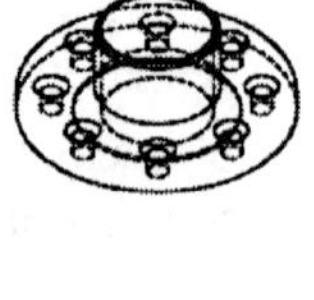

11.2.1A: Creating a Flatshot Detail

7. Let's try another one. Repeat Steps 2 and 3, but this time, don't **Show** the **Obscured lines** and **Include tangential lines.**

8. Insert the flatshot block below the previous insertion …
   Units: Inches Conversion: 1.0000
   Specify insertion point or [Basepoint/Scale/X/Y/Z/Rotate]:

9. … at half scale …
   Enter X scale factor, specify opposite corner, or [Corner/XYZ] <1>: *.5*
   Enter Y scale factor <use X scale factor>: *[ENTER]*

10. … and at a 0° angle.
   Specify rotation angle <0>: *[ENTER]*
   Your drawing looks like the figure at right.

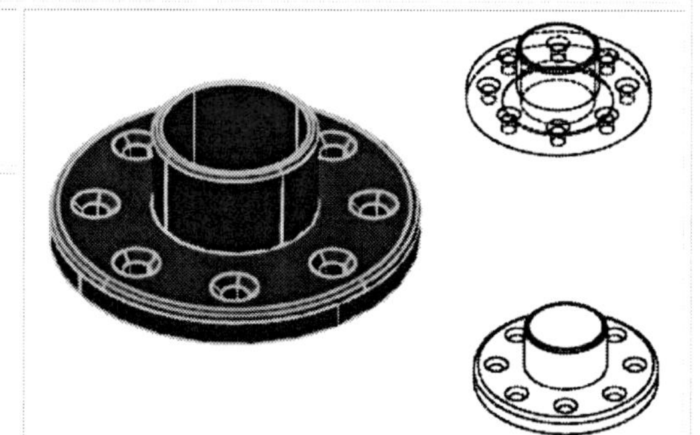

11. Save the drawing.
   Command: *qsave*

You can plot this drawing in Model space, but a better approach (as paper space advocates will insist) follows!

11.2.2 Setting Up the Plot – the *Solview* Command

Paper space plotting tools for 3D solids include: ***Solview***, ***Soldraw***, and ***Solprof***. Of the three, ***Solview*** is the most complex command. It creates viewports according to your input. You can define the viewports by the XY-plane or a user-defined UCS, or by calculating orthographic projections, auxiliary projections, and cross sections from a UCS viewport. ***Solview*** places the viewports on the **VPorts** layer (which it creates if necessary). It also creates viewport-specific layers for visible lines (*Viewname-vis*), hidden lines (*Viewname-hid*), dimensions (*Viewname-dim*), and hatching (*Viewname-hat*).

Where to Find It:	
Command Line:	*Solview*
Ribbon (Tab/Panel):	Home – Modeling (subpanel) – **Solid View**
Menu:	Draw – Modeling – Setup – **View**

You must enter the ***Solview*** command while a **Layout** tab is active; otherwise, AutoCAD will automatically open a layout tab. AutoCAD responds with the initial ***Solview*** prompt:

Command: ***solview***
Enter an option [Ucs/Ortho/Auxiliary/Section]:

Let's consider each option.

- Use the **UCS** option to create the first viewport – usually a front view or plan view of the 3D solid. It creates a two-dimensional profile view of the object. It uses the XY-plane of a user-specified UCS to define the profile. AutoCAD responds to selection of the UCS option with the prompt:

   Enter an option [Named/World/?/Current] <Current>: ***[ENTER]***
   Enter view scale <1.0000>: ***[Enter the scale for the view (if incorrect, you can rescale the view later).]***
   Specify view center: ***[Pick a point where you'd like to place the center point of the view.]***

Specify view center <specify viewport>: ***[Reposition the view, if necessary, or hit ENTER to continue.]***

Specify first corner of viewport: ***[Define the viewport by specifying opposite corners.]***

Specify opposite corner of viewport:

Enter view name: ***[Give the viewport a unique name.]***

The same sequence applies to the **Named** and **World** options, except that AutoCAD will precede the **Named** option sequence with a request for the name of the UCS to use. Use the **?** option to list the named UCSs available for use.

- Use the **Ortho** option (back at the first ***Solview*** prompt) to create orthographic projections from an existing view. The command sequence is the same as with the **UCS** option.
- Creating an auxiliary view – a view perpendicular to an inclined face – is as easy as drawing an orthographic projection for AutoCAD. AutoCAD responds to selection of the **Auxiliary** option with

  Specify first point of inclined plane: ***[Specify two points that define the inclined plane.]***

  Specify second point of inclined plane:

  Specify side to view from: ***[Pick a point from where you wish to see the inclined plane.]***

  AutoCAD continues with the options to size and locate the viewport.
- The **Section** option creates a cross section complete with section lines. (It takes the ***Soldraw*** command to actually create the section.) The sequence is

  Specify first point of cutting plane:

  Specify second point of cutting plane:

  Specify side to view from:

  Again, AutoCAD continues with the options to size and locate the viewport.

All of these options will become much clearer with an exercise.

> You'll notice in our next exercise that, in fact, none of these options creates a drawing. What they do is *set up* a viewport for the orthographic, auxiliary, or cross sectional drawings that you'll create later with the ***Soldraw*** command. You'll see the actually 3D solid in each viewport until you use the ***Soldraw*** command.

Let's begin.

Do This: 11.2.2A	**Using *Solview* to Set Up a Layout**

I. Open the *Sol1* file in the C:\Steps3D\Lesson11 folder. It looks like the figure at right.

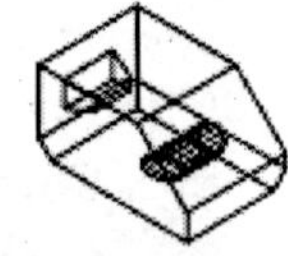

II. Activate the **Layout1** tab. Erase any viewports that appear.

11.2.2A: Using Solview to Set Up a Layout
1. Enter the ***Solview*** command .   **Command: *solview***
2. Select the **UCS** option Ucs.   **Enter an option [Ucs/Ortho/Auxiliary/Section]: *u***
3. Tell AutoCAD to use the **Named** UCS option Named, and use the UCS called ***Front***.   **Enter an option [Named/World/?/Current] <Current>: *n***   **Enter name of UCS to restore: *front***

4. We'll use a three-quarter scale for our viewport.

   Enter view scale <1.0000>: *.75*

5. Center the viewport on Paper Space coordinate 3,3.5.

   Specify view center: *3,3.5*

   Specify view center <specify viewport>: *[ENTER]*

6. Size the viewport as indicated (you can pick approximate coordinates).

   Specify first corner of viewport: *1,4.5*

   Specify opposite corner of viewport: *@4,-2*

7. Call the view ***Front***.

   Enter view name: ***Front***

8. Complete the command.

   Enter an option [Ucs/Ortho/Auxiliary/ Section]: *[ENTER]*

Your drawing looks like the figure at right.

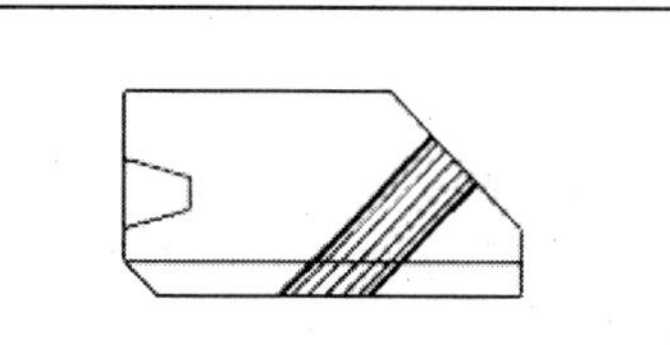

9. Repeat the command.

   Command: *[ENTER]*

10. Now tell AutoCAD to create an orthographic projection Ortho.

    Enter an option [Ucs/Ortho/Auxiliary/Section]: *o*

11. Pick a point on the right side of the existing viewport (notice that AutoCAD automatically uses the midpoint OSNAP) …

    Specify side of viewport to project:

12. … and center the viewport about 3 units to the right of the UCS view.

    Specify view center: *@3<0*

    Specify view center <specify viewport>: *[ENTER]*

13. Locate the viewport at about the coordinates indicated.

    Specify first corner of viewport: *6.5,4.5*

    Specify opposite corner of viewport: *@3,-2*

14. Call the viewport ***Right***, and complete the command.

    Enter view name: ***Right***

    Enter an option [Ucs/Ortho/Auxiliary/Section]: *[ENTER]*

15. Repeat Steps 11 through 14 to create a **Top** viewport as shown in the following figure.

    Command: *[ENTER]*

11.2.2A: Using Solview to Set Up a Layout

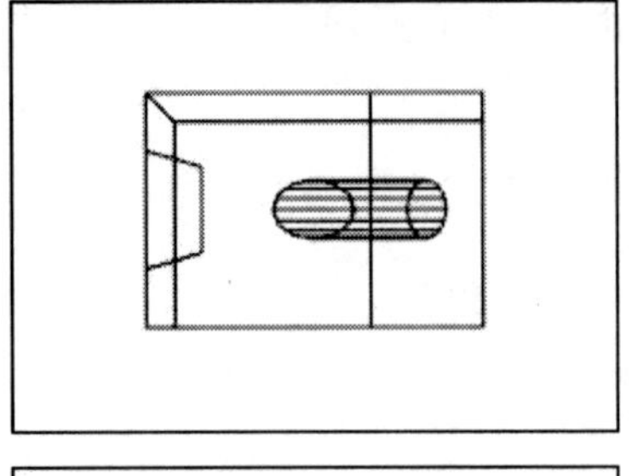

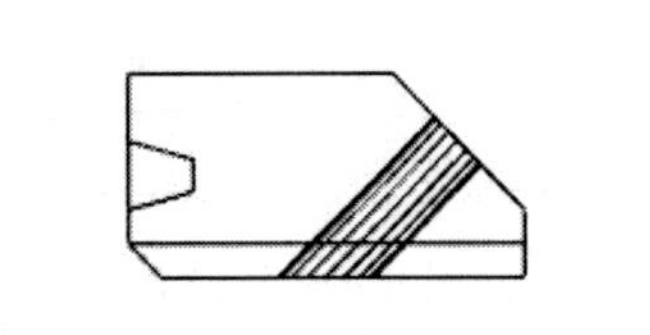

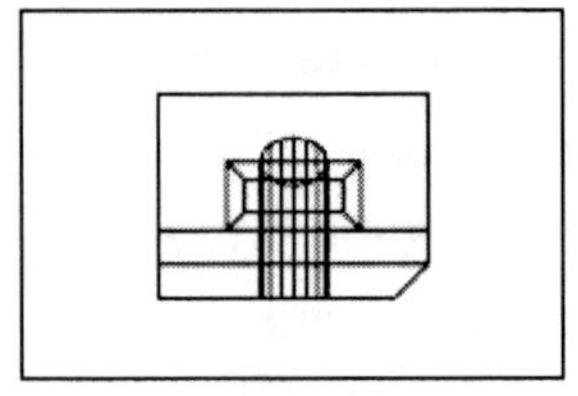

16. Now we'll create an **Auxiliary** view [Auxiliary] of the inclined surface.

    Command: ***[ENTER]***

    Enter an option [Ucs/Ortho/Auxiliary/Section]: ***a***

17. In the original viewport (*Front*), pick the endpoints of the inclined surface. (Pick anywhere in the Front viewport to activate it.)

    Specify first point of inclined plane:

    Specify second point of inclined plane:

18. Tell AutoCAD that you wish to view the surface from the upper-right corner of the viewport.

    Specify side to view from:

19. Pick a point about even with the center of the upper viewport (*Top*).

    Specify view center:

    Specify view center <specify viewport>: ***[ENTER]***

20. Place the viewport around the auxiliary image (don't worry that the viewports overlap), and name the view ***Aux***.

    Specify first corner of viewport:

    Specify opposite corner of viewport:

    Enter view name: ***Aux***

    Complete the command.

21. Move the new viewport to the position shown in the following figure.

    Command: ***m***

11.2.2A: USING SOLVIEW TO SET UP A LAYOUT

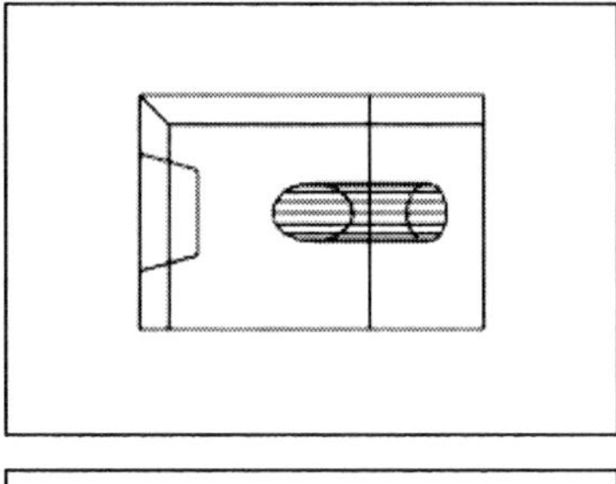

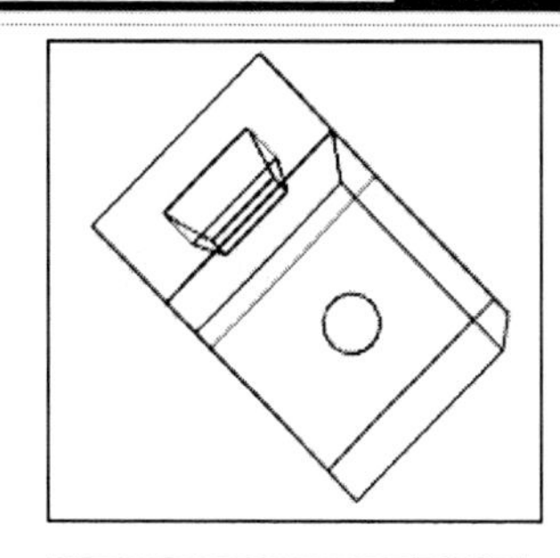

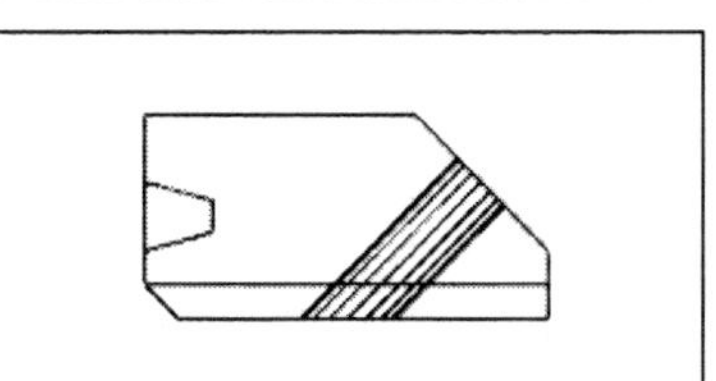

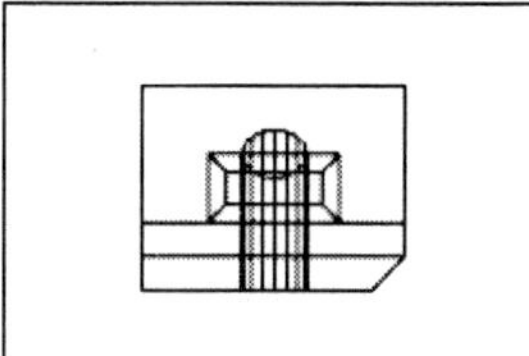

22. Save the drawing.

Command: ***qsave***

23. Repeat the ***Solview*** command.

Command: ***solview***

24. Select the **Section** option.

Enter an option [Ucs/Ortho/Auxiliary/Section]: ***s***

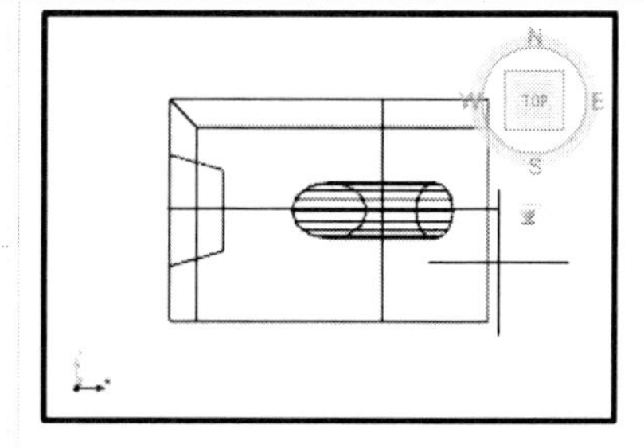

25. In the upper-left viewport (*Top*), specify the cutting plane as shown (use Ortho).

Specify first point of cutting plane:

Specify second point of cutting plane:

26. View the object from the lower half of the viewport, and accept the default view scale.

Specify side to view from:

Enter view scale <0.7500>: ***[ENTER]***

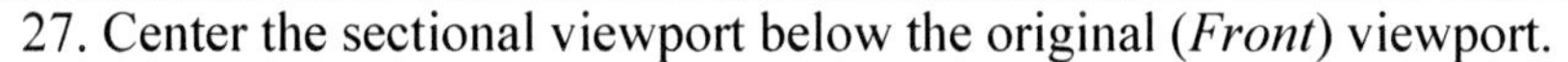

27. Center the sectional viewport below the original (*Front*) viewport.

Specify view center:

Specify view center <specify viewport>: ***[ENTER]***

28. Place the viewport around the image, and call the view ***Sect***.

Specify first corner of viewport:

Specify opposite corner of viewport:

Enter view name: ***Sect***

29. Complete the command.

Enter an option [Ucs/Ortho/Auxiliary/Section]: ***[ENTER]***

Your drawing looks like the following figure.

11.2.2A: Using Solview to Set Up a Layout

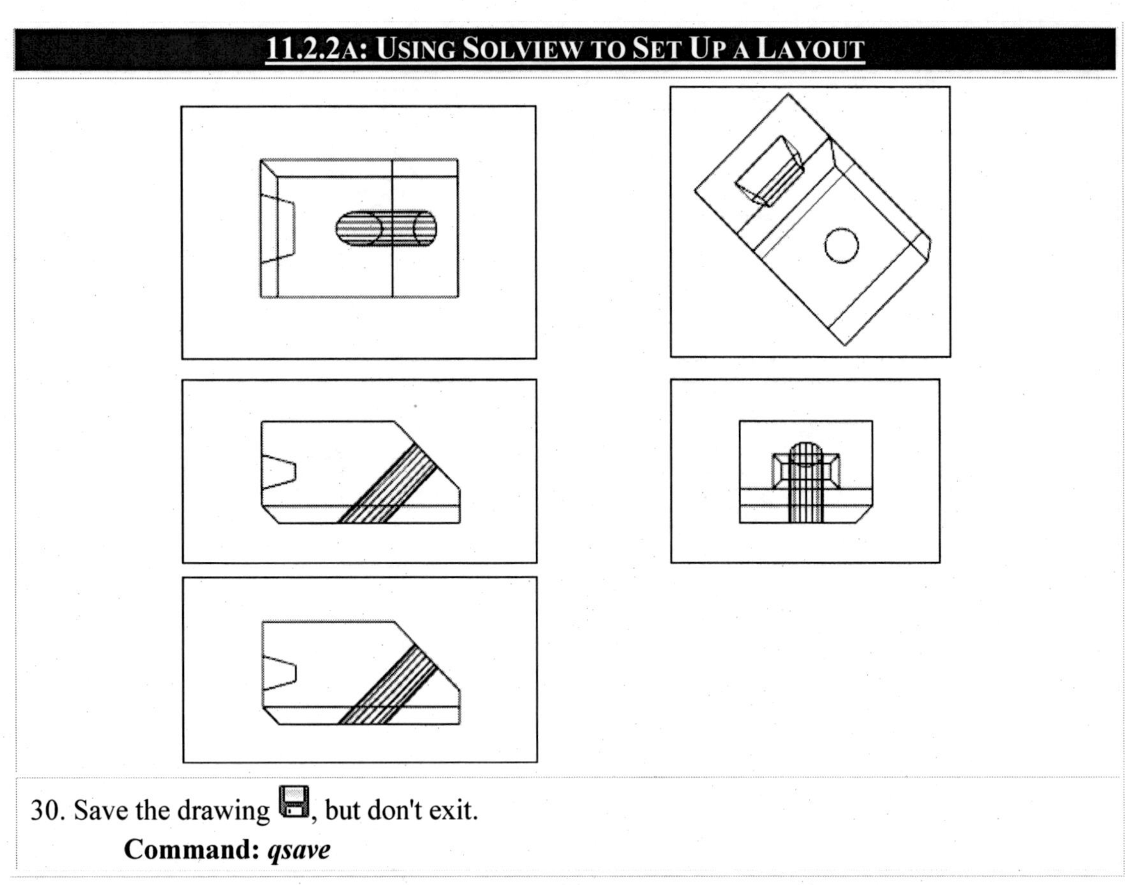

30. Save the drawing, but don't exit.

    Command: ***qsave***

How's that for a quick way to set up several viewports? If you look a little further than your immediate screen, you'll find that AutoCAD has also set up layers specific to each viewport.

Obviously, however, the drawing isn't yet ready to plot – each viewport still shows the full 3D solid. To complete the drawing, we need to show profiles in each viewport. Let's look at the ***Soldraw*** and ***Solprof*** commands next.

11.2.3 Creating the Plot Images – The *Soldraw* and *Solprof* Commands

Both ***Soldraw*** and ***Solprof*** create profiles in a viewport. The biggest difference is that the programmers designed ***Soldraw*** to work specifically with viewports created by the ***Solview*** command. ***Solprof*** will create a profile (see insert) in a viewport created by the ***MView*** or ***MVSetup*** commands. Additionally, ***Soldraw*** will create cross sections where they were set up with the ***Solview*** command.

> A profile shows only those edges and/or silhouettes of a 3D solid that are visible in the specified viewport when hidden lines are removed.
>
> An important thing to remember about the Sol... commands is that they were designed to work only with solids. They won't work with surfaces, meshes, or blocks.

The command sequence for ***Soldraw*** is one of AutoCAD's simplest:

Command: ***soldraw***

Select viewports to draw..

Select objects:

AutoCAD does the rest automatically. Try it.

Where to Find It:	
Command Line:	*Soldraw*
Ribbon (Tab/Panel):	Home –Modeling (subpanel) – **Solid Drawing**
Menu:	Draw – Modeling – Setup – **Draw**

Do This: 11.2.3A	Using *Soldraw* to Create the Profiles

I. Be sure you're still in the *Sol1* file in the C:\Steps3D\Lesson11 folder. If not, please open it now.

II. Follow these steps.

11.2.3A: USING SOLDRAW TO CREATE THE PROFILES

1. Enter the ***Soldraw*** command.

   Command: *soldraw*

2. Select each of the viewports.

   Select viewports to draw…

   Select objects:

   Select objects: *[ENTER]*

AutoCAD creates the profiles and sections.

Your drawing looks like the following figure.

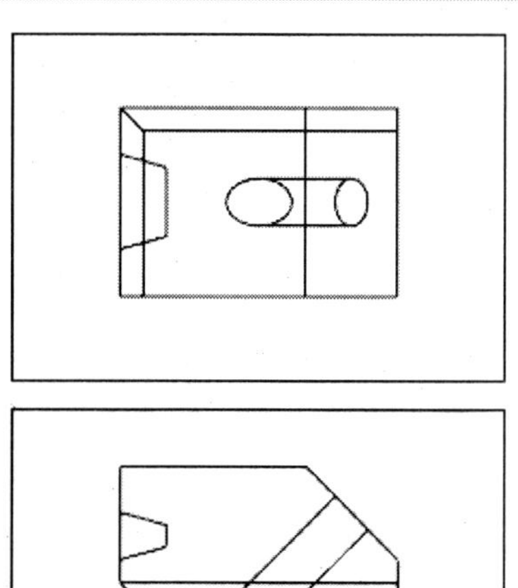
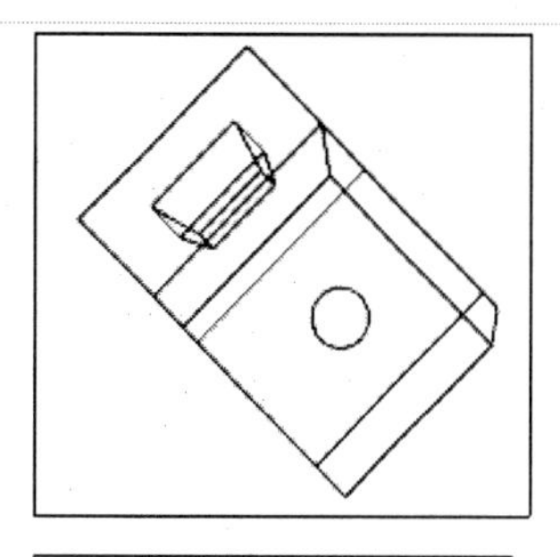
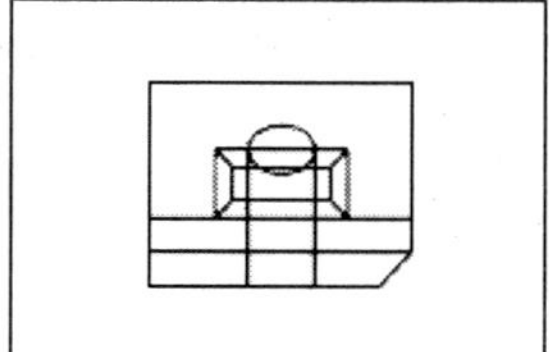
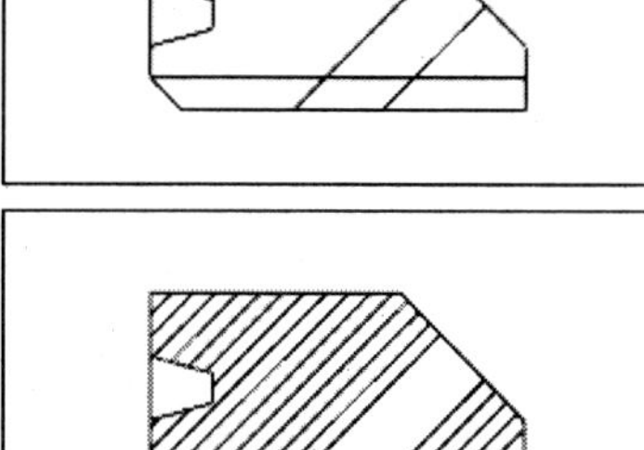

3. Not quite satisfied? Freeze all of the **[Name]-hid** layers except **Front-hid**.

4. Load the **Hidden** linetype and assign it to the **Front-hid** layer.

Your drawing looks like the following figure. (You may need to regenerate.)

11.2.3A: Using Soldraw to Create the Profiles

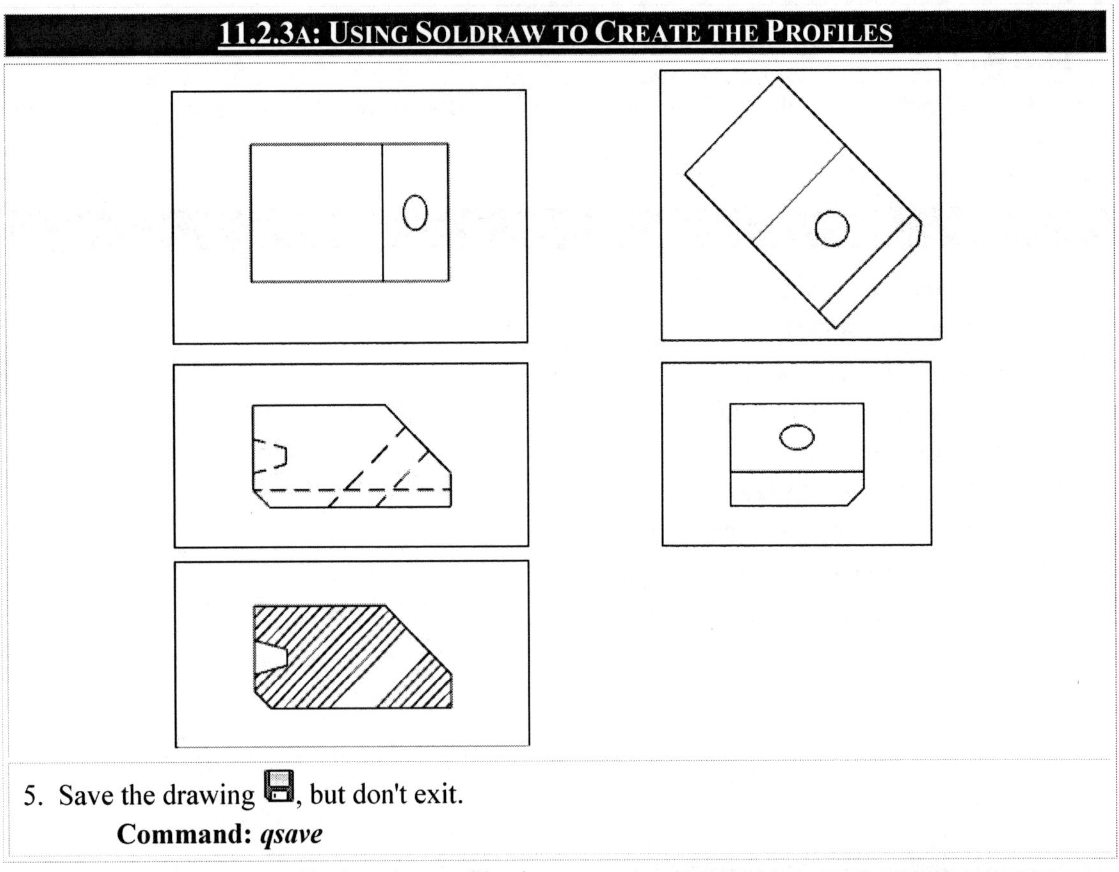

5. Save the drawing, but don't exit.

    Command: ***qsave***

Solprof works almost as easily, but the profiles it creates are actually blocks. Here's the command sequence:

Where to Find It:	
Command Line:	***Solprof***
Ribbon (Tab/Panel):	Home – Modeling (subpanel) – **Solid Profile**
Menu:	Draw – Modeling – Setup – **Profiles**

Command: ***solprof***

Select objects:

Select objects:

Display hidden profile lines on separate layer? [Yes/No] <Y>:

Project profile lines onto a plane? [Yes/No] <Y>:

Delete tangential edges? [Yes/No] <Y>:

Let's look at the options.

- The first option – **Display hidden profile lines on separate layer** – asks if you'd like to place all profile lines on one layer or place the hidden lines on a separate layer. The default is to place hidden lines on a separate layer.

  When you accept the default (generally a good idea), AutoCAD places visible lines on layer *PV-[viewport handle]* and hidden lines on layer *PH-[viewport handle]*. By using the AutoCAD-assigned viewport handle as part of the layer, AutoCAD assures you of a unique layer name. This way, you can freeze the layer or change the linetype of the hidden lines.

- The next option – **Project profile lines onto a plane** – allows you to create a two-dimensional profile by projecting the lines onto the view plane (the default), or to create three-dimensional lines.

- The last option – **Delete tangential edges** – allows you to remove tangential lines. These are objects (lines) that show the transition between arcs or circles. They're essentially the same things as isolines, except that they're actual objects.

Let's use the ***Solprof*** command to create an isometric view of our object.

Do This: 11.2.3B	Using *Solprof* to Set Up a Layout

I. Be sure you're still in the *Sol1* file in the C:\Steps3D\Lesson11 folder. If not, please open it now.

II. Create a new viewport in the lower-right corner of the layout. (Use the ***MView*** command.)

III. Activate the new viewport and set up an isometric view (1,1,1),

IV. Follow these steps.

11.2.3B: USING SOLPROF TO SET UP A LAYOUT

1. Enter the ***Solprof*** command.

   Command: ***solprof***

2. Select the 3D solid in the new viewport.

   Select objects:

   Select objects: ***[ENTER]***

3. Accept the defaults for the next three prompts.

   Display hidden profile lines on separate layer? [Yes/No] <Y>: ***[ENTER]***

   Project profile lines onto a plane? [Yes/No] <Y>: ***[ENTER]***

   Delete tangential edges? [Yes/No] <Y>: ***[ENTER]***

4. Freeze the **PH-[viewport handle]** layer. (The viewport handle will vary.)

5. Freeze the **obj1** layer in the active viewport.

   The viewport looks like the figure at right.

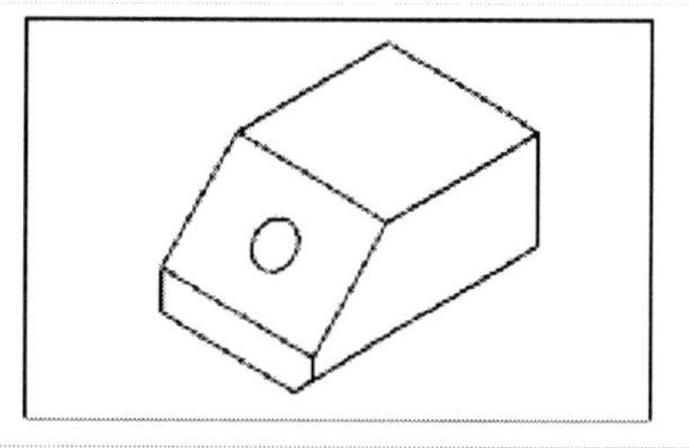

6. Save the drawing.

   Command: ***qsave***

Now you can use the other techniques you've learned to dimension each view, add appropriate text, and otherwise complete the drawing.

11.2.4	Sending a Model to a 3DPrinter

What exactly is a 3D Printer?

In the truest sense of the word, it isn't a printer at all. It's something of a CAM system (Computer Aided Manufacturer) – that is, it actually creates a prototype of the model you send it. Generally, the printer service will return the model to you in wax or plastic, but it should be as functional as the final product (if not as durable).

Where to Find It:	
Command Line:	*3DPrint*
Hotkey(s):	*3dp*
Ribbon (Tab/Panel):	Output – 3D Print – **Send to 3D Print Service**
Application Menu:	Publish – **Send to 3D Print Service**

You'll send your model to a 3D Print Service before

manufacturing because it costs less to produce a sample model out of plastic or wax than it does to produce the product. And if you've made a mistake, you'll be a lot happier to find out before having a client retool his shop!

Where do you find a 3D Printer Service?

The simple (and obvious) answer is the Internet! But I'll give you a couple address to start.

- **Redeye RPM** (http://www.redeyeondemand.com/) – Autodesk uses this company in its tutorials.
- **Dimension** (http://www.dimensionprinting.com/) – These folks operate out of a garage somewhere, but they have the best demo videos of a 3D printer I've found. Take a few minutes and check out the demos.

Several companies produce or will soon produce the equipment for 3D printing (prototyping) at your own company. The prices I've found run from around $5-20k.

You'll find getting your model ready for a 3D printer anticlimactic after all you put into creating it. The command begins a wizard. Let's walk through it now.

Do This: 11.2.4A	Preparing for a 3D Print Service

I. Open the *Flange11a* file in the C:\Steps3D\Lesson11 folder.

II. Follow these steps.

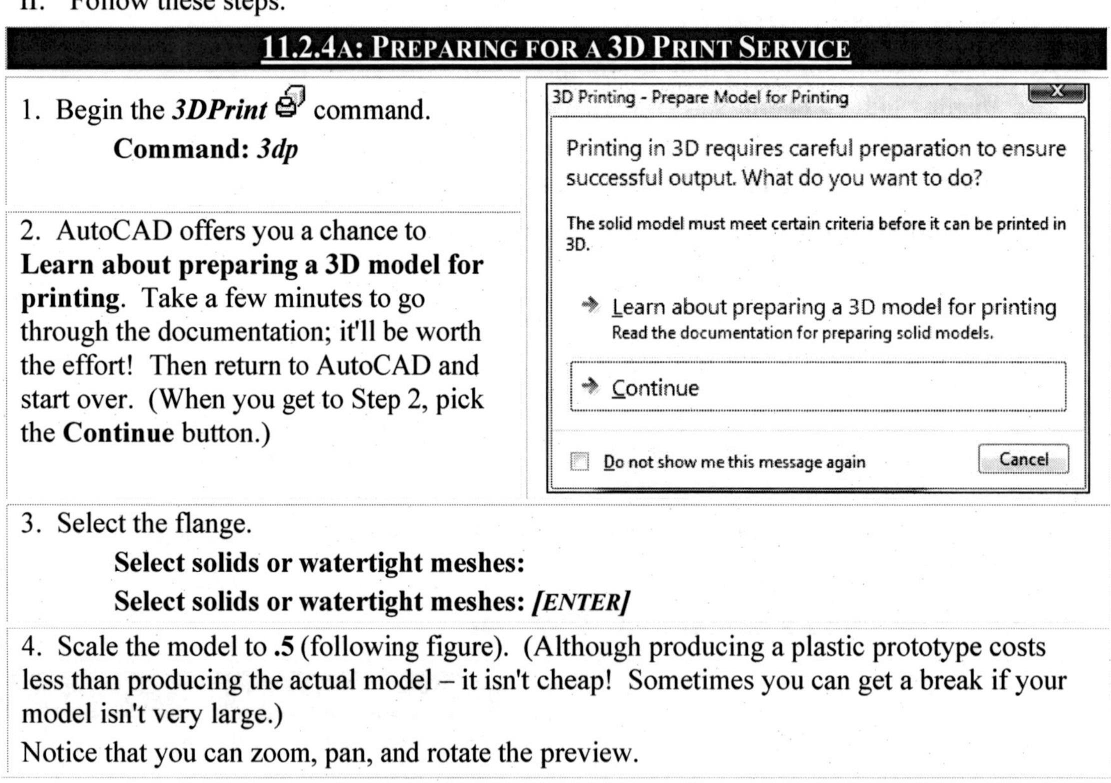

11.2.4A: Preparing for a 3D Print Service

1. Begin the ***3DPrint*** command.
   Command: *3dp*
2. AutoCAD offers you a chance to **Learn about preparing a 3D model for printing**. Take a few minutes to go through the documentation; it'll be worth the effort! Then return to AutoCAD and start over. (When you get to Step 2, pick the **Continue** button.)
3. Select the flange.
   Select solids or watertight meshes:
   Select solids or watertight meshes: *[ENTER]*
4. Scale the model to **.5** (following figure). (Although producing a plastic prototype costs less than producing the actual model – it isn't cheap! Sometimes you can get a break if your model isn't very large.)
   Notice that you can zoom, pan, and rotate the preview.

11.2.4A: Preparing for a 3D Print Service

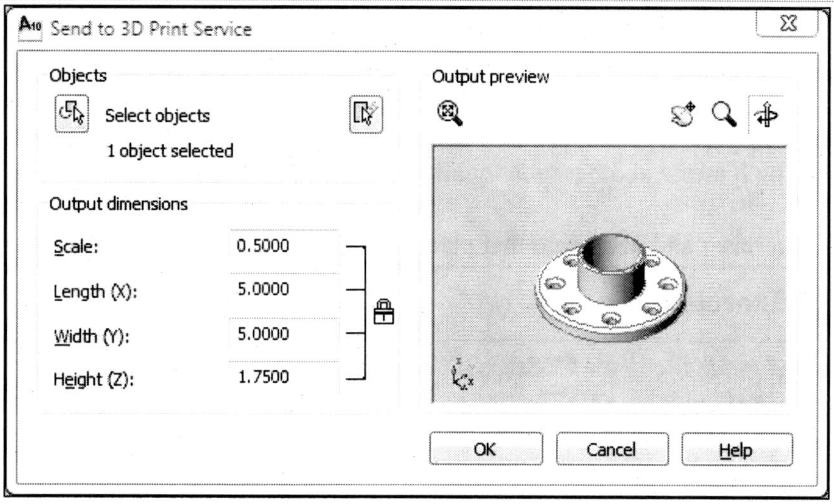

Send to 3D Print Service

Objects

Select objects

1 object selected

Output dimensions

Scale: 0.5000

Length (X): 5.0000

Width (Y): 5.0000

Height (Z): 1.7500

Output preview

OK Cancel Help

5. Pick the **OK** button to continue.
6. Now AutoCAD displays a standard Save file dialog box. It defaults to a Lithography (*.stl) file type, which you can transmit to the printer service.

Call the file, *ModelFlange* and save it in the C:\Steps3D\Lesson11 folder.

Now you can send the file to your printer service! (In fact, if you're on line when you've finished the stl creation, AutoCAD will automatically open its own website with a list of partners who provide 3D printer services.)

Some things you should remember when getting set up for a 3D printer service include:

- Each printer service will have its own requirements. Check with them first before you create the stl file.
- Depending upon what your production will be, you might want to discuss a company investment in your own 3D printer. Prices are coming down and it's a skill that might benefit you in the long run. If you do purchase your own 3D printer, read through the documentation on how to set up your model before creating the stl file.

11.3 Extra Steps

Return to any of the drawings you created in the exercises at the end of Lessons 9 and 10. Create the plotting layouts shown in the exercises (if you created layouts, open **Layout2** and recreate them). This will work best if you use drawings for which you've already created layouts – it'll help you compare the method you used previously with the ***Sol...*** commands.

11.4 What Have We Learned?

Items covered in this lesson include:

- *Three-dimensional uses and techniques for blocks*
- ***Flatshot***
- ***3DPrint***
- *The **Sol...** tools used to set up plots for 3D solid*
  - ***Solview***
  - ***Solprof***
  - ***Soldraw***

In this lesson, you discovered some easier ways to set up a Paper Space plot and some new (and useful) techniques for working with blocks. And you wrapped up your study of three-dimensional drafting and modeling techniques. You can relax for two minutes and pat yourself on the back for having accomplished quite a lot of often-difficult material. Then tackle the exercises at the end of the lesson.

In Lesson 12, you'll learn some presentation tricks. Presentation takes you one step further – to adding material qualities to your objects. This means having a table that shows wood grain or a glass lamp that appears transparent. You'll see how to show your drawing in perspective rather than isometric mode. You'll create photographic-quality images suitable for brochures or posters, and much more!

So, complete the exercises and hurry into that place where AutoCAD meets computer graphics!

11.5 Exercises

1. Open the *slotted guide* file in the C:\Steps3D\Lesson11 folder. Create the layout.
   1.1. Hint: The ***Sol...*** commands work best when the object is viewed through a 2D visual style.
   1.2. Save the drawing as *MySG* in the C:\Steps3D\Lesson11 folder.

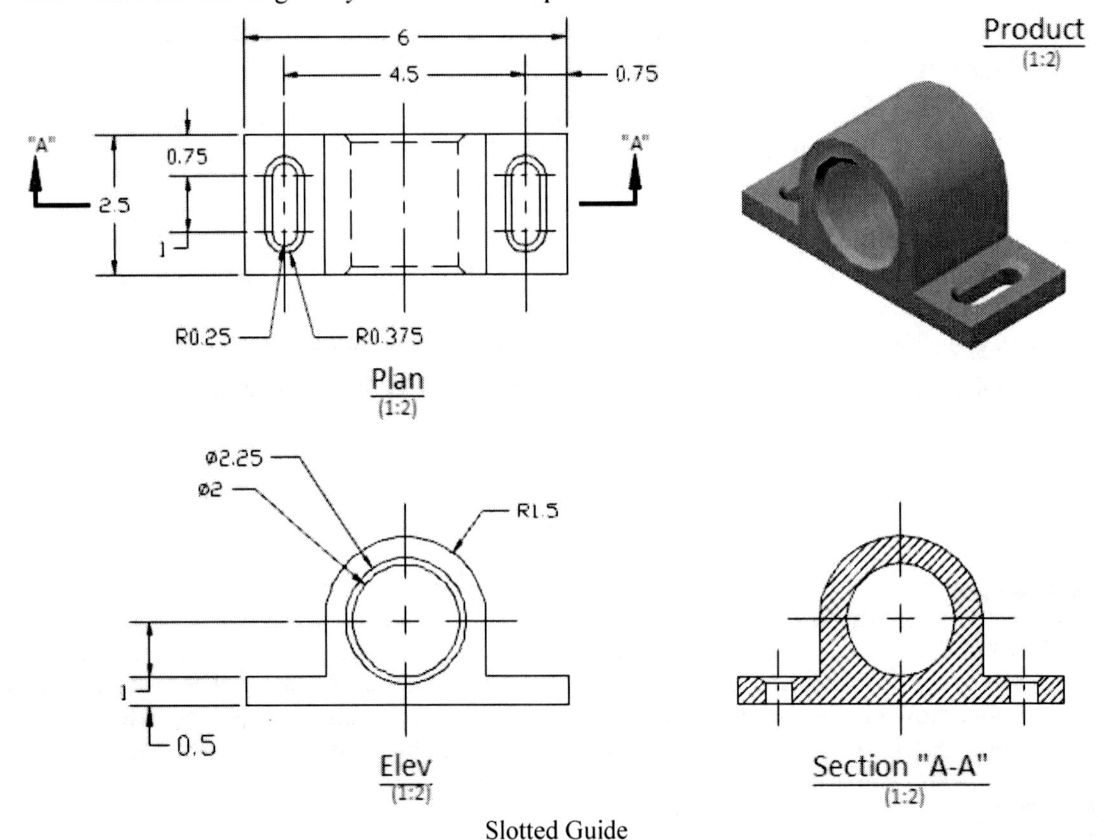

Slotted Guide

2. Open the *ExFlange* file in the C:\Steps3D\Lesson11 folder. Create the layout.
    2.1. Most of the centerlines already exist on layer **Cl**.
    2.2. Save the drawing as *MyFlg* in the C:\Steps3D\Lesson11 folder.

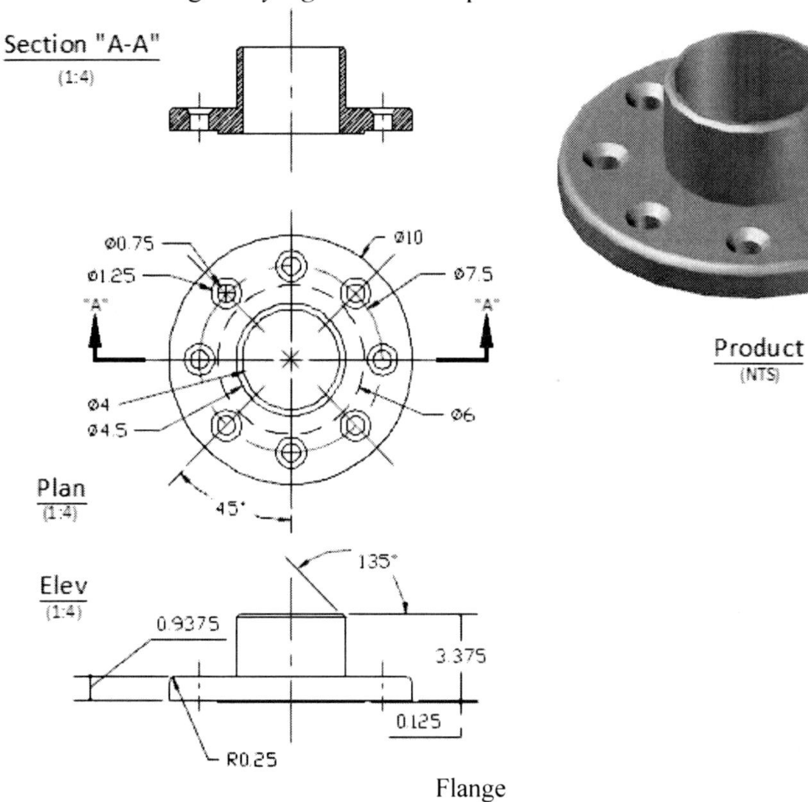

Flange

3. Create the *Bike Rack*.
    3.1. Use the *1_2Ell* drawing found in the C:\Steps3D\Lesson11 folder.
    3.2. Use the *1_2Pipe* drawing to provide the pipe between the elbows (just as you did in Exercise 11.1.2A, p.253).
    3.3. Be sure the UCS = WCS when you finish.
    3.4. Save the drawing as *MyBikeRack* in the C:\Steps3D\Lesson11 folder.

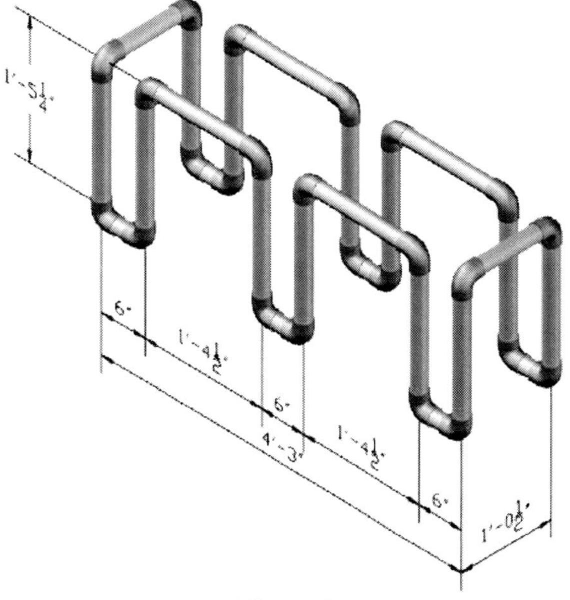

Bike Rack

4. Open *Jig1* in the C:\Steps3D\Lesson11. Create the layout.
   4.1. Set up the drawing on a 17" x 11" sheet of paper.
   4.2. Save the drawing as *MyJig* in the C:\Steps3D\Lesson11 folder.

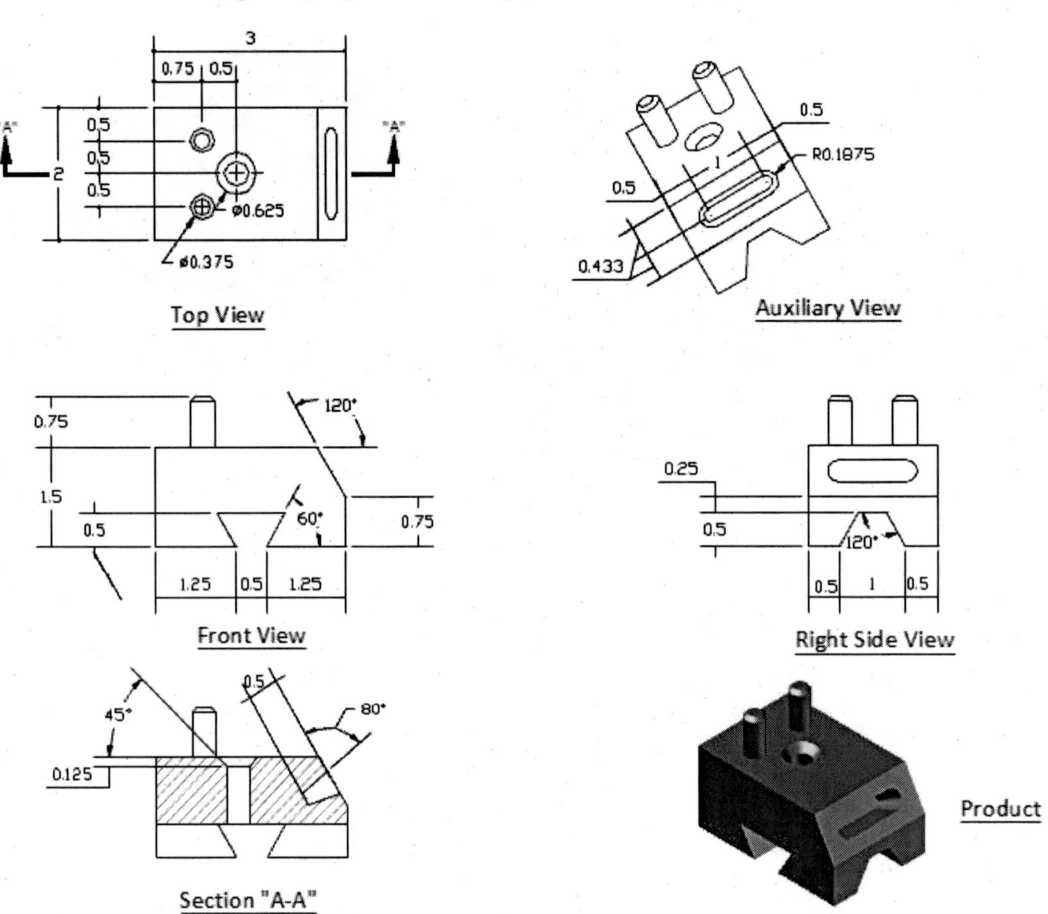

Jig

5. Open the *Thermometer* file in the C:\Steps3D\Lesson11 folder. Create the layout.
    5.1. Set up the drawing on an 11" x 17" sheet of paper.
    5.2. Save the drawing as *MyThermometer* in the C:\Steps3D\Lesson11 folder.

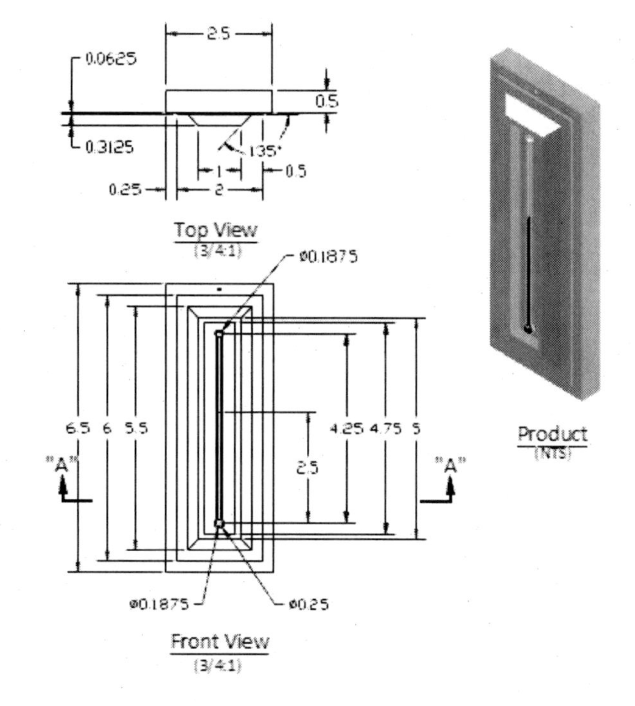

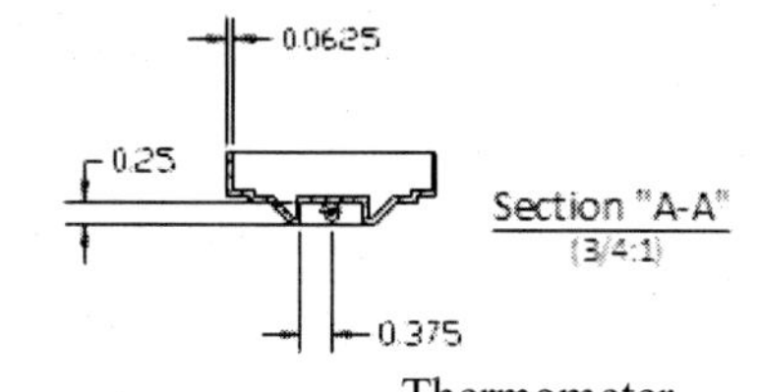

Thermometer

6. Create the *Lawn Chair*.
    6.1. Use the *1_2Ell* drawing found in the C:\Steps3D\Lesson11 folder.
    6.2. Use the *1_2Tee* drawing found in the C:\Steps3D\Lesson11 folder.
    6.3. Use the *1_2Pipe* drawing to provide the pipe between the elbows (just as you did in Exercise 11.1.2A,p.253).
    6.4. Be sure the UCS = WCS when you finish.
    6.5. Save the drawing as *MyLawnChair* in the C:\Steps3D\Lesson11 folder.

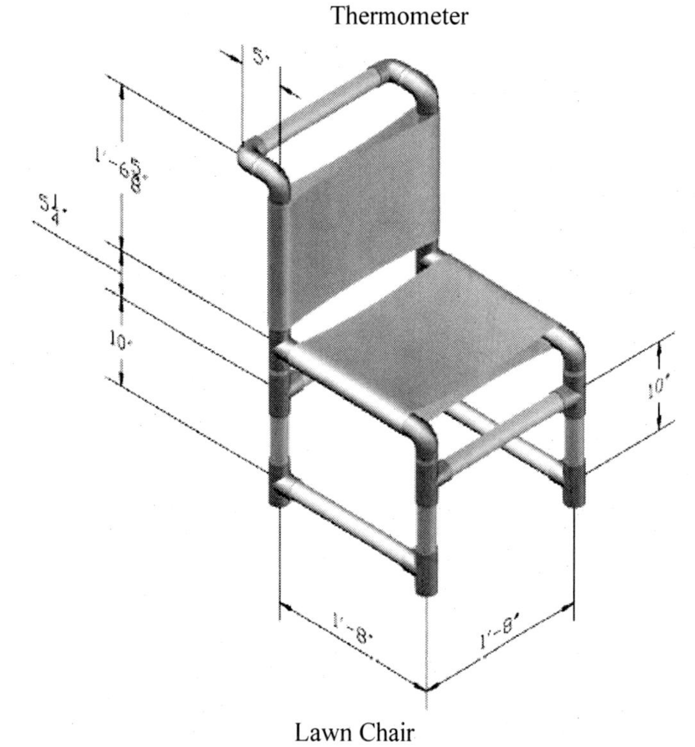

Lawn Chair

7. Create the *service cart* drawing shown below.
   7.1. Create the layout on an 11" x 17" sheet of paper.
   7.2. Adjust the Z-Space and UCS as needed to use a single 2" x 2" block to build the frame.
   7.3. Use the caster you created in Lesson 4 (or the *Caster* file in the C:\Steps3D\Lesson11 folder) for the caster block.
   7.4. Use attributes to create the cutting list.
   7.5. You'll notice that the ***Sol…*** commands won't work properly on blocks, so you'll have to use the ***MView*** or ***MVSetup*** command to create your viewports.
   7.6. Save the drawing as *MyCart* in the C:\Steps3D\Lesson11 folder.

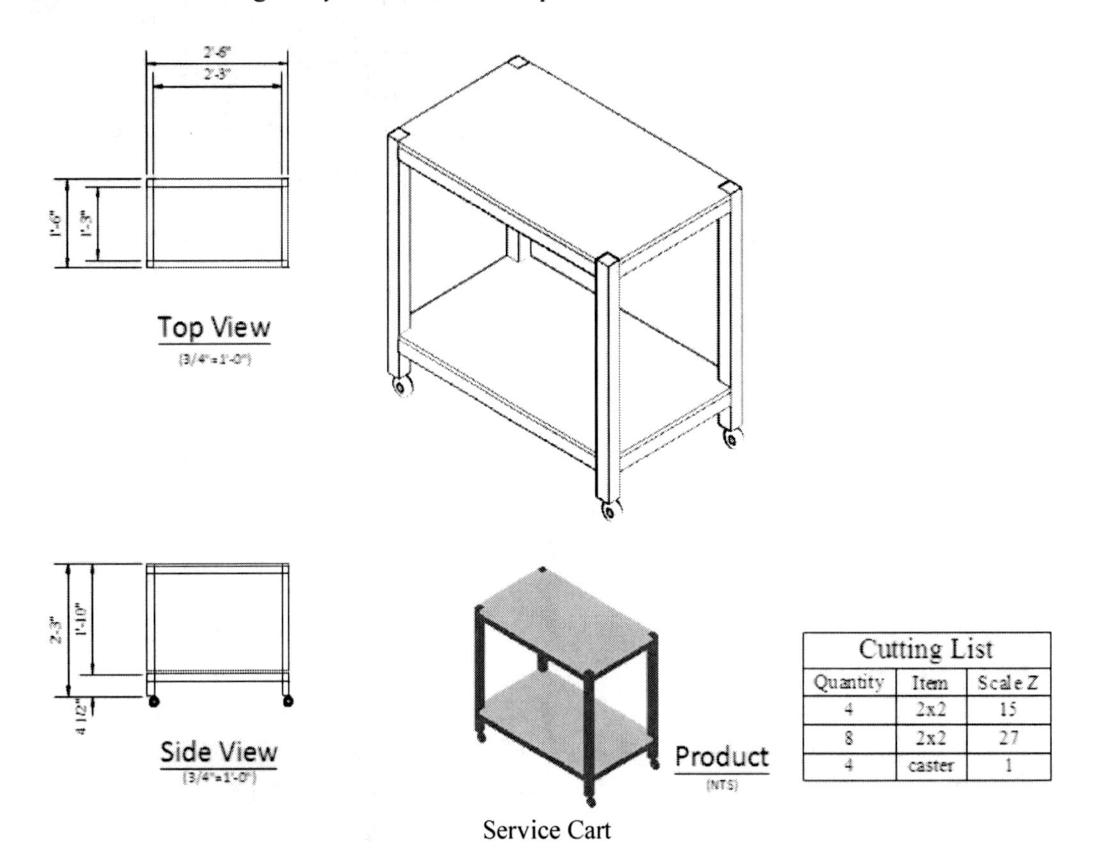

Cutting List		
Quantity	Item	Scale Z
4	2x2	15
8	2x2	27
4	caster	1

Service Cart

8. Create the *½" Ell* drawing.
    8.1. Make sure the base point of the elbow is as indicated. (Use the ***Base*** command to move it, if necessary).
    8.2. Be sure the UCS = WCS when you finish.
    8.3. Save the drawing as *My1_2Ell* in the C:\Steps3D\Lesson11 folder.

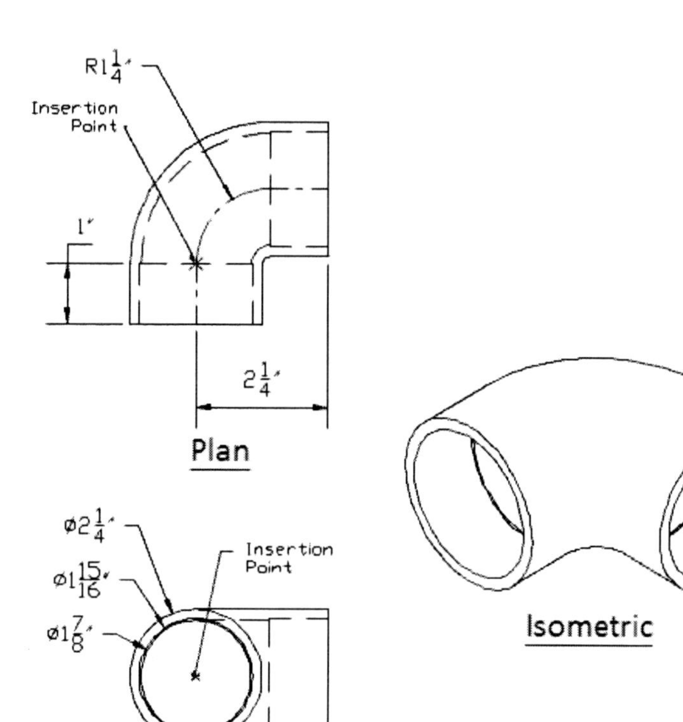

½" Ell

9. Create the *½"Tee* drawing.
    9.1. Make sure the base point of the tee is as indicated.
    9.2. Be sure the UCS = WCS when you finish.
    9.3. Save the drawing as *My1_2Tee* in the C:\Steps3D\Lesson11 folder.

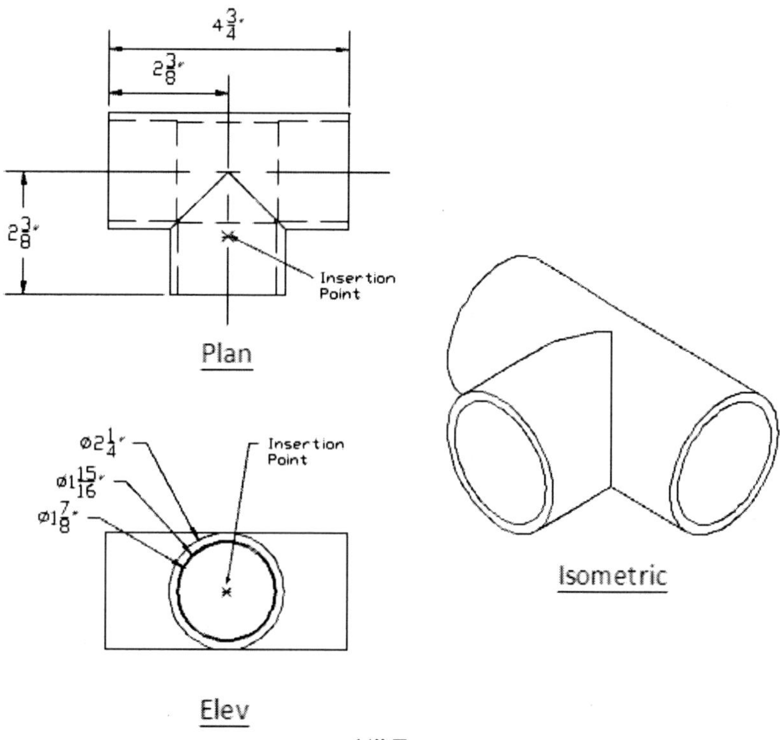

½" Tee

10. Create the *Patio Scene*.
    10.1. Use the equipment you created in Lesson 11's Exercises. (If these aren't available, use the corresponding drawing found in the C:\Steps3D\Lesson11 folder.)
    10.2. You created the garden fence in Lesson 8 and the fountain in Lesson 9. (Both are provided in the C:\Steps3D\Lesson11 folder if you didn't save your drawings.)
    10.3. Save the drawing as *MyPatioScene* in the C:\Steps3D\Lesson11 folder.

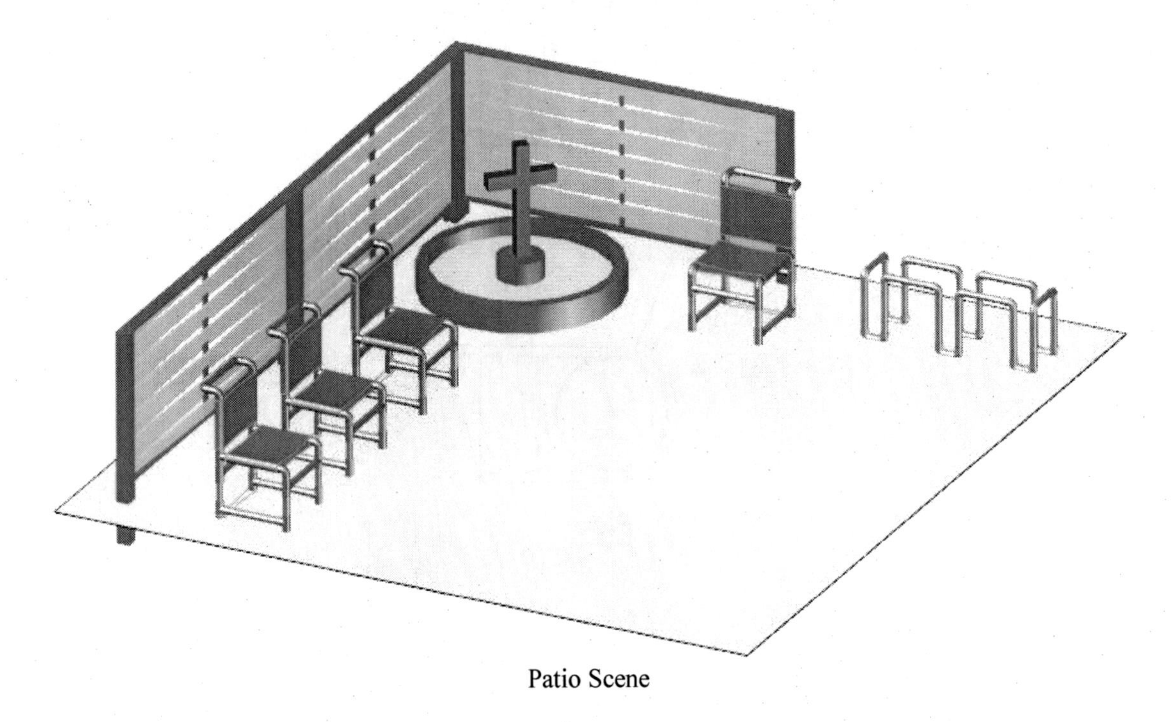

Patio Scene

11.6	**For Web-Based Review Questions, visit: http://foragerpub.com/AcadFiles/2011/2011.htm**

Lesson 12

Following this lesson, you will:

✓ *Know how to render an AutoCAD drawing*
- *Rendering*
- *Assigning materials*
- *Adding lights*

Presentation Tools

From childhood's hour I have not been *As others were – I have not seen* *As other saw.* *Alone* – Edgar Allan Poe
Far better it is to dare mighty things, to win glorious triumphs, even though checkered by failure, than to take rank with those poor spirits who neither enjoy much nor suffer much, because they live in the gray twilight that knows not victory nor defeat. Theodore Roosevelt
The measure of one's soul is calculated *not in successes or failures, but in the number of* *attempts one is willing to make.* Anonymous

As you might guess from the three preceding quotes, you now face the most challenging of the lessons you'll undertake in our One Step at a Time series. So before you start, think back to what you knew when you began Lesson 1 of AutoCAD 2010: One Step at a Time. You began each lesson with anticipation and a touch of anxiety, but you finished each knowing more than you did when you started. It hasn't always been easy, but you've persevered (or else you wouldn't be here). Consider your accomplishments. And take Teddy's advice and "dare mighty things" in Lesson 12.

12.1 What Is Rendering and Why Is It So Challenging?

Rendering is a procedure that takes the objects you've created and gives them properties to make them appear "real." The degree to which they appear real depends on a host of user-defined settings and assignments, including materials, types and positions of lights, and light intensity.

Why is rendering so challenging? Consider what Edgar Allan Poe said in the quote that began this lesson. Every individual will "see" a scene in a different way. Translating what your mind sees to what appears on the screen involves often subtle manipulation of several variables.

Remember Lesson 5 – I told you that we'd reached the edge between CAD operating and CAD programming. Well, in Lesson 12, you've reached the edge between CAD operating and art. Just as not every whittler is a sculptor, not every draftsman is an artist. (A fact I found myself repeating … and repeating … to my employer back when I designed those nifty little houses that Santa sits in down at the mall.) This is where you face the challenge.

12.2 Beyond Visual Styles – The *Render* Command

Where to Find It:	
Command Line:	***RenderPresets***
Hotkey(s):	***rp***
Ribbon:	Render – Render – Render Presets control – **Manage Render Presets**
Dialog Box:	Advanced Render Settings palette control – **Manage Render Presets**

It may fortify you to know that you've been using a rudimentary form of rendering all along when you used visual styles. But here again, consider the whittler and the sculptor. Whereas visual styles have a few settings from which to choose, rendering presents (quite literally) infinite possibilities. Fortunately, we'll navigate the possibilities using tool palettes and dialog boxes. (This should make you appreciate the fact that you're not using one of the earlier releases of AutoCAD!)

Let's begin with the Render Presets (refer to the Render Presets dialog box in Figure 12.001, p.279). Make yourself comfortable; we're going to spend some time with this one.

Let's get started.

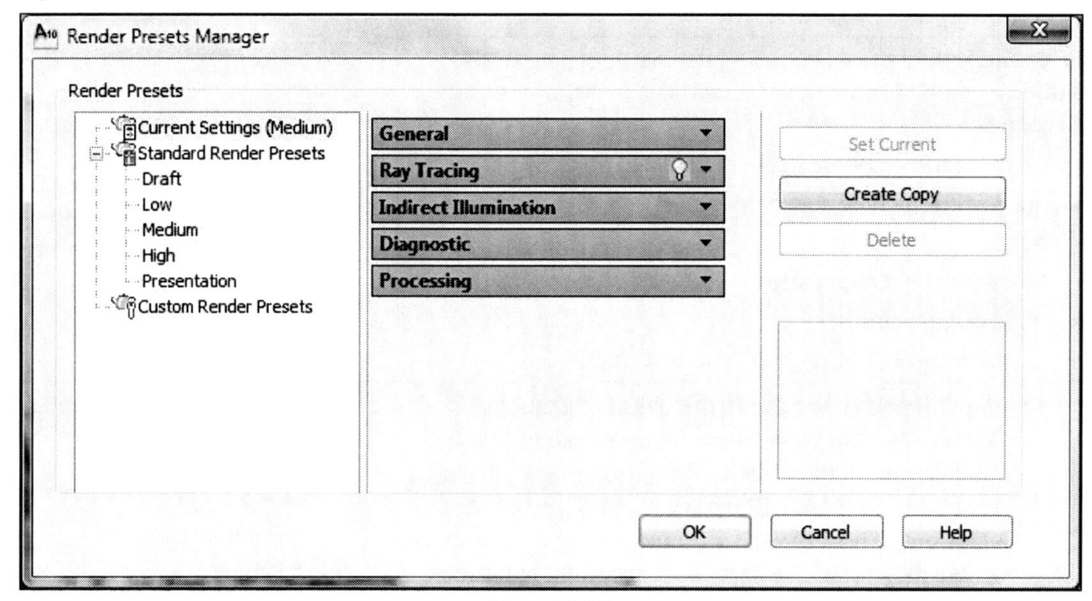

Figure 12.001

- We'll start with the selection box on the left.
    - It begins with an informational item giving you the **Current Settings**. (In this case, it shows the default – **Medium**.)
    - Below the informational item, you'll see the list of **Standard Render Presets**. Here, the list graduates from **Draft** quality renderings to **Presentation** quality. Until you have some experience rendering, you'll find it easier to begin with **Medium** quality renderings and adjust from there.
    - The last item in the list reads **Custom Render Presets**. These are the rendering setups you'll create. Begin with a preset, **Create** [a] **Copy** (use the button on the right), and then make your adjustments on the property panel in the middle of the dialog box.
- The property panel in the middle of the dialog box can (and will) be frightening the first time you see it (unless you're a closet photographer or artist). We're going to spend some time here, but if you don't fully absorb it all, it won't be the end of the world. Remember, the folks at Autodesk have worked hard to keep their baby user-friendly. The default list provides just about everything you'll need in the way of presets. You'll use that scary settings list more for tweaking than anything else.

You'll find similar settings on the Advanced Render Settings palette. Access it with the ***RPref*** command. Using the palette, you can change the current preset or even the settings of your rendering on the fly.

Where to Find It:	
Command Line:	***RPref***
Hotkey(s):	***rpr***
Ribbon (Tab/Panel):	Render – Render – [Settings button]
Menu:	View – Render – **Advanced Render Settings**
Toolbar:	Render – **Advanced Render Settings**

- The **General** tab (Figure 12.002) of the **Property** panel contains settings for **Materials**, **Sampling** and **Shadows**.

The **General** tab of the Advanced Render Settings palette also contains a group of Render Context settings (right) where you can control:

- what's rendered (**View, Crop,** or **Selected** objects)
- the destination (**Window** or **Viewport**)
- the **Output file name**
- the resolution (**Output size**)
- **Type** of **Exposure**
- And brightness of the output (**Physical Scale**)

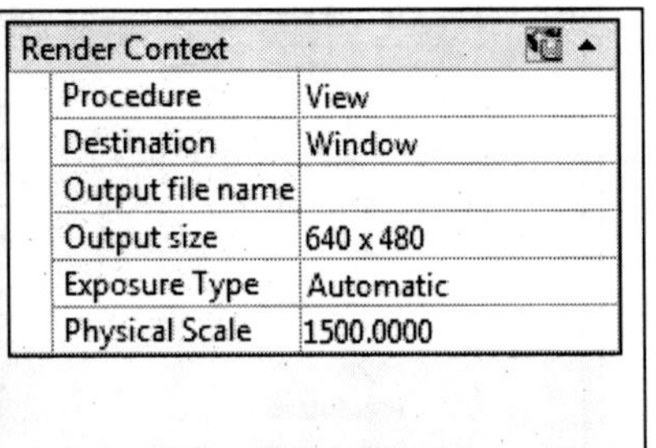

- We'll spend time later in this lesson discussing **Materials**. For now, know that you can toggle their application on or off here. More important, perhaps, is the ability to force materials to appear on one or two sides of the host object. Forcing materials to appear where they won't be seen wastes system resources, while not seeing materials where you should can ruin your presentation.
- Let me try to make **Sampling** simple. Sampling speeds the rendering process by rendering only a ratio of the pixels on your screen –more samples (pixels rendered) mean finer (and slower) renderings. The number of **Min samples** can't exceed the number of **Max samples**. Lower numbers produce faster but lower quality renderings.

  Filters determine how AutoCAD expresses multiple samples in a single pixel. You have a choice of five **Filter types** (listed from fastest and lowest quality): **Box**, **Triangle**, **Gauss, Mitchell**, and **Lanczos**.

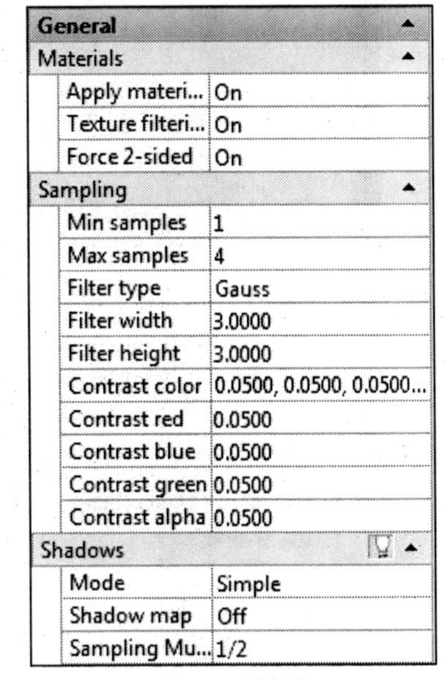

Figure 12.002

  Larger **filter width** and **height** settings slow rendering and soften the image.

  The **Contrast** settings can be confusing, but simply put, they control the color and amount of contrast your rendered image will have. **Zero** settings indicate black; **One** settings indicate fully opaque colors.

- Notice the light bulb atop the **Shadows** tab? This sneaky toggle determines whether or not your rendering will show shadows.

  Shadow **Modes** include: **Simple** (random shadows), **Sorted** (shadows generated in order from the object to the light), and **Segment** (generates shadows along the ray of light). As usual, the simpler shadows are faster but don't produce as high a quality image.

  A **Shadow map** controls the accuracy of a shadow. Using shadow maps can greatly slow a rendering, but you'll almost always like the results.

- **Sampling multipliers** determine how many samples AutoCAD will use when adjusting the shadow quality. As usual, lower is faster and higher is better.

- **Ray Tracing** (Figure 12.003) also involves shadows, and by inference, light. Luckily, this complicated tool has only three options beyond the on/off toggle (the light bulb) on the tab itself.

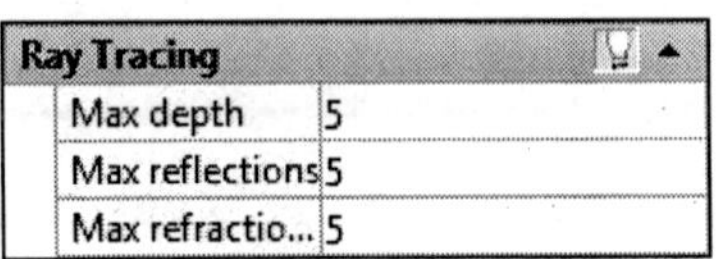

Figure 12.003

  - The **Max depth** setting limits the total reflections and refractions allowed during the rendering.

- **Max reflections** controls how many times light will bounce off of a surface.
- **Max refractions** controls how many times light will bend as a result of collision with a surface.

o **Indirect Illumination** (Figure 12.004) also contains three tabs that deal with light.

- Toggle **Global Illumination** on or off with the light bulb on its tab. Then use its settings to control your scene's lighting.

  Photons/sample sounds Star Truckkie – but it isn't quite that terrifying. It controls the intensity of global light (ambient light). Higher numbers slow renderings and make the scene more blurry but less noisy. (The dictionary defines *Photon* using words like *particle* and *antiparticle* – let's avoid that discussion, shall we?)

  Use radius works with **Radius** to control the use and size of your photons. Each photon is a percentage of the size of the scene. Larger sizes mean slower renders (and possible detection by Romulan war birds!)

  The last three options are the same as those found on the **Ray Tracing** tab.

Indirect Illumination	
Global Illumination	
Photons/sam...	500
Use radius	Off
Radius	1"
Max depth	5
Max reflections	5
Max refractio...	5
Final Gather	
Mode	Auto
Rays	200
Radius mode	Off
Max radius	1.0000
Use min	Off
Min radius	0.1000
Light Properties	
Photons/light	10000
Energy multi...	1.0000

Figure 12.004

- **Final Gather** (when you've toggled the **Mode** on) continues AutoCAD's efforts in controlling global illumination.

  The **Rays** setting determines how many rays of light AutoCAD will use in a final "gathering" before rendering. Higher numbers mean slower renders but less noise.

  The **Radius mode** (**On**, **Off**, or **View**) determines whether or not AutoCAD will use the **Max radius** setting for the final gather. **View** let's you set the mode in terms of pixels or world units.

  Increasing the **Max radius** here will give you a better rendering, but it'll take longer.

  Use min and **Min radius** are similar to **Radius mode** and **Max radius**.

- **Light Properties** affect how indirect light behaves. You only have a couple settings about which to concern yourself.

  Photons/light tells AutoCAD how many photons to emit from each light.

  The **Energy Multiplier** multiplies the global illumination, ambient light, and intensity. Put more simply, it can make your image sharper.

o **Diagnostic** settings (Figure 12.005) can help you figure out why the renderer isn't doing exactly what you expected it to do.

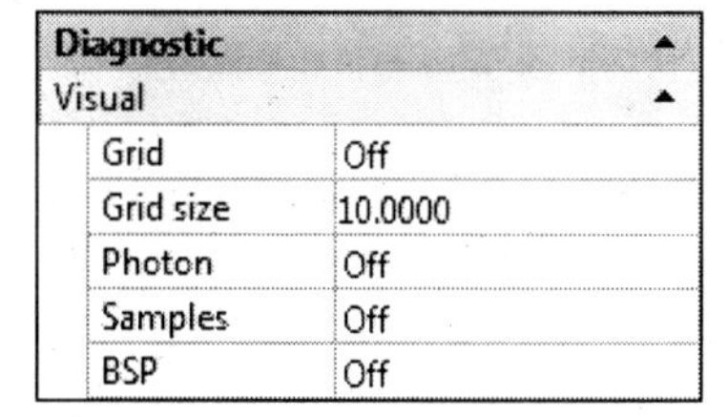

Diagnostic	
Visual	
Grid	Off
Grid size	10.0000
Photon	Off
Samples	Off
BSP	Off

Figure 12.005

Grid toggles a grid on or off. The grid shows coordinate spacing of **Objects**, the **World**, or a **Camera**. **Grid size** determines how large a grid to use.

Use **BSP** to check for **Depth** or **Size** problems when your renderer seems to be going particularly slow.

o When you render, you'll notice a "tile" moving about the screen producing an image. Use options on the **Processing** tab (Figure 12.006) to adjust how this works.

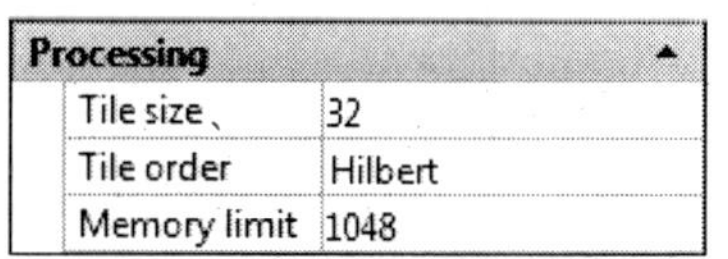

Processing	
Tile size	32
Tile order	Hilbert
Memory limit	1048

Figure 12.006

- Larger tiles (**Tile size**) decrease render time but produce fewer image updates.

- **Tile order** controls the render order AutoCAD uses to render individual tiles. Most of the options are obvious, but **Spiral** spirals outward from the center, and **Hilbert** (the default) moves to whichever tile will be quickest.
- Finally, the **Memory limit** is how much of your system memory (RAM) AutoCAD will use to do the render. If you reach your limit, AutoCAD will discard some geometry in favor of others.

- The three buttons on the right side of the Render Presets Manager require little explanation. But understand that, in order to create a new preset, you must make a copy of an existing one with which to begin.

There's one more thing we have to view before we create our first rendering – the Rendering Window (Figure 12.007). This window appears for each rendering.

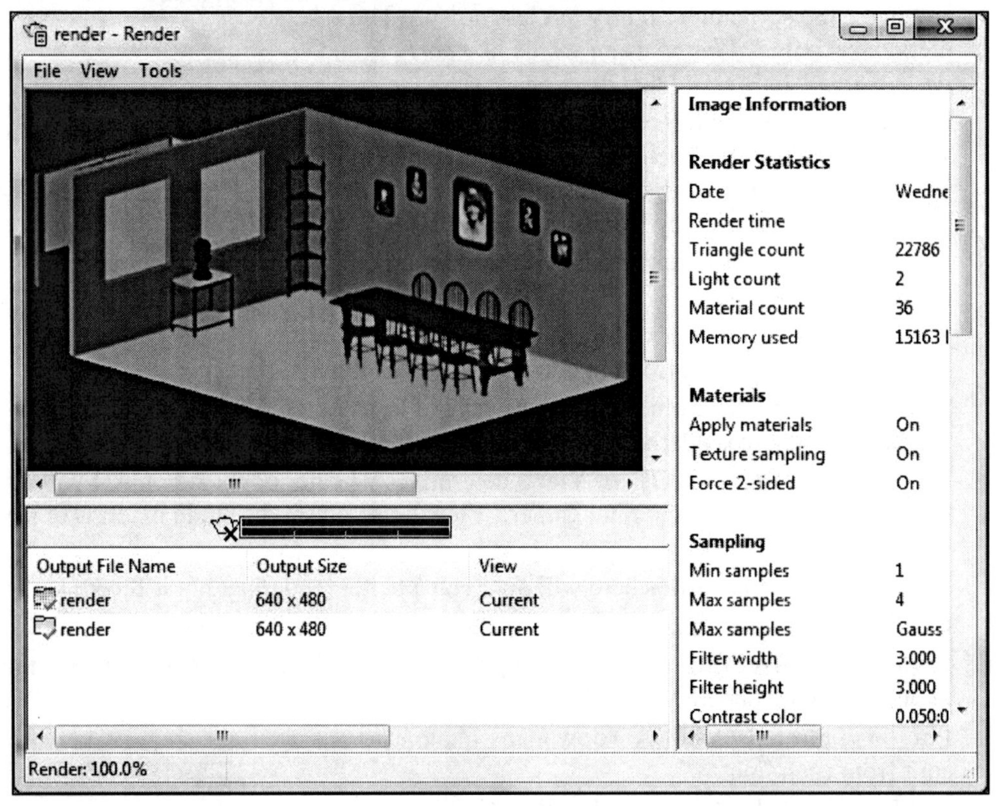

Figure 12.007

This window is an anomaly – one that's a lot simpler than it looks! (I love it when they're easy!) The rendering appears in the upper left corner (the dominant feature of the window). Below, you'll find an informational pane about the rendered image. To the left, you'll find the **Statistics** pane containing information about how the rendered image was created.

- Save the rendered image via the **Save** commands in the File pull down menu.
- Hide the Statistics pane and/or the Status bar with options in the View pull down menu.
- Zoom in or out with options in the Tools pull down menu.

Where to Find It:	
Command Line:	***Render***
Hotkey(s):	***rr***
Ribbon (Tab/Panel):	Render – Render – **Render**
Menu:	View – Render – **Render**
Toolbar:	Render – **Render**

Well, we have to begin someplace. Let's do a basic

rendering just to see how it works. We'll use the ***Render*** command to get things started.

Do This: 12.2A	Discovering Rendering

I. Open the *rendering project* file in the C:\Steps3D\Lesson12 folder. The drawing looks like the figure at right.

II. Follow these steps.

12.2A: DISCOVERING RENDERING

1. Enter the ***Render*** command.

   Command: *rr*

AutoCAD presents the rendered image seen in Figure 12.007, p.282.

2. Save the rendering as *MyFirstRender.jpg* in the C:\Steps3D\Lesson12 folder. (Use the JPEG Image type and accept the default Image Options.)

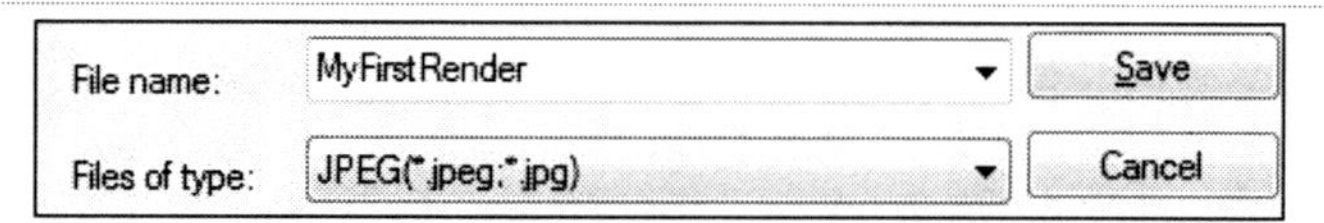

3. Now let's try it again at a higher resolution. Select **Presentation** Presentation in the **Render Preset** control on the ribbon's **Render** panel.

4. Repeat Step 1.

AutoCAD presents the following rendered image. Compare it to the image you saved in Step 2. (An easy way to compare two renderings is to double click on their **Output File Names** at the bottom of the rendering window.) You should find it to be a higher quality.

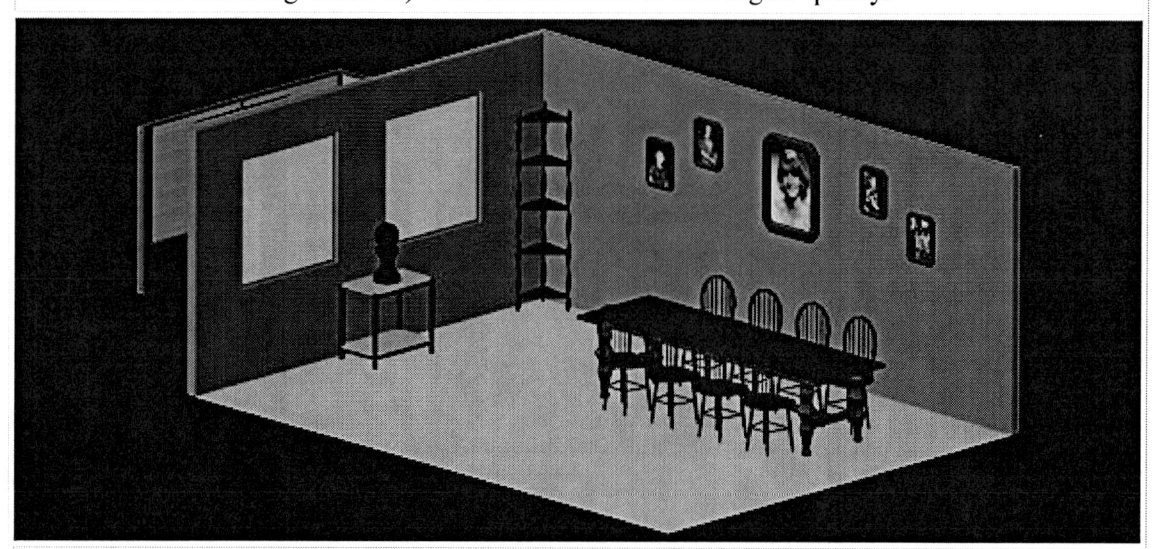

5. Take a few minutes to experiment with some of the other settings before continuing.

We could easily spend a hundred pages exploring the rest of the possibilities, but you should have the general idea. Take some time (once you complete the lesson) to continue exploring on your own.

You've seen the basics of the ***Render*** command, but so far, the rendered drawing is fairly unimpressive. It still looks like a cartoon – bright and colorful but not real. Next we'll begin to add reality to our image by assigning material values to the various objects.

I used the ***ImageAttach*** command to attach the family photos in the frames on the wall.

Let's proceed.

12.3 Adding Materials to Make Your Solids Look Real

12.3.1 Adding Materials via Layers

As you'll soon see, adding materials to an object can mean the difference between colorful cartoon images and images that come close to photographic realism. And luckily, you can accomplish it quite easily.

To understand materials, think of them as paint or wallpaper. The object doesn't actually become wood (or granite, etc.). Rather, it has the image of wood painted onto it. AutoCAD achieves this by attaching an image file to the surfaces of the objects. The only trick involved for you, then, is to know which image file to use. AutoCAD provides a library full of possible images from which to choose. If these don't satisfy your needs, however, AutoCAD helps to create new images (or modify old ones)!

AutoCAD comes with a library of existing materials accessed via the tool palettes. Just pick the **Properties** button on the tool palettes, and then select **Materials Library** ✓ Materials Library from the menu that appears. The Materials Library (Figure 12.008) functions like most palettes – pick the tab you wish on top. Right click on one of the overlapping tabs to produce a convenient menu listing the other tabs available on this palette.

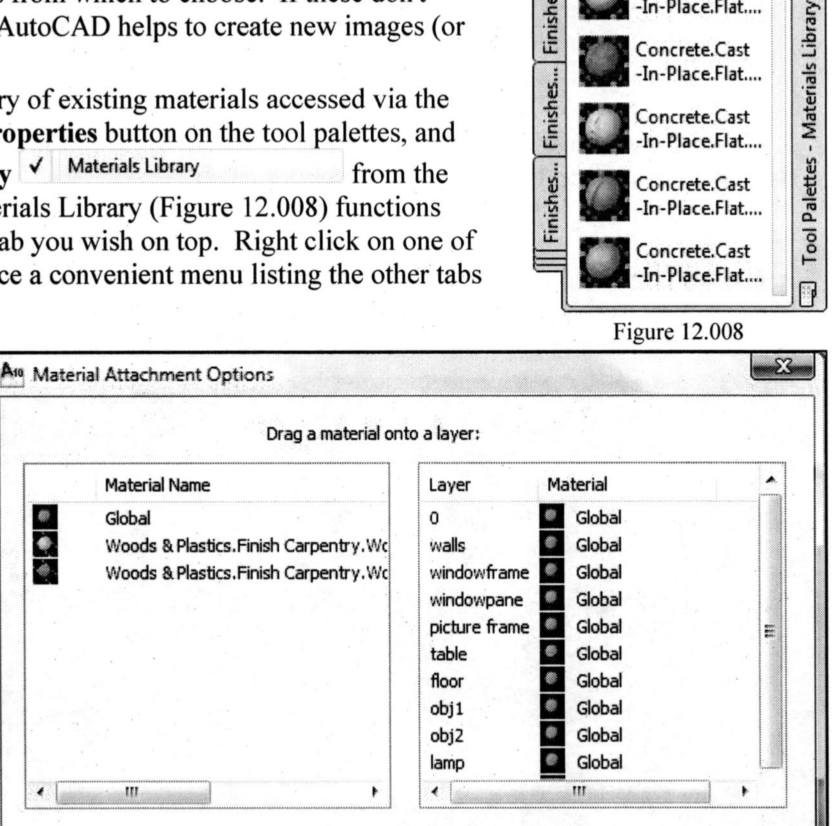

Figure 12.008

Figure 12.009

Attaching a material to an object can be as easy as dragging and dropping a material from the palette to the desired object in your drawing. This procedure actually "loads" the material into the drawing and attaches it to the object at the same time. You can also right click on a material and select **Add to current drawing** from the cursor menu to load the material without actually attaching it. We'll look at why this might be useful after the next exercise.

Another approach for attaching materials (the preferred approach) covers more distance at once that having to individually select objects. Use the ***MaterialAttach*** command to attach materials to entire layers. This command calls the Material Attachment Options dialog box (Figure 12.009, p.284). Just follow the instructions at the top – drag a material onto a layer – and AutoCAD does the rest! Let's attach some materials before we look any deeper.

Where to Find It:	
Command Line:	***MaterialAttach***
Ribbon (Tab/Panel):	Render – Materials (subpanel) – **Attach by Layer**

Do This: 12.3.1A	Attaching Materials

I. Be sure you're still in the *rendering project* file in the C:\Steps3D\Lesson12 folder.

II. If tool palettes aren't open, open them now.

III. Follow these steps.

12.3.1A: Attaching Materials

1. We'll begin by loading several materials. Pick the **Properties** button on the tool palette and select **Materials Library** ✓ Materials Library from the menu. AutoCAD opens the Materials Library (Figure 12.008, p.284).

2. Right click on the bunched up tabs at the bottom of the palette and select **Doors and Windows** Doors and Windows - Materials Library from the menu. AutoCAD places that palette atop the others.

3. Right click on **Doors – Windows.Glazing.Glass. Clear** and select **Add to current drawing** Add to Current Drawing from the menu.

4. Repeat this procedure to add the following materials to your drawing.

Material	Tab
Doors - Windows. Metal Doors - Frames.Steel.Galvanized	Doors and Windows
Woods - Plastics.Finish Carpentry.Wood.Paneling.1	Woods and Plastics
Finishes.Flooring.Carpet.Loop.5	Finishes – Carpet and Rugs
Finishes.Plaster.Stucco.Troweled.White	Finishes – Paint, Plaster, Walls, …

5. Now you have several materials available, let's attach them to your layers. (We'll still have to do some tweaking, but this will complete most of the material assignments for your model.)

Enter the ***MaterialAttach*** command.

Command: *materialattach*

AutoCAD presents the Material Attachment Options dialog box (Figure 12.009, p.284).

6. Drag the materials from the left window to the layer in the right window. Use the following chart to assist you.

12.3.1A: ATTACHING MATERIALS	
MATERIAL	**LAYER(S)**
Doors - Windows. Metal Doors and Frames.Steel.Galvanized	Windowframe
Woods - Plastics.Finish Carpentry.Wood.Paneling.1	Picture frame Table Obj1 Lamp Corner shelf
Finishes.Flooring.Carpet.Loop.5	Floor
Finishes.Plaster.Stucco.Troweled.White	Walls
Doors - Windows.Glazing.Glass.Clear	Windowpane

7. Pick the **OK** button [OK] to complete the procedure.

8. Save the drawing.

9. Render the drawing.

   Command: *rr*

It now looks like the following figure. (Not exactly photographic quality, is it – even though you can see through the windows now!? Not to worry; we haven't finished yet!)

10. Save the rendering as *12.3.1A #9.jpg* to the C:\Steps3D\Lesson12 folder. You can use it for comparison later.

We still need to tweak the materials to see the floor properly, wood grains and glass. We'll create a blue glass for the globe and lamp table shelving. That'll help – but it won't quite finish our efforts. Still, we'll get there *one step at a time!*

12.3.2 Creating and Tweaking Materials – and Another Method for Assigning Them to Objects

The title of this section makes it sound very busy, but these things all work together. It won't be as difficult as it sounds – we'll do the work from the Materials tool palette (Figure 12.010, p.287) – shown with panels closed).

Where to Find It:	
Command Line:	***Materials***
Hotkey(s):	***mat***
Ribbon (Tab/Panel):	Render – Materials – [settings button]
Menu:	View – Render – **Materials**
Toolbar:	Render – **Materials**

Access the Materials tool palette with the ***Materials*** command. AutoCAD divides the palette into an upper frame for creation and manipulation of materials, a selection area, and five panels for modifying materials.

Figure 12.010

- Starting in the top frame, you'll find a large area listing (pictorially) the materials available in the drawing. To the right of the title bar, you'll see a toggle ■ that will provide larger images if you need them.

  Several buttons reside below the images. From the left, these include:

  - A **Swatch Geometry** toggle – use this to view materials
  - attached to a sample sphere, box, or cylinder.
  - A **Checkered Underlay** toggle – use this to toggle the checkered background in the sample images on or off.
  - A **Preview Swatch Lighting** toggle – this toggles lighting on or off for the selected swatch.
  - The **Create New Material** button – this will open the Create New Material dialog box (Figure 12.012) where you'll name and (optionally) provide a description of your new material. We'll use this procedure to create the blue glass for our lamp's globe.

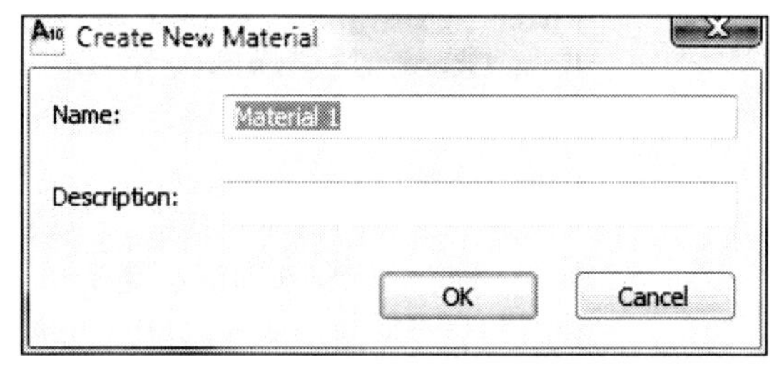

Figure 12.011

  - A **Purge from Drawing** button which comes in handy when your drawing gets too cramped with unused materials.
  - The **Indicate Materials in Use** button updates the display. Materials in use will display an icon in the lower right corner of the image.
  - The **Apply Material to Objects** button allows you to apply the currently selected material to selected objects in the drawing. It prompts with a **Select objects** prompt and replaces the cursor with this one:
  - Use the **Remove Materials from Selected Objects** button to remove materials from selected objects. This also prompts with a **Select objects** prompt and uses the same cursor as the Apply Material procedure uses.

- The Selection Area includes a couple buttons and a control box.

  - Use the **Home to Global Material Settings** button to return to the global level of the selected material's settings.

- The control box lists nested maps (more on maps shortly) associated with the selected material. You can pick the nested map to change its settings.
- Use the **Up One Level** button to move up a level in the nesting.

- You'll find five modifying panels below the selection area. These provide the creation/modification tools you'll need to work with your materials.
  - The **Material Editor** panel appears in Figure 12.012. Let's look at its parts.
    - The **Type** control box presents four options: **Realistic** and **Realistic Metal** provide material options based upon physical properties. **Advanced** and **Advanced Metal** provide more special effect properties.
    - The **Template** control box lists various surface templates available for the **Realistic** or **Realistic Metal** Types. The **Template** control box isn't available for **Advanced** or **Advanced Metal** Types.
    - You can set the diffuse (primary) **Color** by object (the default) or remove the check and AutoCAD will provide a button that will call the Select Color dialog box.
      - **Ambient** allows you to set the color for faces lit by ambient light only. (Available for **Advanced** and **Advanced Metal** only.)
      - **Diffuse** allows you to set the primary color for an object. (**Advanced** and **Advanced Metal** only.)
      - **Specular** allows you to set the color of a highlighted area of a shiny object. (**Advanced** only.)

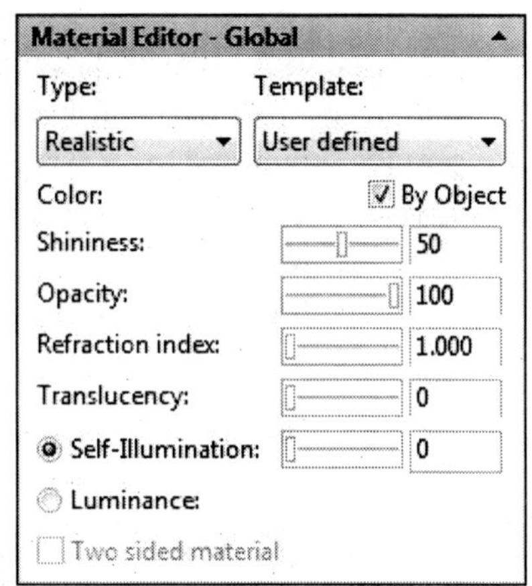

Figure 12.012

    - Shiny material reflects light in fewer directions. The lower the **Shininess** setting, the softer the reflected light. (**Realistic, Realistic Metal, Advanced, Advanced Metal** only.)
    - Set **Opacity** from 1 (transparent) to 100 (opaque). (**Realistic, Advanced** only.)
    - **Reflection** controls the amount of reflection in a surface. When set to 100, the surface will reflect the surroundings. (**Advanced** and **Advanced Metal** only.)
    - **Refraction** is the bending of light. Use this setting on non-metal templates to control how much refraction you'll get through transparent materials. (**Realistic, Advanced** only.)
    - A translucent object both scatters and transmits light. The higher the value, the lower the translucence. **Translucency** isn't available for metal templates. (**Realistic, Advanced** only.)
    - **Self-Illumination** (when set above 0) controls the amount of light an object emits. (**Realistic, Realistic Metal, Advanced, Advanced Metal** only.)
    - **Luminance** controls how much light an object reflects. You can't have values above zero in both **Luminance** and **Self-Illumination**. (**Realistic, Realistic Metal** only.)
  - Below the **Material Editor** panel, you'll find the **Maps** panel (Figure 12.013). If you're using colors in the Materials Editor, there's no need to use maps (and vice versa). Maps replace the diffuse colors.

    Maps assign a pattern or texture in place of the object's diffuse color. AutoCAD provides four possible mapping

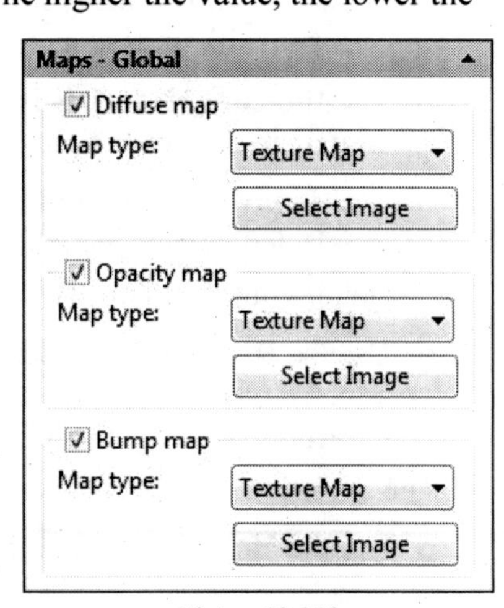

Figure 12.013

channels/sections depending upon the **Type** selection in the Materials Editor.

- **Diffuse map** – provides a color map/pattern for the material.
- **Reflection map** – reflects a scene on a shiny object. (**Advanced**, **Advanced Metal** only)
- **Opacity map** – provides the illusion of transparency/opacity. This can be useful to create the appearance of a hole in an object. I suggest using black and white images for this map – the black area will represent the hole. You can use color images, but AutoCAD will substitute gray scale for the colors.
- **Bump map** – simulates an irregular surface (bumps or waves). Again, I suggest black and white bump maps – black recedes while white protrudes.

Each map allows you to select a **Map type** or an **Image** for textural maps. Other (procedural) map types include: **Checker**, **Gradient Ramp**, **Marble**, **Noise**, **Speckle**, **Tiles**, **Waves**, and **Wood**. If you select one of these types, AutoCAD will replace the **Select Image** button with options similar to those shown in Figure 12.014. These include:

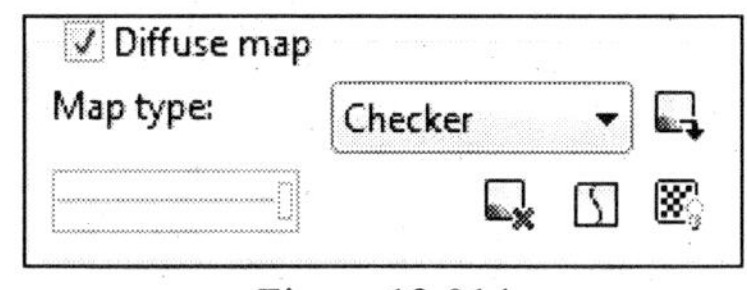

Figure 12.014

- A **Settings** button which calls a panel where you can adjust the settings for the selected map type.
- A **Delete map information from material** button.
- A **Synchronize** button for synchronizing the changes you've made with objects bearing that material.
- A **Preview** button for previewing your changes.

o Use the tools in the **Advanced Lighting Override** panel (Figure 12.015) to override indirect, global, or final gather lightning. (We'll see more on lightning in Section 12.4, p.294.) Using these tools, you can adjust lighting by the material rather than the light source but only when you're using **Realistic** or **Realistic Metal** types.

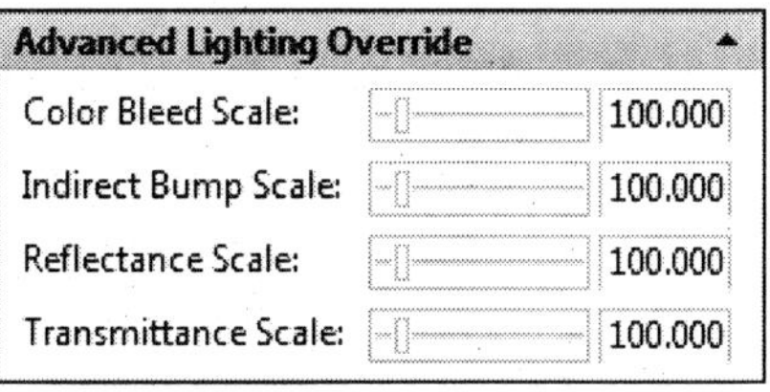

Figure 12.015

You have scale tools with which you can increase/ decrease the following:

- **Color Bleed** – for color saturation (amount of color).
- **Indirect Bump** – the effect of bump mapping.
- **Reflectance** – the amount of diffuse light reflected by a material.
- **Transmittance** – the amount of light that can go through a transparent material.

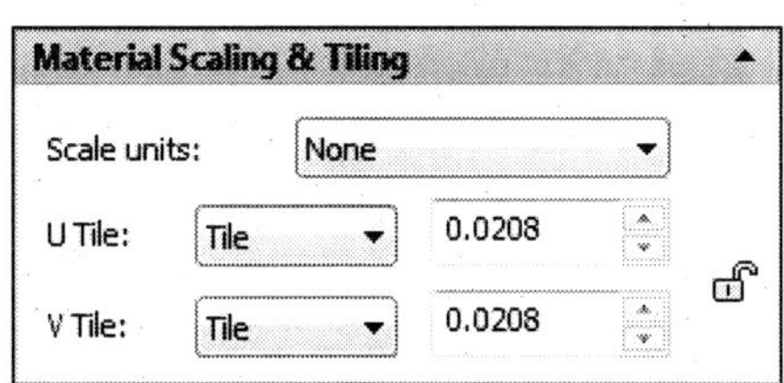

Figure 12.016

o The **Material Scaling and Tiling** panel (Figure 12.016) provides tools for scaling and positioning your material on the object.

- **Scale units** is self-explanatory. (Okay, **Fit to Gizmo** just means to make it fit on the object.)
- **U Tile** and **V Tile** refer to theoretical X and Y axes, but the axes move with the object. Each has three options: **Tile** (which tiles or repeats the image over the surface), **None** (which stretches a single tile over the entire surface) and **Mirror** (which tiles but mirrors each adjoining tile).

The number boxes (spinners) next to **U Tile** and **V Tile** adjust the size of the tiles on the U/V axis.

Use the **Lock Map Shape** button to lock/unlock the shape of the map.

- The **Material Offset and Preview** panel (Figure 12.017) has similar tools to the **Material Scaling and Tiling** panel – but this one adjusts the position of the tile … and it includes a preview to assist you!
  - A check next to **Auto-regen** tells AutoCAD to adjust the preview swatch as you make changes in the panel. If you remove the **Auto-regen** check, use the **Update** button to update the swatch.
  - The **Preview size** buttons zoom in/out of the preview.
  - To position the tile, pick and drag the image inside the **Preview** box much as you pan a drawing.
  - Use the **V Tile** and **U Tile** sliders to adjust the position of the map along the appropriate axis. These are only available on textural or checker map types.
  - **U Offset** and **V Offset** do the same thing as "panning" the map inside the **Preview** box, but you'll have more precise control here.
  - **Rotation** provides a useful means for rotating the map around the "W" (theoretical Z) axis.

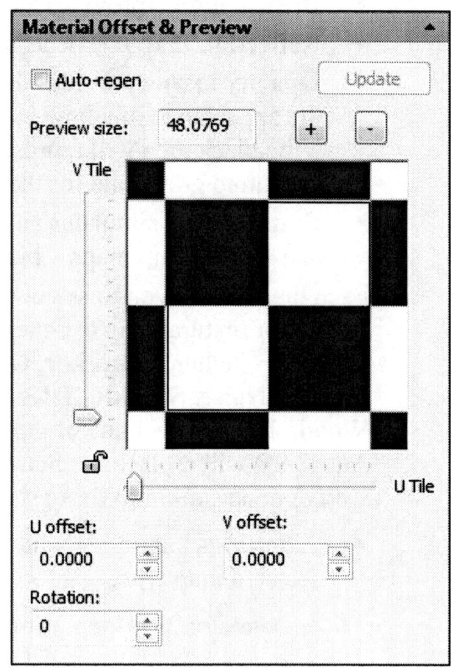

Figure 12.017

That's a lot of power in one little palette!

In our next exercise, we'll create and attach a blue glass material for our lamp globe, and make some needed adjustments to the materials we've already assigned.

Do This: 12.3.2A	**Creating and Tweaking Materials**

I. Be sure you're still in the *rendering project* file in the C:\Steps3D\Lesson12 folder.

II. If tool palettes aren't open, open them now.

III. Follow these steps.

12.3.2A: ATTACHING MATERIALS

1. Open the Materials tool palette.
   Command: ***mat***

2. Pick the **Create New Material** button. AutoCAD presents a Create New Material dialog box (Figure 12.011, p.287).

3. Call the new material *Blue Glass* and give it a description if you wish. Pick the **OK** button OK to continue.

4. Notice that AutoCAD creates a new material and highlights it in the upper section of the Materials palette. We'll define it now.

12.3.2A: Attaching Materials

5. Use these settings on the **Material Editor** panel:

- Use the **Glass-Clear** template.
- Set a **Diffuse** color of blue. (Diffuse by color.)
- Use default settings for **Shininess**, and **Refraction** index.
- Set **Opacity** to ***15***.
- Set the **Translucency** to ~75 and the **Self-Illumination** to ~15. This will make our glass shine a touch brighter than normal.

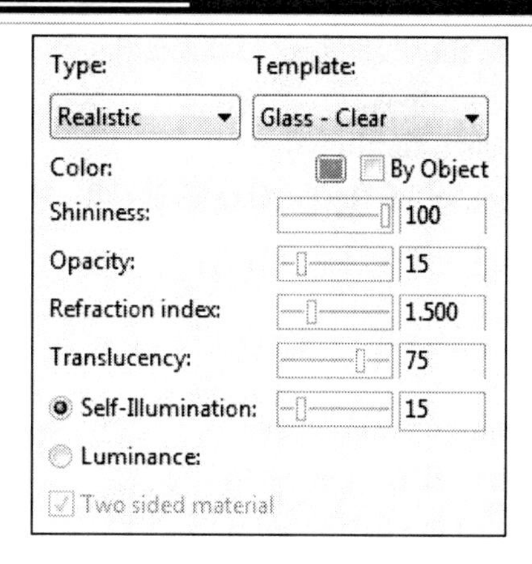

6. We won't need bit maps for glass, so clear all the check boxes on the **Maps** panel.

Congratulations, we don't need to make any other changes! You've created a material! Now let's attach it to something.

7. Pick the **Apply Material to Object** button.

AutoCAD returns to the drawing and asks you to select object.

8. Select the lamp's globe and the upper and lower shelf of the lamp table (right). Complete the command.

Select objects:

Select objects or [Undo]: ***[ENTER]***

9. Render the drawing to check your settings.

Command: ***rr***

The lamp and table should look like the figure at right. (Notice the transparency of the glass.)

10. Save the rendering as you did previously. Compare it to previous renderings. Notice the changes?

11. Save the drawing, but don't exit.

12. Have you noticed that the windows don't appear to have any glass? Have you noticed that the wall doesn't look textured, the floor looks empty and no wood grain shows on the table (or anything else)?

We'll do something about these problems now using the Materials palette.

Select the **Paneling.1** material.

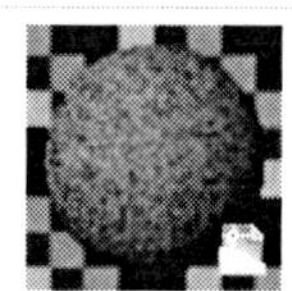

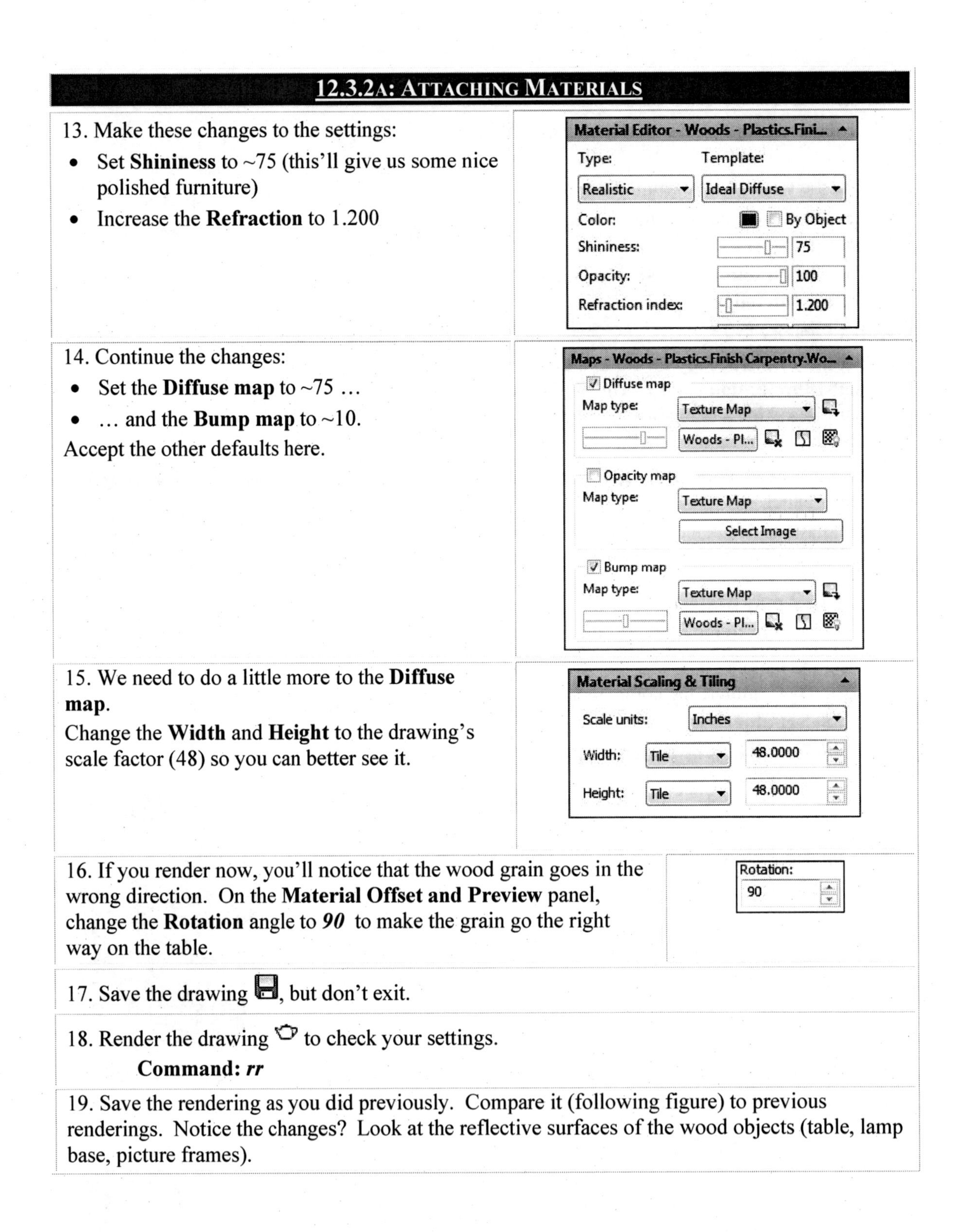

12.3.2A: Attaching Materials

13. Make these changes to the settings:

- Set **Shininess** to ~75 (this'll give us some nice polished furniture)
- Increase the **Refraction** to 1.200

14. Continue the changes:

- Set the **Diffuse map** to ~75 …
- … and the **Bump map** to ~10.

Accept the other defaults here.

15. We need to do a little more to the **Diffuse map**.
Change the **Width** and **Height** to the drawing's scale factor (48) so you can better see it.

16. If you render now, you'll notice that the wood grain goes in the wrong direction. On the **Material Offset and Preview** panel, change the **Rotation** angle to ***90*** to make the grain go the right way on the table.

17. Save the drawing, but don't exit.

18. Render the drawing to check your settings.

Command: *rr*

19. Save the rendering as you did previously. Compare it (following figure) to previous renderings. Notice the changes? Look at the reflective surfaces of the wood objects (table, lamp base, picture frames).

12.3.2A: ATTACHING MATERIALS

20. Make the following changes to the floor (the carpet material):

- Set the **Shininess** to 50.
- Set the **Refractive** index to 1.200.
- We've changed our minds about using carpet; let's use tile. Pick the **Select Image** button in the **Diffuse map** frame (it'll say, "Finishes.Flooring…") and pick the *Finishes.Flooring.VCT.Diamonds.jpg* image file. (You'll find it in the C:\Steps3D\Lesson12 folder.)
- Do the same for the **Bump map** (use the same image file).
- Make the **Width** and **Height** of the tiled image 48" (the drawing's scale factor). (Be sure you're using **Inches** in the **Scale units** control box.)

21. Make the following changes to the walls (the stucco material):

- Set the **Shininess** to 35.
- Select the image file *Finishes.Plaster.Stucco.Medium.White.jpg* for use as a **Bump map**.
- Set the Bump map scale to ~100 (use the slider).
- Set the **Material Scaling** to **Fit to Gizmo**. (Be sure **U Tile** and **V Tile** are set to **1**.)

22. Make the following changes to the windows (the clear glass material):

- Set the **Opacity** to **5**.

23. Render the drawing.

Command: *rr*

How's it look now (see the following image)?

12.3.2A: ATTACHING MATERIALS

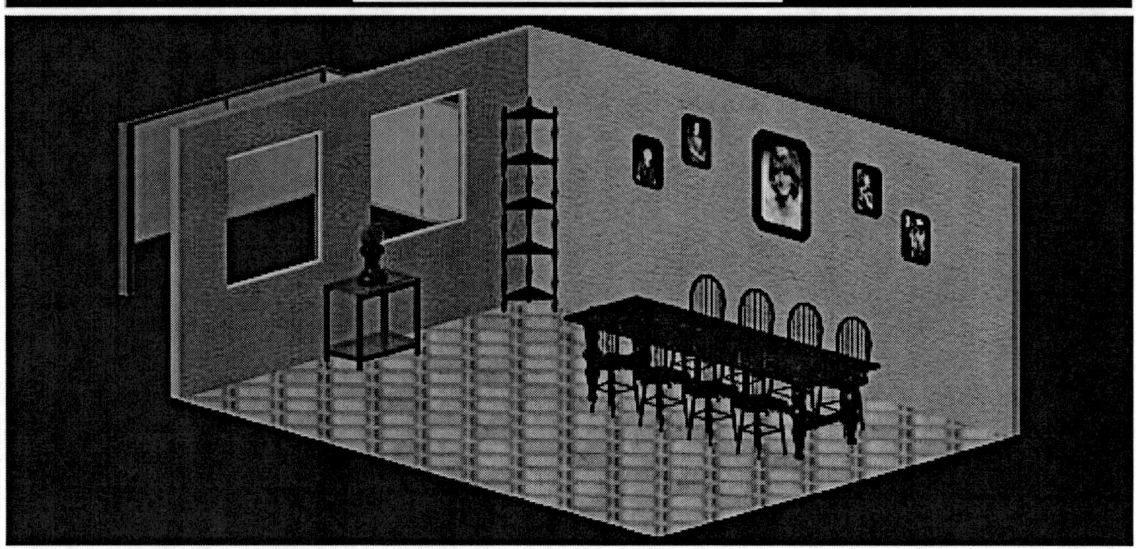

24. Save the drawing, but don't exit.

What's that? It doesn't look real yet? Something's still missing? Well, aren't you particular! What's missing (you are the intuitive one!) is *light*! And that's the topic of our next section!

12.4 Lights

Then God said, "Let there be light," and there was light. God saw how good the light was. ... - the first day.

Genesis 1:3

Perhaps we can better understand the importance of light when we consider that it was the first thing He created.

AutoCAD provides several types of lighting. These include Ambient Light (aka Default Lighting) **Point Light**, **Spot Light**, **Web Light**, **Target Point Light**, **Free Spot Light**, **Free Web Light**, and **Distant Light**. Then it tackles the sun!

- **Ambient Light** lights all surfaces with equal intensity. AutoCAD has two distant sources which light the visible surfaces of a 3D model.
- A **Point Light** works like a light bulb. It spreads rays in all directions from a single source. It dissipates as it moves away from the source, and it casts shadows. A **Point Light** doesn't require a target, a **Target Point Light** does.
- A **Spotlight** works like a **Point Light**, except that it can be pointed in a single direction. A **Spot Light** requires a target; a **Free Spot Light** does not.
- A **Web Light** provides a far more accurate point-type light distribution based upon data provided by light manufacturers. The **Web Light** uses a target whereas the **Free Web Light** does not.
- A **Distant Light** is an even light coming from a distant location.
- The **Sun Light** works much like a distant light except that you can define its location in terms of geography, calendar and time of day.

You can use any combination of light in your rendering except Ambient Light – *which AutoCAD turns off when another light is active*.

12.4.1 The Basic Lights – Point, Spot, Web, and Distant

You can work with lights on the command line or using the ribbon's **Lights** panel.

The command line prompts depend upon the value of the **LightingUnits** system variable. Refer to the following table. (We'll discuss photometric lighting on p.296.)

LightingUnits Value	Description
0 (default)	AutoCAD will use generic lighting (no lighting units).
1	AutoCAD will use photometric lighting with American units.
2	AutoCAD will use photometric lighting with international units.

We'll use the command line approach to lighting and a **LightingUnits** value of 1 to discuss the various prompts. But be aware that some prompts may not be available for generic lighting.

The ***Light*** command begins like this:

Where to Find It:	
Command Line:	*Light*

> **Command: *light***
> **Enter light type [Point/Spot/Web/Targetpoint/Freespot/freeweB/Distant]**
> **<Point>: *[Tell AutoCAD what kind of light you want – your choice will determine the prompts that follow.]***

The prompt asks for the type of light you want to create. Your choice determines the prompts that follow.

> Many of the ***Light*** command's option lights have their own commands which accomplish the same thing. We'll use WTFI tables here to show the various methods of accessing the option/command.

- Selecting **Point** (or using the ***PointLight*** command) brings up this prompt:

  > **Specify source location <0,0,0>: *[Locate your light.]***
  > **Enter an option to change [Name/Intensity factor/Status/Photometry/shadoW/ Attenuation/filterColor/eXit] <eXit>: *[Manage your light's properties – more on this in a moment.]***

Where to Find It:	
Command Line:	***PointLight***
Hotkey(s):	***freepoint***
Ribbon (Tab/Panel):	Render – Lights – **Point**
Menu:	View – Render – Light – **New Point Light**
Toolbar:	Lights – **New Point Light**

- Selecting **Targetpoint** provides the same prompt, except that it includes a prompt for the target following the prompt for the source location. The prompt looks like this:

  > **Specify target location <0,0,-10>:**

- Selecting **Spot** (or using the ***SpotLight*** command) brings up this prompt:

  > **Specify source location <0,0,0>:**
  > **Specify target location <0,0,-10>:**
  > **Enter an option to change [Name/Intensity factor/Status/ Photometry/Hotspot/Falloff/shadoW/ Attenuation/filterColor/eXit] <eXit>:**

Where to Find It:	
Command Line:	***SpotLight***
Ribbon (Tab/Panel):	Render – Lights – **Spot**
Menu:	View – Render – Light – **New Spotlight**
Toolbar:	Lights – **New Spotlight**

  Notice that the prompts are almost identical to the point light prompts except for the inclusion of **Hotspot** and **Falloff** options. We'll discuss these in a moment, too.

- Selecting **Freespot** provides the same prompt except that **Freespot** won't prompt for a target location.
- Selecting **Web** brings up this prompt:

  Specify source location <0,0,0>:

  Specify target location <0,0,-10>:

  Enter an option to change [Name/Intensity factor/Status/Photometry/weB/ shadoW/filterColor/eXit] <eXit>:

  Web's options also reflect those of the point light except that it includes a **weB** option. This option allows you to set an intensity for the light @ X, Y, and Z points within a spherical grid.
- Selecting **freeweB**, of course, provides the same prompts without prompting for a target.
- Selecting **Distant** (or using the ***DistantLight*** command) prompts like this:

  Specify light direction FROM <0,0,0> or [Vector]: ***[As this light will be a distance from your drawing area, locate it by direction rather than a specific point.]***

  Specify light direction TO <1,1,1>: ***[The second prompt can be confusing – the easy way to understand this one is simply to pick a point in the direction you want your light to shine.]***

  Enter an option to change [Name/Intensity factor/Status/Photometry/shadoW/ filterColor/eXit] <eXit>: ***[Manage your light's properties.]***

  Using distant lighting with other lights can cause some unpleasant (overexposure) side affects, so use these with caution.

Where to Find It:	
Command Line:	***DistantLight***
Ribbon (Tab/Panel):	Visualize – Lights – **Distant**
Menu:	View – Render – Light – **New Distant Light**
Toolbar:	Lights – **New Distant Light**

The various lights have several options to manage their properties. These properties, of course, are also available on the Properties palette when you edit the light. Let's look at each. (Refer to Figure 12.018, p.297.)

- **Name** – Give the light a specific name to make it easy to identify from a list of lights. Prompts for a light name.
- **Intensity** – Sets the intensity (brightness) of the light. Your system's resources determine the maximum setting.
- **Status** – Turns the light on or off. You can change this setting in the Properties palette for the specific light.
- **Photometry** – The measurement of the luminous intensity of light. Okay, what's that mean in real words? Simply put, photometric lighting means that you can set light intensity. Unfortunately, you have to set it in *Candelas* (international standard of luminous intensity; one candela = 1 lumen), *Lux* (international standard of luminance; one lux = 1 lumen per square meter), or *Foot Candles* (American unit of luminance; the amount of light one foot from a candle – aka, candela)!

  I know; you want wattage. But wattage refers to the amount of energy used by a lamp rather than the amount of light it produces. Use the following chart as a guide.[*]

[*] These figures come from: http://www.omafra.gov.on.ca/ and http.www.simplyhydro.com/why_hydro.htm

BULB	FOOT CANDLES	LUMENS
incandescent	15-20	10-15
halogen	18-25	20
fluorescent	81-98	68
Mercury Vapor	40-50	
metal halide	80-92	80-120
high pressure sodium	90-110	90-150

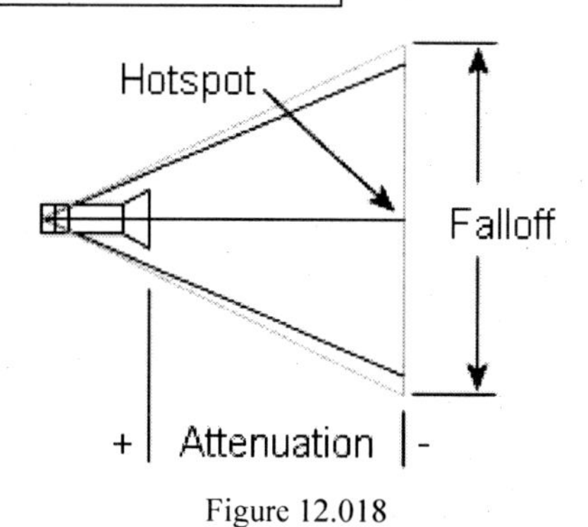

Figure 12.018

- **Shadow** – Do you want your light to cast shadows? **Shadows** includes options for turning the shadows **Off**, as well as **Sharp** (hard edges), **Soft Mapped** (soft edges), or **Soft Sampled** (softer shadows) options.
- **Hotspot** – The hotspot is the focal angle of a spotlight – in other words, it's the center of the spotlight's cone (Figure 12.018).
- **Falloff** – Falloff determines the actual size of the focal point of the spotlight. You'll determine this by angle (from 0° to 160°). The default is 45°, but the falloff angle must be greater than the hotspot angle.
- **Attenuation** –**Attenuation** refers to how light dissipates over distance. You'll be prompted with these options:

  Enter an option to change [attenuation Type/Use limits/attenuation start Limit/attenuation End limit/eXit] <eXit>:

  - **Attenuation Type** provides these options:
    - **None** means that distant objects will reflect as much light from this point light as closer objects reflect.
    - **Inverse Linear** is normal reflectivity. In other words, an object twice as far away as another object will reflect half the light the closer one reflects. This is the default setting and appropriate for most situations.
    - **Inverse Square** is similar to **Inverse Linear** but works with squared numbers. That is, in our previous example, the distant object will reflect one-quarter the light that the closer one reflected.
  - **Use limits** tells AutoCAD to use attenuation limits. You can set the limits using **attenuation start Limit** and **attenuation End limit**.

> The OpenGL video driver doesn't support attenuation limits at the time of this writing. The Direct 3D supports end limits but not start limits. Enter ***3dconfig*** at the command prompt and pick the **Manual Tune** button to see which driver you're using. (AutoCAD running under Windows Vista doesn't support the OpenGL video driver at all – there goes an investment in GEForce graphics cards!)

- **weB** – Web prompts you for the intensity of light at X, Y, and Z points on a spherical grid.
- **filterColor** – This option uses standard AutoCAD procedures to allow you to control the color of the light you're creating.

> With the **LightingUnits** system variable set to 0, AutoCAD won't prompt for **Photometry** and will substitute a **Color** prompt for the **filterColor** prompt. But you may find it easier to set the intensity of your lights.

Let's see what we can do with these lights before we tackle the sun!

Do This: 12.4.1A	Basic Lighting

I. Be sure you're still in the *rendering project* file in the C:\Steps3D\Lesson12 folder.

II. Change the **Opacity** of the **Blue Glass** material to 0. (What we had worked fine for a lamp with no light in it. We're going to put a light in it now.)

III. Be sure the **LightingUnits** system variable is set to 0 (no sense complicating matters.)

IV. Follow these steps.

12.4.1A: BASIC LIGHTING

1. Let's start with a simple spot light. We'll start with the ***SpotLight*** command (although you can use the **Spot** option of the ***Light*** command as well).

    Command: *spotlight*

2. AutoCAD will ask you if you wish to turn the default lighting off or not. Select the **Turn off the default lighting** option.

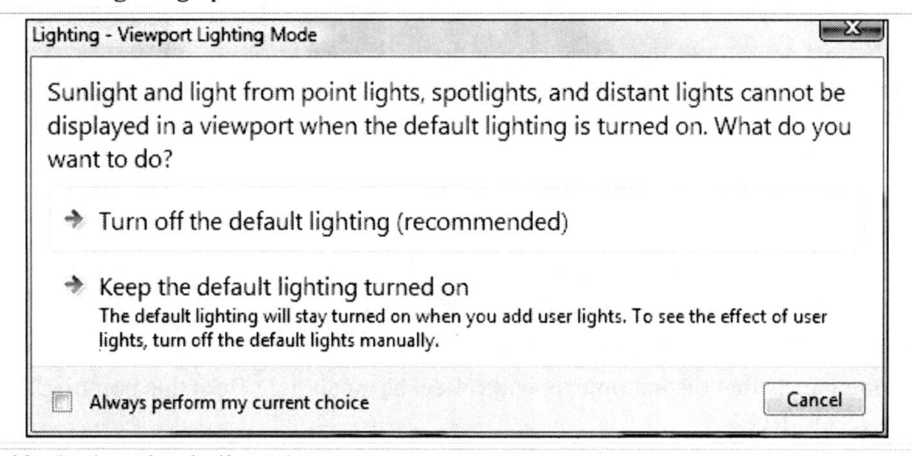

3. Specify the location indicated …

    Specify source location <0,0,0>: *20'6,10'6,3'*

4. … and the target location (about the center of Barbara's picture).

    Specify target location <0,0,-10>: *20'6,16'6,5'6*

5. **Name** the light something convenient.

    Enter an option to change [Name/Intensity/Status/Hotspot/Falloff/shadoW/ Attenuation/Color/eXit] <eXit>: *n*

    Enter light name <Spotlight10>: *Spot*

6. Lower the **Intensity** a bit so we don't outshine our picture.

    Enter an option to change [Name/Intensity/Status/Hotspot/Falloff/shadoW/ Attenuation/Color/eXit] <eXit>: *i*

    Enter intensity (0.00 - max float) <1.0000>: *.5*

7. Set the **Hotspot** and **Falloff** to spotlight just the one photograph.

    Enter an option to change [Name/Intensity/Status/Hotspot/Falloff/shadoW/ Attenuation/Color/eXit] <eXit>: *h*

    Enter hotspot angle (0.00-160.00) <3'-9">: *30*

    Enter an option to change [Name/Intensity/Status/Hotspot/Falloff/shadoW/

12.4.1A: Basic Lighting

Attenuation/Color/eXit] <eXit>: ***f***

Enter falloff angle (0.00-160.00) <160>: ***50***

8. Complete the command.

Enter an option to change [Name/Intensity/Status/Hotspot/Falloff/shadoW/ Attenuation/Color/eXit] <eXit>: ***[ENTER]***

Notice that AutoCAD places a spotlight glyph at your spotlight's location.

9. Render the drawing.

Command: ***rr***

Notice that AutoCAD eliminates the ambient (default) lighting in favor of the user-defined light.

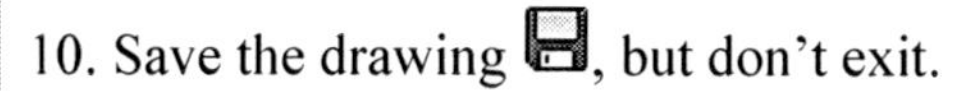

10. Save the drawing, but don't exit.

Command: ***qsave***

11. Let's create a couple point lights – one for the lamp and another for a ceiling light. Use the ***PointLight*** command.

Command: ***pointlight***

12. Locate the first light (the lamp) at the center of the lamp's globe.

Specify source location <0,0,0>: ***cen***

13. **Name** [Name] the light something convenient.

Enter an option to change [Name/Intensity/Status/shadoW/Attenuation/Color/ eXit] <eXit>: ***n***

Enter light name <Pointlight1>: ***lamp***

14. Change the light's color [Color] to blue. (I used AutoCAD's index color 5.)

Enter an option to change [Name/Intensity/Status/shadoW/Attenuation/Color/eXit] <eXit>: ***C***

Enter true color (R,G,B) or enter an option [Index color/Hsl/colorBook] <0,0,0>: ***i***

Enter color name or number (1-255): ***5***

15. We'll accept the other defaults. Complete the command.

Enter an option to change [Name/Intensity/Status/shadoW/Attenuation/Color/eXit] <eXit>: ***[ENTER]***

AutoCAD has placed a point light glyph in the center of the lamp's globe.

16. Create a ceiling light (also a point light) at the coordinates indicated. (Accept the other defaults.)

Specify source location <0,0,0>: ***19'10,2',7'10***

17. Turn the spotlight off. (Select the spotlight glyph and use the Properties palette.)

12.4.1A: Basic Lighting

18. Render the drawing.

Command: ***rr***

Notice (following figure) the shadows. What has blue tints from the blue point light? Notice the double shadows where both lights have shone. Notice the reflections on the table and in the windows. Save the rendering and compare it with earlier efforts.

19. Change to the Perspective ***View***. (Be sure the **ceiling** layer thaws.)

Command: ***view***

20. Render the drawing again. (Notice that, with each new setup, the rendering takes longer.)

Command: ***rr***

This is more of an inside look. Notice the shadows outside – as though it's night. We need some sunshine!

21. Save the drawing, but don't exit.

Command: ***qsave***

Did I leave you in the dark? Of course not, now you can create light! And remember, you can adjust the intensity of the light, shadows, and so forth as you create the lights or using the Properties palette.

Of course, if you're into the science (the art?) of lighting, you can use parametric lighting rather than the easier, basic stuff we used in our exercise.

But we've left out the most import light – the sun. Let's look at that one next.

12.4.2 The Sun

Despite its size, the sun is the easiest light to manage – although you have some options to consider!

You'll use two simple tools – the ***GeographicLocation*** and the ***SunProperties*** commands.

The ***GeographicLocation*** command immediately presents you with a few options in the Define Geographic Location dialog box (Figure 12.019).

Where to Find It:	
Command Line:	***GeographicLocation***
Hotkey(s):	***geo***
Ribbon (Tab/Panel):	Render – Sun & Location – **Set Location**
Menu:	Tools – **Geographic Location**
Toolbar:	Lights – **Geographic Location**
	Render – (Light flyout) **Geographic Location**

- Google Earth creates KML and KMZ files. If you have one, you can use it to define your location.
- You must have Google Earth open and located at the desired location to use the **Import the current location from Google Earth** option.
- **Enter the location values** presents the Geographic Location dialog box (Figure 12.020) where you'll set the longitude and latitude for your model. This can get fairly complicated, but you can pick the **Use Map** button for an easier solution. This presents a Location Picker dialog box (Figure 12.021) where you can tell AutoCAD where you are with the **Nearest City** option? We'll locate our dining room in our next exercise.

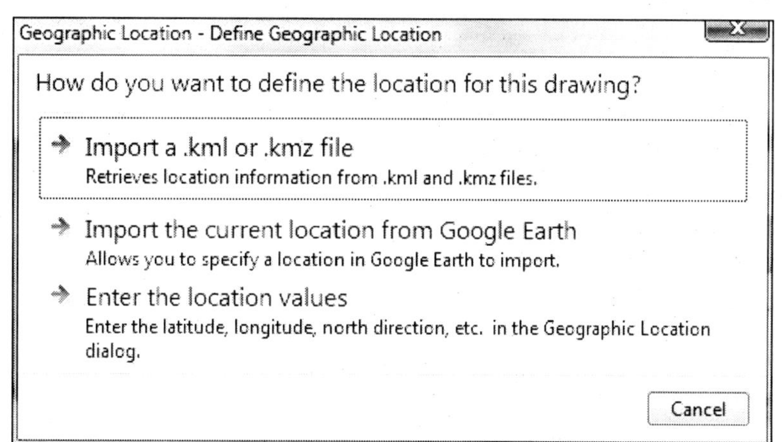

Figure 12.019

> If you've already defined a location for the current drawing, you may see a dialog box other than the Define Geographic Location dialog box.
>
> - The Location Already Exists dialog box gives you the opportunity to edit or change that location.
> - The Coordinate System Defined dialog box gives you the opportunity to convert or not convert the system to AutoCAD's *LL84* system.

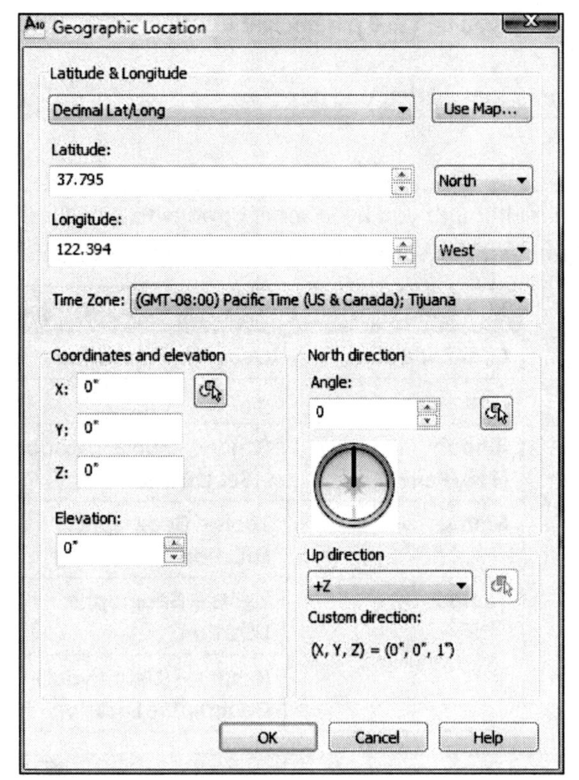

Figure 12.020

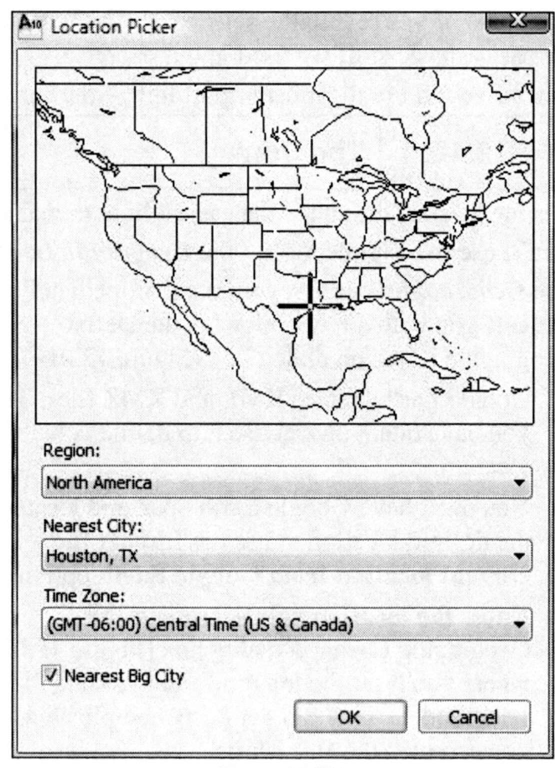

Figure 12.021

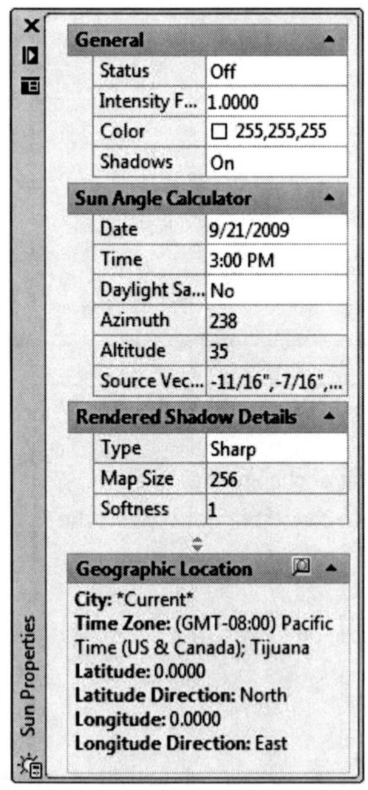

Figure 12.022

Where to Find It:	
Command Line:	***SunProperties***
Ribbon (Tab/Panel):	Render – Sun & Location – [**Settings** button]
Menu:	View – Render – Light – **Sun Properties**
Toolbar:	Lights – **Sun Properties**
	Render – (Sun flyout) **Sun Properties**

The ***SunProperties*** command presents the Sun Properties palette (Figure 12.022). Here you have several tabs to "control" the sun.

- The **General** tab provides options for turning the sun on or off (**Status**), adjusting the brilliance (**Intensity factor**), adjusting the **Color**, and turning **Shadows** on or off.
- The **Sun Angle Calculator** provides options for changing the **Date** and **Time** from your location, and it provides the location of the sun in the sky (**Azimuth** and **Altitude**) for that location and time.
- Finally, you can control the **Type** of **Rendered Shadow Details**.
- The **Geographic Location** tab provides useful information.

Change this information in the Geographic Location dialog box.

Don't you wish you could control the sun like this on those afternoons when your grass is too tall and the neighborhood association wants it to look just like everyone else's?

Well, let's give it a try!

Do This: 12.4.2A	Controlling the Sun

I. Be sure you're still in the *rendering project* file in the C:\Steps3D\Lesson12 folder.

II. Follow these steps.

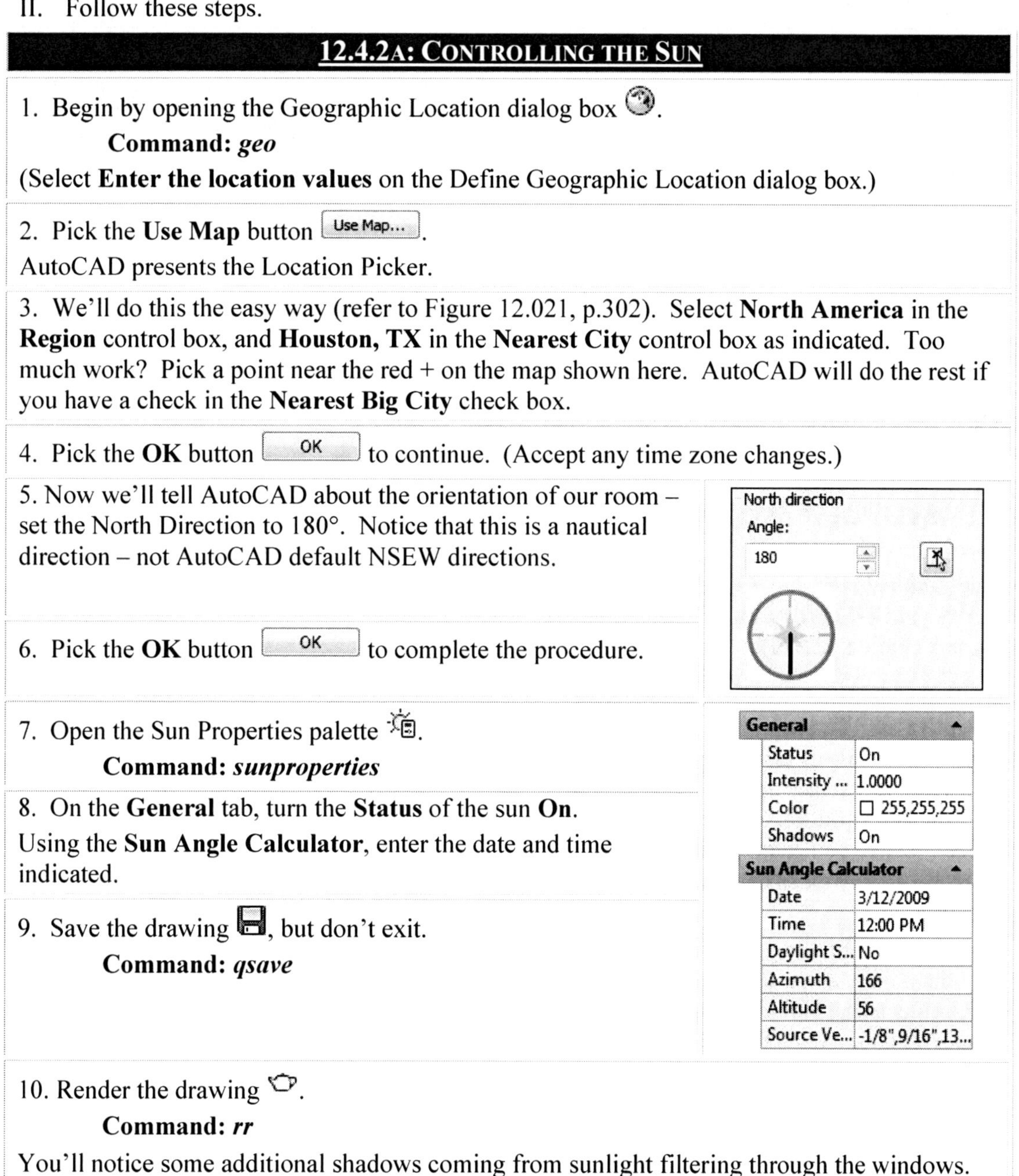

12.4.2A: CONTROLLING THE SUN

1. Begin by opening the Geographic Location dialog box.

   Command: *geo*

(Select **Enter the location values** on the Define Geographic Location dialog box.)

2. Pick the **Use Map** button [Use Map...].

AutoCAD presents the Location Picker.

3. We'll do this the easy way (refer to Figure 12.021, p.302). Select **North America** in the **Region** control box, and **Houston, TX** in the **Nearest City** control box as indicated. Too much work? Pick a point near the red + on the map shown here. AutoCAD will do the rest if you have a check in the **Nearest Big City** check box.

4. Pick the **OK** button [OK] to continue. (Accept any time zone changes.)

5. Now we'll tell AutoCAD about the orientation of our room – set the North Direction to 180°. Notice that this is a nautical direction – not AutoCAD default NSEW directions.

6. Pick the **OK** button [OK] to complete the procedure.

7. Open the Sun Properties palette.

   Command: *sunproperties*

8. On the **General** tab, turn the **Status** of the sun **On**.

Using the **Sun Angle Calculator**, enter the date and time indicated.

General	
Status	On
Intensity ...	1.0000
Color	255,255,255
Shadows	On
Sun Angle Calculator	
Date	3/12/2009
Time	12:00 PM
Daylight S...	No
Azimuth	166
Altitude	56
Source Ve...	-1/8",9/16",13...

9. Save the drawing, but don't exit.

   Command: *qsave*

10. Render the drawing.

   Command: *rr*

You'll notice some additional shadows coming from sunlight filtering through the windows. You should also be able to see the outside fence a little better.

12.4.2A: Controlling the Sun

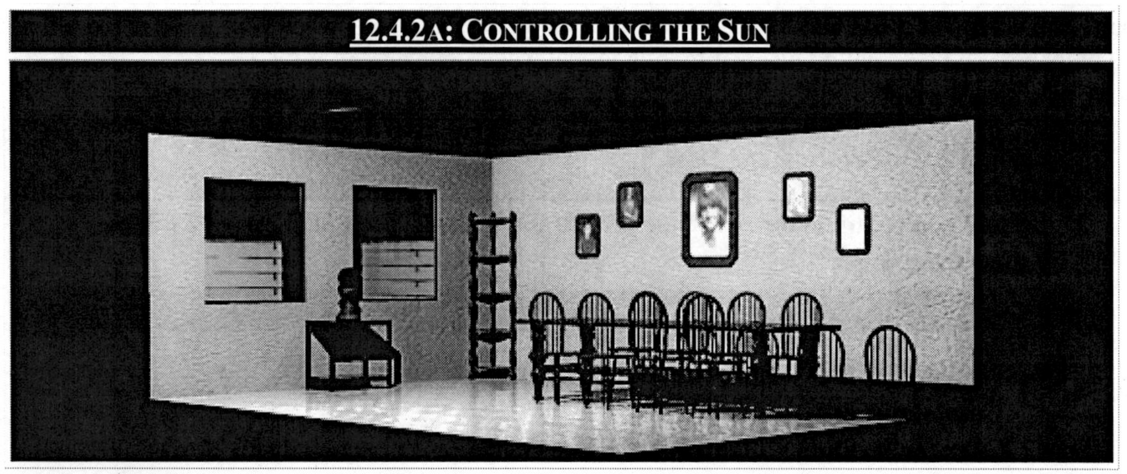

We've almost finished!

12.5 Some Final Touches – Fog and Background

In our "real" world, nothing is perfectly clear – from politics to religion to that stuff that flows from our kitchen faucets. Despite a photographer's best efforts to cut through it, pictures reflect the "haze" that permeates our environment robbing it of the clarity for which we long.

In its efforts to recreate reality, AutoCAD has even provided for the haze!

Use the ***RenderEnvironment*** command to help you de-clarify your efforts. AutoCAD assists with a dialog box (Figure 12.023).

Where to Find It:	
Command Line:	***RenderEnvironment***
Hotkey(s):	***fog***
Ribbon (Tab/Panel):	Render – Render (subpanel) – **Environment**
Menu:	View – Render – **Render Environment**
Toolbar:	Render – **Render Environment**

You'll find a single tab with which to work – **Fog/Depth Cue**. To keep it simple, think of **Fog** and **Depth Cue** as essentially the same thing. (They're actually opposite extremes of "haze"; fog is a white haze while depth cue is a black haze.) Here you can fog (tint) the objects being rendered as well as the background of the rendering.

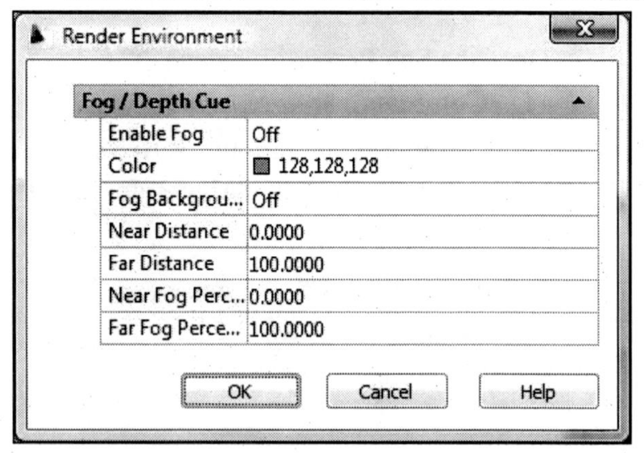

Figure 12.023

- The **Enable Fog** toggle controls whether or not the objects in the drawing will be tinted according to the **Color** defined in the **Color** option. When **Fog Background** is also toggled on, the entire image will be tinted.
- The **Color** option provides AutoCAD's standard color selection methods. In this case, you'll define the tint of the haze in your drawing.
- The **Near distance** and **Far distance** options provide values AutoCAD uses to determine where to begin and end the fog. Values are percentages of the distance from the camera to the back working plane.

- The last options provide percentage values AutoCAD uses to determine how much fog to place at the **Near** and **Far Distance** points.

You'll find the other aspect of the Rendering Environment – Background – accessible through the ***View*** command. We discussed this is Lesson 4 of our basic text, but we'll set up our drawing with a gradient background to refresh your memory.

Let's finish our dining room.

Do This: 12.5A	**Making It "Real"**

I. Be sure you're still in the *rendering project* file in the C:\Steps3D\Lesson12 folder.

II. Follow these steps.

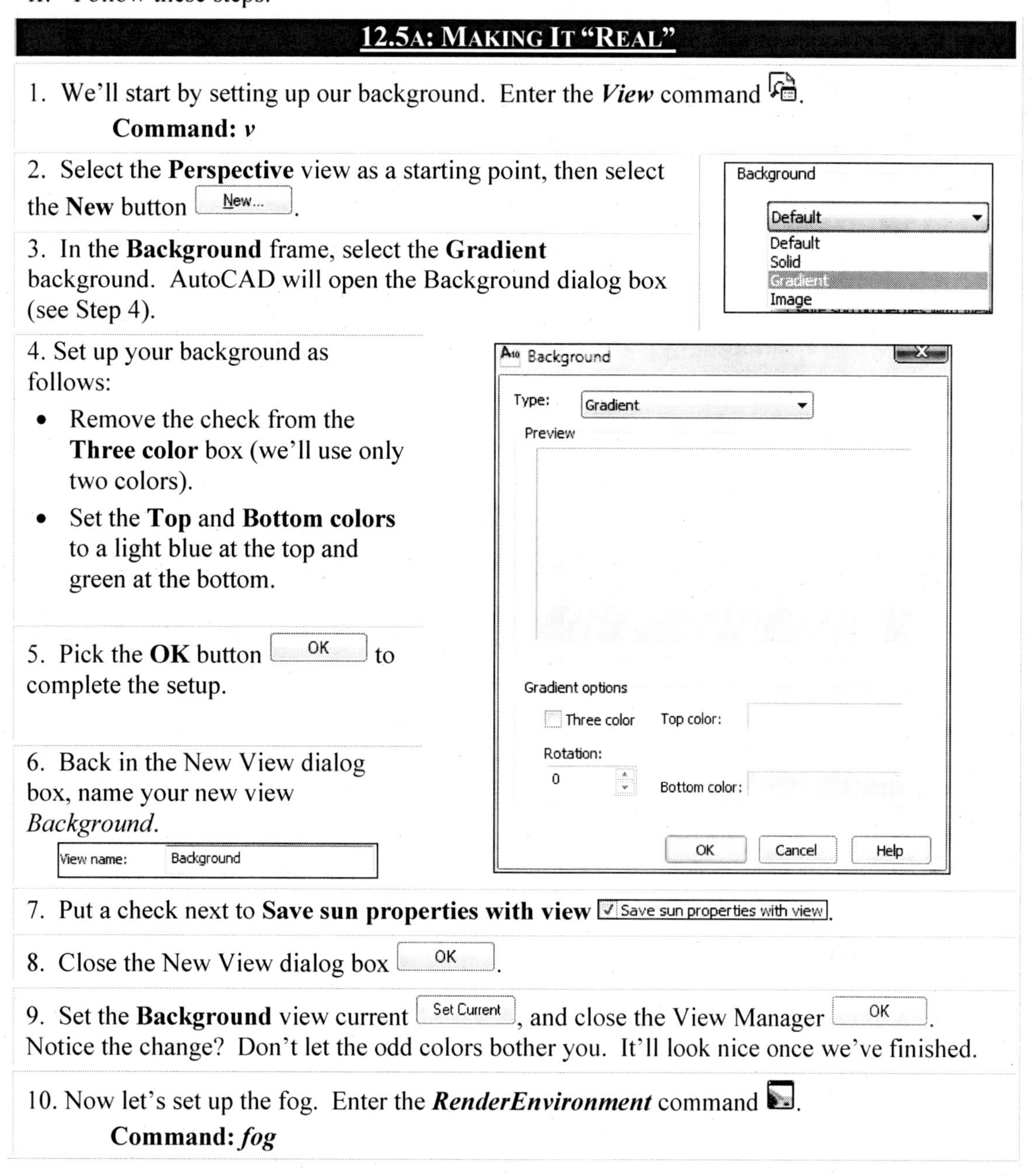

12.5A: MAKING IT "REAL"
1. We'll start by setting up our background. Enter the View command.
Command: v
2. Select the Perspective view as a starting point, then select the New button New...
3. In the Background frame, select the Gradient background. AutoCAD will open the Background dialog box (see Step 4).
Background
Default
Default
Solid
Gradient
Image
4. Set up your background as follows:
Remove the check from the Three color box (we'll use only two colors).
Set the Top and Bottom colors to a light blue at the top and green at the bottom.
Background
Type:
Gradient
Preview
Gradient options
Three color
Top color:
Rotation:
0
Bottom color:
OK
Cancel
Help
5. Pick the OK button OK to complete the setup.
6. Back in the New View dialog box, name your new view Background.
View name:
Background

7. Put a check next to **Save sun properties with view** ☑ Save sun properties with view.

8. Close the New View dialog box [OK].

9. Set the **Background** view current [Set Current], and close the View Manager [OK]. Notice the change? Don't let the odd colors bother you. It'll look nice once we've finished.

10. Now let's set up the fog. Enter the ***RenderEnvironment*** command.

 Command: ***fog***

12.5A: MAKING IT "REAL"

11. Set up the Render Environment as shown:

- Enable both **Fog** and **Fog Background**
- Make the **Color** a soft gold to simulate incandescent lights.
- Accept the **Distance** defaults, but set the **Far Fog Percentage** at about 15%.

(Feel free to experiment with these settings.)

12. Close the Render Environment dialog box OK.

13. Save and render the drawing. The results look something like the figure below.

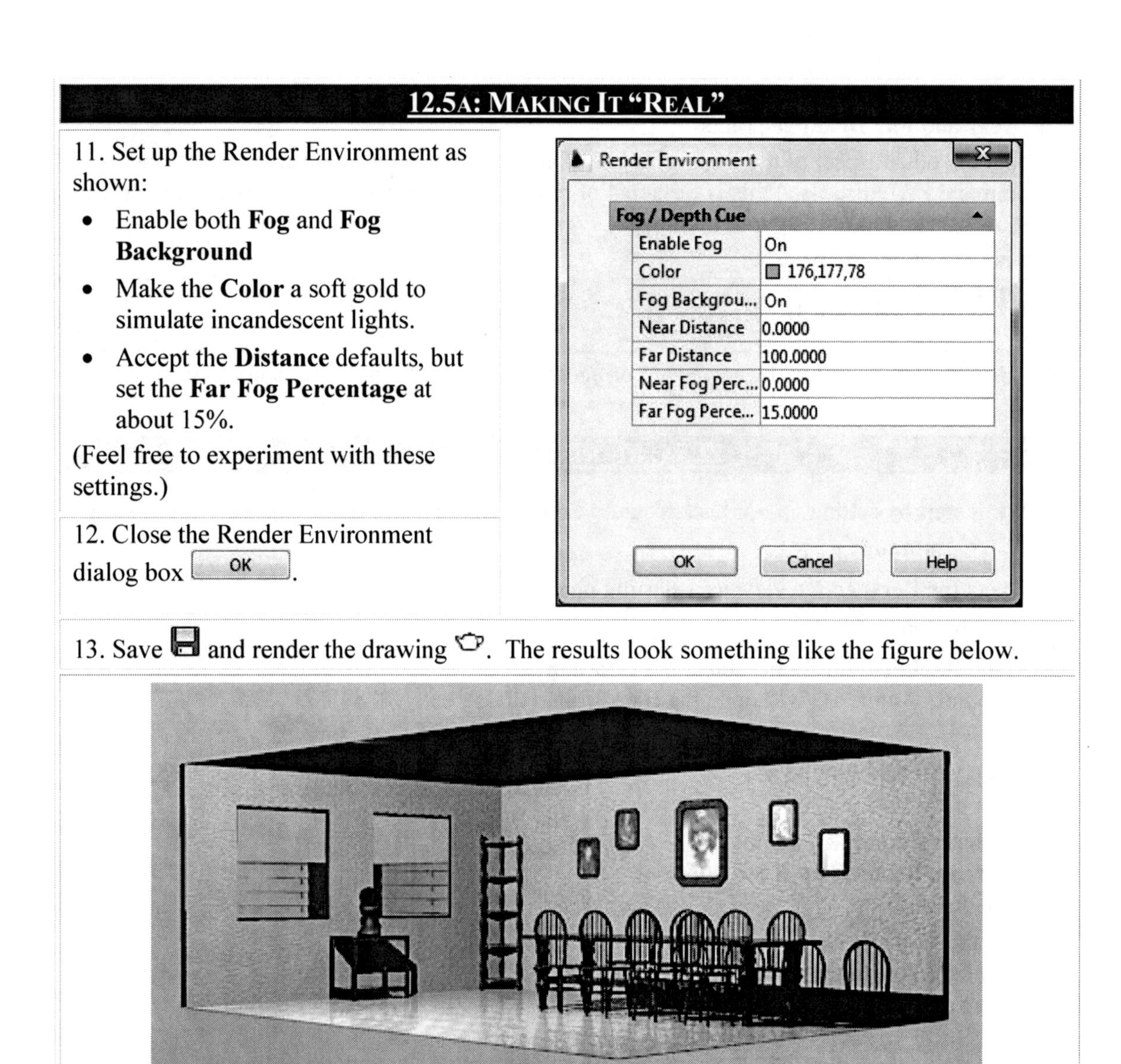

14. Save the rendering and compare it with earlier efforts. What do you think?

Use the rendered image in your client presentations, or include them as image inserts to show the final product in your drawings. Clients are always impressed by these efforts!

12.6 Extra Steps

As if you need anything more to do!

Okay, do this. You may have notice that the ribbon's **Render** tab contains several additional tools beyond what we covered. Take some time to experiment with these. You'll find most of them have alternatives that we discussed … some don't. Frankly, if we tried to cover everything involved in rendering, we'd have to include a course in artistic lighting, photography (black and white and color), and even the weather! But we hit the basics … the rest is there for the curious to discover.

12.7 What Have We Learned?

Items covered in this lesson include:

- *Rendering commands and techniques*
  - *Assigning materials*
  - *Adding lights*
  - *Rendering*
- *Commands*
  - ***Render***
  - ***Materials***
  - ***MaterialAttach***
  - ***Light***
  - ***PointLight***
  - ***SpotLight***
  - ***DistantLight***
  - ***Geographic-Location***
  - ***SunProperties***
  - ***Render-Environment***
  - ***RPref***
  - ***LightingUnits***
  - ***RenderPresets***

You've accomplished a great deal with the completion of Lesson 12. In this lesson, you've conquered the bridge between CAD operation and art. This is as far as AutoCAD goes toward the creation of design imagery. (Beyond this, you'll have to get some tools and build the objects!)

Let's tackle some final exercises before we say goodbye.

12.8 Exercises

1. Open the *emerald12* file in the C:\Steps3D\Lesson12 folder. My rendering appears here.

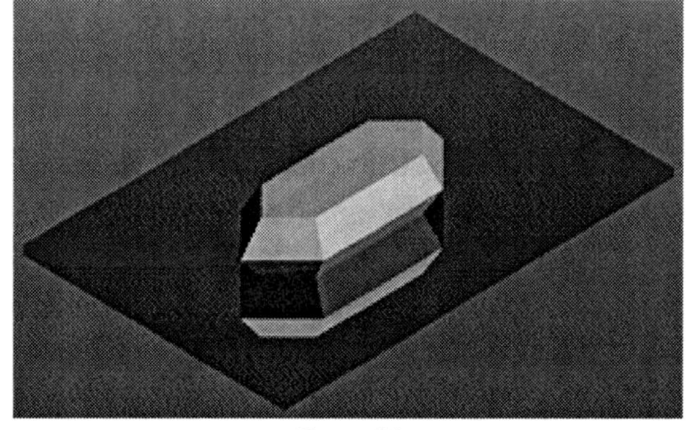

Emerald

   1.1. Assign materials to the emerald and tabletop. (I used **Green Glass** and **Mirrored Glass**. I changed the transparency of the glass to a value of **0.5**.)
   1.2. Assign a background. (I used a two-color gradient.)
   1.3. Place a light(s). (I used a single-point light located at coordinates **18,4,9**.)
   1.4. Render the drawing.
   1.5. Repeat Steps 1.1 to 1.4 using a different set of assignments.

2. Open the *caster12* file in the C:\Steps3D\Lesson12 folder. My rendering appears here.

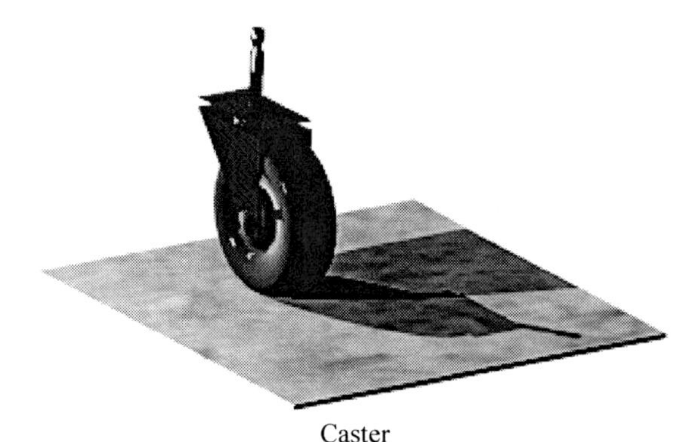

Caster

   2.1. Assign materials to the objects. I used
      2.1.1. Copper (bearings)
      2.1.2. Concrete (flooring)
      2.1.3. Plastic (wheel)
      2.1.4. Steel (axle, spindle, bearing case)
   2.2. Place a light(s). (I used a single-point light with an intensity of **2** located at coordinates **0,-10,4**.)
   2.3. Render the drawing.
   2.4. Repeat Steps 2.1 through 2.3 using a different set of assignments.

3. Open the *pipe12* file in the C:\Steps3D\Lesson12 folder. My rendering appears here.
    3.1. Assign materials to the objects. I used colored plastic templates for the following materials:
        3.1.1. Steel (tank and nozzle)
        3.1.2. Copper (large pipe and fittings)
        3.1.3. Concrete (model platform)
        3.1.4. Steel (pipe rack)

Pipe Model

    3.2. Place a light(s). (I used the sun. The unit is located in a plant in New Orleans. The graphic was created at 3:00 P.M. in late September.)
    3.3. Render the drawing.

4. Open the *train12* file in the C:\Steps3D\Lesson12 folder. My rendering appears here.
    4.1. Assign materials to objects. I used:
        4.1.1. Urethane (front of train)
        4.1.2. Urethane (cab, base, top of cattle guard)
        4.1.3. Urethane (wheels, smokestack, bell)
        4.1.4. Stucco (backdrop)
        4.1.5. Urethane (water tank, undercarriage, base of cattle guard, flag)
        4.1.6. Hickory (bell frame, box, flagpole)
    4.2. Assign a background. (I used a solid green background.)
    4.3. Place a light(s). (I used a single spotlight located at 0,-60,0, and a target location of **4,1,4.5**. Sun light intensity is **0.5**.)
    4.4. Render the drawing.

Train

You have quite a variety of exercises from which to choose in this lesson. Your next assignment is to return to any of the exercises you've completed in the text, assign materials, graphics, and lights, as you deem necessary and then render the drawings. Following are several exercises you can do. (If you haven't completed these drawings, they're available in the C:\Steps3D\Lesson12 folder.)

Wagon

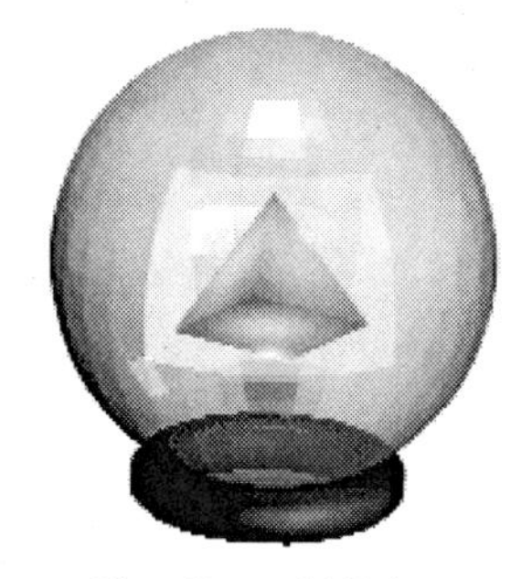

Glass Pyramid Sphere

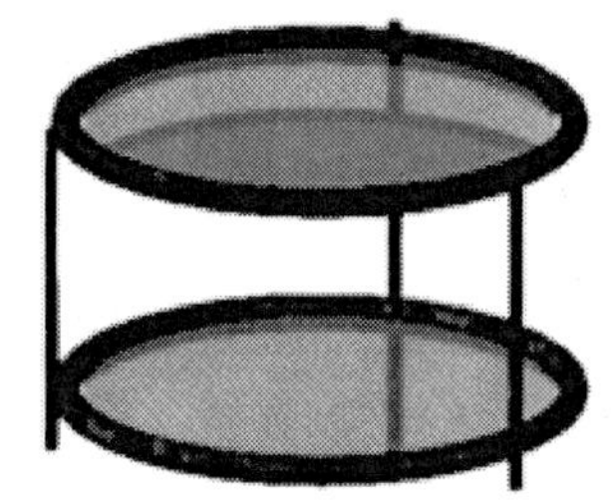

Coffee Table

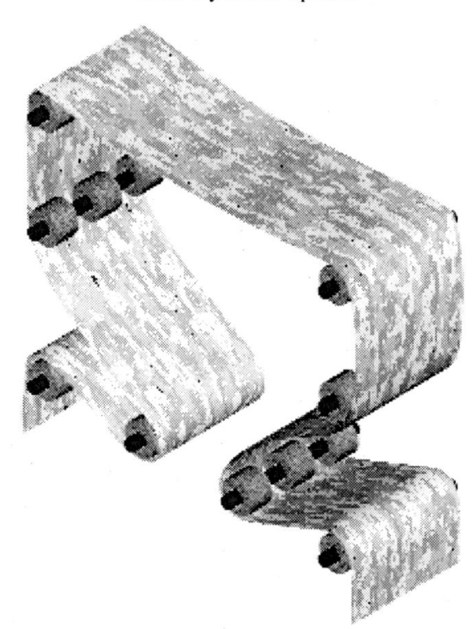

Conveyor Belt

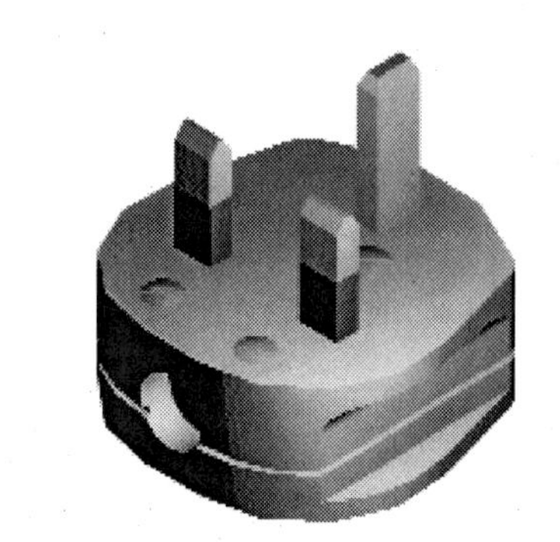

Plug

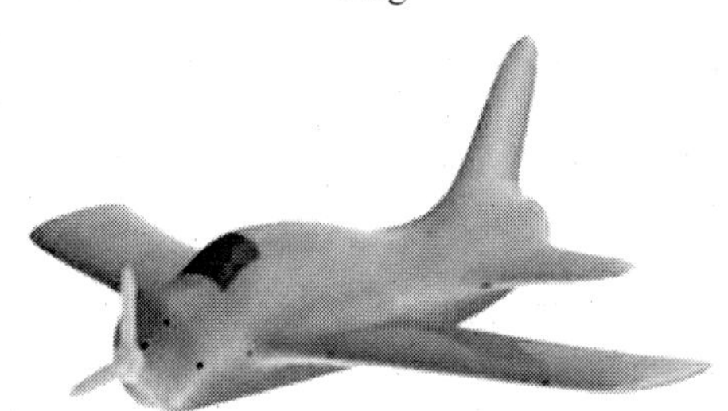

Airplane (mesh)

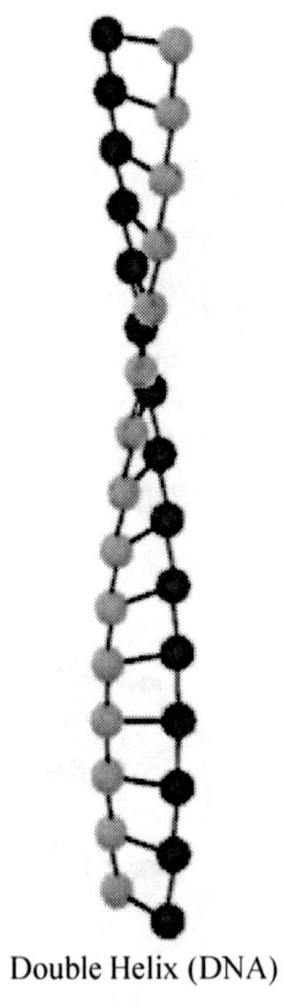

Double Helix (DNA)

Wheel

3D Chess Board

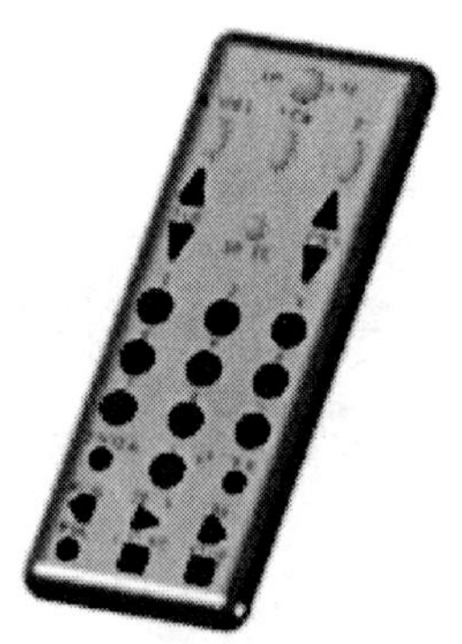

Remote

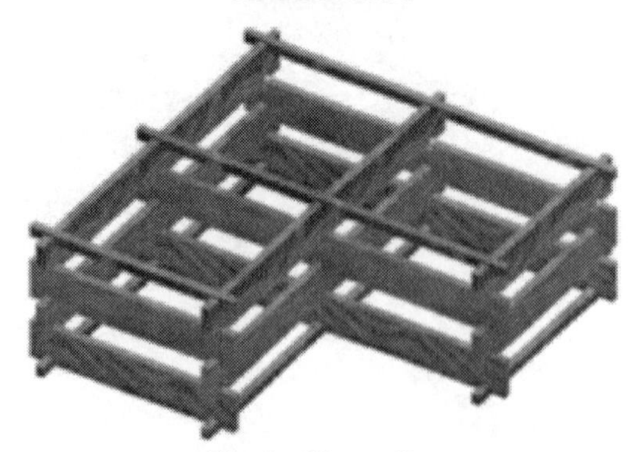

Wooden Planter Box

Switch

Level

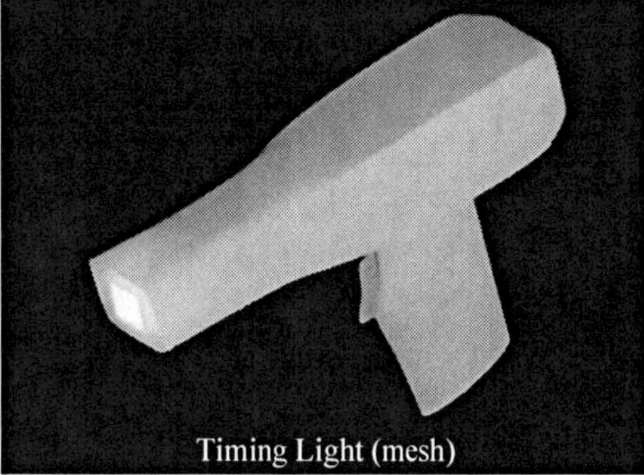

Timing Light (mesh)

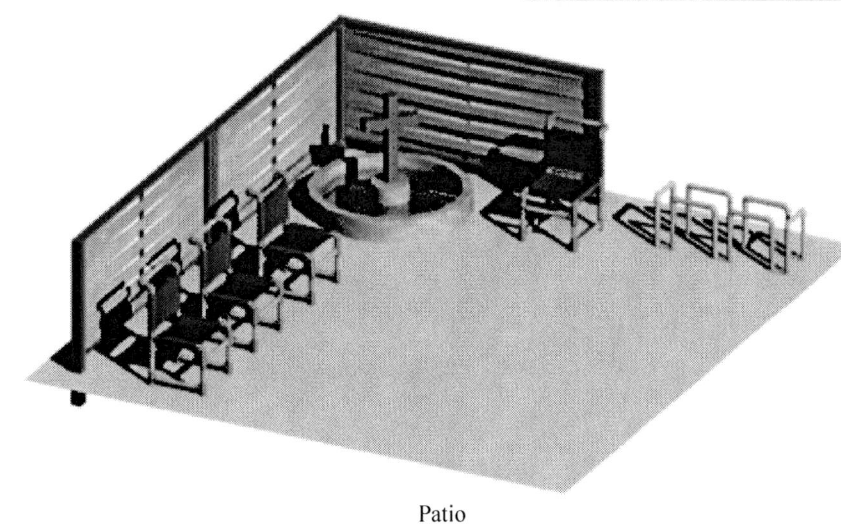

Patio

20. Using other objects you've created in Lessons 4 through 11, redesign the patio scene. Some suggestions follow:
    20.1. Replace the fountain with the planter box.
    20.2. Place planters in the planter box.
    20.3. Add a sidewalk or some decking.
21. Using other objects you've created, redesign the room we created in our rendering project. Some suggestions follow:
    21.1. Replace the corner shelf with the standing lamp (Lesson 4).
    21.2. Put the 3D chess set on the table.
    21.3. Change the pictures on the wall.

12.0 For Web-Based Review Questions, visit: http://foragerpub.com/AcadFiles/2010/2010.htm

Lesson 13

Following this lesson, you will:

- ✓ *Be Able to Create Simple Animations*

Animation Tools

Lights! Action! Camera! Unknown Hollywood Director

We've taken AutoCAD to the edge that separates CAD Operation from programming (in the basic text) and to the edge that separates CAD Operation from art (in Lesson 12). This final lesson will take you to the edge between CAD Operation and motion pictures (the cinema).

Before we begin, however, you should understand that AutoCAD doesn't provide true animation tools. You won't be able to create cartoons or the fancier new animé (virtual reality animation) with AutoCAD; you'll need 3DS Max, Inventor, Poser, Lightwave, or any of a host of other applications designed for that purpose. But that doesn't mean that you can't produce some nice walk–through animations for your clients without the investment required for those other packages.

Let's see how AutoCAD does it.

13.1 Animated Views – The *NavsMotion* Command

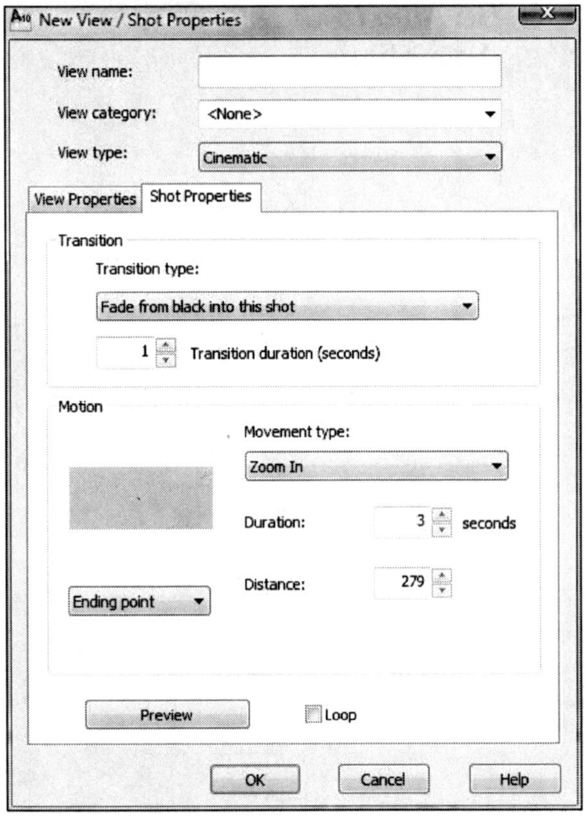

Figure 13.001

Where to Find It:	
Command Line:	***NavsMotion***
Menu:	View – **Show Motion**
Status Bar:	**Show Motion**

Where to Find It:	
Command Line:	***NewShot***
Hotkey(s):	***nshot***
Toolbar:	Show Motion (below the thumbnails) – **New Shot**

NavsMotion (aka. **Show Motion**) provides the easiest methods for creating an animated walk–through. It utilizes a series of views with controlled transitions between each.

We discussed Show Motion in Lesson 4 of our basic text (along with the ***View*** command in Section 4.2) and we won't repeat the basic stuff here, but let's take a look at what it can do in terms of a three dimensional walk-through.

First we need to look at the New View / Shot Properties dialog box (Figure 13.001) accessible using the **New** button New... in the View dialog box or the ***NewShot*** command. We'll concentrate on one of the options we find above the tabs (the **View type** option) and on the **Shot Properties** tab.

- The three **View types** control the type of animation you'll create.
  - A **Cinematic** view type results from a single camera position, although the camera may zoom, crane, track or orbit.
  - A **Still** view type results from a single, stationary camera.
  - **Recorded Walk** (Model Space only) gives you the most freedom in animation creation. With it, you can pick and drag your way through the model.
- The options you find on the **Shot Properties** tab depend upon what **View type** you've selected.

- With a **Still** view type selected, the **Shot Properties** tab looks like Figure 13.002. The simplest setup, this view type still has a few options.
  - The **Transition** frame provides a couple tools to enhance your animation. Using the **Transition type**, you can **Fade from black** [or **white**] **into this shot**, or simply **Cut to shot** without fanfare. The **Transition duration** controls how long it takes to switch to this view. (The **Transition** frame's tools remain consistent for all three view types.)
  - **Motion Duration** determines how long the slide will last before the next transition begins.

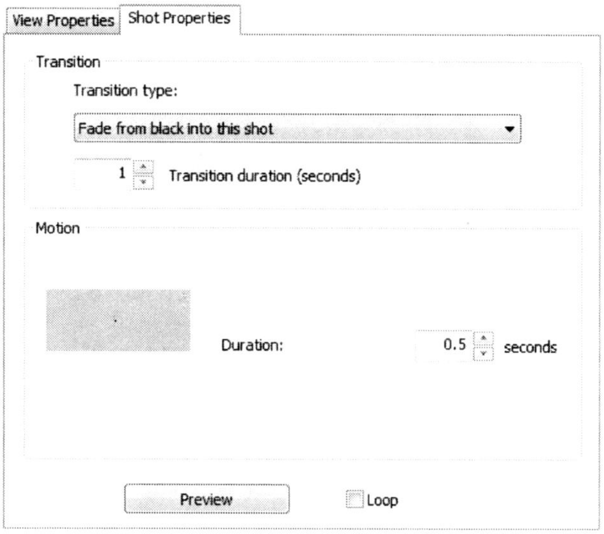

Figure 13.002

- The **Cinematic Shot Properties** tab looks like Figure 13.003. Here, the **Motion** frame has some different options.
  - You'll find several **Movement types** in the drop down list including: **Zoom In / Out**, **Track Left / Right**, **Crane Up / Down**, **Look** (rotate the camera around a stationary position), and **Orbit** (move the camera around the model).
  - **Duration** tells AutoCAD how long to spend performing the movement selected from the **Movement type** list.

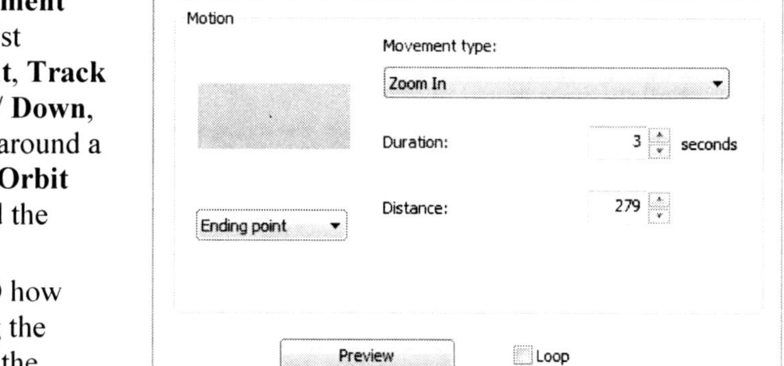

Figure 13.003

  - **Distance** changes with the **Movement type** selection. Use it to limit the distance AutoCAD will zoom, track, or crane.
  - The **Starting** (or **Ending** or **Half-way**) point button determines the current location of the camera.
  - **Preview**, of course, allows you to preview the settings, while **Loop** tells AutoCAD to continually repeat the animation. (These tools are consistent for the three view types.)

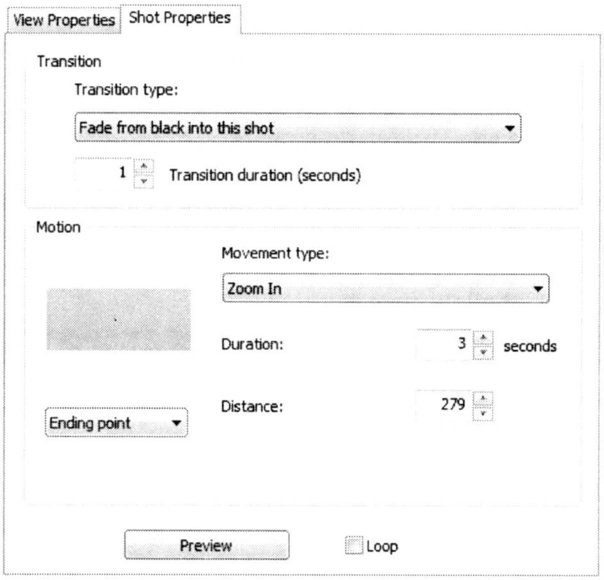

- The **Recorded Walk Shot Properties** tab looks like Figure 13.004.
  - You'll find the **Start recording** button the only tool available in the **Motion** frame. When you pick it, AutoCAD will hide the New View / Shot Properties dialog box and allow you to move about the view using an arrow–type cursor. You'll notice that the view automatically pans as you hold down the left mouse button and move the cursor. When you release the button, AutoCAD returns to the dialog box and lists how long the motion took in the **Duration** display box.

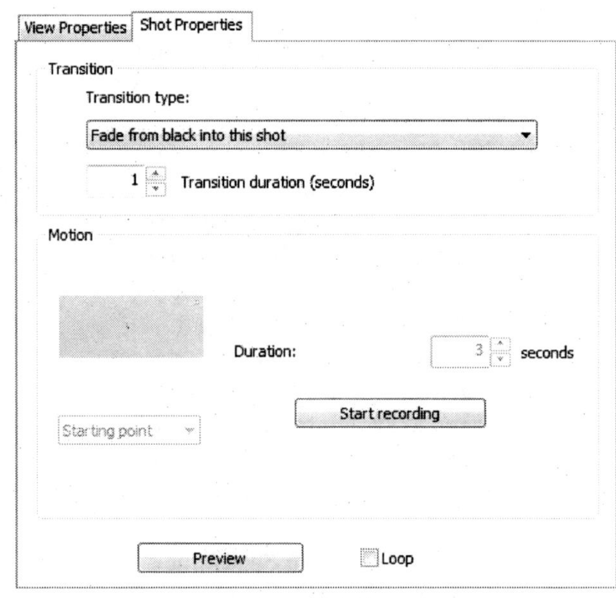

Figure 13.004

As you'll see in our exercise, AutoCAD doesn't limit you to a single view in your animated walk-through. Just remember to *save* your view once you've set it up (no sense wasting all that work)!

Let's see what we can do now.

Do This: 13.1A	Animation Setup and Viewing

I. Open the *Piperack* file in the C:\Steps3D\Lesson13 folder.

II. Follow these steps.

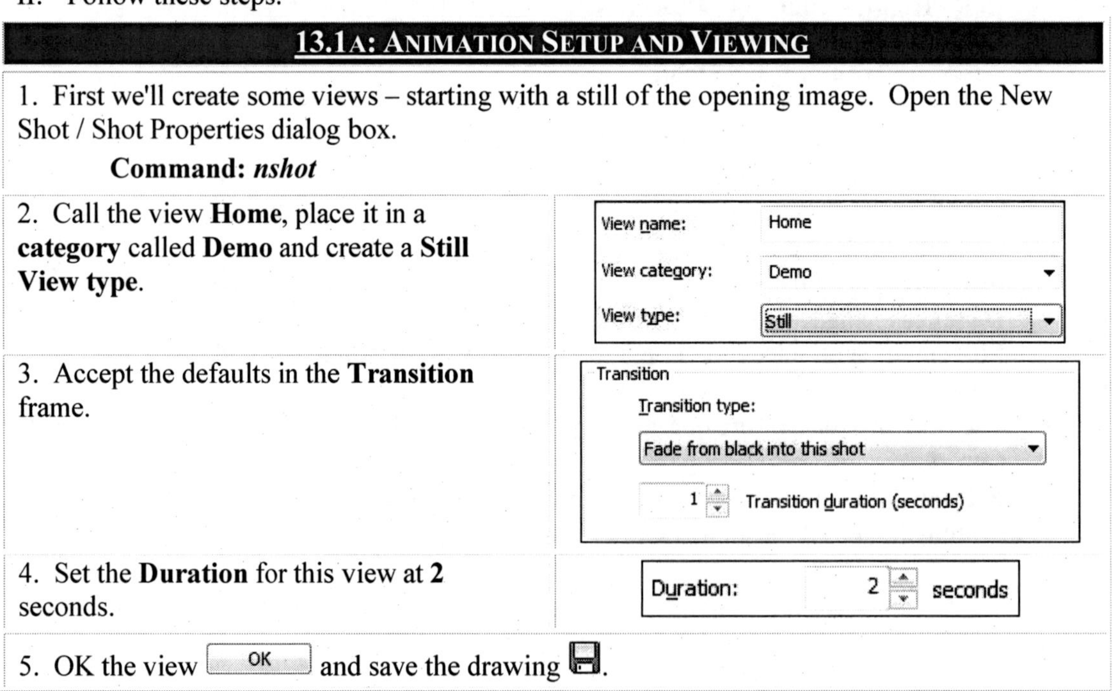

13.1A: ANIMATION SETUP AND VIEWING

1. First we'll create some views – starting with a still of the opening image. Open the New Shot / Shot Properties dialog box.

   Command: *nshot*

2. Call the view **Home**, place it in a **category** called **Demo** and create a **Still View type**.

   View name: Home
   View category: Demo
   View type: Still

3. Accept the defaults in the **Transition** frame.

   Transition
   Transition type:
   Fade from black into this shot
   1 Transition duration (seconds)

4. Set the **Duration** for this view at **2** seconds.

   Duration: 2 seconds

5. OK the view [OK] and save the drawing.

13.1A: Animation Setup and Viewing

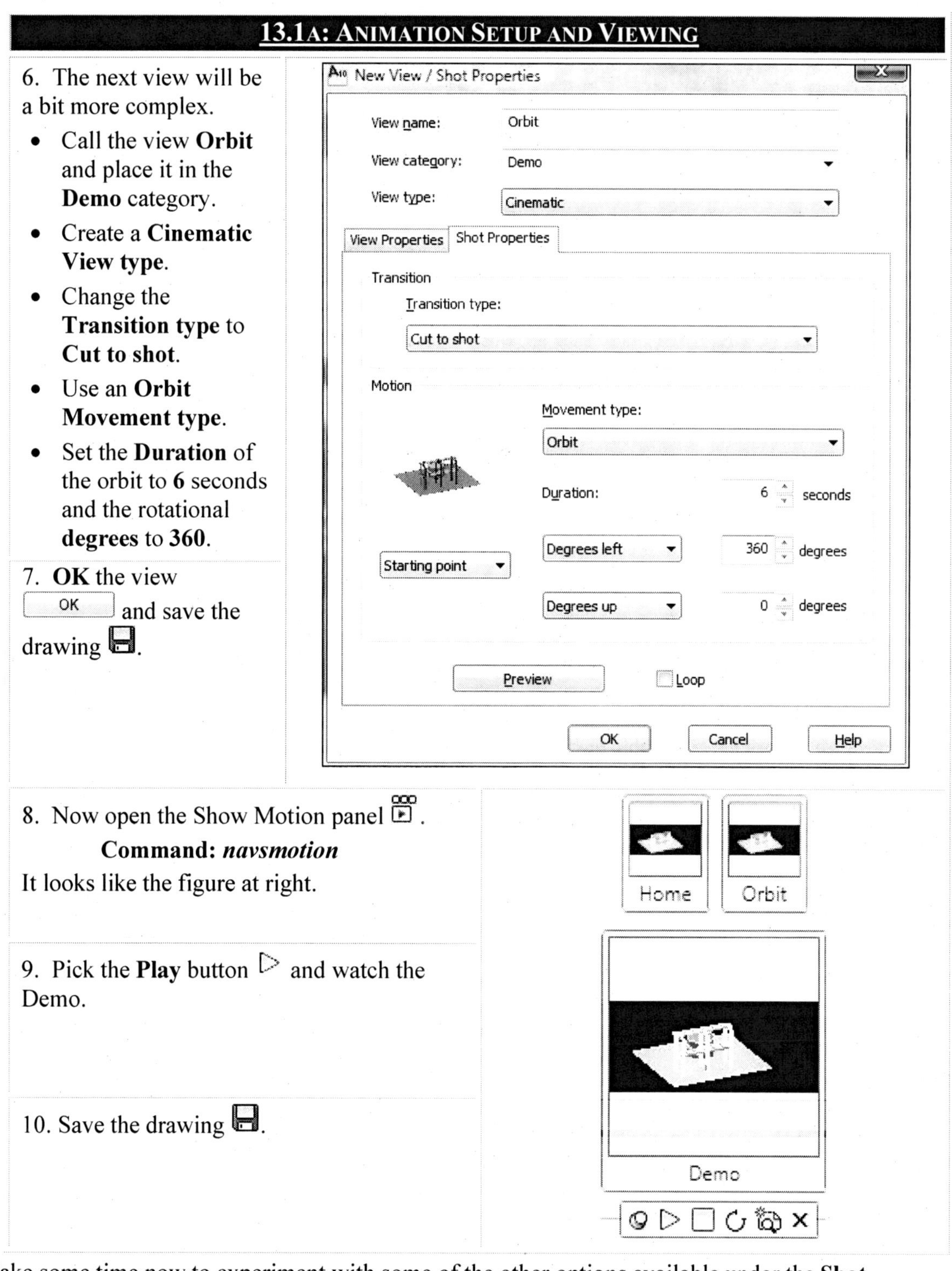

6. The next view will be a bit more complex.
   - Call the view **Orbit** and place it in the **Demo** category.
   - Create a **Cinematic View type**.
   - Change the **Transition type** to **Cut to shot**.
   - Use an **Orbit Movement type**.
   - Set the **Duration** of the orbit to **6** seconds and the rotational **degrees** to **360**.

7. **OK** the view [OK] and save the drawing.

8. Now open the Show Motion panel.

   Command: *navsmotion*

   It looks like the figure at right.

9. Pick the **Play** button and watch the Demo.

10. Save the drawing.

Take some time now to experiment with some of the other options available under the **Shot Properties** tab of the New Shot / Shot Properties dialog box. You're near the end of the text so enjoy yourself!

You may have noticed one glaring omission in this fantastic demonstration tool – AutoCAD has provided *no way to save the demo outside AutoCAD!* To use this procedure to show off to a client or employer, then, you'll need access to AutoCAD and the actual drawing file. But don't fret, we have other tools to allow demonstrations without the application.

Read on!

13.2 3DWalk(ing) and 3DFly(ing)

Okay, it's not like you need to be sold on AutoCAD any further, right? Well, this section might be considered the part of the commercial where a hyper-announcer demands that you, "WAIT! Don't order yet!"

Once your new home (or plant or part etc.) has been built, you still demand that final walk-through before signing the papers. Well, now you can give your client that final walk-through before he buys the first nail *and not require that he purchase a very expensive application to do it!*

We'll examine two approaches – the Manual Approach and the Super-Terrific-Automatic-Realistic (STAR) Approach. (Okay, I just made that up, but you'll see that it really is the Star in AutoCAD's crown!)

You'll want to remember a couple things before we look at these animation tools.

- First, AutoCAD utilizes a very basic form of animation designed to allow simple walk (or fly) throughs. You won't be able to open and close doors and have a hummingbird fly past. If you want that sort of thing, invest in one of the true animation packages available on the market today (Lightwave, 3DS Max, Maya, Poser, etc). Most will have at least some ability to use AutoCAD drawings (or to translate them into something they can use).
- Second, to create a nice, rendered animation takes time – buckets of it! It also takes a healthy investment in system resources. I'll let you know how long it takes me to create the animations we'll be doing in our exercises. FYI: I'm using an HP Pavilion D4100Y with a GEForce 7600 graphics card (513Mb), 3.2Gb of RAM, and an Intel P4 processor with a speed of 3GHz. My operating system is Windows Vista Home Premium, and I'm using Hardware Acceleration. (To use Hardware Acceleration, follow this path: enter ***3dconfig*** at the command line, pick the **Manual Tune** button, and place a check next to **Enable hardware acceleration** in the **Hardware settings** frame.) You can figure your approximate time by how much faster or slower your system is than mine.

13.2.1 The Manual Approach – 3DWalk and 3DFly

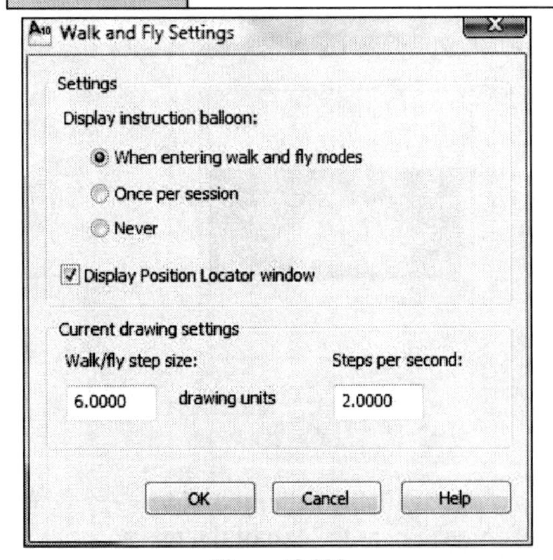

Figure 13.005

Where to Find It:

Command Line:	***WalkFlySettings***
Ribbon (Tab/Panel):	Render – Animations* – (Walk flyout) **Walk and Fly Settings**
Menu:	View – Walk and Fly – **Walk and Fly Settings**
Toolbar:	3D Navigation – (Walk flyout) **Walk and Fly Settings**

What's the difference between ***3DWalk*** and ***3DFly***? You walk on the ground; you fly in the air. (It's really that simple.) Use ***3DWalk*** to do a "walk-through" and ***3DFly*** to do a "fly-by". Settings and tools are shared.

Begin by setting up the walk-through or fly-by with the ***WalkFlySettings*** command, which opens the Walk and Fly Settings dialog box (Figure 13.005).

* If the **Animations** panel doesn't appear, refer to Appendix D, p.334.

- Two **Settings** can save your sanity. The two boxes in the **Current drawing settings** frame control how large the steps are you'll take. The default settings mean you'll be taking 6" steps every two seconds. These small steps allow you time to view, but really cost you in terms of system resources and the size of the animation file you may want to create.
- By default, the Communication Center opens (right end of the title bar) every time you enter the ***3DWalk*** or ***3DFly*** command. It tells you how to walk (or fly) through the drawing. Once you've learned to walk (so to speak) pick **Don't show me this again** so it doesn't become an irritant. To keep it simple, use these tools to walk/fly:

TOOL	ACTION	TOOL	ACTION
Up Arrow or **W**	Move forward	Right Arrow or **D**	Move Right
Down Arrow or **S**	Move backward	Drag Mouse	Look around or Turn
Left Arrow or **A**	Move Left	**F**	Fly/Walk mode toggle
Tab	Reopens the instruction window		

- **Display Position Locator window**, when checked, tells AutoCAD to display the Position Locator (Figure 13.006, p.320) while in the ***3DWalk*** or ***3DFly*** command.

The Position Locator window makes the manual procedure a lot easier – but you have to remember that easy tends toward making a designer think he can make it longer. Unfortunately, *easier doesn't mean faster!*

Where to Find It:	
Command Line:	***3DFly***
Ribbon (Tab/Panel):	Render – Animations* – (Walk flyout) **Walk and Fly Settings**
Menu:	View – Walk and Fly – **Fly**
Toolbar:	3D Navigation – (Walk flyout) **Fly**

- Use the three buttons across the top of the Position Locator to **Zoom in** or **out** on the display window, or to **Move** the locator. Note that you don't have to use the **Move** button here to move the locator.
- The window below the buttons provides a very dynamic method of creating a walk-through.
  - Pick on the indicator's green lines to move it about on the drawing. (Your cursor will become a "pan" hand.
  - Pick on the green arrowhead at the wide end of the indicator and drag it to adjust the target of your camera.
  - Pick on the red ball at the pointed end of the indicator and drag it to adjust the camera itself.

Where to Find It:	
Command Line:	***3DWalk***
Hotkey(s):	***3dw***
Ribbon (Tab/Panel):	Render – Animations* – (Walk flyout) **Walk and Fly Settings**
Menu:	View – Walk and Fly – (Walk flyout) **Walk**
Toolbar:	3D Navigation – **Walk**

You can set the **Camera** and **Target** positions with more precision using the coordinate boxes on the **View** subpanel of the ribbon's **Home** tab.

Camera	21'-2 1/4"	-52'-3 11/	60'-3 15/1
Target	32'-6 7/16	29'-7 13/1	7'-1 13/16

* If the **Animations** panel doesn't appear, refer to Appendix D, p.334.

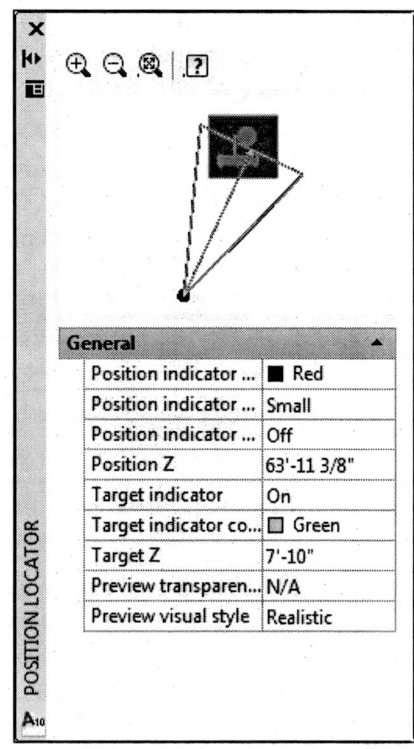

Figure 13.006

- The **General** tab below the display allows you to adjust some of the particulars of the display – mostly position and target colors on the indicator. You can even make the indicator blink in the display. (I don't recommend this for the high-strung among my readers.) More importantly, it allows you to adjust the **Preview visual style**. Set this back to **3D Wireframe** to speed things up a bit. (You don't need realistic in this window.) If you insist on a higher resolution, you can ease the image by increasing the **Preview transparency**.

Once you've set up your 3D Walk (Fly) tools, you might want to consider recording your walk-through. After all, unless your client is at your elbow, you'll need to provide him access to the results.

Use the standard multimedia controls provided on the **Animations** panel **(Tools** tab) to manage your recording. (These are: **Play** ▷, **Record** ○, **Pause** ▯▯, and **Save Animation** ☐.)

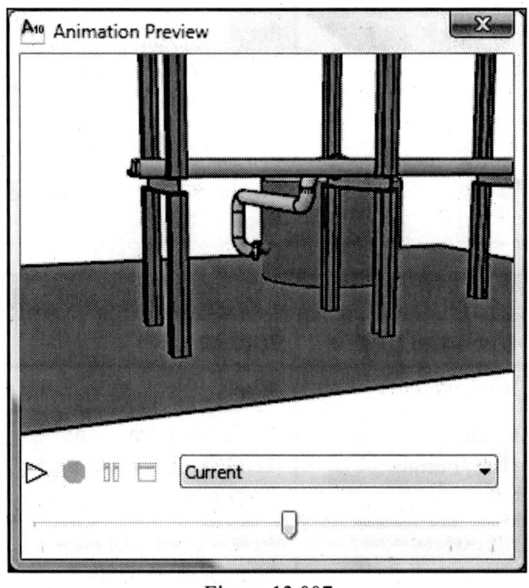

Figure 13.007

During the recording, you can pause and preview what you've done with the **Pause** and **Play** buttons. AutoCAD will display the preview in a window (Figure 13.007). Notice that the same multimedia buttons are available in the preview window, so you can see your walk through as you record it. You can also adjust your position in the preview using the slider rod, and you can adjust how you view the preview. (You'll find the visual styles you've already studied in the control box.)

You may find animations created using AutoCAD defaults to be somewhat disappointing. But let's take a look at them and then we'll see how to customize the setup a bit using AutoCAD's automatic animation tools.

We'll create and record a simple walk-through.

Do This: 13.2.1A	**A Simple Walk-Through**

I. Open the *rendering project* file in the C:\Steps3D\Lesson13 folder.

II. (You must be in a perspective view to create a 3D walk-through or a 3D fly-by.)

III. To hide the glyphs for this procedure, turn off the **Light glyphs display** (you'll find the toggle on the **Lights** subpanel – **Render** tab).

IV. Be sure the **LightingUnits** system variable is set to zero.

V. Follow these steps.

13.2.1A: A Simple Walk–Through

1. First, we'll do some set up. Open the Walk and Fly Settings dialog box (Figure 13.005, p.318).

   Command: ***walkflysettings***

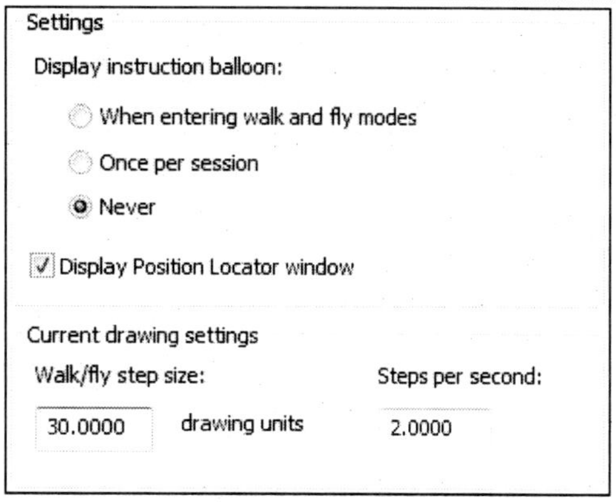

2. Create the setup shown.
   - Don't display the instruction window.
   - Set your **step size** to something closer to normal.
   - Accept the other defaults..

3. Close the Walk and Fly Settings dialog box [OK].

Settings
Display instruction balloon:
When entering walk and fly modes
Once per session
Never
Display Position Locator window
Current drawing settings
Walk/fly step size: 30.0000 drawing units
Steps per second: 2.0000

4. Set up a new visual style starting with the **Realistic** style and removing isolines. Save the style as **WalkDemo**.

   Command: ***vssave***

5. Begin the 3D Walk procedure. (Enter the command, but don't do anything yet!)

   Command: ***3dw***

AutoCAD opens the Position Locator (Figure 13.006, p.320).

6. Using the Position Locator, move the target to the center photo on the wall. Then move the camera to just over the southeast corner of the floor. (I'm using the **3D Hidden Preview visual style**.)

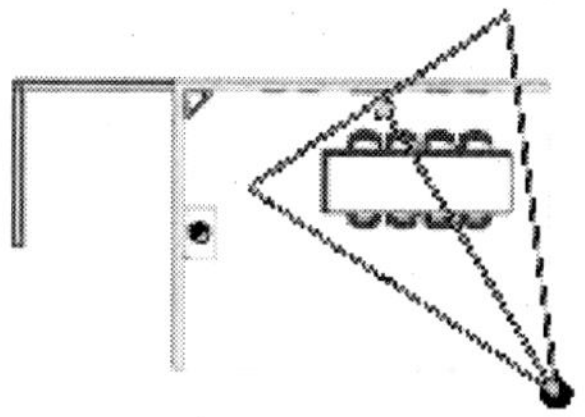

7. Pick the **Record** button on the **Animations** panel (**Tools** tab).

8. Using the Position Locator, move the camera to the other southern corner of the room.

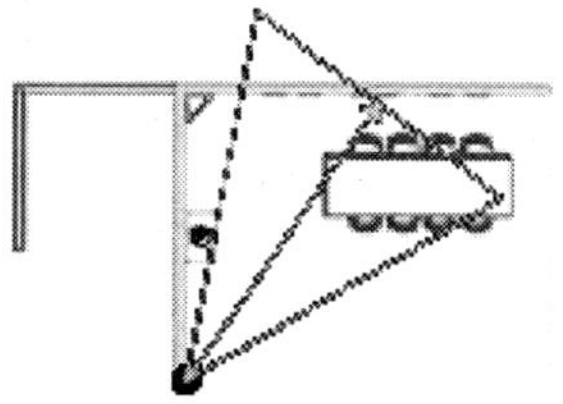

9. Pick the **Pause** button on the **Animations** panel.

10. Now pick the **Play** button and watch your animation. Experiment with the different visual styles during the preview.

11. You can save the animation if you wish. (It took ~3 minutes on my system. It takes considerably more time when you use rendered images rather than a visual style.) I've included the animation file as *First Animation* in the C:\Steps\Lesson\Steps3D\Lesson13 folder if you'd like to view it.

12. Use the ESC key [Esc] to close the procedure. Zoom previous to restore the drawing to its original view.

13.2.1A: A Simple Walk-Through

13. View the animation using Windows Media Player to get a better understanding of what you can do. There, you'll see the rendered animation. (It looks a lot better than the preview!)

Of course, you could have used arrow keys to establish a more interesting path, but you'll find a much easier approach when you create an animation using the automatic process we'll discuss next. Personally, I find it more versatile as well.

13.2.2 The Automatic Approach – Motion Path Animation

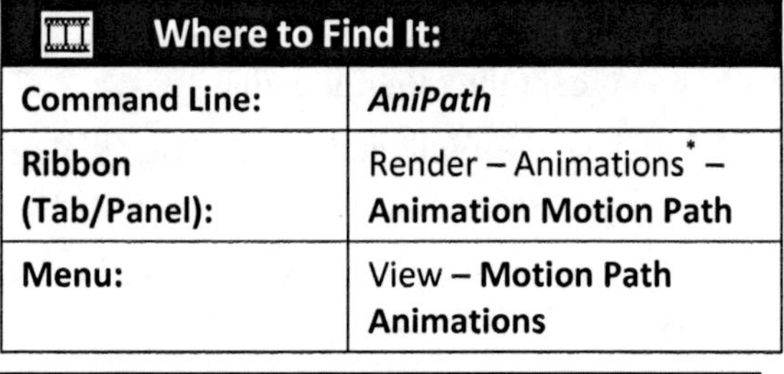

Where to Find It:	
Command Line:	***AniPath***
Ribbon (Tab/Panel):	Render – Animations* – **Animation Motion Path**
Menu:	View – **Motion Path Animations**

When you use the automatic approach – Motion Path Animations – you can create the path you wish your camera and/or target to follow. Then you let AutoCAD handle the repositioning.

Access the Motion Path Animations dialog box (Figure 13.008) with the ***AniPath*** command. The dialog box contains everything you'll need for this effort.

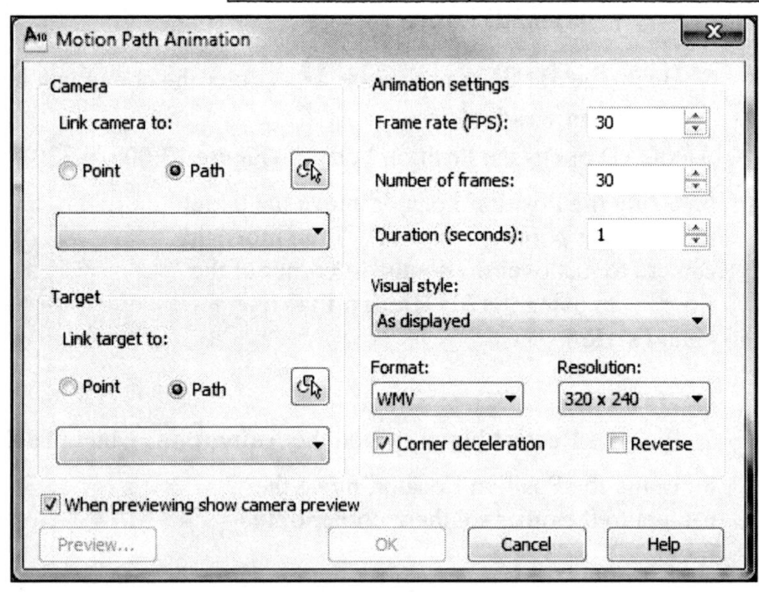

Figure 13.008

- Define your camera's location (**Point**) or **Path** in the **Camera** frame. (You can also use the tools on the **View** subpanel of the **Home** tab to select a point.) Tell AutoCAD which link you want to use for the camera with the radio buttons, then use the **Select Path / Pick Point** button to return to the drawing for your selection. A path can be a line, arc, elliptical arc, circle, polyline, 3D polyline, or spline. Once you've made your selection, AutoCAD will ask you to name the point or path. It then places the name in the **Camera** frame's control box for later selection if you need it again.
- You'll do the same thing for the **Target** that you did for the camera. Both can have paths if you wish the animation to move that way. Both cannot, however, have points. (Forgive the pun, but what would be the point? Nothing would move!)
- **Animation Settings** work together. Enter values into any two of the three numbered control boxes (**Frame rate**, **Number of frames**, and **Duration** of animation) and AutoCAD will fill out the third. (Remember, higher frame rates make better animations but take long to produce and result in larger animation files.) You'll also set the **Visual Style** or rendering option here, as well as the **Resolution** of the animation. A check next to **Corner deceleration** slows the camera at the ends of the animation, while a check next to **Reverse** reverses the animation.

* If the **Animations** panel doesn't appear, refer to Appendix D, p.334.

Movies run at 24 frames per second (FPS); computer animations typically run at about 30 FPS. A frame rate of 13 isn't unusual, especially for web animations, but they can appear choppy. Experiment with different rates to determine what works best for you, but be aware that experimenting with animation can be an absorbing and time-consuming task.

Notice that you have a **Format** control in the **Animation Settings** frame with several options.

- WMV – popular windows format; plays in the Media Player that comes with Windows.
- MPG – good for web play, small size, also plays in the Windows Media Player.
- AVI – older file type, much larger, also plays in the Windows Media Player.
- MOV – Quicktime movie, player is downloadable and free, works on both IBM compatibles and Mac computers.

- With a check next to **When previewing show camera preview**, AutoCAD shows the same Animation Preview window (Figure 13.007, p.320) you saw in our last exercise as it creates the animation for you.

Once you've filled out the Motion Path Animation settings, pick the **OK** button to begin creating the animation.

On my system, AutoCAD took about three minutes to complete the animation sequence in the following exercise. If you haven't that much time, read over the exercise and view the results (the *MP Animation.mpg* file in the C:\Steps3D\Lesson13 folder).

Do This: 13.2.2A	Motion Path Animation

I. Be sure you're still in the *rendering project* file in the C:\Steps3D\Lesson13 folder.
II. Be sure that light glyphs are still toggled off.
III. Thaw the **animation path** layer. The object you see is a spline I created to speed up the exercise. Normally, you'll create your own path.
IV. Follow these steps.

13.2.2A: MOTION PATH ANIMATION

1. Begin the command.

   Command: ***anipath***

2. Create the setup shown in the following figure.
   - Select the spline for the **Camera Path** and accept the default name.
   - Your target should be the center of the large photograph on the wall. Use OSNAPs! (You don't want to spend the time to create an animation to find it focused on the wrong target!)
   - Give it a reasonable frame rate for your animation. (For a client, you might want to set the FPS at 24 – 30, but that isn't necessary for our movie.)
   - Set the **Duration** to 10 seconds. (Any longer and you'd be here all day!)
   - Use the **WalkDemo** rendering (or use a **Presentation** rendering for a really nice – but slowly produced – video).
   - Use the same 800 x 600 **Resolution**. It creates a smaller movie but doesn't take as long as larger imaging.
   - Create an MPG to save on the size of the final movie.
   - Finally, Use **Corner deceleration** to avoid sharp starts and stops.

13.2.2A: Motion Path Animation

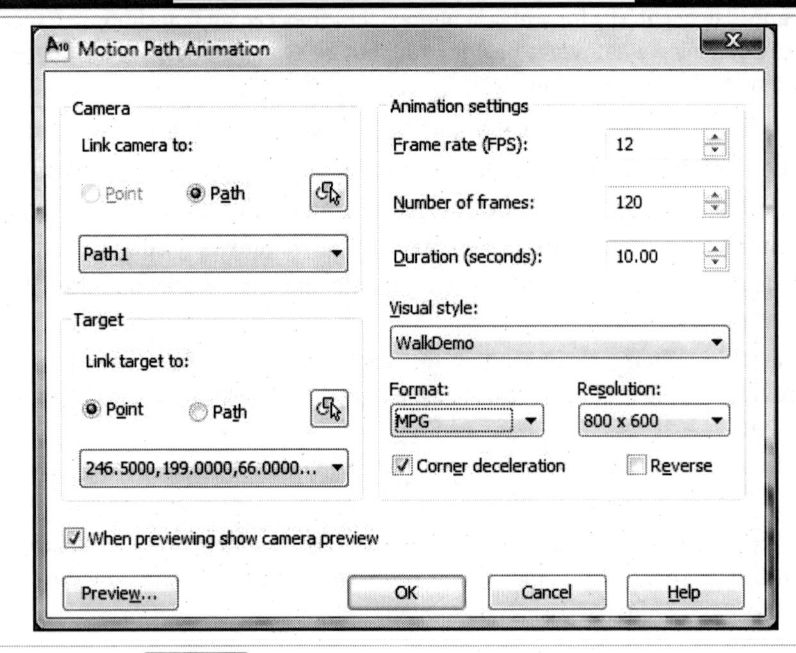

3. Begin the animation OK. (Call it, *My MP Animation* and save it to the C:\Steps3D\Lesson13 folder.) If you checked **When previewing show camera preview**, you can watch AutoCAD creating your animation. (Relax, this is going to take awhile.)

4. View your animation using Windows Media Player. (Let the video loop several times while you study your results.) Pay particular attention to the smoothness of the video (controlled by the **FPS** setting), the detail (**Visual Style**), and the size of the video on your screen (**Resolution**). Notice the wood grain pattern in the picture frame and table, and the reflection of the light on the photo and table top. These are the result of careful material controls. Finally, notice the difference when the camera moves outside and you see the photo through the window.

13.3 A Final Word

Well, I have to leave you here. We've come a long way together and you're ready to go out and make your fortune (sniffle), so I suppose I should leave you with some words of wisdom and some thoughts on where to go from here.

AutoCAD serves as a backbone – almost an operating system – for several other programs. These include many industry-specific applications, such as *Architectural Desktop*, *Mechanical Desktop*, and a host of third-party applications. You can continue your training into any of those.

If you wish to continue with the animation side of things, I highly recommend NewTek's *Lightwave*. It's hard to get better bang for your design bucks. Autodesk has a cool package it calls *3DS Max*. Use *Max* or *Lightwave* for your cartooning efforts. Smith Micro sells an animation package – *Poser* – for those on a budget. I've been very pleased with it as a tool for creating cartoons for some of my other books.

Oh, yes; final words of wisdom. I've racked my brain and can't come up with anything better than the immortal words of Dr. Sidney Freedman (MASH psychiatrist). So, from an old piper to all you kids out there looking to get ahead in the world, here's my advice: *pull down your pants and slide on the ice.*

13.4 What Have We Learned?

Items covered in this lesson include:

- *Creating animations*
- *Commands*
  - ***3DWalk***
  - ***3DFly***
  - ***WalkFlySettings***
  - ***Anipath***

Let's tackle a final (big) exercise before we say goodbye.

13.5 Exercises

5. Open the *2DFlrPln* file in the C:\Steps3D\Lesson13 folder.
   - 5.1. Create 3D walls from those shown.
   - 5.2. Add some furniture – feel free to block and copy furniture, fixtures, photos, and/or frames from the exercises in this text. Create other items as you desire. (Block your new creations for use later.)
   - 5.3. Assign some materials to the walls, floor, furniture, etc.
   - 5.4. Add some lights, lamps, the sun, etc.
   - 5.5. Add a ceiling.
   - 5.6. Render each room as you complete it.
   - 5.7. Create views and view presentations to discuss possibilities with your instructor/boss/significant other …
   - 5.8. Now create a final animated walk-through to show off your house! Save it as an mpg.
6. Go get one of those top–paying jobs I hear so much about (and next Christmas, don't forget lowly authors who helped you along the way)!

13.6 For Web-Based Review Questions, visit: http://foragerpub.com/AcadFiles/2010/2010.htm

Appendix – A: Drawing Scales

SCALE (= 1')	SCALE FACTOR	DIMENSIONS OF DRAWING WHEN FINAL PLOT SIZE IS:				
		8½"x11"	11"x17"	17"x22"	22"x34"	24"x36"
1/16"	192	136'x176'	176'x272'	272'x352'	352'x544'	384'x576'
3/32"	128	90'8x117'4	117'4x181'4	181'4x234'8	234'8x362'8	256'x384'
1/8"	96	68'x88'	88'x136'	136'x176'	176'x272'	192'x288'
3/16"	64	45'4x58'8	58'8x90'8	90'8x117'4	117'4x181'4	128'x192'
¼"	48	34'x44'	44'x68'	68'x88'	88'x136'	96'x144'
3/8"	32	22'8x29'4	29'4x45'4	45'4x58'8	58'8x90'8	64'x96'
½"	24	17'x22'	22'x34'	34'x44'	44'x68'	48'x72'
¾"	16	11'4x14'8	14'8x22'8	22'8x29'4	29'4x45'4	32'x48'
1"	12	8'x6'11	11'x17'	17'x22'	22'x34'	24'x36'
1½"	8	5'8x7'4	7'4x11'4	11'4x14'8	14'8x22'8	16'x24'
3"	4	34"x44"	3'8x5'8	8'x6'11	7'4x11'4	8'x12'
(1" =)						
10'	120	85'x110'	110'x170'	170'x220'	220'x340'	240'x360'
20'	240	170'x220'	220'x340'	340'x440'	440'x680'	480'x720'
25'	300	212'6x275'	275'x425'	425'x550'	550'x850'	600'x900'
30'	360	255'x330'	330'x510'	510'x660'	660'x1020'	720'x1080'
40'	480	340'x440'	440'x680'	680'x880'	880'x1360'	960'x1440'
50'	600	425'x550'	550'x850'	850'x1100'	1100'x1700'	1200'x1800'
60'	720	510'x660'	660'x1020'	1020'x1320'	1320'x2040'	1440'x2160'
80'	960	680'x880'	880'x1360'	1360'x1760'	1760'x2720'	1920'x2880'
100'	1200	850'x1100'	1100'x1700'	1700'x2200'	2200'x3400'	2400'x3600'
200'	2400	1700'x2200'	2200'x3400'	3400'x4400'	4400'x6800'	4800'x7200'

Appendix – B: Project Drawings

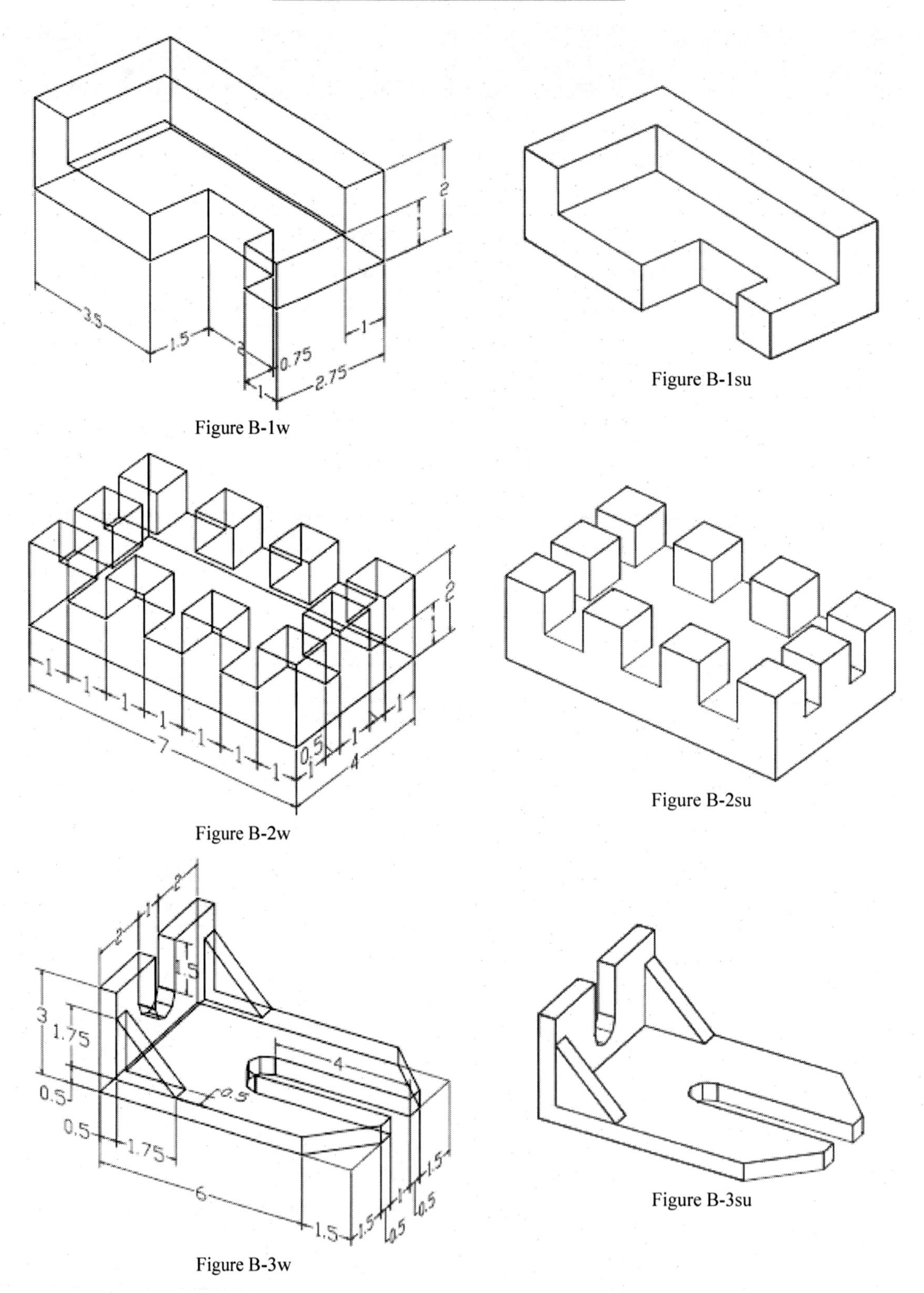

Figure B-1w

Figure B-1su

Figure B-2w

Figure B-2su

Figure B-3w

Figure B-3su

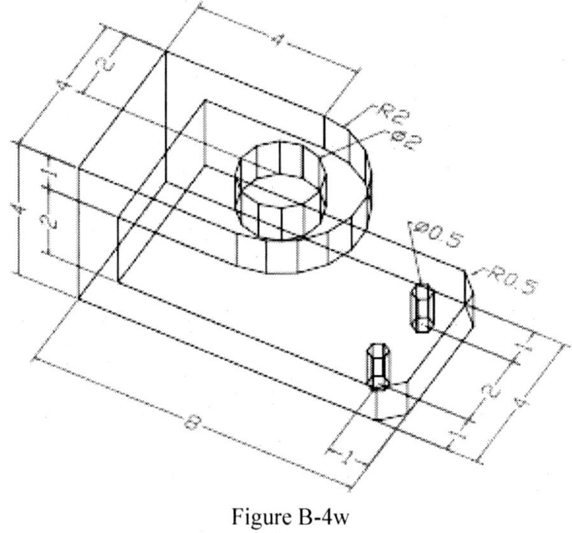

Figure B-4w

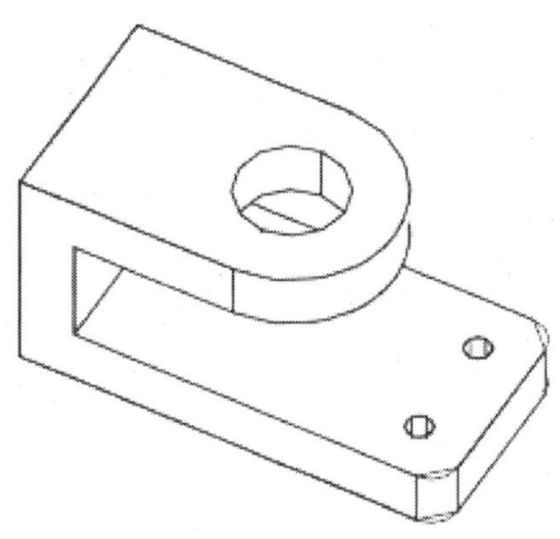

Figure B-4su

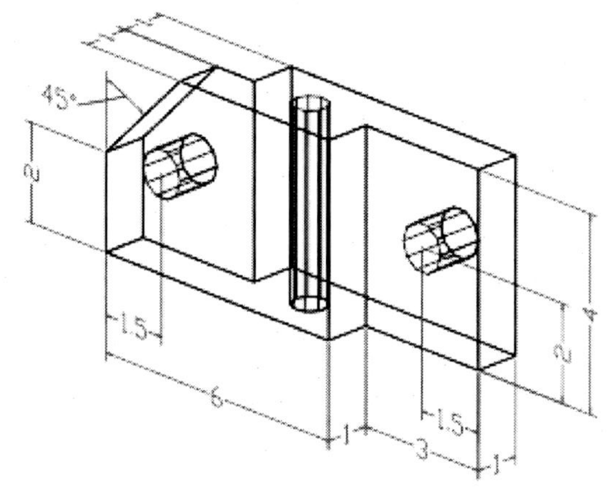

Figure B-5w

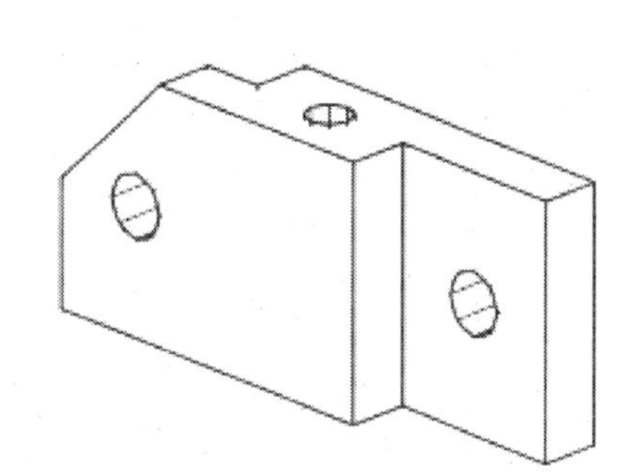

Figure B-5su

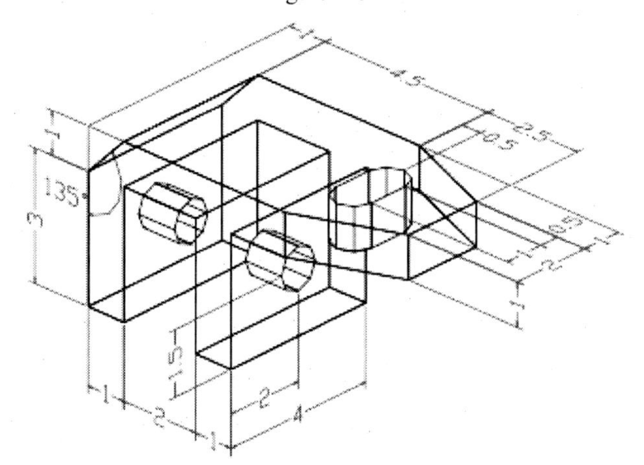

Figure B-6w

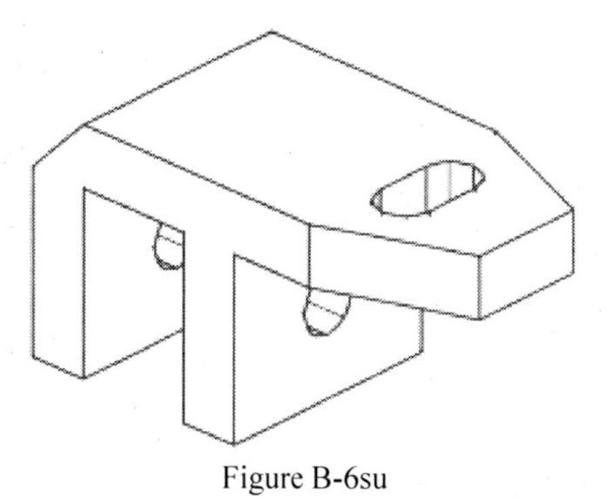

Figure B-6su

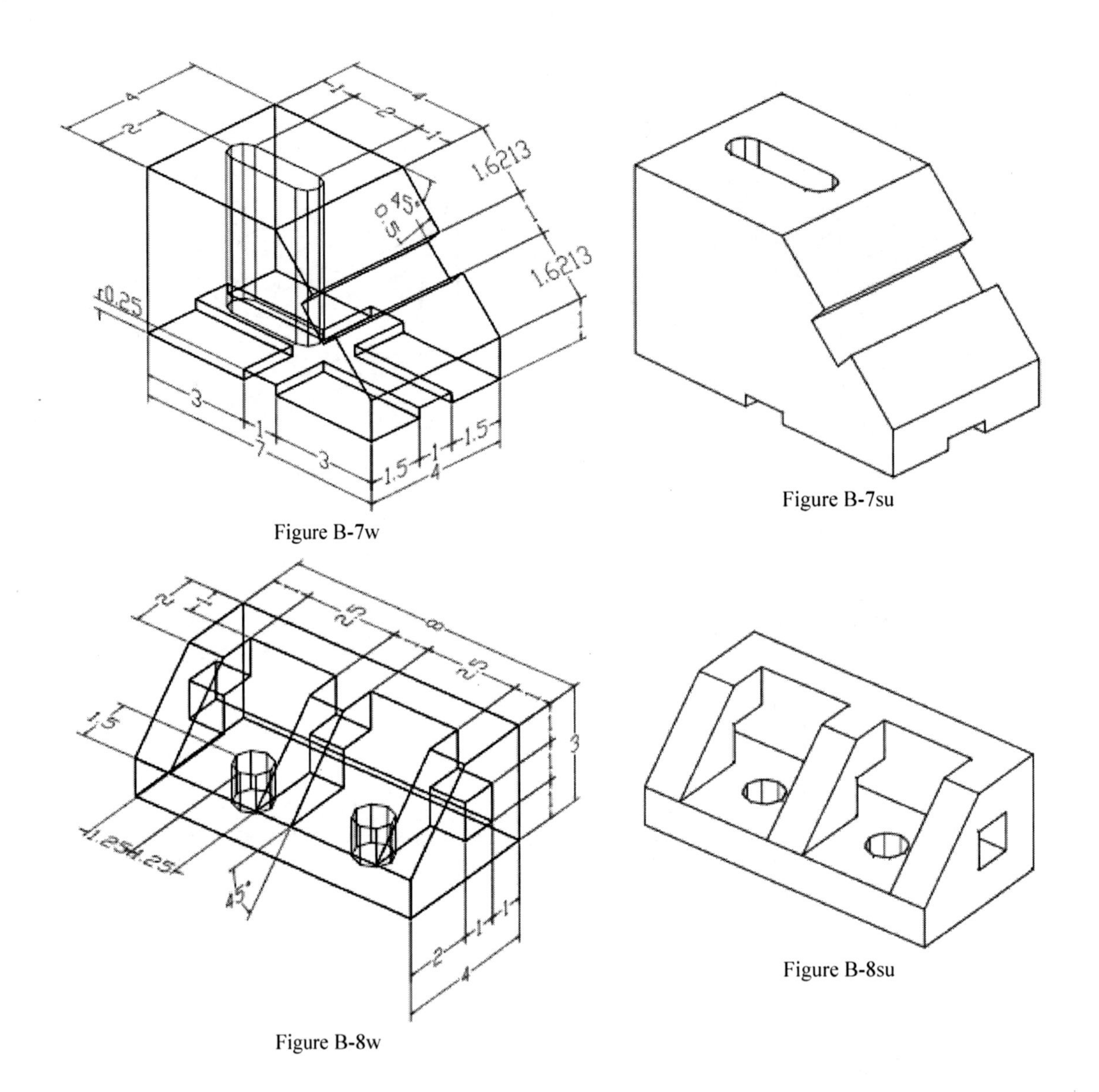

Figure B-7w

Figure B-7su

Figure B-8w

Figure B-8su

Appendix – C: Additional Projects

Here are two drawings that might challenge you.

1. The pieces for Aloysius' Cabin can be found in C:\Steps3D\Lesson\Cabin. They include:
    1.1. *longlog.dwg*
    1.2. *medlog.dwg*
    1.3. *shortlog.dwg*
    1.4. *longhalflog.dwg*
    1.5. *medhalflog.dwg*
    1.6. *roof.dwg*
    1.7. *gable.dwg*
    1.8. *HS Logo.bmp*

2. The bookshelves are 7'-6" square x 11½" deep, with a ½" plywood back inserted into rabbets on the sides. Use your own spacing for the internal shelves.

3. Search the web for additional projects. Try to draw some of them. I used keywords like "3D render" and "solid model" and found these:
    3.1. http://www.pleione.com/pithouse/ [This site has some cool renderings.]
    3.2. http://www.gsmmedia.com/cad/renderings.html [Fair renderings.]
    3.3. http://www.3drender.com/ [renderings from the technical director at Pixar – not necessarily AutoCAD, but most of these things can be done in AutoCAD]
    3.4. http://www.cadalyst.net/ [leading industry mag]
    3.5. http://www.smlib.com/gallery.html

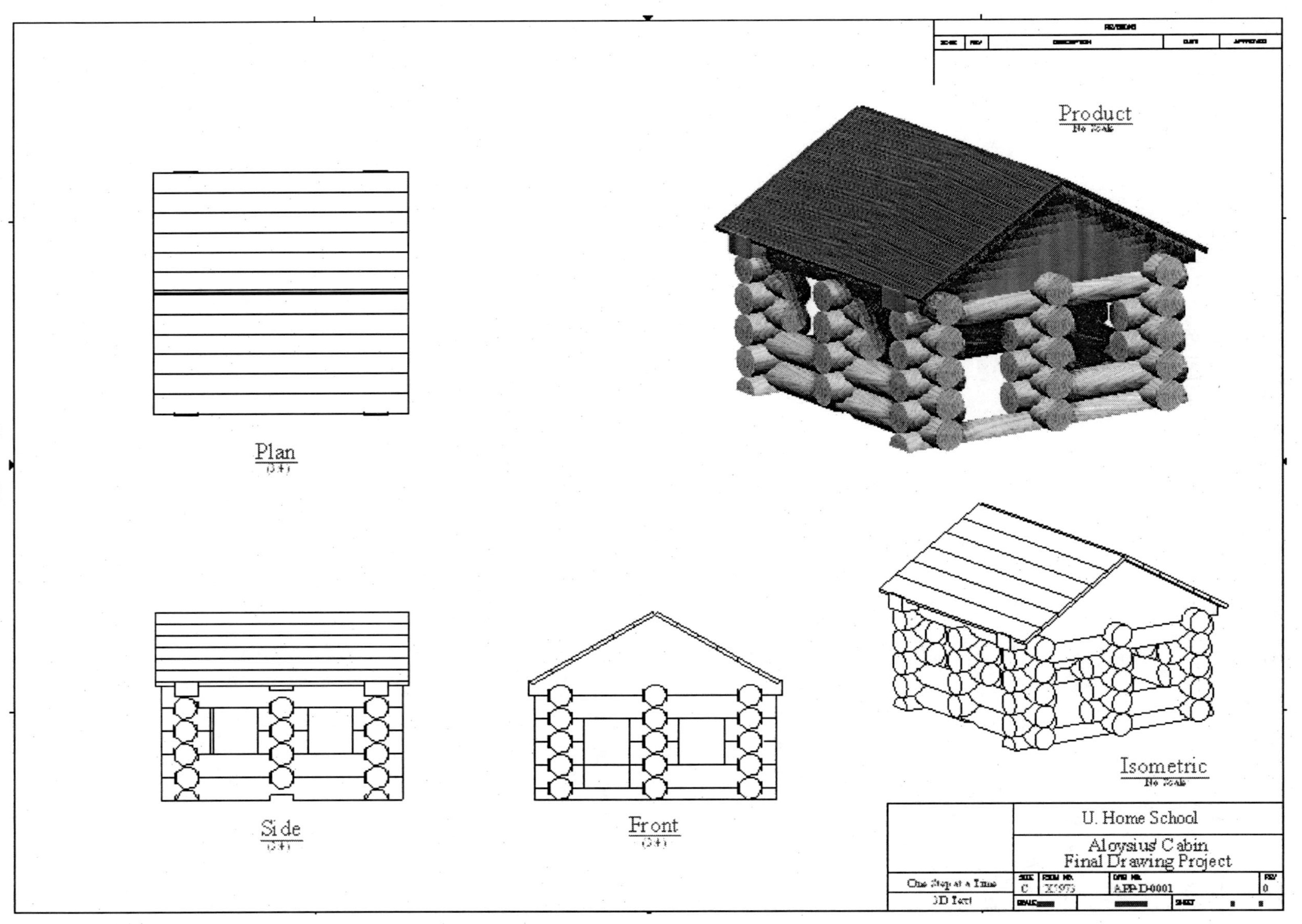
Product
No Scale
Plan
Side
Front
Isometric
No Scale
U. Home School
Aloysius' Cabin
Final Drawing Project
One Step at a Time
3D Text
C
APP-D-0001
0

Figure C1

Figure C2

Appendix – D: Adding Ribbon Panels

AutoCAD 2010 began offering the option of selecting what you wish to see on the ribbon panel during their initial installation. Occasionally, the options you select won't provide the panel(s) you need.

Here's how to display a panel after the initial setup.

1. Right click on the tab and select **Show Panels** from the menu. (You'll get a different list of panels for each tab you use.)
2. Place a check next to the panel you wish to display.

And you thought it was going to be difficult!

Index

Lightning Source UK Ltd.
Milton Keynes UK
UKOW03f1919110615

253363UK00006B/203/P